Contemporary Ergonomics 2002

Contemporary Ergonomics 2002

Edited by

Paul T. McCabe

WS Atkins, UK

THE **Ergonomics**
society

First published 2002
by Taylor & Francis
11 New Fetter Lane, London EC4P 4EE

Simultaneously published in the USA and Canada
by Taylor & Francis Inc,
29 West 35th Street, New York, NY 10001

Taylor & Francis is an imprint of the Taylor & Francis Group

This book is produced from camera-ready copy supplied by the Editor
Printed and bound in Great Britain by TJ International Ltd, Padstow, Cornwall

British Library Cataloguing in Publication Data
A catalogue record for this book is available from the British Library

Library of Congress Cataloging in Publication Data
A catalogue record for this book has been requested

ISBN 0-415-27734-5

CONTENTS

MANUAL HANDLING

MUSCULOSKELETAL DISORDERS

INTERFACE DESIGN

WORKPLACE DESIGN

WORK DESIGN

TEAM WORKING

WARNINGS

SLIPS TRIPS AND FALLS

PREFACE

Contemporary Ergonomics 2002 is the proceedings of the Annual Conference of the Ergonomics Society, held in April 2002 at Homerton College, Cambridge, UK. The conference is a major international event for Ergonomists and Human Factors Specialists, and attracts contributions from around the world.

Papers are chosen by a selection panel from abstracts submitted in the autumn of the previous year and the selected papers are published in *Contemporary Ergonomics*. Papers are submitted as camera ready copy prior to the conference. Each author is responsible for the presentation of their paper. Details of the submission procedure may be obtained from the Ergonomics Society.

The Ergonomics Society is the professional body for ergonomists and human factors specialists based in the United Kingdom. It also attracts members throughout the world and is affiliated to the International Ergonomics Association. It provides recognition of competence of its members through its Professional Register. For further details contact:

The Ergonomics Society
Devonshire House
Devonshire Square
Loughborough
Leicestershire
LE11 3DW
UK

Tel: (+44) 1509 234 904
Fax: (+44) 1509 235 666

Email: ergsoc@ergonomics.org.uk
Web page: http://www.ergonomics.org.uk

HOSPITAL ERGONOMICS

HOSPITAL ERGONOMICS:
ORGANISATIONAL AND CULTURAL FACTORS

Sue Hignett[1] and John R. Wilson[2]

[1]Ergonomics and Back Care Advisory Department
Nottingham City Hospital NHS Trust, Nottingham NG5 1PB
[2]School of 4M, Management and Human Factors Group
University of Nottingham, Nottingham NG7 2RD

This paper describes an exploration of the organisational and cultural factors in the practice of hospital ergonomics in a qualitative interview-based study with twenty-one ergonomists (academics and practitioners). The analysis identified three themes: organisational; staff; and patient issues.

It is suggested that hospitals present a particularly complex setting in which to practice ergonomics. This is partly due to the organisational structure (with multiple professional and managerial lines), but also to the core business of health care.

The area of female workers in hospital ergonomics was found to be under-researched both with respect to the type of work and to social and cultural issues about gender stereotyping.

Introduction

Ergonomics has been described as a socially-situated practice (Hignett, 2001a & b). In order to achieve the goals of 'design' and/or 'change' the ergonomics practitioner has to have an understanding of the culture of the industry or organisation in which they are working. Hospital ergonomics is a relatively new area of practice, but has an enormous potential scope of practice. The National Health Service (NHS) is the biggest civilian employer in Europe, employing more than 1.1 million people, 5% of the UK working population. It is the largest employer of women, with approximately 75% female workers, with nurses accounting for 50% of all staff.

The following questions were explored:

- What are the characteristics of the health care industry with respect to the organisational and cultural factors?
- How do these characteristics impact on the practice of ergonomics in hospitals?

The paper is intentionally written in the first person to emphasise the use of a qualitative (or interpretative) approach throughout the study. Establishing one's position by writing in the first person is supported by a tradition in the social sciences and education (Wolcott, 1990; Webb, 1992). The literature review is embedded throughout the findings and discussion to give a more interactive analysis (Wolcott, 1992) and to facilitate the testing of the data against the literature (inductive analysis).

Methodology

Qualitative methodology was chosen as a suitable approach to explore the two aims. This approach enabled the questions to be explored by giving:

1. Access to information through interactive interviews, with the flexibility to develop the questionnaire both during an individual interview and throughout the study.
2. An inclusive position to reflect on the diversity of the perspectives held by academics and practitioners involved in ergonomics.

The choice of qualitative methodology was supported with a middle ground philosophical stance, giving an ontological position of subtle or transcendental realism. This allows that there is a physical structure beyond our minds, '..things exist and act independently of our descriptions ... objects belong to the world of nature' (Bhaskar, 1975), but also that different people will have different perceptions of them, the idea of non-competing multiple realities (Murphy et al, 1998). This accepts the view of Hammersley and Atkinson (1995) that '.. there is no way in which the researcher can escape the social world in order to study it'. So two people may interact with the same situation or product and have very different experiences and perceptions of it, and both can be equally valid.

Methods

Twenty-one semi-structured interviews were carried out with academics and practitioners using a questionnaire proforma which developed iteratively over the 18 months of the project. The interviews were audio-taped and transcribed verbatim. The transcripts were returned to the interviewee for an accuracy and confidentiality check before analysis.

Contact data sheets were completed after each interview to capture my immediate thoughts and summarise the main points from the interview. These were used as the first stage of data reduction. As the study progressed the new/target questions started to develop into questions for the data analysis rather than the interviewees.

Sampling strategies

A progressive four stage sampling strategy was used starting with purposive sampling to spread the net. Suggested contacts were then followed up (snowball sampling), before the third stage of intensity sampling to focus on subjects with specific experience in hospital ergonomics. A final strategy of analysis sampling sought extreme and deviant cases to test the analysis and interpretation.

All the interviewees agreed to participate but unfortunately for three I was unable to arrange a convenient time. Other interviews were booked to try and achieve saturation from different discipline areas although input from psychology remained limited.

Analysis

The analysis used the three steps of data reduction, data display and conclusions drawing/verification (Miles and Huberman, 1994).   The interview transcripts were imported into a qualitative data management tool, NUD*IST N₄ (Gahan & Hannibal, 1998).   The data were summarised, coded and broken down into categories using qualitative classification (Miles and Huberman, 1994; Sanderson and Fisher, 1997).

Findings

The analysis resulted in three categories: organisational, staff and patient issues.

Hospital Themes

Organisational issues	**Workers**	**Caring for People**
Size	**(Staff issues)**	**(Patient issues)**
Complexity	*Multiplicity of professions*	*Dirty and emotional work*
	Gender	*Patient expectations*
		Life, death and mistakes

Figure 1.   Hospital Themes

Organisational Issues

The organisational issues included both the size and complexity of the National Health Service.   For example, three hierarchical lines were identified in the management structure: an administrative line, a professional line and a patient-focused clinical management line.

> *'..there are three sources of power from the management which is, of course, connected to the health authorities, and you have the doctors, and then you have the nursing staff which is also a source of power and you cannot do anything if you can't make agreement with all these three...'*

The three-way hierarchy adds to the complexity with respect to accountability, authority and power.   Quantitative measures of performance are applied at the level of units of provision, or 'cost centres' (directorates).   This means that meeting the clinical targets may not be the responsibility of individual nurses.   NHS managers may not directly control the work of nurses through performance.   But they probably control the supply of other things, for example: (1) the context in which the nurses carry out their work and exercise their professional autonomy; and (2) the number of patients and therefore the amount of time (staff-patient staffing ratio).

The recurring themes of the complexity and interface with the patient seem to be fundamental in how health care differs from other industries.

> *'you start to try and draw your person, equipment interaction and always there's another person in the picture as well, so it's actually a people-people interactions are, are quite a big focus '*

Van Cott (1994) called this difference 'people-centred and people-driven' in contrast to other industries which are technology-centred where the human role is to monitor the equipment or supervise small numbers of other staff.   This relates to the core business of the hospital providing the public service of health care (to include both public and private sector organisations).

'the product that the hospital has, as a business, is caring for patients, and caring for patients isn't seen as a 'product'. So whereas, whereas in industry or in commerce you're producing something which you're selling or a service that you're providing..'

Implementing change is often a key part of ergonomics projects and it was suggested that 80% of the effort when working in hospital ergonomics was needed to progress the project and with only 20% on understanding the problem. The reverse was perhaps the more usual model for ergonomics projects, with 80% of the time spent on understanding or solving the problem and only 20% on progressing the project.

Staff Issues

Two main areas were explored. The first related to the multiplicity of professions found in health care, and the second to the high proportion of female workers. The literature points to evidence for gender stereotyping for care tasks, but the case study generated very little data in support.

'..there may be cultures that are specific to predominantly female professions and semi-professions which may be about sacrifice, and all of that stuff, that actually may not be true of, I don't know, car workers in the Midlands...'

Paid care work has been considered to be a low status occupation and almost an extension of housework (Giddens, 1993; Miers, 1999). This has led to dubious assumptions, for example that 'women are equipped to deal with bodily substances and that they enjoy this work as an extension of their 'natural' role and engage in it by choice' (Lee-Treweek, 1997).

There was a general feeling that there was a lack of data or information on women within ergonomics.

'if you look at any of the standard texts there's really, there has never been, in my view, sufficient general data gathered on either females, or anything more than the fit population which is invariably youngish ..'

Patient Issues

The patient issues incorporated three dimensions associated with the caring role: the type of work; expectations; and possible outcomes. The work tends to be dirty and emotional, with a professional subculture to allow the handling of other peoples' bodies.

Other aspects of handling people include the emotional impact of other people's nakedness. Lawler (1998) looked at the nurses' first experience of suffering, disfigurement and death, and suggested that speed was often used as a method to manage difficult or potentially embarrassing situations. Technical vocabulary or jargon is also used to cope with the full significance of handling bodies.

'in many, many situations you have to deal with an interaction between people which both parties have to really have very high belief in, where there are, can be, very strong emotional influences at a level which is just about as sharp as you can get I think in terms of interactions between people'

This subculture was linked to a 'coping' attitude where staff put the patients' needs and well-being before their own, and has resulted in staff taking risks. The change in patient expectations (from being apologetic through to demanding their rights) is mirrored in the cultural move from paternalism to partnership (Boseley, 2000). This change might also fit the two models of care described by Miller and Gwynne (1972) with respect to risk-taking. A minimum risk environment was called the 'warehousing model of care', whereas a more stimulating, riskier environment was described as the 'horticultural model of care'. In order to provide both care and cure there are different type of service provision required.

There is a growing field of application of human factors in medicine, especially in the area of human error. This growth was discussed by Caldwell (1996) and a parallel was drawn between medical practice and 'other technologically dynamic, error-critical systems', e.g. aviation, nuclear and petrochemical industries. At the moment approximately one in ten patients are known to suffer adverse consequences as a direct result of their admission to hospital (Department of Health, 2000) and there are initiatives to change this through audit, further research and education.

Discussion and Conclusion

It is suggested that hospitals are different to all other industrial organisations. Health care is a service industry like banking, but additionally it is also a public service (like the railways). The difference for ergonomics practice may lie in the definition of the 'user' in the context of a user-centred design or task analysis. Every member of the United Kingdom population is a potential user of the NHS so the definition of the user group is difficult for many areas. As a service industry the clients (patients) are not paying at the point of contact (unlike banking or transport services) and they do not have to be there (unlike education or the prison service). For banking and transport services the 'users' are all either paid employees or paying customers. A closer comparison might be education, but here the 'users' are paid employees (teachers and support staff) or children, who are legally required to attend the school. The prison service again has a complex user definition with the inmates giving an additional interface to the employees, but again the prisoners have not chosen to be there, that choice is made for them. This makes the definition of 'user' very difficult and creates complex interface. My conclusion is to suggest that the complexity of the organisation and culture needs to be taken into account for ergonomics practice in the health care industry.

References

Bhaskar R. 1975, *A realist theory of science* (Leeds, Leeds Books)

Boseley S. 2000, *Doctors and nurses to be trained together to relax the elitist divide,* The Guardian, Monday 15 May 2000, 7

Caldwell B.S. 1996, Organisational bridges from research to practice: cases in medical practice, Proceedings of the Human Factors and Ergonomics Society 40th Annual Meeting, (Philadelphia), 530-532

Department of Health. 2000, *An organisation with a memory*, (Norwich, The Stationary Office)

Gahan, C. and Hannibal, M. 1998, *Doing Qualitative Research using QSR NUD*IST*, (London, Sage Publications)

Giddens, A. 1993, *Sociology*, Second Edition, (Cambridge: Polity Press)

Hammersley, M. and Atkinson, P. 1995, *Ethnography: Principles in Practice*, (London and New York, Routledge)

Hignett, S. 2001a, Embedding ergonomics in hospital culture: top-down and bottom-up strategies, *Applied Ergonomics*, **32**, 61-69

Hignett, S. 2001b, *Using Qualitative Methodology in Ergonomics: theoretical background and practical examples*, Ph.D. thesis, University of Nottingham

Lawler, J. 1998, Body Care and Learning To Do for Others. In M. Allott and M. Robb (eds.) *Understanding Health and Social Care. An Introductory Reader,* (London, Sage Publications and The Open University), 236-245

Lee-Treweek, G. 1997, Women, resistance and care: An ethnographic study of nursing auxiliary work, *Work, Employment and Society*, **11**, 1, 47-63

Miers, M. 1999, Nurses in the labour market: exploring and explaining nurses' work. In G. Wilkinson and M. Miers, (eds.), *Power and Nursing Practice*, (Basingstoke: Macmillan), 83-96

Miles, M.B. and Huberman, A.M. 1994, *Qualitative Data Analysis: An Expanded Source Book,* Second Edition. (Thousand Oaks, CA: Sage Publications)

Miller, E.J. and Gwynne G.V. 1972, *A life apart: A pilot study of residential institutions of physically handicapped and the young chronic sick*, (London, Tavistock).

Murphy, E., Dingwall, R., Greatbatch, D., Parker, S. and Watson, P. 1998, *Qualitative research methods in health technology assessment: a review of the literature*, Health Technol Assessment, 2, 16

Sanderson, P.M. and Fisher, C. 1997, Exploratory Sequential Data Analysis: Qualitative and Quantitative Handling of Continuous Observational Data, Chapter 44. In G. Salvendy (ed.), *Handbook of Human Factors and Ergonomics*, Second Edition, (NY, John Wiley and Sons), 1471-1513

Van Cott, H. 1994, Chapter 4. Human Errors: Their Causes and Reduction. In M.S. Bogner, (ed.), *Human Error in Medicine*, (New Jersey, Lawrence Erlbaum Associates)

Webb, C. 1992, The Use of the First Person in Academic Writing: Objectives, Language and Gate keeping, *Journal of Advanced Nursing*, **17**, 747-752

Wolcott, H.F. 1992, Posturing in Qualitative Inquiry. In M.D. LeCompte, W.L. Milroy and J. Preissie (eds.), *The Handbook of Qualitative Research In Education*, (New York, Academic Press), 3-52

Wolcott, H.F. 1990, *Writing up qualitative research. Qualitative Research Methods Series 20.* A Sage University Paper, (Thousands Oaks, CA, Sage Publications Inc.)

SIMPLE SOLUTIONS REDUCE MSDs IN HOSPITALS

Tamara James, Sabrina Lamar, Tony Alleman

Duke University and Health System
Box 3834
Durham NC 27710 (USA)

This paper describes three case studies that involve musculoskeletal disorders affecting radiology technicians, histopathology technicians, and nursing assistants. The purpose of this paper is to provide examples of unique problems and solutions for reducing musculoskeletal disorders so that others may benefit from these ideas. The emphasis here is on *simple* solutions. Although they may seem insignificant, these simple solutions can have a tremendous impact on an organization just starting an ergonomics improvement program. Small successes such as those created by implementing simple solutions can help build credibility and pave the way for future success. The use of multi-disciplined teams to develop solutions appears to be a good approach for greater likelihood of success.

Introduction

Many groups of hospital workers experience musculoskeletal discomfort and injuries as a result of their occupational activities. The following case studies exemplify a multi-disciplinary approach and application of ergonomic principles to simple redesigns of equipment and work practices that result in a more comfortable work environment. Many of the solutions described are simple, low cost modifications that can be incorporated rapidly and routinely to reduce injuries to health care workers. The case studies presented here illustrate improvements made at a major medical center in the southeastern United States.

Case summaries

Radiology technologists
Radiology workers have reported musculoskeletal discomfort that they attribute to their work tasks (Wright and Witt, 1993). Patients who are not medically stable to transport from their hospital room must have procedures performed in their room. Radiology

technologists use portable x-ray equipment to perform the x-ray exams in these patients' rooms. In these cases, the technologist slides the film cassette under the patient. The current method of positioning portable x-ray cassettes under patients causes ergonomic stressors such as poor body mechanics and excessive force while pushing cassettes under a patient between the bed sheets and the mattress ticking (Figure 1). Now that hospitals in the United States are required to cover mattresses with a non-permeable material that is less slippery, the task is even more difficult. The sharp corners of the cassettes and the increased friction from new mattress ticking, contribute to the difficulty of the task. This results in significant force required to push the cassette under patients. There are no handles on the cassettes, so an awkward and forceful grip is also required. These factors result in symptoms in the hands, wrists, shoulders, and back in this group of hospital workers. It is estimated that over 60,000 portable radiographs are performed each year at this medical center.

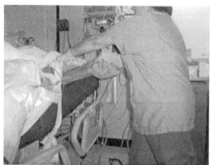

Figure 1. Posture used to place radiology cassettes

For this particular project, a multi-disciplinary team approached the portable x-ray problem with the goal of decreasing physical stressors to employees and increasing patient comfort during the procedure. A new cassette holder was designed with the following characteristics (Figure 2):

- Light weight materials used
- Broad, smooth, curved front edge to reduce friction
- Handle incorporated into the design to improve grip and improve posture during use

The device design transfers the required pushing force from the weaker, upper extremities to the stronger, lower extremities. In addition, the smooth rounded surface of the device reduces the friction between the sheets and the mattress ticking, thereby reducing the push force. The new device requires an average of 15% less force to slide cassettes under a patient. In addition, the change in users' body mechanics has resulted in less stress to the upper extremity and a significant reduction in awkward body postures (Figure 3).

The reduction of force required to slide the cassettes using the new holder is likely due to the combination of reduced friction and providing a handle with which to grasp the cassette while pushing. The cassette holders are relatively inexpensive to fabricate and are a simple solution to removing ergonomic stressors from this type of activity.

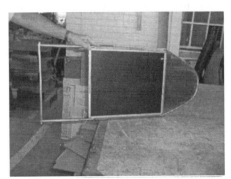 

Figure 2. Prototype cassette holder   **Figure 3. Using new cassette holder**

Microtome workstation
Studies have shown that histologists who perform microtomy may be exposed to awkward postures and repetitive tasks (Herman *et al*, 1995a and Herman *et al*, 1995b). Histopathology technicians have been treated at the employee health unit of this medical center for various musculoskeletal disorders ranging from wrist pain to epicondylitis. The ergonomists at this medical center were asked to evaluate the work practices and work environment to determine if any ergonomic improvements might decrease risk of developing these disorders.

Employees in histology typically use a microtome at least four to six hours each day to prepare tissue specimens for pathologists. The use of the microtome involves repetitively turning two handles, one on either side of the equipment (Figures 4 and 5). It was determined that turning the handles for several hours each day was contributing to elbow and wrist pain of one operator in particular. The equipment was poorly designed from an ergonomic standpoint, forcing the operator to use the wrist and arm in awkward positions.

Figure 4. Typical microtome use   **Figure 5. Typical microtome use**

Through ergonomic analysis using a multi-disciplined approach (ergonomics and occupational medicine), it was determined that an adaptation of the microtome handles would improve some of the ergonomic stressors to the wrists, elbows, and arms. The handle was lengthened and the diameter was increased to allow the operator to use it with a flat palm motion rather than a twisting or turning motion (Figure 6). The microtome workstation was modified using readily available household items. Tubular foam was used to modify the handle. The entire unit was elevated with plywood and elbow pads were added to prevent contact stress (Figure 7). These changes have increased the employee's comfort and should help reduce ergonomic stressors that are associated with developing musculoskeletal disorders. Other microtome workstations in histology will likely be modified as well. Automated microtomes have been recommended for future implementation here and elsewhere (Esinger, 2000) for further reduction in ergonomic stressors in this environment.

Figure 6. Modified handle   **Figure 7. Pads and elevated microtome**

Linen and trash chutes

For this project, a job was analyzed involving environmental services employees responsible for removal of trash and linen from floors of the multi-level hospital. There had been numerous reports of acute injuries as well as reports of musculoskeletal injuries related to removal of bags of trash and laundry from patient care units.

The hospital uses a Pneumatic Linen Trash Transport System (PLTTS) to move trash and soiled linen from patient units to the loading dock where it is transported off site (Figure 8). As required for accreditation, the doors to chutes must be kept closed when not in use. Therefore, they are designed with a side-hinged door and a self-closing mechanism.

Because of the spring mechanism on the doors and the pull of the vacuum when the inner door opens, employees must hold the door open with one hand while placing trash or linen in the chutes with the other hand. This has resulted in numerous injuries to the hands from doors accidentally slamming shut. It has also resulted in musculoskeletal injuries since employees are required to lift bags of linen and trash with one hand and twist their torso to place items in chutes.

Figure 8. Laundry and waste disposal vacuum system

A multi-disciplinary team (engineering, ergonomics, employee health, safety, and facilities maintenance) was assembled to evaluate the hazards associated with the laundry and waste disposal chutes. The goal was to determine potential solutions in an attempt to decrease the frequency of injuries and to prevent a more serious injury from occurring.

The evaluators discovered that a significant amount of negative pressure is generated in the chute to facilitate movement of trash and linen. This negative pressure may have been partly responsible for some of the acute injuries that have occurred. The design of these doors requires the operator to use one hand to prop open the door, thereby increasing the horizontal distance from the chute opening. If the operator is using one extremity to hold the door open and the door accidentally closes on the other extremity, an acute injury may occur. In addition, there is increased risk with the operator using one extremity to prop open the door and the other to dispose of items. Laundry and trash bags can be heavy and typically should be handled with both upper extremities. When the load is not distributed evenly between both upper extremities there is a greater risk for musculoskeletal injuries.

The team recommended the use of a wedge (Figure 9) for holding chute doors open. (This type of device is used by firefighters around the country for temporarily propping doors open to aid in moving hoses and equipment.) Such a device can be used to hold these doors open in the negative pressure situation so the operator is less likely to be injured.

These devices are not permanent, are inexpensive and are simple to fabricate. They can be labeled with operator names, identification numbers or color-coded, and issued to each operator. If a device were to be accidentally left in place, keeping the doors propped open, the responsible operator could be identified. This particular wedge was not an exact fit with the existing laundry/waste chutes, but a similar device was fabricated in-house to serve the same purpose (Figure 10). The injuries can be reduced through the use of simple wedges to temporarily prop doors open at the hinges.

Figure 9. Prototype door wedge

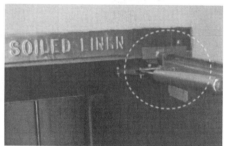

Figure 10. Final door wedge design

Conclusions

Hospital work is extremely unique and offers physical challenges that may result in musculoskeletal injuries. With proper evaluation and the implementation of some simple solutions these challenges can be reduced to effectively increase employee safety and reduce occupational injury rates in health care environments.   Success with implementation of simple solutions can help lead to a successful ergonomics program. A successful ergonomics program could lead to a more positive attitude towards ergonomics in general.

All of the case studies presented in this paper utilized a multi-disciplinary approach to analyze jobs and work environments and identify solutions. This approach appears to be very effective in resolving these issues and may actually increase the overall success of implementing solutions.

References

Esinger, W. 2000, The role of motorized microtomes in histotechnology. *American Clinical Laboratory*, **19**(8), 18

Herman, G.E.[a], Schork, M.A., Shyr, Y., Elfont, E.A., and Arbit, S. 1995, Histologists, microtomy, chronic repetitive trauma, and techniques to avoid injury. 1. A statistical evaluation of the job functions performed by histologists, *Journal of Histotechnology,* **18**(2), 139-143

Herman, G.E.[b], Arbit, S., Hyman, S.C., Currie, M.K., and Elfont, E.A. 1995, Histologists, microtomy, chronic repetitive trauma, and techniques to avoid injury. 2. A physical and rehabilitation medicine physician's perspective, *Journal of Histotechnology,* **18**(4), 327-329

Wright, D.L and Witt, P. 1993, Initial study of back pain among radiographers, *Radiologic Technology*, **64**(5), 283-9

CHANGING PRACTICE – IMPROVING HEALTH: AN INTEGRATED BACK INJURY PREVENTION PROGRAMME IN NURSING HOMES

Emma Crumpton[1], Carol Bannister[1] & Nico Knibbe[2]

[1]*Royal College of Nursing, 20 Cavendish Square, London, W1G ORN,*
[2]*LOCOmotion, Nijenbeek 20, 3772 ZE Barneveld, The Netherlands*

An integrated back injury prevention programme was put in place in 3 nursing homes. The programme comprised various strategies based on both management and care issues. Outcome measurements included care plan details, staff perception of their health, and exposure to manual handling risk. The results indicate that an integrated approach to the prevention of musculo-skeletal pain amongst nursing home staff is essential. Staff exposure to manual handling risk decreased at stage 2 due to risk management and problem solving interventions only. The decrease continued to stage 3 with new equipment and training interventions. No major changes in staff health perception were recorded, perhaps due to the brief time between completing all the interventions and measuring the outcome. A further period of time is needed to fully assess the effects of the project interventions.

Introduction

The original intention was to select homes for the study that would show a range of organisational set ups from small independent homes to large organisations with a number of homes. However, the choice was limited as the response rate to letters was low. All leads were followed up, but in the end the sampling strategy was convenience, purely based on whether or not the homes were willing to be part of the study.

The first home recruited was a 54 bedded home in London, which was part of a large chain. The second home was also in London, and part of a chain, but was a larger 200 bedded home. Attention was focused on just two of the heaviest and most problematic units. The third home was recruited several weeks into the project from outside London to ensure that the project was applicable to all nursing homes in the UK. It had 40 beds, and was part of a charitable trust, with a large chain of homes.

Method

Initially each home was visited, and management and workers were interviewed informally. Time was spent working alongside the nursing staff and health care assistants, and an initial view on what the general problems might be was made.

Initial objectives were written for each home under the following headings, and interventions were put in place:

Change management
- Raising staff awareness by creating focus groups and ensuring staff participation

Identifying problems and solutions
- Examination and updating residents care plan using a risk assessment / problem solving approach
- A comprehensive Occupational Health Audit

Occupational health - Management Systems
- A leadership course
- Guidelines and information provision
- Manual handling policy review

Occupational health - Staff health and wellbeing
- Occupational health provision for staff

Improving staff skills - training
- Train the Trainers Manual Handling Course
- Training in manual handling techniques

Equipment provision
- Equipment Audit recommending equipment provision based on the database of information for each work area, including the residents dependency levels

Monitoring progress

Outcome measures

Measurements were taken at 3 stages throughout the project. *Stage 1* was the baseline information, taken at the beginning of the project before any interventions. *Stage 2* was roughly halfway through once some interventions had started. The occupational health and equipment audits had taken place, and problem solving sessions were on going. Training had started but was not complete. *Stage 3* was at the end once all interventions were complete and a period of working without weekly input from the project manager had been undertaken. The outcome measures that were used are detailed in the results section.

Results

A large amount of information was collected and recorded over the duration of the project, and a brief over view is given in this paper

Database
Information from the residents care plans was taken in order to review exactly what was going on in the work areas and to be able to comment on any changes. Information was

gathered, and recorded through out the course of the project giving background information and evidence of some improvements in recording

Table 1. Background information from nursing home care plans

Nursing Home	Average Length of Stay	Average Age of Residents
home 1	2.6 years	88 years
home 2	3.4 years	84 years
home 3	2.8 years	89 years

The dependency levels of the residents were assessed by the staff at the first leadership course, and were recorded at stage 2.   The ARJO residents gallery was used to determine the levels. The residents gallery is a classification system that identifies five levels of dependency ranging from A (independent to E (dependent). The high numbers for home 2 are probably due to the fact that they have an EMI unit and the confusion levels led to a higher dependency level.

Table 2. Dependency levels of residents

Nursing Home	Percentage of high dependency residents (D&E – residents gallery)
home 1 (n=54)	31%
home 2(n=40)	60 %
home 3(n=40)	42 %

The dates of the latest moving and handling assessment for each resident were recorded at each stage. They show that all of the nursing homes are carrying out assessments and recording them more frequently.   The biggest difference was shown at nursing home 2 where at stage 1 the average time of the last assessment was over 14 months, by stage 3 the figure was reduced to just over 4 months.   This is an important finding as it shows that the homes are assessing more frequently and documenting the results for communication to staff. This should result in an improvement in resident care.

Figure 1 : Average time since last M & H assessment.

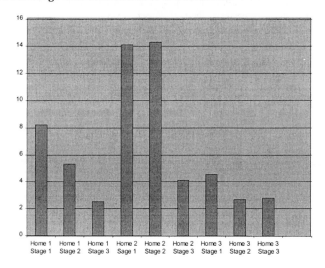

StaDyMeter

The StaDyMeter is a frequency oriented log to monitor the exposure of staff to manual handling risks. This measurement was necessary for the present study to monitor the number and duration of the manual handling tasks that were being carried out by staff. Knibbe and Friele (1999), stated that a change in back pain prevalence can only be attributed to an intervention if it is paralleled by a change in exposure. The StaDyMeter was used with the assistance of LOCOmotion who developed the tool, to monitor the exposure of staff to manual handling risk. Results show an increase in the number of tasks being done safely

Table 3. Activities performed safely

Transfers	stage one (n=2116)	stage two (n=1113)	stage 3 (n=1381)
Transfer bed <-> chair, wheelchair, toilet	45.7%	60.2%	85.9%.
Transfer wheelchair / chair <->chair, toilet	59.8%	68.7%	81 %
Transfers within the bed	15.9%	16.5%	44.9%

Figure 2. Percentage of safe activities comparing stage 1,2 and 3

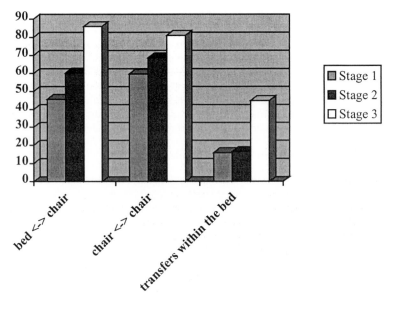

These results show an increase in safe practice between stage 1 and 2, when only management systems and change management strategies were implemented, with a small amount of training. The increase is sustained and further improvement is seen to stage 3 when the training programme was complete and equipment was introduced.

SF 36 and back Pain Questionnaire

The SF-36 (Ware1993) is a recognised tool for measuring health status outcomes. It measures outcomes that are functional, rather than clinical causes. This questionnaire was distributed with the back pain questionnaire which was designed to get staff to state clearly if they are suffering from musculo-skeletal aches and pains

The staffs perception of their own health was not affected very much. This was probably due to the length of time available for the project, and further time is needed to show any effects on health.

However, according to the back pain questionnaire the numbers of respondents experiencing pain more than once a week and more than once a month both decreased slightly by 1 and 3% respectively, whilst numbers experiencing pain less than once a month increased from 17 to 25%. This seems to indicate that respondents are experiencing pain less frequently.

Conclusions

The results show that the approach taken has been successful in some ways. Examination of the resident care plans shows that the care of the residents has improved according to the measures used for the purposes of this project. Records were more complete and more regularly updated, also the errors in equipment use seem to have decreased.

The exposure to manual handling risk seems to have decreased as more tasks are being performed 'safely' according to the criteria for the purposes of this project.

These improvements occurred between stage 1 and 2. That is before the introduction of any equipment, and with incomplete training programmes. Therefore it seems that they may be, at least partly, due to the change management strategy of making staff aware of problems and assisting them to find their own solutions. Staff were given the information about where to make improvements in the form of the occupational health report. Then, on-going support and problem solving in the work place. Also they had time, in the form of the leadership course, allocated to consider the issues and find solutions. It would appear that these management issues were the main reason for a successful out come at stage 2. Once equipment and training were introduced at stage 3 the improvements were maintained and continued to increase.

The idea that many of the standard approaches to back injury prevention have had limited success has been demonstrated in the review of the literature. Richardson and Hignett (1994) suggest there are many features of working life that may contribute to the risk of the build up of musculoskeletal stress and that they will not be simply solved by management strategies that aim to eliminate a single handling operation, or to automate, or to create procedures to decrease the load.

The results and process of the present study lead to the suggestion that resident care is inextricably linked to staff health. A very simplistic explanation for this statement is that if residents are rehabilitated they will do more for themselves and so decrease the load on staff. Resident care refers to all aspects of life in a nursing home, including the allocation of tasks between staff, the order in which things are done over a shift, the passing of information between staff, other professionals and the resident, effective problem solving for all the residents needs, as well as, a good physical environment, adequate equipment, appropriate policies and procedures and management support. It seems that by ensuring that there is good communication amongst all staff and residents, the working environment will be more conducive to better care for the resident as it is an environment that is receptive to change and identifying and finding solutions to problems.

Effective management is an absolute prerequisite for a change process to be effective in reducing back symptoms in care staff. A range of interventions is required, none of which can improve staff health individually, but collectively they are able to both improve staff health and the standards of resident care. There is no single correct method which can be imposed on an organisation from the outside. The present evidence seems to point towards a need for open dialogue and communication across the organisation. A collaborative management style which will provide a forum for culture change and better resident care is needed, as well as a sound basis of ergonomics knowledge in order to begin to effectively tackle the problem of back pain in nurses.

References

Arjo Limited (2000) Arjo *Residents Gallery*. Gloucester: Arjo

Knibbe J and Friele R, 1999, The use of logs to assess exposure to manual handling of patients, illustrated in an intervention study in home care nursing, *International journal of industrial ergonomics 24 445- 454*

Richardson, B., and Hignet, S., 1994 Risk Assessment- Myth or Method. In: *Ergonomics and Health and Safety. Proceedings of the meeting held on 20th October 1994, Bristol. The Ergonomics Society, in affiliation with the Institute of Occupational Safety and Health and the Health and Safety Executive*

Ware JE, Snow KK, Kosinski M, Gandek B. 1993.. *SF 36 health survey manual and interpretation guide.* Boston, MA; New England Medical Center, Health Institute,

DIAGNOSTIC ULTRASOUND: THE IMPACT OF ITS USE ON SONOGRAPHERS IN OBSTETRICS AND GYNAECOLOGY

Joanne O. Crawford[1], Jo McHugo[2] and Rachel Vaughan[3]

Institute of Occupational Health,
The University of Birmingham, Birmingham, B15 2TT[1]

Department of Radiology, Birmingham Women's Hospital
Birmingham Women's Health Care NHS Trust, Birmingham B15 2TG[2]

Industrial Ergonomics Group, School of Manufacturing and
Mechanical Engineering, The University of Birmingham, B15 2TT[3]

The aim of the following study was to identify contributing factors in the development of musculoskeletal pain and discomfort in Ultrasound Sonographers involved in Obstetric and Gynaecological scanning. The methodology involved a cross-sectional study of Sonographers in one hospital. The methodology included the use of the Rapid Upper Limb Assessment (RULA) to identify the exposure to postural risk, static muscle work and repetition, and the use of an adapted Nordic Musculoskeletal Questionnaire to evaluate the frequency and distribution of musculoskeletal problems,. Fourteen participants were assessed in the workplace. The RULA analysis identified that the task element with the highest risk factor within this sample was scanning patients; it was found that the participant spent between 31% and 39% of their working time doing this in a 26-hour week. The results from the questionnaire found that 64% had experiences one or more combined physical problems over the previous 12 months, with the shoulder joint (57%), cervical spine (50%) and wrist and hands (50%) identified as the most frequently reported problems. The study highlights the prevalence of self-reported symptoms among sonographers and the postural constraints individuals have to adopt while working. Suggestions are proposed to reduce the risks via adjustable work equipment, education in using adjustable equipment, work organisation changes and the use of arm supports.

Introduction

Diagnostic ultrasound (DU) was first introduced to the medical world in 1942 by Austrian Physician Dr Karl Dussik (Levi, 1997). It was initially used to detect gall stones in 1950 by Ludwig and Stutler, but was developed for use in gynaecology where there was a profound need to develop safer imaging techniques (Levi, 1997). DU was first used in the applied setting in 1958 by Professor Ian Donald from Glasgow who used ultrasound to examine gynaecological patients for pelvic lesions (Levi, 1997). It was used on a more regular basis within the clinical environment during the 1960's gaining recognition by the American Medical Association in 1974 (Vanderpool, 1993).

In modern day medicine, the use of ultrasound, in both gynaecology and obstetrics has become a normal part of the diagnostic and foetal assessment process. However, the increased use of ultrasound equipment has been identified as a source of workplace pain and discomfort in Sonographers. Craig (1985), surveyed 100 sonographers with 5-20 years experience. The results showed that that majority of respondents had experienced symptoms of musculoskeletal problems including wrist and shoulder problems. The study however, did not document exact figures concerning work related problems nor did it give details pertaining to the methodology or response rate of the study.

A more recent study by Vanderpool (1993), surveyed 225 Cardiac Sonographers. A 47% response rate was achieved with 72% of respondents female. Results found that 63% of respondents had experienced wrist problems during their career and 3% had been diagnosed specifically with Carpal Tunnel Syndrome.

Wihlidal and Kumar (1997), surveyed 156 Sonographers in a postal survey in Alberta. A 61.5% (N=96) response rate was achieved and 88.5% of respondents reported work related symptoms either historically or ongoing. Clusters of symptoms included neck and intrascapular pain (54%), shoulder or upper arm pain (53%), low back pain (37.5%) and elbow pain (23.5%). Respondents were asked about absence from work and 16% reported that they had been forced to take absence due to symptoms (Wihlidal and Kumar, 1997).

In comparison with others involved radiography work, May *et al* (1994) surveyed breast screening radiographers in a UK national survey of 800 participants. There were 320 respondents to the survey. This study used two control groups including clerical staff (N=400) and general radiographers not involved in screening (N=400). Preliminary results found that those involved in general radiography reported most muscular complaints (94.4%), 76% of those involved in breast screening reported pain and 70% of clerical staff reported muscular discomfort. Although only descriptive data is reported in the study, it highlights the level of complaints within general and breast screening radiography.

Habes and Baron (2000) presented a case study of ergonomic evaluation of ultrasound testing. The study highlighted the postural extremes sonographers had to adopt while using ultrasound equipment, the static loading from holding the scan heads and the biomechanical loading on the sonographers. Several recommendations from this study included the use of adjustable chairs including sit/stand seats and beds, the provision of elbow support, customising one room for specific scanning types and a secondary monitor in the line of sight of the sonographers.

The results of the previous studies suggest that the use of ultrasonography equipment is accompanied by physical musculoskeletal problems. The following study was carried out after an initial ergonomics evaluation of the work carried out in the radiography department of a hospital. The aim of the study was to identify the prevalence of musculoskeletal pain and discomfort, to identify postural risk factors when carrying out scanning tasks and to evaluate the workplace and equipment design.

Methodology

The participants in the study all worked at one hospital specialising in Gynaecology and Obstetrics. To become familiar with the working environment and the scanning process, a period of time was spent observing Sonographers at the Neonatal unit. To further identify the principle components of the scanning process, a talk-through was carried out based on the method in Kirwan and Ainsworth (1992).

The Rapid Upper Limb Assessment (RULA), methodology developed by McAtamney and Corlett (1993) was used to identify whether the postures used when carrying out sonography tasks were high risk. The RULA analysis was carried out with 6 participants scanning obstetric patients and 6 participants scanning gynaecological patients including transvaginal scanning during 90 minute observation periods.

A modified version of the standardised Nordic Musculoskeletal Questionnaire (NMQ) was used in the form of a structured interview during the initial stages of the study. The structured interview was based on the standardised and validated questionnaire developed by the Nordic Group (Kuorinka, *et al*,1987).

Results

A total of 14 Sonographers took part in the study, 11 were registered Radiographers and three Medical Doctors. The age range of the participants was 35 to 52 years, with time working with ultrasound equipment ranging from 6 months to 23 years. The time spent using equipment ranged from 6 hours per week to 35 hours per week with a mean of 26 hours per week scanning patients.

The observation data identified the different types of scans carried out within the department, familiarisation with the working environment and the scanning process. Two main types of scans were identified, firstly the obstetric scan and secondly the gynaecological scan. Nine rooms were routinely used for scanning each fitted out with the relevant equipment including computers for record keeping. Four different types of ultrasound equipment were used within the department. Eight of the consulting rooms contained non-adjustable stools of varying heights

The talk-through process identified the task elements involved in scanning patients. Table 1., identifies the task elements from A to H that the sonographer carries out.

Postural observations made during the RULA analysis identified that when scanning a patient, the sonographer is required to twist the neck and trunk in order to view the monitor while at the same time maintaining probe contact with the patient. The cervical spine is also held in moderate side flexion usually when the sonographer is pointing out features on the display screen. Table 1., presents the data for the RULA analysis and indicates that although the majority of tasks carried out are scored at action level 2, the scanning of patients for both types analysed was scored at action level 3, requiring investigation and change in the near future. The degree of static loading during the scanning element was also analysed visually and it was approximated that when carrying out an obstetric scan, the maintenance of static posture was required for 84% of the time and for a gynaecological scan, 74% of the scanning time.

Table 1. Task Elements and RULA Scores of the Scanning Process

Task Element Code	Task	Obstetric Scan Mean Grand Score	Action level	Gynae. Scan Mean Grand Score	Action Level
A	Reading the patient's notes	3	2	3	2
B	Walking	2	1	2	1
C	Computer Work	4	2	4	2
D	Setting up the patient/equipment	4	2	3	2
E	Performing the scan	5	3	6	3
F	Talking to the patient/relatives	3	2	3	2
G	Clearing the plinth	3	2	3	2
H	Communicating with colleagues	3	2	3	2

The percentage of time spent on each of the task elements was also calculated and the data is presented in Figure 1. What is highlighted from this is that the individuals surveyed, spent between 31% and 39% of their time with patients carrying out the task with the highest RULA grand score.

Figure 1. Percentage of Time on Scanning Tasks

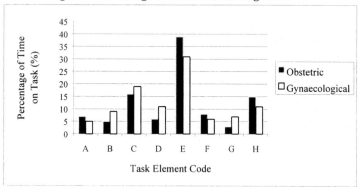

The NMQ identified that 13 (93%) of respondents had previously or were currently experiencing at least one or more physical symptoms. Table 2., presents the summary results of the prevalence of musculoskeletal pain and discomfort. The data presented show that pain and discomfort is most commonly reported for the neck, shoulder, wrist/hands, upper and lower back. Two of the respondents felt that the symptoms they experienced did encroach upon their leisure time but none felt that their work activities had been reduced. Of those who complained of shoulder problems, 60% attributed this to sonography work including working with the arm elevated for long periods. Moving patients and heavy equipment were also implicated to a lesser degree.

Table 2. Incidence of Musculoskeletal Problems in the past 7 days or 12 months

Body Segment	Last 7 days		Last 12 months	
	N	%	N	%
Neck	5	36	7	50
Shoulders	3	21	8	57
Elbows	0	0	1	7
Wrists / Hands	3	21	7	50
Upper back / Thoracic Spine	5	36	5	36
Lower back	5	36	7	50
One or both: hips, thighs, buttocks	0	0	1	7
One or both knees	1	7	2	14
One or both ankles or feet	0	0	0	0

Discussion

The main limitation on this study was the small sample size. Although there were only 14 respondents, this did represent all the staff working at the time of the study. In comparison with previous work, the results are similar to those within sonography.

There are a number of potential factors that are thought to contribute to the pattern of symptom reporting including the nature of sonography work and the equipment in use, where approximately one third of the working time is spent interacting with the ultrasound equipment. This particular task has been identified by the RULA analysis as being Action Level 3 where investigation and change are required in the near future. The repetitive nature of the work is also a risk factor where scans are at times carried out at 15 minute intervals during the working day.

The ultrasound equipment design is also an issue in terms of risk. Sonographers use different ultrasound machines in the course of the working week in a variety of standing or sitting postures. All users were seen to conduct the scan in postures which place the joints under stress. They often had to stretch the upper limb to reach the console and controls while maintaining physical contact with the patient. In this case it appears that although some equipment has been designed with the user in mind, this is not often the case as the ultrasound equipment is cumbersome to move and use. This may be a concern as ultrasound equipment should be considered under the Display Screen Equipment Regulations (HSE, 1992) where there are different requirements for the screen and the keyboard as currently exist in many hospital departments.

In terms of the workplace and setting up of equipment, the risk factors included non-adjustable seats of varying heights. This did not allow the sonographers the opportunity to adjust the workplace to fit themselves. Other factors that should be considered are the use of support for the upper limb as suggested by Habes and Baron (2000)

Conclusion

Sonography is a relatively young profession that became prominent in the 1960s. Its characteristic scanning technique involves a great deal of static muscle work in the upper body. A consequence of this rapid growth is starting to emerge and research over the past 10 years has associated sonography with a number of musculoskeletal problems. Implicated in the symptom profile of this study were the postures adopted when using ultrasound equipment, the non-adjustability of workplace seating and the ultrasound equipment design.

References

Craig, M. 1985, Sonography. An occupational health hazard? Focusing on the issues. *Journal of Diagnostic Medical Sonography,* **1**, 121-126

Habes, D.J., Baron, S. 2000, Case Studies: Ergonomic evaluation of antenatal ultrasound testing procedures. *Applied Occupational and Environmental Hygiene,* **15**, 521-528

HSE 1992, *Display Screen Equipment Work: Health and Safety Regulations 1992, Guidance on the Regulations L26,* (HMSO, London)

Kirwan, B., and Ainsworth, L.K. 1992, *A Guide to Task Analysis,* (Taylor and Francis, London)

Kuorinka, I., Jonsson, B., Kilbom, A., Vinterberg, H., Biering-Sorensen, F., Anderson, G., Jorgensen, K. 1987, Standardised Nordic questionnaire for the analysis of musculoskeletal symptoms. *Applied Ergonomics,* **18**, 233-237

Levi, S, 1997, The history of Ultrasound in Gynaecology 1950-1980. *Ultrasound in Medicine and Biology,* **23**, 481-522

May, J., Gale, A.G., Haslegrave, C.M., Castledine, J., Wilson, A.R.M. 1994, Musculoskeletal problems in breast screening radiographers. In S.A.Robertson (ed.) *Contemporary Ergonomics 1994* (Taylor and Francis, London), 247-252

McAtamney, L., and Corlett, E.N. 1993, RULA: A survey method for the investigation of work related upper limb disorders, *Applied Ergonomics,* **24**, 91-99

Vanderpool, M.P.T. 1993, Prevalence of Carpal Tunnel Syndrome and other work related musculoskeletal problems in cardiac sonographers. *Journal of Occupational Medicine,* **35**, 604-610

Wihlidal, L.M., Kumar, S. 1997, An injury profile of practising diagnostic medical sonographers in Alberta. *International Journal of Industrial Ergonomics,* **19**, 205-216

USE OF ERGONOMIC CHAIRS IN A DENTAL CLINIC

Christy Turner, Beth Rohrer, Tamara James

Duke University and Health System
Box 3834
Durham NC 27710 (USA)

The purpose of this field study was to identify an appropriate chair for the activities performed in a dental clinic and to evaluate the effects of using the chair. Before new chairs were purchased, five full-time employees rated the frequency and severity of any body discomfort. They repeated the survey after using the ergonomic chairs for eight months. The data showed an overall decrease in both frequency and severity of discomfort in the head, neck, low back and shoulders after using the ergonomic chairs. Although the backrest of the new chairs could be used to support the chest while leaning forward, the employees in this study found that using the chairs in the traditional way was ideal for their tasks. The armrest shape and saddle seat make this chair more comfortable and easier to use than a traditional dental stool.

Introduction

Studies have shown that dental workers are at risk for developing musculoskeletal discomfort and disorders as a result of non-neutral postures adopted during work. In a survey of dental hygienists, Osborn *et al* (1990) found that 68% had experienced some type of musculoskeletal pain within the past year. Similarly, Shugars *et al* (1987) found that 60% of general dentists experienced pain over a one-year period. The lower back, neck and shoulder were the most common sites of pain for both groups, consistent with other studies of dentists (Finsen *et al*, 1998; Lehto *et al*, 1991; Rundcrantz *et al*, 1990).

The tasks of a dentist have high visual demands, which require dentists to assume fixed postures for extended periods of time (Marshall *et al*, 1997). In addition, non-neutral postures are often required to gain manual and visual access to the patient's mouth. Although dentists' working positions change depending on the part of the mouth and tooth surface on which they are working (Rundcrantz *et al*, 1991), flexion and rotation of the neck and trunk are maintained throughout most tasks (Visser and Straker, 1994). These postures can cause the worker to experience increased physical discomfort throughout the workday (Visser and Straker, 1994).

Because dental workers often assume a forward leaning posture, traditional chairs do not provide adequate back support. This lack of support may contribute to increased musculoskeletal discomfort. The purpose of this study was to identify an appropriate chair for the activities performed in a dental clinic and to evaluate the effects of using this chair.

Methodology

This field study was conducted at a pediatric dental clinic within a major medical center in the southeastern United States. Five full-time female employees participated in the project. The average age of the employees was 26 years old, and the average number of years in the dental profession was 5.2 years.

Using discomfort surveys, the employees rated the frequency and severity of discomfort in any of sixteen body areas. Frequency was rated on a scale from 0 to 5 (0 = never, 3 = frequently, 5 = all of the time). Severity was also rated from 0 to 5 (0 = none, 3 = uncomfortable, 5 = painful and prevents my working).

Shortly after the survey was conducted, the dental clinic moved to a new facility. Before the move took place, an ergonomist made recommendations to assist with the design of the new clinic. Recommended heights for tool manipulation, sinks, computer workstations and waste receptacles were provided. In addition, appropriate locations for equipment storage and proper lighting were addressed in the recommendations.

The ergonomist also suggested that new chairs be purchased. Several non-neutral postures are required to perform dental work, including trunk flexion, neck flexion and neck rotation. The existing chair (Figure 1) did not support the employees' back in these positions. Hag Capisco chairs (Figure 2) were recommended and were purchased after the move to the new facility.

Figure 1. Traditional chair

Figure 2. Hag Capisco

The Hag Capisco was selected because the backrest can be used to support the chest when the user leans forward, reducing stress on the back. Many of the activities performed in the dental clinic require this forward leaning posture (Figure 3), and it was thought that upper body discomfort would decrease with use of a more supportive chair (Figure 4). When the chairs arrived, an ergonomist demonstrated how to use and properly adjust them. Approximately eight months after the facility change occurred, the employees completed the discomfort surveys again.

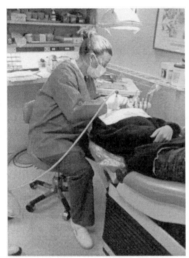

Figure 3. Use of traditional chair

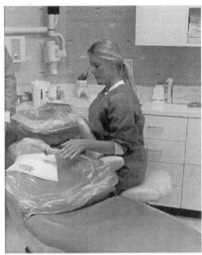

Figure 4. Possible use of ergonomic chair

Results

The discomfort data prior to use of the new chairs showed that the majority of the employees had the most body discomfort (where frequency or severity was 3 or greater) in the head (frequency 40%, severity 40%), neck (severity 40%), low back (frequency 40%, severity 40%), and shoulders (frequency 20%, severity 20%). After using the new chairs for 8 months, there was an overall decrease in both frequency and severity of discomfort in these areas. Only the neck (severity 20%) and low back (frequency 20%, severity 20%) had ratings of 3 or greater. An interesting finding is that overall low back discomfort did not increase even though employees reported an average increase of 22% more time spent seated per day than at the time of the first survey.

Table 1. Frequency of discomfort: percent of employees reporting discomfort frequency greater than or equal to 3 "frequently"

Area	Pre-intervention	Post-intervention
Head	40%	0%
Neck	0%	0%
Low back	40%	20%
Shoulders	20%	0%

Table 2. Severity of discomfort: percent of employees reporting discomfort severity greater than or equal to 3 "uncomfortable"

Area	Pre-intervention	Post-intervention
Head	40%	0%
Neck	40%	20%
Low back	40%	20%
Shoulders	20%	0%

Discussion

The body areas in which employees experienced the most discomfort (head, neck, low back and shoulders) correspond with those ranking highest in previous studies of dental workers (Lehto et al, 1990; Rundcrantz et al, 1990; Shugars et al, 1987). This suggests that the decrease in discomfort experienced by these employees after using the new chairs may also occur in other dental employees.

It was originally thought that supporting the chest while leaning forward would reduce body discomfort. After the survey results were gathered, however, it was discovered that the dental clinic employees did not use the backrest to support the chest. Instead, they used the chairs in a traditional manner (Figure 5). The employees stated that the height and width of the backrest interfered with the arm positions that are required for working in a patient's mouth when the chair is turned around. They did, however, find several other advantages of the ergonomic chair as compared to the traditional dental stool.

1. *Saddle-shape seat pan*: The employees stated that the saddle seat allows for more postural variety and helps them maintain a more neutral back posture while working. Additionally, they feel much more secure in this seat as compared to the traditional stool. This is important because they often swivel in the chair to obtain other items or to reach different areas of the patient's mouth.

2. *Arm rests*: The employees like having arm rests on the chair, both for resting their arms while performing a procedure and for leaning on when taking short breaks. The dental assistants prefer this chair over a traditional assistant's chair, which does have an arm rest, because the arm rest is at the back of the chair (Figure 6) instead of curved around the side.

There are several limitations of this study. One is the small sample size, as the five subjects may not be representative of most dental workers. Additionally, the employees moved to a new facility at the same time they received the new chairs. It is possible that facility improvements also contributed to the decrease in discomfort and it is difficult to tease out what portion of this decrease is due solely to the ergonomic chairs. However, because these employees spend more than half of the workday sitting, it is likely that the ergonomic chairs had a greater impact on their level of comfort than did facility changes. Another limitation of this study is the fact that the data are based on subjective feedback.

Future studies could collect more objective data, such as the differences in working postures or EMG measurements assumed by dental clinic employees when using a traditional stool or the Hag Capisco. Additionally, the feasibility of using these chairs should be investigated in similar work environments such as operating rooms or ambulatory surgery.

Figure 5. Use of ergonomic chair

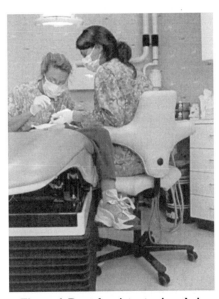

Figure 6. Dental assistant using chair

Conclusions

The fact that the increase in sitting was combined with a decrease in discomfort implies that the ergonomic chairs had a positive influence in the workplace. Although the backrest of the new chairs can be used to support the chest while leaning forward, the employees in this study found that using the chairs in the traditional way was ideal for their tasks. The employees report armrest shape and saddle seat make this chair more comfortable and easier to use than a traditional dental stool. The study's findings indicate that this chair may be useful not only for dental workers, but also other occupational groups that use a forward leaning posture to perform their work, such as microscope users, surgeons, and other health care clinicians. These users may find it beneficial to have a saddle seat as well as an option to use the backrest as a chest support.

References

Finsen, L., Christensen, H., and Bakke, M. 1998, Musculoskeletal disorders among dentists and variation in dental work, *Applied Ergonomics,* **29** (2), 119-125

Lehto, T.U., Helenius, H.Y.M., and Alaranta, H.T. 1991, Musculoskeletal symptoms of dentists assessed by a multidisciplinary approach, *Community Dental Oral Epidemiology*, **19**, 38-44

Marshall, E.D., Duncombe, L.M., Robinson, R.Q., and Kilbreath, S.L. 1997, Musculoskeletal symptoms in New South Wales dentists, *Australian Dental Journal*, **42** (4), 240-6

Osborn, J.B., Newell, K.J., Rudney, J.D., and Stoltenberg, J.L. 1990, Musculoskeletal pain among Minnesota dental hygienists, *Journal of Dental Hygiene*, **64**, 132-138

Rundcrantz, B-L., Johnsson, B., and Moritz, U. 1990, Cervical pain and discomfort among dentists. Epidemiological, clinical and therapeutic aspects, *Swedish Dental Journal*, **14**, 71-80

Rundcrantz, B-L., Johnsson, B., and Moritz, U. 1991, Occupational cervico-brachial disorders among dentists, *Swedish Dental Journal*, **15**, 105-115

Shugars, D., Miller, D., Williams, D., Fishburne, C., and Strickland, D. 1987, Musculoskeletal pain among general dentists, *General Dentistry*, **4**, 272-276

Visser, J.L., and Straker, L.M. 1994, An investigation of discomfort experienced by dental therapists and assistants at work, *Australian Dental Journal*, **39** (1), 39-44

PATIENT HANDLING IN THE AMBULANCE SERVICES: CASE STUDY INVESTIGATIONS

Mark G. Boocock[1], Mike I. Gray[1] and Sally Williams[2]

[1] *Health & Safety Laboratory,*
Broad Lane,
Sheffield S3 7HQ

[2] *Health & Safety Executive,*
Health Services Unit,
14 Cardiff Road,
Luton LU1 1PP

Accidents and injuries resulting from the moving/handling of patients remain a primary cause of ill health and early retirement amongst ambulance personnel. Such statistics appear to reflect the high physical demands of the job and psychosocial aspects of the work. Whilst in recent years there have been considerable advancements in ambulance design and patient handling equipment, recent case study examples reveal that poor design and disparity between equipment and operator requirements are still a cause for concern, which can have a detrimental effect on good handling practices. Recognition of the need to reduce musculoskeletal injuries within the ambulance services has led to the formation of an HSE-led Ambulance Services Working Group which brings together many different professions and disciplines.

Introduction

Epidemiological studies suggest that amongst the ambulance profession employees are at high risk of suffering health problems and an excess level of health-related early retirement (Rodgers, 1998; Safas, 1993). Rodgers (1998) identified the three most common causes of early retirement as being due to musculoskeletal disorders/injuries, circulatory disorders, or mental disorders - musculoskeletal disorders/incidents accounting for 41% of all early retirements. The reason for such high morbidity rates amongst ambulance workers may stem from many contributory factors, including both long-standing and more recent changes to working practices. Long standing risk factors include cumulative effects, such as years of lifting and handling patients in uncontrolled working environments, shift working and disturbed sleeping and eating patterns, and the mental stressors of an emergency environment. More recent changes to working practices, such as target setting (e.g. Ambulance Performance Standards set 8 minute call response times for category 'A' patients), shorter and more frequently interrupted rest periods and increased public expectations, may be increasing the risks even further.

In recognition of the growing concerns for the high prevalence of musculokeletal disorders amongst ambulance workers, coupled with the high social and economic costs of such injuries, HSE, through the tripartite Health Service Advisory Committee (HSAC), has established an Ambulance Service Working Group (ASWG) challenged with the task of:

- identifying the main processes and tasks giving rise to musculoskeletal injuries amongst ambulance workers;
- determining the causal factors of such injuries, taking into account possible underlying failings in risk management systems; and
- directing future strategy of the HSAC targeted towards reducing the incidence of musculoskeletal disorders.

Accident data and associated costs

The first job of the ASWG was to establish the current extent of musculoskeletal problems throughout UK ambulance services in order to facilitate and identify patterns or trends in accident / incident reporting. In order to elicit reliable and comparable data, ambulance services were approached directly, either through WG members or representatives of the National Ambulance Risk and Safety Forum (NARSF). In all, accident and injury data was provided by seven UK ambulance Trusts

The format and categorisation of data provided by trusts was generally considered poor and varied considerably from one Trust to another. In some instances, full years' data could not be provided due to recent mergers between Trusts or changes in IT systems. Consequently, it was necessary in some cases to extrapolate partial year figures over a 12 month period, which may well have introduced a measure of error as there is some evidence of seasonal variations in reporting rates.

Results showed that between 30 and 51% of all recorded incidents (not just RIDDOR reportable) involved the moving/handling of loads and resulted in some form of musculoskeletal injury. The mean incidence rate was 178 per 1000 employed, representing an 18% risk of musculoskeletal injury due to the moving/handling of loads. Where accident data could be subdivided according to Accident and Emergency crews (A&E) and Patient Transfer Service (PTS) staff, 90% of injuries were assigned to A&E staff. When expressed as a ratio of injury to call response, A&E staff sustained one injury per 965 calls, compared to one injury per 18,600 calls for PTS staff.

Sickness absence data was provided by three trusts and average sickness absence due to all causes ranged from 7 to 20 days per employee per year. One trust identified an average sickness absence rate due to back and musculoskeletal injury of 8.25 days lost per employee per year. Another trust estimated that the annual cost of absence due to injury and ill health amounted to £764,400. One trust supplied statistics of early retirement through ill health/injury, which showed that the average age for early retirement was 49 years (representing an average loss of 16 years working life), with musculoskeletal injury assumed to be the major contributory factor. Information provided by the Trade Union showed that the average settlement for a compensation claim/award was £5653. Analysis of 1039 incidents identified 3 main tasks linked to accident/injury causation: the use of stretchers; the use of carry chairs; and patient transfers (e.g. floor-to-bed, bed-to-chair).

Risk factors and design issues

Regardless of the weight of load being handled, ambulance work is extremely hazardous due to the varied and complex nature of the risk factors present. These pose a constant threat to the health of the operator, particularly when handling tasks are performed in compromising and risky positions, and are complicated by constraints imposed by the work environment. The working environment is probably the most difficult area to control, as its often unpredictable and inhospitable (Collins, 1997). However, improvements can be made to the regular, more controlled interface between ambulance worker and his/her equipment, e.g. the ambulance vehicle.

The following examples are intended to illustrate where failings in the decision making processes during the selection and purchasing of new equipment, coupled with external influences (e.g. financial and targeting setting), may have led to inappropriate or poor handling procedures. Of particular concern are those examples where the driving force for change was based on the need to improve manual handling practices.

Case study 1: Ambulance vehicle and equipment compatibility
Concerns were raised by ambulance crews at one ambulance Trust following difficulties reported by some operators (particular short operators) when loading/unloading stretchers (referred to as an 'Easy-Load' stretcher) to and from ambulances (with a fixed tailgate height of 770 mm). This followed the recent purchase of new ambulances by the trust as part of their overall programme of modernisation.

An ergonomic investigation attempted to simulate stretcher loading tasks performed by operators at the extremes of the stature distribution curve. The main focus of the study was to investigate the effects of vehicle tailgate height and operator variables on the physical capabilities for carrying out the task. Video recordings, force measurements and biomechanical modelling techniques were employed as part of the investigation.

Outcomes of the study showed clearly that the existing configuration of vehicle and patient stretcher placed high physical demands on both the 'short' and 'tall' operator, with the short operator being at a particular disadvantage. The human factor most influencing the ability to carry out the task appeared to be arm strength; this placed the operator at risk of injury and compromised the stability afforded to the stretcher. Measurements of forces required to support the weight of the trolley prior to it being pushed into the back of the ambulance showed that operators were required to support approximately 47% of the combined weight of the stretcher and patient. For a patient weighing 70 kg, this meant that the operator was required to exerted a vertical upward force of approximately 530 N (trolley weight = 45 kg). Observations of the task also revealed that the operator was required not only to support the weight of the trolley, but also to operate a control handle to release the locking mechanism of the trolley's undercarriage, thereby allowing it to be raised/lowered to the ground by a second operator. Both placement and positioning of this handle were considered poor and led to awkward hand/wrist postures. Other aspects of the operator-equipment interface considered detrimental to the handling task included: door seals at the back of the ambulance which created a double step (each 40 mm in height) and impeded the movement of the stretcher into the ambulance; and a saloon height which imposed awkward postures on those standing inside the vehicle.

As a consequence of this investigation a number of recommendations were forthcoming. In the short-term, there was an immediate need to reduce tailgate heights throughout these vehicles and the fitting of drop-down suspension was seen as the most viable and cost-effective approach. A reduction in tailgate height of only 160 mm, as simulated by releasing air from the rear tyres of the ambulance, was shown to have a significant benefit on the ease of the handling operation. Although not an approved document at the time, a BSI standard for the design of ambulance vehicles specifies a maximum tailgate height of 750mm (BS EN 1789, 2000). This study suggests that 750mm may be too high for most combinations of ambulance and stretcher loading systems. A requirement was placed on the Trust to re-examine its purchasing and procurement policy, with particular emphasis placed on improved methods of evaluation and user and specialist involvement. Furthermore, it was suggested that work should be undertake to assess the impact of any proposed modifications on other handling operations, e.g. carry chair transfers.

Case study 2: Designing to meet the needs of service care

In response to government-led Ambulance Performance Standards to meet 75% of Category 'A' calls (immediately life threatening) one Trust set about the development and construction of a 'fast response' ambulance vehicle. Building of the ambulance in-house was seen as an attractive proposition, as it facilitated the use of in-house expertise, as well as offering significant financial savings. The design team began the conversion of three standard LDV Pilot vehicles fitted with the necessary equipment to respond to Category 'A' calls. This included the fitting of a 'self loading' stretcher, storage facilities for equipment and specially designed passenger seats. The rationale behind opting for the LDV chassis, as opposed to other vehicles, was that its compact size would facilitate easier manoeuvrability through traffic.

Following complaints raised by users of the vehicle it was decided to carry out a risk assessment of the vehicle design. Outcomes from this assessment identified some 14 design issues that were felt to compromise the health and safety of ambulance crews, including problems associated with: loading/unloading patient stretchers via the rear tailgate; the manual handling of equipment from storage areas inside the vehicle; and access within the vehicle to attend to the needs of the patient. Whilst operational requirements and patient needs had been given necessary consideration during vehicle construction, it appeared that there had been little evaluation of the work carried out by ambulance staff as part of its everyday use. The major concern was the confined space within the passenger compartment, presenting a significant risk to ambulance staff when administering emergency care, such as CPR. It was clear that the design team failed to recognise a number of ergonomic design issues, presumably due to a lack of experience amongst the design group, and whilst some retrofit modifications were possible the intended use of the vehicle needed to be reassessed.

Case study 3: The design of alternative patient transfer systems

Methods of loading/unloading patient stretchers to/from ambulances is one area where considerable progress has been made over recent years. Currently available methods include 'easy-load' and 'self-loading' patient stretchers, tailgate lifts and ramp systems, often in conjunction with a towing winch. Following evaluation of the available methods, one Trust decided upon a ramp system and purchased a fleet of ambulance vehicles incorporating a manually retractable ramp. Following concerns raised by both staff and HSE inspectors, an investigation of the methods of patient transfer using this system was carried out.

Following a detailed study of the handling procedures and forces required to operate the ramp, a number of factors were found to severely limit the benefits of this approach over more conventional methods. In particular, the manual task of deploying the ramp resulted in high pushing and pulling forces. Maximum peak forces when removing the ramp from the ambulance were found to rise steadily to a peak of approximately 25 kg and when returning it to the ambulance, a constant force of approximately 21 kg was required over a period of approximately 4-6 s. These requirements appeared to stem mainly from the overall weight of the ramp (34 kg) and aspects of its design, e.g. friction within the ramp housing. Other concerns centred on the risks associated with: slipping and/or tripping when walking on the ramp; overall ramp angle and marked changes to the angle over its length; a poorly designed handle for deploying the ramp; and the narrow width of the ramp and the potential for misalignment with the stretcher trolley.

The overall conclusion of this investigation was to suggest a major redesign of the ramp in order to reduce the manual force during deployment, ideally by using a power operated system.

Discussion

Analysis of the accident and incident reported data from seven ambulance Trusts revealed considerable variation in the nature and type of data recorded, which led in turn to difficulties in providing reliable indicators of accident/incident causation, as well benchmarking across the service. This will inevitably influence the effectiveness of assessing future preventative strategies and/or of directing change. Despite these difficulties, the data did show that UK ambulance personnel are sustaining a high number of injuries at work, with A&E crews most at risk. It also showed that there is a resultant high cost due to sickness absence, early retirement and civil claims. Amongst the hidden costs are the loss of staff expertise and the costs of recruiting and training replacements.

With regard to preventative strategies for reducing the incidence of musculoskeletal accidents/incidents, the existing situation, particularly with regard to hardware solutions (e.g. the design of ambulance vehicles and patient transfer equipment), is currently undergoing considerable developments. However, as these case studies have shown, successful interventions require the involvement of all concerned: employers, users and suppliers, with ergonomists also having an important role to play.

Within the context of the ambulance work environment, incompatibility between worker and workspace remains a major impacting factor on safe handling practices. As Collins (1997) suggested there are areas where improvements can be made: the removal of constraints imposed on the work area due to design limitations; improved access and handling of equipment within the passenger compartment; optimisation of saloon heights to allow unrestricted movements; and assurance that tailgate heights conform to user requirements. Advancements in patient stretcher loading and unloading systems for ambulances continue at great pace with, more recently, the introduction of tailgate lifts and ramp and winch systems. The suitability and practicality of these systems still need to be evaluated and other possible approaches developed. More importantly, the work of ambulance staff needs to be considered within a much wider context in order to include all aspects of the job, including physical, psychosocial and individual factors relevant to the work.

Employers and manufacturers are now aware of the need for improved communication between the various parties concerned and a number of discussion groups have been created, e.g. vehicle and equipment design groups, safety and manufacturer forums. The formation of an HSE-led Ambulance Services Working Group bringing together operatives, ambulance and equipment manufacturers, health and safety professionals, union representatives and ergonomists, will hopefully also serve to facilitate change and ultimately lead to a reduction in injuries to employees within this profession.

Acknowledgements

The authors wish to acknowledge the considerable work of the members of HSAC ASWG and, in particular, staff and representatives from the ambulance Trusts who assisted in this work.

References

BS EN 1789: 2000, Medical vehicles and their equipment – Road ambulances. BSI 01-2000

Collins B. 1997, An ergonomic approach to manual handling in the ambulance service. *Ambulance UK*.

Rodgers L.M. 1998, A five year study comparing early retirements on medical grounds in ambulance personnel with those in other groups of health service staff. Part II: Causes of retirement. *Occupational Medicine*, **48**, 119-132.

Safas H. 1993, Ill health retirement in health care workers. *Occupational health*, **45**, 101-103.

MEASURING WORKING POSTURES OF MIDWIVES IN THE HEALTHCARE SETTING

Dianne Steele[1] and David Stubbs[2]

[1]*University of Greenwich, School of Health & Social Care*
E-mail D.R. Steele@greenwich.ac.uk
[2]*University of Surrey, Robens Centre for Health Ergonomics. EIHMS*
E-mail d.stubbs@surrey.ac.uk

This paper presents an overview of how the working postures of midwives were measured when assisting women to breast-feed. Focus group interviews were conducted with midwives, to ascertain their views and perception of musculoskeletal discomfort in the midwifery working environment. The Quick Exposure Check (QEC) observational paper assessment tool, was used to gather quantifiable data by measuring the working postures of midwives when assisting women to breast-feed. Following the observational study, midwives were requested to note their perceived musculoskeletal discomfort using self-reporting discomfort scores. Using triangulation between the methods the findings suggest midwives do experience musculoskeletal discomfort in the back and neck.

Introduction

The project was developed in response from midwives, who reported a high incidence of backache when assisting women to breast-feed. Second, the study was prompted by the dearth of literature and consequently its impact on clinical practice. The literature reviewed, found little or no reference directed at midwifery practice. Recognition that musculoskeletal disorders may occur within the midwifery workforce was first reported by Hignett (1996) and later the (Royal College of Midwives 1997).

Today, midwifery practice is reliant on the knowledge of postural practices drawn from the nursing literature. However it would be beneficial for the midwifery profession to address this area.

The methodology selected was drawn from the quantitative and qualitative paradigms. These methods were chosen to enrich the research experience, an area which has been neglected. The study of working postures has been consumed in scientific aura, but if used in isolation can be flawed. In this study apart from measuring working postures in a controlled and systematic way, the search for meaning in which individual's interact within their environment was also obtained. Whilst we can obtain crude measurements of body postures in order to enrich the findings, the need to study human behavioural patterns is also evident. As a method, qualitative research studies the patterns of behaviour within a social group, it was worth considering, as issues of rituals, traditions and relationships are innate in this cultural group and have not been studied. This study selected both research methods for its design because it is widely reported to be successful (Parahoo 1997).

Focus Group Interviews

Focus group interviews were conducted in order for midwives to express their feelings, thoughts and attitudes regarding their physical work environment.

Midwives were recruited into two focus groups from two postnatal wards from two hospitals. The group size was six and eight respectively. Both groups were asked questions about their perceived experiences on the nature of midwifery work in relation to musculoskeletal discomfort. They were also asked to identify areas of midwifery practice that may cause such discomfort. The data from the groups were recorded on paper using spider diagrams, and later the groups could confirm a true recording of events. The data was coded and developed into themes using Swanson & Cheniez (1986) outcomes model. The data from the two groups were analysed together as it was not intended to compare groups.

From a total of thirty-seven themes, eight categories were perceived to be contributory factors to musculoskeletal discomfort in the workplace these were: labour ward, breast feeding, furniture, ward environment, neonatal care, working conditions, work attitude, education and training. Four of the eight are worth further discussion.

Working postures

Midwives cited the positions they often had to adopt in the labour ward as problematic. Backache and muscle stiffness was widely reported. Many felt static bending, stooping and twisting was the cause. Many midwives stated working around a bed was a problem, even though beds were height adjustable. Whether bed design and or the positioning of women on the bed remains a problem for the midwife, by not being able to get close to the work without compromising back flexion and torsion, is unclear? Midwives referred to assisting with breast-feeding as a cause of significant backache in the lumbar region. Other contributing factors such as women sitting in low chairs and choosing to breast feed in bed were problematic for the worker. One midwife stated " *Which ever way I stand, sit, I always get backache, its because I know I twist and bend my back*"

Contributing factors may be the mismatch of working height between the worker and the woman who is breast feeding, a factor, which was considered during the observational study. Some midwives felt they could not tolerate musculoskeletal discomfort in the back when assisting with breast-feeding and chose to kneel. When questioned, they stated that by kneeling they could get closer to the woman and baby, however they could not maintain this posture for long periods due to knee discomfort. One midwife felt she would rather have knee discomfort than backache; she could accommodate this position by displacing pressure from one knee to another and keep her back in a neutral position.

Later, it was interesting to note in the observational study those midwives who chose to kneel spent the same amount of time with the woman as those who chose to stand or sit. Whether this position was adopted for short periods only may be due to the task being uncomfortable for the worker and therefore continued to change position to avoid or alleviate discomfort.

Furniture

Beds appeared to be a continual problem for a variety of reasons. Few midwives admitted they adjusted the bed to a comfortable working height. This was perceived to take too long and was yet another task to do. However, during the observational study it was noted it took less than ten seconds to adjust the height of a profile bed. A considerable amount of time was spent moving beds, cots and wheelchairs. Staff reported difficulty with mobility blaming poor wheel design, wheels jamming and ineffective brakes. When questioned the midwives were unaware of any equipment maintenance programme and none had personally reported defective equipment. On exploring this further the workers felt they would rather work with defective equipment than have the equipment removed for repair. Nobody thought equipment could be replaced in the interim period. It was concluded that workers were prepared to compromise their physical well being, and the tendency to display compliant behaviour was overwhelming. As the midwives were unable to confirm how and what maintenance provision was in a place in the respective hospitals, then the problems were likely to be compounded further.

Ward environment

The main discussion centred on lack of space in ward areas, including individual rooms. As a result, midwives felt this impeded their ability to work, restricting access to the work task. Clutter, lack of space between beds, bed curtains and lack of storage space for property, led to items stored on the floor and under beds.
" It's like an assault course, you have to clamber over bags, dodge the flowers and the cards, before you can get near the woman to give care. "
The problem was compounded if the bed curtains were drawn around the bed.
A change in social trends was interesting. Midwives had noticed, with the introduction of early transfers, compared to six to ten day stays, increased women's social isolation. Here, women kept the bed curtains closed around the bed for prolonged periods, despite no procedural reason.
When questioned, midwives seemed to accept this problem as part of the job. Nor did they seek to resolve the problem of poor housekeeping in order to improve their working environment.

Work attitude

A significant part of the discussion focused on psychological rather than physical aspects. This information was useful as it was used to clarify behaviour noted in the observational study. One midwife did not consider 'backache' to be part of her job and would not do a task if it caused discomfort.
" If a get backache I will leave the situation and go back to it later. I won't compromise"
Other midwives did state they would compromise their physical well being if it was in the interest of the client especially in an emergency situation. This was only cited by a few, others felt they conformed to the client wishes without considering their own

welfare. This may be perceived as compliant behaviour similar to the actions noted with the continual use of defective equipment.

Midwives were asked how they might improve their working conditions. Some thought additional knowledge of working postures would be beneficial. Few midwives felt work conditions would not change, nor felt empowered to make change to reduce musculoskeletal discomfort. It was perceived as part of the job and a legacy from their 'nursing days'.

Given this information it was useful to draw conclusions from some of the behaviour exhibited by midwives during the observational study. Had the methodology only included postural measurements then the understanding of what was actually happening behind the adopted postures would have been missed.

Quick Exposure Check Assessment Tool

The Quick Exposure Check (QEC) observational paper tool was selected for the study. (Li and Buckle 1998). Its reliability is reported to be 'fair' to 'moderate' (Li and Buckle 1999).

The data gathering was completed on the postnatal wards in three NHS Healthcare Trusts. A total of thirty registered midwives were recruited through purposive sampling.

Midwives were asked to assist the woman to breast-feed in whatever manner she/he thought was appropriate. The observer offered no further instructions. During this period of interaction between the midwife and the woman, the recording of four body measurements along with the 'workers assessment' scores was noted.

A numerical score was calculated to establish an 'exposure score'

The QEC mean scores suggest midwives frequently adopt back postures from the erect position to one of flexion between 20 and 60 degrees. No midwives maintained a fixed posture, but altered their posture frequently. Why midwives change their posture frequently is unknown, however midwives independently cited in the focus group interviews that they admitted to changing posture to alleviate discomfort. This aspect is worth investigating further by using case studies.

As well as flexed back postures, midwives also demonstrated torsion of the lower and mid back regions. This occurred when the midwife assisted women who were positioned in bed. Commonly midwives sat on the edge of the bed facing the woman.

Three midwives chose to kneel to avoid backache. They reported they would rather risk knee discomfort than tolerate back discomfort. This was reinforced from the information already gathered from focus group interviews. The subjects who knelt to assist women were seated in a low chair; all midwives knelt on a hard surface with no protective covering. Despite reporting knee discomfort they all changed their position from one knee to the other and sometimes on both knees. Some midwives, who knelt, did so by placing one knee on the bed. This group reported less discomfort and maintained the back posture in a neutral position without flexion and torsion to the spine. One midwife knew she should place one knee on the bed but failed to understand the principles of appropriate working heights. Her posture caused excessive hip flexion with back flexion greater than 60 degrees.

In some cases extensive stooping was attributed to the midwife failing to adjust the bed to an appropriate working height, citing this activity took too long to execute. They did this knowing they may experience low back discomfort.

There was a correlation between back flexion and extension of arms. Back flexion increased as the arms extended away from the trunk. It was evident that the arms had to be extended in order to get closer to the woman and her baby. This was more noticeable when the midwife was standing and the woman was seated in a chair. Midwives reported discomfort in the shoulders, mid and lower back during this activity.

The QEC tool assesses wrist and hand positions. Deviations of wrist/hand movement from straight to lateral flexion and rotation involving movement of the forearm were noted. In all cases, whether the midwife stood, knelt or sat, the hand supporting the breast always caused ulnar and radial rotation causing the wrist to rotate, but the angle of the wrist did not deviate, remaining almost straight. The thumb in most cases was abducted, thus extending the saddle joint. Despite this posture no discomfort was reported on the Borg Scale for wrists, hand or lower arm. The correlation between back flexion and wrist/hand posture was interesting. It appeared the greater the back flexion the greater degree of deviation in wrist flexion. This was particularly noticeable when women were seated and the midwife standing.

Visual demands and neck posture were noted. The study found no subjects with a neck flexion of greater than 20 degrees, but neck torsion was noted. Two contributing factors may predispose to neck torsion it was noted in:

- midwives who conducted the task to the side of the woman and seated on the bed.
- midwives whose back flexion was greater than 40 degrees.

Whilst the QEC tool measured body postures relating to the work demand, aspects of non-work exposure to risk factors was limited. In order to address the work demand aspects more fully findings from the focus group interviews and self-assessment reporting of discomfort scores were triangulated. Further information about the methods and results are reported by Steele (2001).

Self-Assessment Reporting

In order to correlate the experiences of the midwives, a self-reporting assessment was completed immediately after the observational study using Corlett and Bishop's(1976) body map, and Borg's(1990) discomfort rating scale. The body maps primarily identify the link between work postures and pain. This data being largely subjective, reports the perceived experience of the individual undertaking an activity, the event being unique to the individual. Body maps and discomfort scales have been used with some success, indeed it provides additional qualitative data when used with other quantitative data. The major drawbacks to subjective rating scales are that they are prone to other influences within the workplace or personal lifestyles. To use either methods in isolation is not recommended. Where self reporting assessment is used without any other measures, the validity and reliability is reported to be low (Burdorf & Laan 1991, Wiktorin et al 1993). In this study Borg's scale ratio scores had rank order intervals with a scale 1 to 10. The larger the range the harder it is to define the categories and makes judgements more difficult. As the tool was administered immediately after the observational study and given pain measurement is subjective, it was recognised that this was a crude measure to assess pain when compared to other biophysical measurements.

The self-assessment reporting did confirm the highest amount of discomfort was reported in all the back regions with the low back scoring the most. Conclusions were drawn between the amount of perceived discomfort and what was observed of the body postures. A correlation between increased back flexion and torsion and reported discomfort was confirmed. Reflecting back to comments made by the midwives in the focus groups confirmed the notion that midwives do experience back discomfort not only when helping women to breast feed but during other work activities.

Body regions of the shoulders, lower neck and knees reported less discomfort but should not be ignored.

Conclusion

This paper draws together not just an overview of one study investigating the working postures of midwives when assisting women to breast-feed but encompasses the notion of using a varied research approach. In the healthcare setting, work roles are diverse and may not be comparable. Therefore where midwifery working postures are thought to be a threat to workers health, the research design should include qualitative as well as quantitative methods, in order to understand the relationship between the psychosocial and the physical demands of work.

References

Borg G 1990, Psychophysical scaling with applications in physical work and the perceptions of exertions. *Scandinavian Journal Environmental Health* 16:55-58.

Burgdorf A & Laan J 1991, Comparisons of methods for the assessment of postural load on the back. *Scandinavian Journal of Work & Environmental Health* 17: 425-429.

Corlett E N & Bishop R 1976, A technique for assessing postural discomfort. *Ergonomics* 19(2):175-182.

Hignett S 1996, Manual handling risks in midwifery: identification of risk factors. *British Journal of Midwifery* 4(11):590-596.

Li G & Buckle P 1998, The development of a practical method for the exposure to work related musculoskeletal disorders. *General Report to HSE.* Robens Centre for Health Ergonomics, EIHMS, University of Surrey.

Li G & Buckle P 1999, Current techniques for assessing physical exposure to work-related musculoskeletal risks, with emphasis on posture-based methods. *Ergonomics.* 42 (5): 674-695.

Parahoo J 1997, *Nursing Research: Principles, Process and Issues. MacMillan* Press.

Royal College of Midwives 1997, *Handle with care. A midwife's guide to preventing back injury.* London

Swanson J & Cheniez T 1986, *From Practice to Grounded Theory.* Addison Wesley.

Steele D 2001, A study of working postures adopted by midwives when assisting women to breast-feed. *MSc Dissertation . Unpublished.* University of Surrey. Guildford.

Wiktorin C, Karlqvist L, Winkel J and Stockholm Music 1 Study Group 1993, Validity of self-reported exposures to work postures and manual materials handling. *Scandinavian Journal of Work, Environment and Health* .19:208-214.

EVIDENCED BASED PATIENT HANDLING:
A SYSTEMATIC REVIEW

Sue Hignett, Emma Crumpton, Pat Alexander, Sue Ruszala, Mike Fray and Brian Fletcher

Ergonomics and Back Care Advisory Department
Nottingham City Hospital NHS Trust
Nottingham NG5 1PB

This paper will present a systematic literature review on patient handling activities. It reports selected findings from the search results of over 1,000 papers.

Patient handling activities, as primary or secondary tasks, are carried out by staff throughout the health care industry. Over the last 20 years a number of guidance publications have been published. These have been used operationally throughout the health care industry to guide clinical practice, and additionally as reference texts for expert opinion in legal proceedings. To date there has been no substantial review that places the advice about patient handling techniques and equipment in the context of research rather than expert opinion or anecdote.

Introduction

The aim of this project was to bring together all available research in a systematic literature review framework. It is hoped that this will enable both published and future research to be read and appraised in a context of robust, scientific research and provide the foundation for future guidance publications.

This paper describes the analysis of one task: transferring patients from a sitting position (on a bed, chair etc.). The review of the literature will be described with respect to the level of evidence for current practice.

Methodology

The methodology, inclusion/exclusion criteria and development of the data appraisal/extraction form are described elsewhere (Crumpton et al, 2002).

Search results

The search was extended to include all languages (with translation being arranged as required) and included a number of strategies (Hart, 1998; Hamer and Collinson, 1999) as shown in table one.

Table 1.   Search strategies and results to Sept. 2001

Source	Retrieved	Retained at Sept. 2001
Medline (1960-present)	1130	336
AMED	34	31
Psychinfo	73	11
EMBASE	631	174
CINAHL	493	168
British Nursing Index	12	10
Best Evidence	2	0
Hand searching journals and exploding the reference list of identified papers Screening cited references	47	?
Contacting expert informants and theses	42	?
Personal collections Books, newsletters, conference papers	305	305
TOTAL	2769	1035

A quantitative meta-analysis was not considered to be appropriate for the data synthesis in this review due to the heterogeneity of interventions, settings, participants and outcome measures used.

Data synthesis

The data were synthesised in two stages. The first stage involved grouping papers into the categories of task, equipment or intervention with some papers being allocated to more than one category. This was followed by agreeing evidence levels (table two) based on the quality and practitioner score from extraction/appraisal process.

Table. 2. Evidence levels (Bernard, 1997; Faculty of Occupational Health, 2000)

++	Strong evidence - provided by multiple, high quality ($\geq$ 75% score) studies. High practitioner score (> 4)
+	Moderate evidence- provided by generally consistent findings in fewer, smaller or lower quality (50%-74% score) studies Moderate practitioner score (3-4)
-	Limited or contradictory evidence- provided one study, or inconsistent findings in multiple studies. Lower quality (25%-49% score) studies Low practitioner score (2-3)
- -	Poor or no evidence - no studies or low quality score ($\leq$ 24%). Very low practitioner score (< 2)

Current practice, advice and guidelines

In order to place the findings of the project in the context of current practice, advice and guidelines, the following sources were reviewed to develop inclusive flow charts: Disabled Living Foundation (1994); Essex Group of National Back Exchange (1996); The Disability Information Trust (1996); Royal College of Midwives (1997); National Back Pain Association/ Royal College of Nursing (1997); South London and Kent Group, National Back Exchange, (1998); Health Services Advisory Committee (1998); Chartered Society of Physiotherapy (1998); Oxford Region, National Back Exchange (1999); ASSTSAS (1999); Human Services/Victorian Work cover Authority (1999); Graham et al (2000); The Resuscitation Council (2001). Figure one gives an example the flowcharts by showing options for transferring from sitting for three levels of patient ability (needing no help, some help or a lot of help)

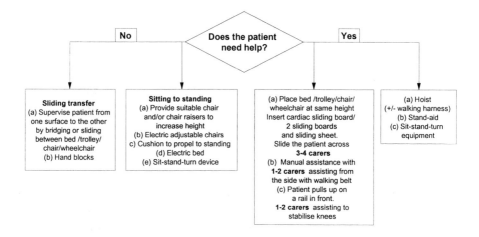

Figure 1. Transferring from a sitting position

Results for specific task (transferring from a sitting position)

Table three shows the summary the papers for the task of transferring a patient from a sitting position, listing the quality and practitioner scores from the extraction/appraisal process (Crumpton et al, 2002).

The first draft of the evidence level for the options in the flow chart is currently being considered by the project team. The proposed interpretation is as follows:

+   Moderate evidence for using mechanical (hoists) lifts for dependent patients, but consideration should be given to the selection of the mechanical equipment

+   Moderate evidence for using a walking belt with two carers for weight bearing patients over lifting with two carers; gait belt with two carers; handling sling with one carer; and walking belt with one carer. The walking belt technique was found to be faster than using mechanical aids (standing and lifting hoists)

Table 3. Summary of papers relating to the task of transferring a patient from a sitting position

Study	Recommendation	Quality Score	Practitioner Score
Garg et al (1991a) Bed ↔ Wheelchair	Walking belt with two carers for weight bearing patients Two of the hoists had a lower rating than the walking belt with two carers	75.5%	4.5
Garg et al (1991b) Wheelchair ↔ shower chair	Walking belt with two carers for weight bearing patients Two of the hoists had a lower rating than the walking belt with two carers	54%	4
Garg and Owen (1994) Toilet ↔ Wheelchair	Walking belt with two carers for weight bearing patients Two of the hoists had a lower rating than the walking belt with two carers	66.5%	4.5
Marras et al (1999)	Found that the single carer hug method resulted in about 10% higher risk than any of the two carer transferring methods. The use of the gait belt only reduced the loading for the carer positioned on the right side of the patient.	72%	5
Ulin et al (1997) Bed ↔ Wheelchair	Mechanical lift should always be used for totally dependent patients regardless of patient weight	66.5%	5
Zhuang et al (1999) Bed ↔ Chair	Basket lift and overhead lift	75.5%	4.5
Zhuang et al (2000) Bed ↔ Chair	Walking belt, two of the stand-up lifts, two of the basket lifts	57.5%	4.5

Discussion and conclusion

It seems, from the data, that many of the currently recommended options have no evidence base. Moderate evidence was only found for the use of mechanical lifts for dependent patients and walking belts (with two carers) for weight bearing patients.

These findings will be put into context when the full review (Hignett et al, 2002) is reported.

References

ASSTSAS. 1999, *Principles for moving patients safely,* L'Association pour la santé et al sécurité du travail, secteur affaires socials (ASSTSAS), Montréal, Canada

Bernard B.P. 1997, (ed.) *Musculoskeletal disorders and work place factors. A critical review of epidemiologic for work-related musculoskeletal disorders of the Neck, Upper Extremities and Low Back.* (NIOSH, US Department of Health and Human Sciences), 1-13 – 1.14

Chartered Society of Physiotherapy. 1998, *Moving and Handling for Chartered Physiotherapists,* (London, Chartered Society of Physiotherapy)

Crumpton, E., Hignett, S., Goodwin, R. and Dewey, M. 2002, Reliability study of a data extraction and quality assurance tool applied to literature on patient handling. In P.T. McCabe (ed.) *Contemporary Ergonomics 2002,* (London, Taylor and Francis)

Disabled Living Foundation. 1994, *Handling People: Equipment, Advice and Information,* (London: Disabled Living Foundation)

Essex Group of National Back Exchange. 1996, *Paediatric Moving and Handling: Report of workshops*

Faculty of Occupational Medicine. 2000, *Occupational Health Guidelines for the management of low back pain at work.* Evidence Review and Recommendations. London: Faculty of Occupational Medicine. ISBN 1-86016-131-6, 27-28

Garg, A., Owen, B., Beller, D. and Banaag, J. 1991a, A biomechanical and ergonomic evaluation of patient transferring tasks: bed to wheelchair and wheelchair to bed. *Ergonomics,* 34, **3**, 289-312

Garg, A., Owen, B., Beller, D. and Banaag, J. 1991b, A biomechanical and ergonomic evaluation of patient transferring tasks: wheelchair to shower chair and shower chair to wheelchair. *Ergonomics.* 34, **4**, 407-419

Garg, A., and Owen, B. 1994, Prevention of back injures in healthcare workers. *International Journal of Industrial Ergonomics,* **14**, 315-331

Graham, J., Hurran, C., and MacKenzie, M. 2000, *Paediatric Manual Handling. Guidelines for Paediatric Physiotherapy,* (Long Eaton, Association of Chartered Physiotherapists)

Hamer, S. and Collinson, G. 1999, *Achieving Evidenced-based practice. A handbook for Practitioners,* (Edinburgh, Ballière Tindall)

Hart, C. 1998, *Doing a literature review. Releasing the Social Science Research Imagination,* (London, Sage Publications)

Health Services Advisory Committee. 1998, *Manual Handling in the Health Services,* (Health and Safety Commission, London. HMSO)

Hignett, S., Crumpton, E., Alexander, P., Ruszala, S., Fray, M. and Fletcher, B. 2002, *Evidenced Based Patient Handling: Techniques and Equipment,* (London, Routledge). In press

Human Services/Victorian Work cover Authority. 1999, *Manual Handling: Reducing the Risk, Reducing Injuries,* (Human Services/Victorian Work cover Authority, Australia)

Marras, W.S., Davis, K.G., Kirking, B.C. and Bertsche, P.K. 1999, A comprehensive analysis of low-back disorder risk and spinal loading during the transferring and repositioning of patients using different techniques. *Ergonomics,* 42, **7**, 904-926

National Back Pain Association/ Royal College of Nursing. 1997, *The Guide to the Handling of Patients,* 4th Ed., (Teddington, Middlesex, National Back Pain Association/ Royal College of Nursing)

Oxford Region, National Back Exchange. 1999, *Generic Safe Systems of work for patient handling and inanimate load management,* (Oxford Region, National Back Exchange)

Royal College of Midwives. 1997, *Handle with Care. A midwife's guide to preventing back injury,* (London, Royal College of Midwives)

South London and Kent Group, National Back Exchange, 1998, *Manual Handling procedures,* (South London and Kent Group, National Back Exchange)

The Disability Information Trust. 1996, *Hoists, Lifts and Transfers*, (The Disability Information Trust, Oxford: Nuffield Orthopaedic Centre)

The Resuscitation Council. 2001, *Guidance for safer handling during resuscitation in hospital,* (London, The Resuscitation Council)

Ulin, S., Chaffin, D.B., Patellos, C. and Blitz, S. 1997, A biomechanical analysis of methods used for transferring totally dependent patients, *Scientific Nursing*, 14, **1**, 19-27

Zhuang, Z., Stobbe, T.J., Hsiao, H., Collins, J.W. and Hobbs, G.R. 1999, Biomechanical evaluation of assisting devices for transferring residents, *Applied Ergonomics,* 30, 285-294

Zhuang, Z., Stobbe, T.J., Collins, J.W., Hsiao, H. and Hobbs, G.R. 2000, Psychophysical assessment of assistive devices for transferring patients/residents, *Applied Ergonomics*, 31, 35-44

ESTABLISHING RELIABILITY FOR DATA EXTRACTION IN A SYSTEMATIC REVIEW ON PATIENT HANDLING

Emma Crumpton[1], Sue Hignett[1], Rob Goodwin[1], Michael Dewey[2]

*[1]Ergonomics and Back Care Advisory Department
Nottingham City Hospital NHS Trust, Nottingham NG5 1PB
[2]Trent Institute for Health Services Research
University of Nottingham, Nottingham NG7 2RD*

This paper describes the development and testing of a data appraisal/extraction tool for use in a systematic review on patient handling (Hignett et al, 2002a). Prior to the main review an inter-rater reliability exercise was carried out. Statistical analysis of the results showed good inter-rater reliability for the overall quality score and so the tool will now be applied, with confidence, to the full systematic review.

Introduction

Since the early 1980s the health care industry has used formal methods of systematically reviewing studies to produce reliable, useful summaries of the effects of health care interventions. This type of information is then used to formulate guidelines and clinical strategies (Ukoumunne et al, 1999, Cameron et al, 2000). The change in approach has been driven by both external (political) and internal (professional) pressures to base clinical decisions on scientific evidence.

Chalmers and Altman (1995) identified a number of applications for systematic reviews, with the most important being the ability to give an accurate assessment of the literature. They suggested that traditional (narrative) reviews could be criticised as being haphazard and biased by the individual reviewer, a more systematic approach would achieve greater reliability by allowing an assessment of how the review was carried out. Additional benefits include: a more explicit definition of exclusion/inclusion criteria and the critical appraisal process; and the possibility of explaining inconsistencies or conflicts in data which might relate to the sampling strategy or outcome measures etc.;

Aim
The aim of this paper is to:
1.   describe the process of developing and testing the data extraction/appraisal tool
2.   provide evidence for the reliability of the data extraction/appraisal process in the systematic review on patient handling.

Method

Hamer and Collinson (1999) describe the key components of a systematic review as being:

1.   Definition of the research question
2.   Methods for identifying the research studies
3.   Selection of studies for inclusion
4.   Quality appraisal of included studies
5.   Extraction of the data
6.   Synthesis of the data

The research question (step 1) is discussed elsewhere in these proceedings (Hignett et al, 2002a), together with details of the search strategy and findings (step 2). This paper will discuss steps 3-5 of the above process, with an example for step 6 also being given in Hignett et al (2002a).

Inclusion/Exclusion Criteria (step 3)

All languages were included in the search, where possible English abstracts were obtained, with translation being arranged as required. The following inclusion/exclusion criteria were applied, whereby a paper was:

1.   Included if it described a named task, piece(s) of equipment or intervention relating directly to patient handling.

2.   Included as a professional opinion if it:

    • Had references.
    • Critically appraised the literature.
    • Provided new interpretation of the literature.

3.   Excluded if it was related to epidemiology of musculoskeletal disorders (usually low back pain) and did not meet criterion (1) for any part of the study.

4.   Excluded if it was not the primary source of a study. The primary source was sought and included.

Appraisal/Extraction Tool (steps 4 and 5)

All the project team (Hignett et al, 2002a) have specialist experience in the area of patient handling and were concerned that there might be limited research which met the highest traditional criteria for systematic reviews (randomised controlled trials). One of the authors (MD) has extensive experience in carrying out systematic reviews and advised that a check list which included both randomised and non-randomised studies would be more appropriate for this area of interest. Downs and Black (1998) have produced such a checklist, and this formed the basis for the appraisal/extraction tool for this project.

Figure one shows the decision-making hierarchy used in the appraisal/extraction process. The Downs and Black (1998) checklist was used for all studies 'with an intervention'. In order to include a range of study types the checklist was modified for 'studies with no intervention' and 'other' types of studies. The modifications were tested within the project group before undergoing the inter-rater reliability trial. In order to include qualitative studies a checklist was developed from the appropriate literature (Grbich, 1999, Whalley-Hammell et al, 2000, Seale, 1999). A full copy of the data appraisal/extraction tool can be found in Hignett et al (2002b).

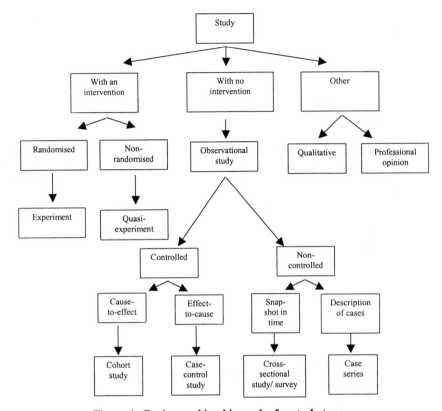

Figure 1. Design-making hierarchy for study type

The data appraisal/extraction tool produces scores in four sections with a final quality score. The four sections are:

1. Reporting: Assessing whether the information presented in the paper would allow the reader to make an unbiased assessment of the findings of the study.

2. External validity: Assessing the extent that the findings from the study could be generalized to the population from which the study subject came.

3. Bias: Assessing the biases in the measurement of the intervention and outcome.

4. Confounding: Assessing the bias in the selection of the study subjects.

An additional reality check was added in the form of a five point subjective practitioner score. This asked the reviewer to consider whether they would be likely to use the findings from the study in their practice regardless of the scientific quality of the paper.

Inter-rater reliability

As the review process would use two members of the project team to appraise and extract data from each paper, before the main part of the review could commence it was important to assess the inter-rater reliability of the reviewers. Ten papers were randomly selected and copies of each were sent to the members of the project team.

Inter-rater reliability was calculated using intraclass correlation coefficients (ICC). From this calculation of single measure intraclass correlation, the ICC for two judges was calculated using the Spearman-Brown prophecy formula (Shrout and Fleiss, 1979).

Results

There was complete agreement, by all reviewers, on the study design type as shown in table one.

Table 1. Results of Study Design Type

Study identification	Study design type
1. Charney *et al*, 1991	Quasi-experimental
2. Coleman, 1999	Cross-sectional study
3. Dehlin and Lindberg, 1975	Cross-sectional study
4. Ellis, 1993	Survey
5. Engels *et al*, 1994	Cross-sectional study
6. Fenety and Kumar, 1992	Survey
7. Finsen *et al*, 1998	Cross-sectional study
8. Fragala, 1993	Quasi-experimental
9. French *et al*, 1997	Cross-sectional study
10. Garg and Owen, 1991	Quasi-experimental

The results of the inter-rater reliability are shown in table two.

Table 2. Results of Quality Assurance Review

Quality assurance component	Pair-wise, Intra-class correlation (r=)
Reporting	0.87
External validity	0.53
Bias	0.72
Confounding	0.76
Overall quality score	0.95

The individual components of the appraisal/extraction tool reliability varied from poor/moderate (external validity) to good (reporting, bias and confounding). The overall quality score showed good reliability (r=0.95).

Discussion and conclusion

Downs and Black (1998) found similar results, with poor reliability for external validity (Spearman correlation coefficient= -0.14). They suggested that this might either be due to the small number of items in this section (three questions compared with six to ten in the other sections) or to the construction of the questions. The overall quality score showed good reliability (r=0.95), which compared well with the findings of Downs and Black (1998), correlation= 0.75.

The data appraisal/extraction tool for this systematic review has been developed to include a wider range of study types and it will be interesting to see how this works for the main review. Inter-rater reliability has been established, giving a level of confidence to the main review process.

References

Cameron, I., Crotty, M., Currie, C., Finnegan, T., Gillespie, L. and Gillespie W. 2000, Geriatric Rehabilitation Following Fractures in Older People: A Systematic Review, Health Technol Assess, 4, 2

Chalmers, I. and Altman, D.G. (eds.) 1995, *Systematic Reviews,* (London. BMJ Publishing Group)

Charney, W., Zimmerman, K. and Walara, E. 1991, The lifting team. A design method to reduce lost time back injury in nursing, *AAOHN Journal,* **39,** 231-234

Coleman, S. 1999, Manual handling in the operating theatre, *Professional Nurse,* **14,** 682-686

Dehlin, O. and Lindberg, B. 1975, Lifting Burden for a Nursing Aide During Patient Care in a Geriatric Ward, *Scand J Rehab Med,* **7,** 65-72.

Downs, S.H. and Black, N. 1998, The Feasibility of Creating a Checklist for the Assessment of the Methodological Quality both of Randomised and Non-Randomised Studies of Health Care Interventions, *Journal of Epidemiological Community Health,* **52,** 377-384

Ellis, B.E. 1993, Moving and Handling Patients: An evaluation of current training for Physiotherapy students, *Physiotherapy,* **79,** 323-326

Engels, J.A., Landeweerd, J.A. and Kant, Y. 1994, An OWAS based analysis of nurses' working postures, *Ergonomics,* **37,** 909-919

Fenety, A. and Kumar, S. 1992, An ergonomic survey of a hospital physical therapy department, *International Journal of Industrial Ergonomics,* **9,** 161-170

Finsen, L., Christensen, H. and Bakke, M. 1998, Musculoskeletal disorders among dentists and variation in dental work, *Applied Ergonomics,* **29,** 119-125

Fragala, G. 1993, Injuries cut with lift use in ergonomics demonstration project, *Provider,* **Oct,** 39-40

French, P., Lee Fung Wah, F., Sum Ping, L. and Wong Heung Yee, R. 1997, The prevalence and cause of occupational back pain in Hong Kong registered nurses, *Journal of Advanced Nursing,* **26,** 380-388

Garg, A. and Owen, B. 1991, A biomechanical and ergonomic evaluation of patient transferring tasks, *Proceedings of 11th Congress of IEA, Paris 1991* **1,** 60-62

Grbich ,C. 1999, Qualitative Research in Health. An Introduction, (Sage Publications. London)

Hamer, S. and Collinson, G. 1999, *Achieving Evidenced-based practice. A handbook for Practitioners,* (Edinburgh, Ballière Tindall)

Hignett, S., Crumpton, E., Alexander, P., Ruszala, S., Fray, M. and Fletcher, B. 2002a, *Evidenced Based Patient Handling: A Systematic Review.* In P.T. McCabe (ed.) *Contemporary Ergonomics 2002,* (London, Taylor and Francis)

Hignett, S., Crumpton, E., Alexander, P., Ruszala, S., Fray, M. and Fletcher, B. 2002b, *Evidenced Based Patient Handling: Techniques and Equipment, (*London, Routledge). In press

Seale, C. 1999, *The Quality of Qualitative Research.* (Sage Publications. London)

Shrout, P.E. and Fleiss, J.L. 1979, Intraclass Correlations: Uses in Assessing Rater Reliability, *Psychological Bulletin,* **86,** 420-428

Ukoumunne, O.C., Gulliford, M.C., Chinn, S., Sterne, J.A.C. and Burney, P.G.J. 1999, *Methods for Evaluating Area-Wide and Organisation-Based Interventions in Health and Health-Care: A Systematic Review*, Health Technol Assess, 3, 5

Whalley-Hammell, K., Carpenter, C. and Dyck I. 2000, *Using Qualitative Research. A Practical Introduction for Occupational and Physical Therapists.* (Edinburgh, Churchill Livingstone)

INTEGRATING NEW TECHNOLOGY INTO NURSING WORKSTATIONS: CAN ERGONOMICS REDUCE RISKS?

Marina Naki and Rachel Benedyk

Ergonomics and HCI Unit, UCL,
26 Bedford Way,
London WC1H 0AP

Within a hospital ward setting, Health Care Professionals are often forced to adopt poor working postures for administrative tasks, particularly in cases where they are required to work at workstations that are not ergonomically designed to accommodate the introduction of computers. An evaluation of nursing workstations at a large NHS hospital was carried out; problems and risks were identified with the current design and recommendations were made for the redesign of the workstations, with a view to accommodating the introduction of new technology and reducing musculoskeletal complaints. The Concentric Rings Model was employed to guide the evaluation, and its effectiveness was evaluated. An enhanced model (the Labyrinth model) was proposed.

Introduction

This paper focuses on inpatient wards in a large London hospital. Each ward has its own nursing workstation, which is the area on the ward which nurses and other members of staff use to perform non-clinical tasks related to the patients. Functions performed at the nursing workstation include: inputting or retrieving patients' information (possibly via a computer), using the phone, writing medical reports, communicating with other staff or visitors, using the patient name board, using storage areas as well as sorting out patients' medication. The furniture and equipment in nursing workstations differs and the position of the computer at the workstation varies, according to need and the availability of space.

The need for this study emerges from the gradual introduction of computers for ward tasks, to be used in parallel with manual tasks such as writing medical notes and drug charts. Health care professionals can be forced to adapt to poorly designed workstations, which have not been ergonomically designed to accommodate the new technology, leading to an increased number of reported musculoskeletal problems among such professionals (McHugh and Schaller, 1997).

This study employed a holistic approach, as ergonomics would recommend that the computer use should not be viewed in isolation. Nursing workstation usage can be influenced and affected by other factors, including: the physical workplace and work environment, such as the wards; the organisation and the social environment, such as the hospital; and the world outside i.e. society. The study used a methodological framework, to enable a comprehensive approach to identifying the risks at workstations to be employed. The Concentric Rings Model (Girling and Birnbaum, 1988) emphasises the interactions and influences that each of the factors bears on each other and on the system as a whole.

The main objectives of this study were: to evaluate the existing nursing workstations on the wards and identify problems that health care professionals have to face, with a view to accommodating the increased demands of new technology and preventing musculoskeletal injuries; and to evaluate the utility of the Concentric Rings Model in this context.

Data collection

Four wards were selected based on ease of access, which according to their managers needed urgent evaluation and attention. Twenty-five voluntary subjects for the study were all employees of these wards and all spent at least 30 minutes per day using the workstations. These health care professionals included ward managers, nurses, midwives, physiotherapists, pharmacists and dieticians.

A convergent techniques approach was adopted (Figure 1) which included structured interviews, direct observation, postural analysis using photographic material and direct measurements of the nursing workstation. A questionnaire was designed to identify how the users perceive problems at the nursing workstations. Interviews were conducted to obtain more details about the different tasks and to discuss possible solutions and future perspectives.

The following areas were investigated:
• Nursing workstation characteristics and limitations as perceived by the user, such as adequate space, working postures, functions, furniture and equipment (especially regarding the use of the computer).
• Ward characteristics, such as location, physical aspects and nurse's workload.
• Hospital procedures, standards and availability of training
• Subject's individual characteristics, such as sex, age, job title, length of employment and health problems.

Problems and recommendations

36% of the subjects reported musculoskeletal problems related to the use of the workstation. Using the telephone (100%) and the computer (56%), as well as reading and writing medical notes (84%) and communicating with other people (56%), were amongst the most frequent tasks and 96% of the staff had to perform concurrent tasks. It was reported that a mean of 7 to 8 people could have been using the workstations at a particular time, with a maximum of 15 people for a short time (for example, at the beginning of the ward rounds).

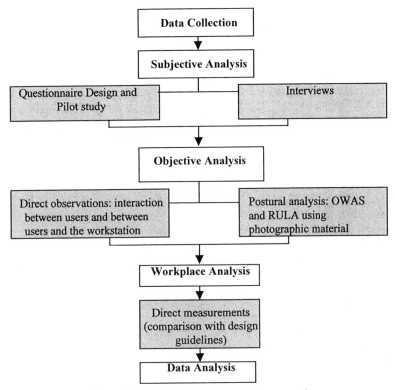

Figure 1. Selected methodological framework.

Amongst the most common complaints were: inadequate leg room when using the bench, inadequate storage area leading to the inappropriate use of the space under the bench to store equipment and patients' property, as well as lack of adjustable working surfaces leading to poor working postures and potentially contributing to musculoskeletal stress. 52% thought that the bench was too high when sitting and 56% that the bench was too low when standing, leading to the conclusion that the present bench cannot function sufficiently in either situation. A lack of adequate working surfaces and properly functioning adjustable chairs was also identified.

In relation to the computer, the main problems that were observed were the increased glare on the computer screens, insufficient wrist support, poor positioning of the screen in relation to the user and the lack of adequate legroom due to objects stored underneath the computer table. From the postural analysis, it can be concluded that the majority of the working postures observed at the workstations needed immediate attention and change.

Overall, the majority of the furniture and equipment at the workstations of the four wards was not designed for the specific tasks performed there.

The whole of each workstation was evaluated, as the implementation of new technology could lead to potential problems and limitations for other tasks in the workstations. Recommendations were made for the adaptation of the current workstations in order to accommodate computers. Solutions were proposed based on problems identified at situational, behavioural and managerial levels, as per the Concentric Rings Model. Examples of the recommended solutions were: repositioning of the telephones, one close to the computer and one to a more private area, availability of more adjustable working surfaces and chairs, reorganisation of the workstation to ensure adequate storage areas. Staff education and training, as well as job re-design and better equipment maintenance were also suggested.

Although a variety of methods were employed in this study to investigate the risks and ergonomic recommendations were proposed, further research is needed to establish long term solutions to the problems and to evaluate the potential implementations.

The Models

The results indicate that at the situational level, many problems existed in the design of the workstation. For example, inadequate size of the workstation for the number of people working at it, poor furniture dimensions and limited equipment. Such limitations are reinforced by other problems at the workplace (the wards) and the organisation (the hospital), such as poor noise control, poor equipment maintenance as well as poor job design.

It follows that more information can be collected and interpreted if managerial and behavioural factors are considered concurrently with situational factors. At the level of management, problems included limited application of design guidelines, lack of adequate risk assessments and feedback to the employees, as well as inadequate training of ward staff.

The results of this study identified the behaviour of the ward staff in response to the above problems and coping strategies adopted by them in order to function at the workstations. Some members of staff passively accepted the existing situation and showed little interest in changing it and others adopted techniques to temporarily overcome the limitations, for example, not using the top shelf often because it was too high.

From the above, it can be concluded that the model used in this study was of particular value in determining a comprehensive range of relevant factors. However, it could be interpreted that the Concentric Rings Model is fairly static and the ability to demonstrate movement through the design process cycle needs to be addressed. We therefore propose a development of the model in order to increase its effectiveness and suitability when applied in future research. This is the 'Labyrinth Model' (Figure 2). The new Model could be described as an 'open loop system' and as such provides more flexibility at each level as well as emphasising the differentiation between levels.

The Labyrinth Model continues to differentiate between situational, managerial and behavioural factors at each level as Girling and Birnbaum (1988) proposed. It suggests that the analysis could be initiated at a global level – the environment/society and the organisation – and then move to more specific levels in the system – the workplace and the workstation –

always in relation to the user and his/her tasks. The Labyrinth Model could be described as a 'top down' approach to system design with the view to identifying the possible origins of the problems, which in some cases lie at the organisational level. If problems are found to be at a lower level (closer to the user) e.g. the workplace, the proposed level allows the researcher to return to a higher level if the need for further investigation arises. We suggest this refined model will have useful wider application in both evaluation and design.

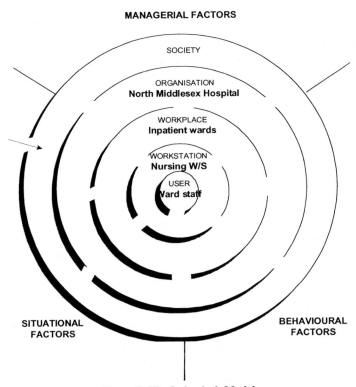

Figure 2. The Labyrinth Model.

References

Girling B. and Birnbaum R. 1988, An ergonomic approach to training for prevention of musculoskeletal stress at work, *Physiotherapy*, **74**, 9, 479-483.
McHugh M.L. and Schaller P. 1997, Ergonomic nursing workstation design to prevent cumulative trauma disorders, *Computers in nursing*, **15**, 5, 245-252.

PROVISION OF ASSISTIVE TECHNOLOGY FOR DISABLED PERSONS

Gary T Clark, Robert H Campbell & Frank Bell

Glasgow Caledonian University
Cowcaddens Road, Glasgow, G4 0BA

Technical aids such as hearing aids, spectacles, walking aids and wheelchairs are common examples of customised assistive technology designed specifically to meet the needs of people with physical impairments. More complex systems for more complex disabilities have also been available for many years. For example some 15 years ago one of the authors (Bell et al) was involved in the investigation of new users of Possum PSU3 environmental control equipment. This equipment was provided through the National Health Service to severely disabled people to enable them to continue to live in the community without the need for constant attendance from relatives, and allowed them to control lighting, telephones, television sets, door and curtain openers and so forth.

Over the last two decades, the provision and use of technology in the general population has greatly increased. Common examples include automatic bank telling machines, web based shopping and commerce, remote controls for audiovisual and other electrical devices, speech recognition software, mobile phones and video conferencing. The concepts of commercial universal design, taking account of the wider population, including people with disabilities and of customised design specifically for disabled people, have attracted considerable debate over the past decade. In particular, by Lewis, 1998, Pell et al 1999 and Vanderhein, 1998.

This current study comprised a review of relevant literature and products, visits to agencies providing technical aids and assistance, and three case studies. Whilst recognising the uniqueness of the individual it was found nevertheless useful to consider five broad groupings of people when reviewing the literature, as the onset and development of impairment can have quite different consequences and technical needs.

The poster presentation presents examples of universal and customised products with comments on use and costs, one model for assessment used in one local rehabilitation centre and summary of design philosophy criteria.

References

Bell, F., Whitfield, E. and Rollet, P. 1987, Investigation of Possum users in Scotland, *International Rehabilitation Medicine*, **8** (34), 105 – 112

Lewis R.B., 1998, Assistive technology and learning disabilities: today's realities and tomorrow's promises, *Journal of Learning Disabilities*, **31** (10), 16 – 26

Pell, S.D., Gillies, R.M. and Carss, M., 1999, Use of technology by people with physical disabilities in Australia, *Disability and Rehabilitation*, **21** (20), 56 – 60

Vanderhein, G.C., 1998, Universal design and assistive technology in communication and information technologies, *Assistive Technology*, **10**, 29 - 36

The ASSESSMENT OF HAND FUNCTION AND ITS RELEVANCE TO PATIENTS WITH ARTHRITIS

Mohssen Hassan-Zadeh[1], Frank Bell[1] , Marc Bransby – Zachary[2], Deborah Hutchinson-Clyde[2]

[1]Glasgow Caledonian University
Cowcaddens Road, Glasgow G4 0BA
[2]Southern General Hospital, Glasgow

Wrist and finger joint mobility are particularly important in arthritic patients because multiple joints are destroyed simultaneously. Arthritis is a systemic inflammatory disease of unknown aetiology which can incapacitate a person in his/her daily living (Boyes, 1979). It is three times more prevalent in women than in men (Brattstrom, 1987) and, in the age group 50-55 years, arthritis affects 7% of all women and 3% of all men (Allander, 1970). Wrist fusion may be performed for; pain relief, restoration of function of the hand, correction of deformity, cosmoses and improvement of motion of the fingers by immobilising the wrist. The purpose of this project was to evaluate appropriate measures of hand function in terms of strength (using a Jamar dynamometer), dexterity (using a Purdue Peg board) and activities of daily living (using a questionnaire) with the wrist fixed in different positions. The dominant hand was splinted at the wrist in randomized positions of: 15^0 of extension, 30^0 of extension, neutral, neutral with15^0 of ulnar deviation (UD), 15^0 of extension with 15^0 UD and 30^0 of extension with 15^0 UD. The sample comprised 23 normal subjects, six males and 17 females, age range 19 to 57 years. All subjects were right hand dominant and used their splinted right hand in the trials. There have been surprisingly few studies which have established normative values for muscle force measurements using hand-held dynamometers and with respect to wrist position for optimum hand function (Andrews, 1996). With respect to grip strength, of the six positions tested, maximum strength was recorded at 30^0 extension with 15^0 UD ($p<0.05$ using analysis of variance). With respect to dexterity, analysis indicated that there were no significant differences between the six wrist positions. Analysis of activities of daily living (ADL) questionnaires revealed differences related to specific activities, eg personal hygiene, but overall no statistically significant differences between the six positions.

References

Allander E. 1970, A population survey of rheumatoid arthritis, *Acta Rheumatology Scandinavian, Supplement* **15**

Andrews, A.W. 1996, Normative values for isometric muscle force measurements obtained with hand-held dynamometers, *Physical Therapy*, **76**, 248-259

Boyes, J.H. 1979, *Tendon Surgery of the Hand*, (Churchill Livingston), 313

Brattstrom M. 1987, *Joint Protection and Rehabilitation in Chronic Rheumatic Disorder*, (Wolfe Medical Publications), 11

PROTOTYPE DYNAMOMETER FOR ANALYSIS OF MANUAL AND HYBRID-POWERED WHEELCHAIRS

Gordon Keary, Robert H Campbell and Frank Bell

Glasgow Caledonian University
Cowcaddens Road
Glasgow, G4 0BA

Until fairly recently wheelchairs were categorised into two types, manual and powered, and this distinction is still acknowledged in current Standards (Bardsley, 1998). During the last decade there has been some interest in hybrid powered or power assisted wheelchairs (Cremers, 1989; Hopping et al, 1993; Hiroaki and Atsushi, 1999). Such wheelchairs allow users to continue to use manual propulsion as appropriate but assist where necessary by sensing, controlling and delivering additional power via a motor. At least one commercially available model of a power assisted wheelchair is now available, the Rea® Vortex developed by Scandinavian Mobility (Invacare® Scandinavian Mobility UK Ltd., Corby).

Ergonomic analysis of this type of wheelchair will play an important part in its evaluation and subsequent prescription also the basic design concept offers scope for measuring key patient-related parameters. For example, instances of injury, including carpal tunnel syndrome and rotator cuff injuries are reported in high numbers in manual wheelchair users and may be due to excessive vertical force application to the handrim, and wrist angles exceeding published values for active range of motion (Veeger et al, 1998).

Our prototype dynamometer comprises a three-beam, $120°$ spaced steel cantilever attached to a standard aluminium handrim as a simplified adaptation of the Smart[Wheels] system (Asato et al, 1993). The beams were instrumented with strain gauges, and the arrangement provided accurate and reliable sensing of moments generated at the hub when calibrated tangential forces were applied to the handrim. Control systems theory was used as a measure of determining the potential output from the system that would be controlled by a tuned PID controller; this was simulated on a SIMULINK software package.

The results of the experiment showed that by varying feedback levels, the assistance from the motor can be predetermined, thus allowing the clinician to 'set the level' of assistance for the client, if required. The basic design and laboratory test data results are illustrated via the poster. The next stage of development is fitting a motor and feed back circuit and relocating the dynamometer from the engineering laboratory to the Occupational Therapy Department for eventual subject trials.

References

Asato. K.T., Cooper, R.A., Robertson, R.N., Ster, J.F., 1993, Smart[Wheels.] Development and testing of a system for measuring manual wheel chair propulsion dynamics, *IEE Transactions on Biomedical Engineering*, **20** (12), 1320 – 1324

Bardsley, G., 1988, European Standards for Wheelchairs, *IEE Engineering in Medicine and Biology*, May/June, 142 – 148

Cremers, G.B., 1989, Hybrid power wheelchair: a combination of arm force and electrical power for propelling a wheel chair, *Journal of Medical Engineering and Technology*, **13** (1/2), 142 – 148

Hiroaki, O. and Atsushi, U., Yamaha Motor Co. Ltd., 1999, Power-assisted wheelchair and method of operating same, *Patent Number* EP0945112

Hopping, J.M., Owen, A.R., Ahsing, T.P. and Stenehem, J.C., Staodyn Inc., 1993, Power-assisted wheelchair, *Patent Number* US5234066

Veeger, D.H.E.J., Meershoek, L.S., Woude, L.H. van der, and Lagenhoff, J.M., 1998, Wrist motion in handrim wheelchair propulsion, *Journal of Rehabilitation Research and Development*, **35** (3), 305 - 313

EVALUATION OF A VISUAL DISPLAY STETHOSCOPE

Karen Drummond, Robert H Campbell & Frank Bell

Glasgow Caledonian University
Cowcaddens Road
Glasgow, G4 0BA

The acoustic stethoscope is a device designed to amplify weak inaudible sounds, such as heart sounds, blood flow sounds, lung sounds and abdominal sounds to audible level. The process requires excellent hearing and the ability to discriminate between slight differences in pitch and timing. The aim of the study was to evaluate the traditional acoustic stethoscope against a visual display stethoscope. A questionnaire study of a variety of clinicians and a laboratory based comparative study were undertaken. The results showed that the acoustic stethoscope is accepted by clinicians as easy to use, but doubt has been cast on the validity and subjectivity of the sounds heard. The visual display stethoscope was perceived as considerably better in both sound reproduction and objectivity of evaluation.

Introduction

The acoustic stethoscope is a medical device that has remained almost unchanged for 200 years. It is designed to amplify weak and inaudible sounds, such as heart sounds, blood flow sounds, lung sounds and abdominal sounds to audible level. The process requires excellent hearing and the ability to discriminate between slight differences in pitch and timing. Many clinicians never acquire these skills or will lose them as they age (Mangione and Neiman, 1997).

Studies have been undertaken to examine the ability of doctors to accurately evaluate the sounds they hear when using a stethoscope. Mangione and Neiman (1997) studied such skills in medical students and physicians in training. They concluded that the skills of the group studied were very poor and showed low levels of accuracy. Other studies show that training in the use of acoustic stethoscopes and the interpretation of the sounds heard may not be highly accurate or effective (Wranne et al. 1999). The topic has been the subject of considerable discussion and debate in the medical literature (Prasad, 1987; Lewis, 1997).

Durand et al (1998) considered that the acoustic stethoscope was ergonomically sound but was poor at aiding the clinician in evaluating faint sounds such as heart murmurs. Callahan et al (1998) commented that high ambient noise may impede the efficiency of the traditional stethoscope, while others comment on the subjective nature of the sound interpretation (Lukin et al, 1996).

An electronic stethoscope attempts to overcome some of the difficulties with the traditional acoustic stethoscopes. They can selectively amplify at different frequency ranges, filter to provide enhanced clarity, do not distort sound due to tubing resonance, are lightweight and have volume adjustment. Furthermore the visual display stethoscope gives the clinician the opportunity to visually check their acoustic interpretation, to identify inaudible arrhythmia's and provide an output record which can be kept for later reference (Caditec.com).

The aim of this study was therefore to compare the visual display stethoscope to the traditional acoustic stethoscope.

Method.

The study comprising a questionnaire survey of clinicians from a variety of specialities and a laboratory comparison test involving males with a predominantly technical background (technicians and final year engineering students), and no prior experience in the use of stethoscopes. The questionnaire was designed in the light of published guidelines and followed a small pilot study. Fifty questionnaires were distributed to clinicians in one local hospital and one large general practitioners surgery. Questions related to the ease of use and perceived performance criteria of the acoustic stethoscope and awareness of alternatives such as electronic and visual display stethoscopes. 20 respondents (40%) completed the questionnaire and their specialities included general medical practitioners, senior and junior medical staff in a variety of ward types, midwives and final year medical students. The comparison tests of one acoustic stethoscope and one visual display stethoscope, (Cadiscope®, CADItec AG, Switzerland) involved 15 male subjects age range from 21 to 50 years. A protocol was used for testing two normal heart sounds (S1 and S2) and three normal lung sounds based on the McGill Virtual Stethoscope web site (mcgil.ca), rating, ease of use, isolation of background noise and amplification of sounds. Comparison test sheets were designed to allow ratings of all these parameters.

Results

The questionnaire results showed that the acoustic stethoscope, was widely used by the targeted groups with 90% using the acoustic stethoscope daily and the remaining 10% several times per week.. The clinicians were therefore experienced in the use of the acoustic stethoscope. Only 15% of the clinicians were aware of the existence of visual display stethoscopes, while 40% knew of electronically assisted stethoscopes.

The clinicians considered acoustic stethoscope (mode ratings) to be, good to excellent for ease of use, good for reliability, good to fair for comfort, appreciation of high and low frequency sounds, amplification of sounds and accuracy and fair to poor for isolation of background noise. Several respondents classed accuracy and reliability as user dependent. Since the clinicians were all experienced in the use of the acoustic stethoscope it is perhaps not surprising that they rated it highly for ease of use. This view is generally accepted and it is described as "straight forward to use" by All Heart [online]. They also rated the acoustic stethoscope as good to fair for sound appreciation and accuracy though several did concede that the evaluation of the sound was open to individual interpretation. This appears to indicate that the clinicians perceptions of the validity of the acoustic stethoscope is at odds with the limited literature on the topic which highlights the subjective nature of such evaluation and the limitations on validity of the interpreted results. The clinicians did agree with the literature on the acoustics stethoscopes limitations with regard to background noise (Durand et al, 1998, Callahan et al 1998, Mangione and Neiman, 1997).

The comparison laboratory tests indicated that the visual display stethoscope was rated significantly better than the acoustic for every test using Mann Whitney tests with confidence intervals set at 95%. The visual display stethoscope was rated 1 to 2 times better for the appreciation of heart sounds, isolation of background noise and the ease of use for heart sounds. The ratings for use on lung sounds were also considerably better but less so than for the heart.

Conclusion

This study has shown that the acoustic stethoscope is a widely accepted diagnostic medical device that is in common use. The acoustic stethoscope is accepted by clinicians as easy to use but doubt has been cast on the validity of evaluation of the sounds heard due to the subjectivity of such evaluation and limitations in the quality of sound reproduction.   The visual display stethoscope was perceived as considerably better in both sound reproduction and objectivity of evaluation.

Continued research into this area is strongly recommended. Further study into the validity of the sounds reproduced by both types of stethoscope undertaken by highly experienced and qualified medical personnel would yield a more diverse evaluation.

References

Callahan, M.G., Callahan, T.F., Foster, K.S., Graber, G.P., Glifort, K., Jones, J.T. and Wodicka, G.R.,   (1998), An adaptive noise reduction stethoscope for auscultation in high noise environments, *Journal of the Acoustical Society of America,* **103** (5), 2483 - 2491

Durand, J., Durand, L., Gagnon, K., Genest, G. and Grenier, M., 1998, Clinical comparison of acoustic and electronic stethoscopes and design of a new electronic stethoscope, *American Journal of Cardiology,* **81** (5), 653 – 656

Lewis, P.R., 1997, Cardiac auscultation skills of physicians in training (Letter), JAMA, 278 (21), 1739

Lukin, A., Polic, S., Rumbolt, Z., Bagatin, J., Rakic, D. and Kuzmanic, A., 1996, Comparison of auscultation findings using a classic stethoscope and electronically amplified stethoscope, *Lijecnicki Vjesnik,* **118** (5-6), 127 - 8

Mangione, S. and Neiman, L.Z., 1997, Cardiac auscultatory skills of internal medicine and family practice trainees, *JAMA.,* **278** (9), 717 – 722

Prasad, C.S., 1987, Teaching stethoscope (Letter), *J. Assoc. Physicians India,* **35** (2), 175

Wranne, B., Ask, P. and Hoh, B., 1999, To use the stethoscope correctly is not easy, The intricate art of auscultation should receive more attention in medical education, *Lakartidningen,* **9 6** (24), 2981 – 4

http://www.Caditec.com (accessed 3/3/00)

http://www..store.yahoo.com/allheart/index.html (accessed18/2/00)

http://www.sprojects.mmip.mcgil.ca/MVS/STFRAME.HTM (accessed 14/3/00)

ERGONOMIC ANALYSIS OF A PODIATRIC WORKPLACE: A PILOT STUDY

Abdullah Al Naseeri, Robert H Campbell & Frank Bell

Department of Physiotherapy, Podiatry and Radiography,
Glasgow Caledonian University,
Glasgow, G4 0BA

Although the examination of a podiatry patient requires the podiatrist to assume a number of work positions (sitting, standing and squatting), treatment tends to require, in the main, the podiatrist to sit at the feet of the supine patient in a fixed or awkward work posture. Despite the apparent ergonomic risk factors there are few published studies in this area. This study was carried out over a 6-week period at a 24-work station hospital based podiatry clinic in Glasgow. The study methods include visual and instrumented examination of the workplace, a questionnaire survey, video filming and photography. This study supports the limited literature available on the ergonomics of podiatry related neck, upper limb and back problems and relates them to poor posture, work station design and user training.

Introduction

The podiatrist work is multi faceted. It may involve sitting at a desk, standing beside the patient, sitting at the patients feet while talking to and treating the patient who may be seated higher than the podiatrist and occasionally squatting to examine and treat parts of the patients foot. It also involves the use of surgical instruments that require a high degree of skill and manual dexterity, reaching for instruments and dressings from the workstation trolley, using therapeutic machinery such as dental type drills and electronic diagnostic testing equipment as well as the chair-side manufacture of therapeutic padding and strapping. The well being and comfort of the patient has always been considered paramount but until fairly recently, scant regard has been paid to the comfort, safety and well being of the practicing podiatrist.

Ergonomic analysis is a mechanism, which may be used to facilitate the identification of particular problems in the workplace (Pheasant, 1991). Through a process of systematic recording of specific aspects of the work done or the effects on the person of that work, ergonomic defects and possible health hazards may be defined and subsequently eliminated (Jacobs, 1999). Consequently this ergonomic study was designed to examine the podiatry work station and the effects of the workstation design on the practicing podiatrist. In particular, the effects of body posture (back, head, neck, arms, shoulder and wrists) related to the basic work activities undertaken by the podiatrist. Environmental factors such as noise and temperature were also recorded.

Body posture

Little work has been published on the work conditions of the podiatrist however there is literature on the ergonomic work factors affecting other health care professionals that are pertinent to the podiatrist.   Murphy (1997) was of the opinion that dental workstations ergonomic risk factors are often triggered by the use of vibrating tools, fixed or awkward work postures or multiple repetitions of the same movement.   Much of podiatry workstation furniture and fittings are adapted from dental designs.

Piggot (1991) linked work conditions and equipment to back pain and observed, in a study on self-reported back pain in chiropodists (podiatrists), that the risk of back pain is increased with the height of the individual.   Lowerson (1990) found that bad posture including bending and squatting to be a significant cause of various complaints including backaches and fatigue.   While the aetiology of back pain is multi faceted and cannot be solely related to work, there is a growing accumulation of evidence to link the workplace with back pathologies (Putten, 1995).

Piggot (1991) also considered the neck to be at risk due to the postures adopted by the podiatrist when working on lesions on the dorsal aspect of the foot.   Chaffin (1973) observed that the average time to reach marked muscle fatigue of the neck musculature, was shortened as the neck/head angle increased.   While Kroemer and Grandjean (1997) concluded that localised muscle fatigue in the neck area can be a preliminary sign of other more serious and chronic musculo-skeletal disorders and that inclination angle of the head should not exceed 30° for any prolonged period of time   Neck flexion while working may be associated with non-specific neck and shoulder problems.   Conversely neck and shoulder problems have been associated with poor hand wrist and arm posture (Pollack, 1996).

Podiatrists use a variety of surgical instruments including nail nippers to reduce pathological nails (which may involve considerable pinch force) and scalpels to debride pathological skin lesions, which involves rapid repeated movements of the wrist and fingers.   Poor arm and wrist postures can lead to serious musculo-skeletal problems including nerve entrapment neuropathies (Pollack, 1996). While repetitiveness in the absence of other factors may increase the risk of upper limb pathologies (Haslegrave et al, 1990).

Working Environment

The selection of a suitable chair is a critical step in preventing health problems in people who work in a sitting position, particularly in situations where the workstation is fixed. (Jacobs, 1999).   An appropriately designed chair allows the user to sit in a balanced position but the actual sitting position adopted may depend on the individual's habits. The users posture will also be influenced by other factors such as work surfaces; footrests, task lighting, document holders etc. and the individual user may have to learn or be taught how to sit correctly in the chair.   Many papers have been published on the design of chairs.   Some novel chairs have been designed such as the saddle, kneeling and sloping chairs which have been shown to have advantages and disadvantages largely dependant on the work task undertaken (Wilson and Corlett, 1990, Rice, 1998).   Similarly adaptations and additions to more conventional type chairs (e.g. armrests, seat angulations and lumbar supports) will be effective in reducing pathology in some situations but may be hazardous in others by impeding the users movements and in some situations may increase the chance of injury (Jacobs, 1999).   Similarly the design of the workstation has been extensively studied.   Pollack (1996) suggested that work surface height should be adjustable and be adapted to suit individual body sizes.   Gasset et al (1996) agreed and also consider that sufficient legroom to allow knee clearance and "stretching the legs", was essential.   Equipment positioning can dramatically affect working posture and contribute to musculo-skeletal disorders and inefficiencies.   Beurger (1997) considered that reach length should not exceed 20 inches, while Hazelgrave et al (1990) considered that equipment and controls should be situated as close as feasible and be positioned directly in front of the worker to prevent extended forward (bent) and side (twisted) body movements

Podiatrists sit on chairs, which are often 'typist' type chairs that normally are seat height adjustable but have no other specific design features.   They treat the patients feet which will be situated directly in front of them and in doing so will use a variety of instrumentation and other equipment, that is situated on (and in) a non-height adjustable trolley at their right hand side.   The patients chair will be height adjustable so that the feet can be raised or lowered as required, however the height range will vary considerably and is dependant on the design of the particular chair.

Methodology

The study was carried out over a six-week period in a 24-work station hospital based podiatry teaching clinic in Glasgow.   The study methods included visual and instrumented examination of the podiatric workplace, a questionnaire survey, photography and video filming.   Following a small pilot study of three podiatry staff, a questionnaire survey of 30 podiatry staff and final year students was undertaken to elicit basic information related to the clinicians posture and seating and the general clinical environment.   The questions were derived from published questionnaire studies undertaken in other relevant healthcare fields.   The general clinical environment was assessed by taking sample temperature, noise and light meter readings at three sites in the clinic and at 23 of the workstations.   Video recording was undertaken at one workstation during examination and treatment of patients.   Using published data as a guide video analysis forms of 20 second time intervals each covering a 5 minute work period were designed and used to analyse positions of back, neck, arms and wrists over a 45 minute work session.

Results and Discussion

24 respondents (80%) completed the questionnaire.   There were 18 females and 6 males, with a mean age of 26 years, ranging from 21 to 50 years   It is generally considered throughout the podiatry profession that low back pain and neck pain are "occupational hazards" and a common cause of lost working hours and general discomfort.   This opinion is confirmed, to some extent, by Lowerson (1990) and Piggot (1987 & 1991).   In this study 30% - 60% of respondents complained of postural or workspace related problems.   Video analysis revealed that for at least 60% of work time the podiatrist worked leaning forward with spine in forward flexion, neck in forward flexion and one or both arms flexed (between 30° and 90°) and with their wrists extended.   The duration of each postural position varied.   Lowerson (1990) and Putten (1995) agreed that such

forwardly flexed postures considerably increased the stresses on intervertebral discs. Furthermore the podiatrists spent 60% of work time with neck flexed more than is considered acceptable (Milerad and Ekenvall, 1990). The podiatrists prolonged spine and neck flexion during work time may imply a risk factor.

40% of respondents also complained of insufficient legroom under the operating (patients) chair.   Measurements indicated that the operating (patients) chair and the podiatrists chair may not be completely compatible to allow the podiatrist sufficient working legroom.   Luopajarvi (1995) considered work surface height, in this case the patient's chair, to be an important ergonomic factor. While Pollack (1996), observed that legroom may be restricted if the working height is too low.

The literature generally agrees that poor arm posture can lead to musculo-skeletal pathologies and nerve entrapment syndromes.   In this study, video analysis demonstrated that over the 45 minute work sample period, the podiatrists arms were flexed between 30° and 90° for 65% of work time and that the wrist were in extension for some 40% of work time.   However a direct causal link between arm and wrist postures and musculo-skeletal pathologies could not be demonstrated.

Between 50% and 70% reported some difficulties related to noise, vibration and ventilation when nail drilling.   The general environment was acceptable but problems with local lighting and shadow effects were reported by about half of the respondents. Measured temperatures ranged from 19° - 22°C (mean 21 °C), illumination from 220 lux – 420 lux (mean315 lux) and noise from 56 dB(A) – 76 dB(A) (mean 67 dB(A)).

Conclusion

This small pilot study supports the limited data available specifically on ergonomics in podiatry related back, neck and upper arm aches and pains to bad posture and consequently to work station design and user training (Lowerson, 1990; Piggot, 1991; Putten, 1995).

References

Beurger, R., 1999, Ergonomics in the operating room, *The American Journal of Surgery*, 171, 385 - 387

Chaffin, K., 1973, cited in Kroemer, K. and Grandjean, E., 1997, *Fitting the Task to the Human: A Textbook of Occupational Ergonomics, Fifth Edition,* (Taylor and Francis, London), 68 – 70

Gasset, R., Hearne, B. and Keelan, B., 1996, Ergonomics and Body Mechanics in the Workplace, *Occupational Disorder Management*, 27 (4), 861 – 879.

Haselgrave, C., Wilson, J., Corlett, E. and Manenica, L., 1990, *Work Design in Practice,* (Taylor and Francis, London)

Jacobs, K., 1999, *Ergonomics for Therapists, Second Edition*, (Butterworth-Heinneman, London)

Kroemer, K. and Grandjean, E., 1997, *Fitting the Task to the Human: A Textbook of Occupational Ergonomics, Fifth Edition,* (Taylor and Francis, London)

Lowerson E., 1990, Posture at work, *The Chiropodist*, March 1990, 60 –70

Luopajarvi, T., 1995, Ergonomic analysis of workplace and postural load, *Ergonomics,* 38(10), 51 – 78

Milerad, E. and Ekenvall, L., 1990, Symptoms of the neck and upper extremities in dentists, *Journal of Work Environmental Health,* 16, 129 - 134

Murphy, D., 1997, Ergonomics and dentistry, *New York State Dental Journal*, 63(7), 31 – 34

Pheasant, S., 1991, *Ergonomics Work and Health*, (McMillan Press, London)

Piggot, A., 1987, Chiropodial questionnaire to the 1985 graduates, *The Chiropodist*, 42 (6), 63 – 67

Piggot, A., 1991, A study into the incidence and prevalence of self reported back pain amongst chiropody students and 1986 graduates, *Journal of British Podiatric Medicine*, 46 (5), 83 – 87

Pollack, R., 1996, Dento-ergonomics: the key to energy saving performance, *Journal of the California Dental Association,* 24 (4), 63 - 68

Putten, M., 1995, Safety and health of the podiatrist at work, *Journal of British Podiatric Medicine*, 50 (1), 6 – 9

Rice, V., 1998, *Ergonomics in Healthcare and Rehabilitation*, (Butterworth Heinemann, London)

Wilson, J. and Corlett, E., 1990, *Evaluation of Human Work: A Practical Ergonomics Methodology,* (Taylor and Francis, London)

MANUAL HANDLING

PROSPECTIVE INVESTIGATION OF THE REVISED NIOSH LIFTING EQUATION

Patrick G. Dempsey[1], Gary S. Sorock[1], M.M. Ayoub[2], Peter H. Westfall[2], Wayne Maynard[1], Fadi Fathallah[3] and Niall O'Brien[1]

[1]*Liberty Mutual Center for Safety Research, Hopkinton, Massachusetts, USA*
[2]*Texas Tech University, Lubbock, Texas, USA*
[3]*University of California at Davis, Davis, California, USA*

The preliminary results of a prospective epidemiological study of the revised NIOSH lifting equation are presented. The baseline evaluations included assessment of lifting and lowering tasks with the revised NIOSH equation, as well as a questionnaire regarding personal variables. Subject follow-up was primarily accomplished through postal questionnaires, telephone interviews, and surveillance for workers' compensation claims for low-back disorders. The preliminary results reported are based on over 600 person-years of exposure. Important findings related to the usability of the revised NIOSH equation across several types of common exposures are also discussed.

Introduction

The revised National Institute for Occupational Safety and Health (NIOSH) lifting equation (Waters et al., 1993), a revision of the equation developed in 1981 (NIOSH, 1981), is one of the more widely discussed materials handling assessment tools. The equation has been factored into a number of regulatory and consensus documents. Perhaps the first use of the equation in this type of application was the use of the equation by inspectors from the Occupational Safety & Health Administration (OSHA) in the US. Although the equation is not part of regulations, the inspectors have used the equation when they believe lifting or lowering tasks pose demands in excess of what the "General Duty Clause" of the Occupational Safety and Health Act of 1970 specify (i.e., "...a place of employment ...free from recognized hazards that are causing or likely to cause death or serious harm").

Aside from the aforementioned use of the equation in the US, several other countries or regulatory bodies have considered or adopted the equation for similar purposes. For example, the primary components of the NIOSH approach have been integrated into draft standards developed by the European Committee for Standardization (Draft prEN 1005-2) and the International Organization for Standardization (ISO/DIS 11228-1.2). The Health Council of the Netherlands (1995) performed a review of the various materials handling assessment tools available, and the revised NIOSH equation was selected as the preferred method for analyzing lifting exposures.

Despite widespread application of the NIOSH equation, little information concerning the ability of the equation to predict, and subsequently control, the incidence or severity of low-back disorders is available. This information is critical if the equation is going to be used as a risk assessment tool, *per se*. Risk assessment requires estimation of the outcome of interest in a population for a given exposure level. In the case of the NIOSH equation, this indicates that an estimate of the incidence of low-back disorders for a given lifting index (LI) is required for a risk assessment.

The goal of the present study was to prospectively investigate the relationship between the 1991 NIOSH equation and the incidence and severity of filing workers' compensation claims. The severity information, in particular, may be critical since a small percentage of low-back disability claims leads to a high percentage of the total costs. The severity information is not reported here since follow-up is continuing. A secondary aim was to evaluate the usability of the equation in a variety of industrial and service sectors.

Methods

An exposure assessment protocol was designed to measure the frequency, duration, and magnitude of exposure to lifting and lowering tasks. These three parameters are the primary determinants of physical exposure. 'Magnitude' of lifting/lowering tasks was assumed to be captured by the NIOSH equation lifting index (LI). Frequency was simply defined by individual task frequencies, and duration was assessed through the hours of daily or weekly exposure to the tasks. Although, the equation has a restriction that no pushing, pulling or carrying tasks be performed, this restriction had to be violated in order to obtain a sufficient sample. These tasks were assessed with psychophysical data. There was a requirement that jobs included in the study to have a primary lifting and lowering component.

Job selection criteria were used to promote selecting jobs with fairly stable exposure to two-handed lifting and lowering tasks. There were numerous difficulties finding suitable jobs. Approximately 30 loss prevention professionals were trained to collect the data according to the protocol, and almost all reported trouble finding jobs suitable for inclusion in the study. It should be noted that all those involved in data collection were trained by the same individual for consistency.

A survey of jobs not meeting the criteria was initiated to gain a more quantitative understanding of why so many jobs failed to qualify for the study. A total of 40 jobs not suitable for the study were included in the survey, and informal feedback was used to develop the reasons that could be recorded. Multiple reasons were allowed. The results confirmed the informal feedback the investigators received up to that point. The most common reason indicated was that the weights lifted and lowered were too variable (63% of the jobs). The next most common reasons (each with approximately 25% of jobs) were job changing often, job rotation, and that the job contained too many lifts to analyze.

In order to increase the number of subjects being entered into the study, the job selection criteria were revised to include jobs such as those typically found in warehouses, jobs with rotation of workers among several exposures, and jobs with variable weight demands. The protocol for warehousing/complex jobs involved a sampling 15 representative tasks. The job rotation protocol was designed to measure the rotations with the most stressful tasks and the longest duration. If all rotations had similar stress levels, the two rotations with the longest durations were assessed. The variable weight protocol was designed to capture the distribution of weights handled.

Aside from the assessment of manual handling tasks, workers were interviewed to gather basic demographic and personal variables (e.g., smoking, regular exercise, back belt use). Participation was voluntary in accordance with ethical guidelines for research. Workers are followed up via postal questionnaire every three months to determine if they have reported a work-related injury or illness and whether or not they are still working the same job as when interviewed. Initial low response rates to the survey were increased by using a lottery consisting of three $50 prizes. All workers returning the survey by a specified date are eligible. Phone calls are made to non-respondents, and site visits or contact with site personnel are sometimes used to locate non-respondents.

The outcome measures used include OSHA 200 reports and workers' compensation claims during the follow-up period. Only the workers' compensation results will be discussed here. The primary outcome measure being studied is compensable low-back disorders, with a secondary focus on the broader class of over-exertion injuries to all body parts. It should be noted that all workers' compensation claims are collected. Low-back claims due to traumatic events (such as the result of a fall or struck-by accident) are excluded, as the etiology of these injuries is quite different. Also, subjects are removed from the study when a low-back claim is reported, including those attributed to traumatic events, when they change jobs, or if they do not wish to participate any longer.

Results

As of the end of data collection, 449 subjects from over 60 facilities throughout the United States completed baseline evaluations. Table 1 provides a summary of the demographics of the sample.

Table 1. Summary of subject demographics.

Gender	Mean Age (SD)	Mean Weight (SD)	Mean Height (SD)
Female	37.0 (9.5)	67.5 (14.1)	163.6 (8.6)
Male	34.1 (9.9)	85.1 (16.3)	177.3 (8.1)

SD = Standard deviation, height in cm., weight in kg.

Table 2 provides the rates of filing workers' compensation claims for low-back disorders during the follow-up period. The method used was to consider 2000 hours as one full-time equivalent. This is the method used by the Occupational Safety and Health Administration.

Table 2. Summary of claims by exposure category.

	LI Category			
	$0 \leq LI < 1$	$1 \leq LI < 2$	$2 \leq LI < 3$	$LI \geq 3$
Exposure Hours	244,479	306,050	227,001	459,404
# WC Claims	1	2	4	5
Rate/100 FTE	0.82	1.31	3.52	2.18

WC = workers' compensation, FTE = Full-time equivalent

Conclusions

The results presented represent only a crude analysis, with no examination of potential confounding, etc. The follow-up of workers is still in progress, and more formal analyses will be conducted when the follow-up ends. The exposure groups used in Table 2 are similar to the groups used by Waters et al. (1999), and the results are similar in that the rate of WC claims increased for the lowest 3 LI categories, and then showed a decrease for the highest category. Waters et al. (1999) reported a similar pattern for self-reported low-back pain.

Regardless of the final outcome of the study, it was apparent that the revised NIOSH lifting equation is not suitable for quite a few MMH jobs in the U.S. Originally, a sample size of 2000 workers was the goal of our study, and this had to be reduced drastically because of the difficulty locating suitable exposures. This is in spite of ignoring several of the primary assumptions of the equations, such as no other materials handling tasks besides lifting or lowering. If this restriction was followed, over 56% of the workers that volunteered for the study would have been excluded (Dempsey, 2001). Thus, this restriction is not realistic.

Aside from the epidemiologic results, a key finding from the study is that more systemic tools to assess and evaluate MMH exposures are needed. Although the NIOSH equation is well-suited for the monotonous tasks that were common in some manufacturing settings, many jobs currently in the manufacturing sector, and particularly the service sector, are not suited to analysis with the equation.

Acknowledgements

The authors would like to thank the participating facilities, workers that volunteered to participate, and the many Liberty Mutual employees that were involved in the study in many different ways. This study was partially supported by NIOSH Grant No. 5R18OH03202

REFERENCES

Dempsey, P.G. 2001, Field Investigation of the Usability of the Revised NIOSH Equation, *Proceedings of the Human Factors and Ergonomics Society 45th Annual Meeting*, (Human Factors and Ergonomics Society, Santa Monica, CA), 972-976

Health Council of the Netherlands 1995, *Risk Assessment of Manual Lifting*, Publication No. 1995/02, (Health Council of the Netherlands, The Hague)

NIOSH 1981, *Work Practices Guide for Manual Lifting*, NIOSH Technical Report No. 81-122, (National Institute for Occupational Safety and Health, Cincinnati, OH)

Waters, T.R., Baron, S., Piacitelli, L.A., Anderson, V.P., Skov, T., Haring-Sweeney, M., Wall, D.K., and Fine, L.J. 1999, Evaluation of the Revised NIOSH Lifting Equation: A Cross-Sectional Epidemiologic Study, *Spine*, **24**, 386-395.

Waters, T.R., Putz-Anderson, V., and Garg, A. 1994, *Applications Manual for the Revised NIOSH Lifting Equation*, (National Institute for Occupational Safety and Health, Cincinnati, OH)

Waters, T.R., Putz-Anderson, V., Garg, A, and Fine, L.J. 1993, Revised NIOSH equation for the design and evaluation of manual lifting tasks, *Ergonomics*, **36**, 749-776

POWER CONSIDERATIONS IN ESTIMATING SAMPLE SIZE FOR EVALUATION OF THE NIOSH LIFTING EQUATION

ADJ Pinder

Health and Safety Laboratory
Broad Lane
Sheffield, S3 7HQ, UK

The efficient design of human studies requires careful consideration of issues of intended analysis and statistical power. A prospective epidemiological study is under way to test the ability of the NIOSH lifting equation to predict work-absence due to low back pain. Power calculations using a Proportional Hazards Model show that a sample of 650 would provide sufficient power for an overall evaluation of the equation. Sample sizes in the tens of thousands would be required to evaluate the correctness of the least influential multipliers in the equation. Due to a need to balance costs and benefits, a sample size of 1000 has been chosen for the study. This will permit the overall evaluation of the equation and of some of the constituent multipliers, but not of the smaller multipliers.

Introduction

Considerations of statistical power are crucial to the efficient design and costing of studies involving human subjects and are of increasing concern to research ethics committees. The Health and Safety Laboratory (HSL) and Liberty Mutual Research Center (LMRC) are collaborating in a prospective epidemiological study to test the ability of the 1991 NIOSH lifting equation (Waters *et al.*, 1994) to predict work absence due to low back pain caused by manual handling.

The importance of the 1991 NIOSH equation as a manual handling risk assessment tool can be gauged from the fact that European (CEN) and International (ISO) committees are developing proposals for standards for evaluating manual handling tasks using equations based it. The forms of the equations are:

$RWL = LC \times HM \times VM \times DM \times FM \times AM \times CM$ (Waters *et al.*, 1994)

$M \leqslant M_{ref} \times k_H \times k_V \times k_D \times k_a \times k_F \times k_G$ (ISO, 2000)

$R_{ML} = M_C \times H_M \times V_M \times D_M \times F_M \times A_M \times C_M \times O_M \times P_M \times A_T$ (CEN, 1999)

The equations consist of a Load Constant (LC, M_{ref}, M_C) and a series of multipliers. The horizontal (HM, k_H, H_M), vertical (VM, k_V, V_M), distance (DM, k_D, D_M), frequency

(FM, k_F, F_M), asymmetry (AM, k_a, A_M) and coupling (CM, k_G, C_M) multipliers cannot exceed 1.0 so the Recommended Weight Limit (RWL) cannot exceed the Load Constant. The NIOSH Lifting Index (LI) is the ratio of the load to the RWL. NIOSH deems that values of LI>1 indicate that the person lifting is at increased risk of low-back pain. The three extra CEN multipliers further reduce the CEN mass limit, R_{ML}, when one-handed tasks (O_M), two-person handling (P_M) or 'additional tasks' (A_T), such as pushing, occur.

When first designing their study to evaluate the NIOSH equation, LMRC planned to collect data on 2000 subjects in the USA. The aims were to determine prospectively the ability of the equation to predict the incidence and severity of work absence due to low-back pain and to test if the predictive capability is modified by workers' sex, age, height, weight, and history of previous low-back pain or injury. HSL offered to collect comparable information on subjects in the UK and to pool the data. HSL had the additional aim of similarly evaluating the proposed ISO and CEN equations.

Due to difficulties finding sufficient subjects in suitable jobs, LMRC stopped recruiting subjects having obtained only 449. Dempsey *et al.* (2000) reported the study when they had recruited 392 subjects, had followed them up for an average of 43 weeks, with a dropout rate of 46%, and had recorded a total of nine lost time back injuries. This paper considers the implications of these numbers for the UK part of the study in terms of the number of subjects required to achieve adequate power.

Statistical models suitable for cohort studies of low back pain

There are three possible types of model that can be constructed to evaluate the relationship between a risk index such as the LI and the true risk of injury. The simplest is a binary split between values below and above a threshold value such as LI=1. The next is to split values of the index into categories, such as LI<1, 1≤LI<2, 2≤LI<3, 3≤LI. The most complex and valid approach is to treat the index as a continuous variable and to attempt to find a linear or non-linear function relating the index to risk of injury.

Methods suitable for analysis of relationships between the risk of low back injury and measured variables are: 1) Logistic regression; 2) Proportional Hazard Models (PHMs); and 3) General Additive Models (GAMs). These all allow covariates such as age, and prior back injury to be accounted for. Logistic regression predicts a binary outcome (e.g., injury or non-injury) from a number of predictor variables (e.g., manual handling task parameters). The PHM predicts the risk of injury and is able to account for loss of subjects due to injury and during follow up. Dempsey and Westfall (1997) favoured GAMs and an additive model extension to the PHM because of their flexibility and ability to deal with non-linear relationships between risk factors and outcomes.

Estimation of the sample size needed for overall evaluation of the equations

Available computer software for calculation of power and sample size lacked options for GAMs or for PHMs more complex than binary comparisons. Therefore, a PHM was used to estimate the power of a comparison of lost-time due to back injury between a control group (LI<1) and an exposed group (LI≥1). A range of sample sizes, two levels of risk and two levels of dropout were used (Table 1). A 5% incidence rate was assumed in the control group. Rates of 26.4% (Marras *et al.*, 1995) and 10% were used for high and conservative estimates of the incidence in the exposed groups. Annual drop out rates of 10% (Hoogendoorn *et al.*, 2000) and 46% (Dempsey *et al.*, 2000) were used.

Power calculations were carried out using the 'advanced log rank two-sided Proportional Hazards Model' in PASS ('Power Analysis and Sample Size', NCSS, Kaysville, Utah). It was assumed that the control group would be equal in size to the exposed group and that subjects would be recruited over 9 months and followed up for 18 months each. Conventional significance and power levels of 5% and 80% were used.

Table 1. Effect of risk level, drop out rate and sample size on power

Risk	Annual drop out	Sample size				
		200	450	650	1,000	2,000
Conservative	10%	0.360	0.670	0.822	0.947	0.999
(Relative Risk = 2.0)	46%	0.268	0.519	0.674	0.848	0.988
High	10%	0.997	1.000	1.000	1.000	1.000
(Relative Risk = 5.3)	46%	0.979	1.000	1.000	1.000	1.000

For the conservative risk estimate, a sample of 200 would provide very little power. A sample of 650 would provide the conventionally necessary 80% power at an annual drop out of 10%, provided the other assumptions are met. If the 46% dropout suffered by LMRC were to occur then a sample of 1000 would provide this power. In the unlikely circumstance that all jobs in the exposed group fell into the high risk category, very great power would be obtained, even with a sample of 200 and an annual drop out of 46%.

Evaluation of multipliers common to the NIOSH, ISO and CEN equations

To evaluate fully multiplicative equations such as the 1991 NIOSH equation, it is necessary to evaluate each of the constituent multipliers. This can be done for the six NIOSH multipliers by using the data to create models that treat variations in all but one multiplier as random errors. Because the multipliers have a range of minimum values, calculations were carried out between 0.9 and 0.1, using both conservative and high estimates of risk (Table 2). It is possible to estimate sample size by assuming that there is a perfect inverse relationship between the multiplier and the relative risk of low back pain, so that a decrease in a multiplier from 1.0 to 0.5, reflects a true doubling of risk.

Table 2. Effect of multiplier value on required sample size

Multiplier	Relative Risk	'Conservative' risk		'High' risk	
		% surviving	Required N	% surviving	Required N
1	1.00	95.0%		73.6%	
0.9	1.11	94.4%	35,342	70.7%	5,417
0.7	1.43	92.9%	2,724	62.3%	403
0.5	2.00	90.0%	615	47.2%	86
0.3	3.33	83.3%	162	12.0%	20
0.1	10.00	50.0%	28	0.0%	0

The minimum value of the coupling multiplier of 0.9 means that many thousand subjects are required to evaluate the difference in risk between a 'good' and a 'poor' coupling, and even more to evaluate the difference between a 'fair' and a 'good' or a 'poor' coupling (CM=0.95). The ease of assessing the other multipliers varies according to their ranges. Thus, the distance multiplier, which can decrease to 0.85, is much harder to evaluate than the frequency multiplier, which can decrease to zero. Moreover, because all except the coupling multiplier are continuous variables, values distributed across their ranges are needed to evaluate each one, which will to increase the sample size required.

Evaluation of the extra multipliers in the CEN equation

The three extra multipliers in the CEN equation are binary (factor present or not present). Table 3 shows the eight possible combinations and the relative risks for tasks that differ only in the presence or absence of these factors. As two combinations of the factors are very unlikely to occur, only six groups are needed to evaluate the multipliers fully.

Table 3. Evaluation of combinations of factors in the proposed CEN equation

	One hand	Two person	Additional tasks	Comments	Multiplier	Relative Risk
1	No	No	No	Standard NIOSH group	1	1
2	Yes	No	No	One-hand	0.6	1.6667
3	No	Yes	No	Two person	0.85	1.1765
4	No	No	Yes	Additional Tasks	0.8	1.25
5	Yes	No	Yes	One-hand × Additional	0.6×0.8	2.0833
6	No	Yes	Yes	Two-person × Additional	0.85×0.8	1.4706
7	Yes	Yes	No	Inherently unlikely		
8	Yes	Yes	Yes	Inherently unlikely		

Table 4 gives estimates of sample sizes required for the five possible comparisons with Group 1 (the control group), using conservative and high estimates of incidence of 5% and 26.4%. It would be possible to compare Group 5 with Groups 2 or 4 since each of these has one of the two risk factors present in Group 5. However, distinguishing Group 2 and Group 5 would take 4326 (conservative risk) or 505 (high-risk) subjects. Distinguishing Groups 3 and 6 would require 6267 or 892 subjects; For Groups 4 and 5, 974 or 127 would be required; For Groups 4 and 6, 11436 or 1605 would be needed.

Table 4. Incidence rates and sample sizes for evaluating the extra CEN risk factors

	Group	'Conservative' risk		'High' risk	
		Incidence rate	Required N	Incidence rate	Required N
1	Standard NIOSH	5.0%	7,231	26.4%	1,097
2	One-handed	8.3%	615	44.0%	89
3	Two-person	5.9%	7,231	31.1%	1,097
4	Additional	6.3%	3,718	33.0%	560
5	One-handed × Additional tasks	10.4%	269	55.0%	38
6	Two-person × Additional tasks	7.4%	1,149	38.8%	169
	Total subjects required		20,213		3,050

Table 4 shows that the total number needed for the evaluation of the six groups depends crucially on the actual injury rate occurring in Group 1. It also shows that testing the two-person multiplier is unrealistic unless very large subject numbers are obtained. It would, however, be fairly easy to evaluate the one-handed × additional tasks combination, but less easy to evaluate the one-handed factor by itself. Numbers almost an order of magnitude larger would be needed to evaluate the additional tasks multiplier.

Modification of project plan

These calculations have been used to inform decisions about subject numbers for the HSL part of the study. A figure of 1000 subjects has been chosen as a compromise between the need to keep the study size within manageable bounds and available resources and to provide meaningful results. With a 10% annual loss to follow-up, power of 95% for an alpha value of 5% will result for the evaluation of the overall predictive abilities of the NIOSH and CEN/ISO equations. Even if the drop out rate reaches the 46% experienced by LMRC, the power will still be 85%. This sample size may be sufficient to evaluate the more important individual factors within the equations, but will definitely not be sufficient to evaluate the least influential. It would only be feasible to do this by collecting data from tens of thousands of subjects. However, within the limitations of the data actually collected, as much evidence as possible will be extracted.

Bibliography

CEN, 1999, *Safety of machinery - Human physical performance - Part 2: Manual handling of machinery and component parts of machinery*, Draft PrEN 1005-2

Dempsey, P.G. and Westfall, P.H. 1997, Developing explicit risk models for predicting low-back disability. A statistical perspective, *International Journal of Industrial Ergonomics*, **19**, 483-497

Dempsey, P.G., Sorock, G.S., Cotnam, J.P., Ayoub, M.M., Westfall, P.H., Maynard, W., Fathallah, F. and O'Brien, N. 2000, Field evaluation of the revised NIOSH lifting equation. In Proceedings of the XIVth Triennial Congress of the IEA, (Human Factors and Ergonomics Society, Santa Monica, CA), Volume **5**, 37-40

Hoogendoorn, W.E., Bongers, P.M., de Vet, H.C.W., Douwes, M., Koes, B.W., Miedema, M.C., Ariens, G.A.M. and Bouter, L.M. 2000, Flexion and rotation of the trunk and lifting at work are risk factors for low back pain, *Spine*, **25**, 3087-3092

ISO 2000, *Ergonomics - Manual handling - Part 1: Lifting and Carrying*, ISO DIS 11228-1.3

Marras, W.S., Lavender, S.A., Leurgans, S.E., Fathallah, F.A., Ferguson, S.A., Allread, W.G., Rajulu, S.L 1995, Biomechanical risk factors for occupationally related low back disorders, *Ergonomics*, **38**, 377-410

Waters, T.R., Putz-Anderson, V. and Garg, A. 1994, *Applications Manual for the Revised NIOSH Lifting Equation*, (U.S. DHHS, Cincinnati, Ohio), Report No. 94-110

MUSCULOSKELETAL DISORDERS

BACK PAIN IN ROYAL NAVY HELICOPTER PILOTS: PART 1. GENERAL FINDINGS

R S Bridger, M R Groom,
R J Pethybridge, N C Pullinger, G S Paddan

Institute of Naval Medicine, Gosport, PO12 2DL, UK

A questionnaire survey of back pain in RN helicopter pilots revealed a high prevalence of back pain (80%). This confirms the findings of previous research. The questionnaire included items from the Nordic back pain questionnaire. Psychosocial factors were accounted for by including the Back Beliefs Questionnaire and the Psychological Aspects of Work Questionnaire. Pilots identified seat design and the posture needed to operate the controls as the main contributing factors. No demographic or psychosocial variables accounted for variance in back pain prevalence or back pain disability. Vibration dose values are unlikely to be exceeded in most flying operations. Back pain was reported significantly more frequently in the flying pilot roles (56 and 54% in hover and forward flying and 72% in instrument flying) than the co-pilot role (24%). Flying the aircraft, rather than sitting in the aircraft, appears to be the key factor underpinning the high prevalence of pain.

Introduction

A recent review of back pain in helicopter aircrew (Raggatt, 2000) revealed a high prevalence of back pain compared to controls, an exposure-response relationship with flying hours and evidence for degenerative changes in the spines of pilots. Helicopter pilots appear not to differ from fixed wing pilots in the prevalence of back pain *not* associated with flight, but *do* appear to suffer more pain when flying (Froom et al., 1986). The pain is located in the mid to low back, normally without radiation, and may persist after the flight. The pain may sometimes interfere with flying activities. Shanahan et al. (1986) canvassed 802 US army helicopter pilots and found that 28.4% admitted to rushing through one or more missions because of back pain. Most (83.5%) shifted in their seats in some way and a third had, at some time relinquished control to the co-pilot. The overall prevalence of pain was 72% which may explain why no differences were found between the back pain and no pain groups in terms of age, weight, height, years of flight experience and total flying hours. Only total rotary wing flying hours differed between cases and controls.

Many surveys demonstrate that the prevalence of back pain in helicopter pilots is high. However there is less agreement about the cause because pilots are exposed to both vibration and to postural stress and both of these might conceivably be responsible for the pain. Table 1 summarises these studies of back pain in helicopter pilots and presents the authors' conclusion about the main risk factor.

Table 1. Summary of helicopter aircrew back pain studies (Raggatt, 2000)

Authors	No.	Subjects Occupation	%Pain	Main Risk Factor
Fitzgerald and Crotty	100	Aircrew	58	Environment
Sheard et al.	196	RN aircrew	82	Posture
Fischer et al.	221	Pilots	61	Vibration
Schulte-Wintrop and Knoche	145	Aircrew	40	Vibration
Froom et al.				Vibration
Shanahan and Reading	11	Pilots		Posture
Bongers et al.	163	Pilots	68	Vibration
Pope et al.	20			Posture
Zimmerman and Cook	30			Vibration
Froom et al.		Pilots, Gunners		Posture

Pope et al. (1986) found that the sitting posture of helicopter pilots elicited complaints of back pain, even in the absence of vibration. Similar results were obtained by Shanahan and Reading (1984). The layout of manual controls in helicopters is asymmetric with the collective to the left and orientated similar to the handbrake of a car and the cyclic directly between the pilot's legs. Both feet operate pedals and direct observation indicates that the knees are flexed by less than 70 degrees.

The present investigation was a follow-up of that of Sheard et al. (1996) in which it was found that 82% of Royal Navy helicopter aircrew experienced back pain. The purpose of the investigation was to investigate ergonomic and task factors, including vibration, and their relationship to the prevalence of pain.

Method

Prevalence of Pain – Risk Factors and Mediating Variables
A questionnaire was developed and sent out to all RN squadrons. Items from the Nordic Questionnaire (Kuorinka et al. 1987), the Back Beliefs and PAW scales of Symonds et al. (1996) and a cartoon-based, posture rating method described in Bridger et al. (2002) were included. As a result of discussions at squadron level and familiarisation flights, it became clear that RN pilots worked in pairs, dividing their flying time fairly equally between the roles of "flying pilot" and "non-flying pilot". In the flying pilot role, the main flight regimeswere: visual forward flying; prolonged hover and instrument flying. Back pain prevalence in each of these roles/regimes was determined. Further details can be found in Bridger et al. (2001). Severity of low back pain, LBP, was rated between 1 (mild discomfort) to 7 (severe pain).

Vibration Measurements
Acceleration measurements were made in the pilot's and co-pilot's (non-flying) seats in 4 Sea King helicopters (Mk 2, 4, 5 and 6); the data have been reported elsewhere (Paddan, 2000). Measurements were made of five channels of acceleration at the seats: vertical (z-axis) on the floor beneath the seat, three translational axes on the seat (fore-and-aft (x-axis), lateral (y-axis) and vertical (z-axis)) and fore-and-aft (x-axis) on the seat backrest. The vibration measurements were made with the helicopters operating in the general modes of flight (forward flight at set speed, hover) and typical modes of flight specific for the different types of helicopter (for example, search and rescue, anti-surface warfare activities, different radar positions).

The accelerations measured were assessed in accord with British Standard BS 6841 (1987); the human response to vibration frequency weightings are also shown in the standard. The data were analysed in many formats including the estimation of the discomfort likely to be caused by the vibration, the effect of vibration on the health of the exposed personnel and the vibration isolation efficiency of the seats. The discomfort likely to be caused by the vibration can be put into 5 categories ranging from "*not uncomfortable*" to "*extremely uncomfortable*". The standard states that a vibration dose value, VDV, be calculated; a VDV of 15 ms$^{-1.75}$ will "... *usually cause severe discomfort*" and increase the "... *risk of injury*". The vibration isolation efficiency of the seats can be evaluated by calculating a Seat Effective Amplitude Transmissibility, SEAT, value.   The SEAT value is calculated from vertical vibration measured on the floor beneath the seat and vertical vibration on the seat.   A SEAT value greater than 100% implies that a person sitting on the seat experienced more discomfort than sitting on the floor (Griffin, 1990).

Results

Completed questionnaires were received from 185 pilots (response rate 75%). 63% of pilots were between 25 and 34 years of age and 26% were aged 35 or over. One third of the pilots had less than 1000 flying hours, one third had 1000-2000 hours and the remaining third had more than 2000 hours. Almost all pilots flew between 100 and 300 hours per year. Back pain and back pain disability were not related to any of the demographic or psychosocial factors, nor to flying hours. Pilots over 35 years of age reported significantly less back pain. The 12-month prevalence of back pain was 80% for pilots of all RN helicopter types.

Influence of task and postural factors
Due to the high prevalence of back pain, the sample was split, approximately, into two halves - a group of "LBP3 cases" (LBP ratings of 3 or more) and a group of non-cases (ratings lower than 3). Table 2 gives LBP3 prevalence in different flying regimes/roles.

Table 2. Back pain prevalence (%LBP3) by helicopter in different flying regimes

Helicopter	Pilots(n)	VFF	IF	PH	NFP*
Gazelle	15	60	67	47	20
Lynx	36	53	69	50	22
Merlin	12	33	42	42	8
Sea King	118	60	79	58	27
Other	4	25	25	0	0
All	185	56	72	54	24

*VFF = Visual forward flying, IF = Instrument flying, PH = Prolonged hover, NFP = Non-flying pilot role

LBP3 prevalence was greatest in instrument flying and least in the non-flying pilot role.

Table 3. Self-reported forward flexion of the trunk in different flying regimes (Sea King Pilots only)

Trunk inclination	VFF	IF	PH	NFP*
	%	%	%	%
Noticeably forward	4	48	5	0
Slightly forward	75	49	77	25
Up straight	19	3	16	54
Reclining	2	1	3	21

*VFF = Visual forward flying, IF = Instrument flying, PH = Prolonged hover, NFP = Non-flying pilot role

Table 3 shows the percentage of self-reported forward flexion of the trunk in different flying operations for the 118 Sea King Pilots. Application of the Friedman test on the responses from 114 pilots providing information for all 4 regimes, revealed that median reported trunk inclination differed significantly between the flight regimes ($p<0.01$). Testing for contrasting regimes revealed that forward flexed sitting postures were most frequently reported in instrument flying. A significantly less "forward" posture was reportedly adopted during the visual forward flying and prolonged hover. The reported posture for these latter regimes was significantly more "forward" than in the NFP (non-flying pilot) role. Similar findings were obtained for lateral tilting and twisting of the trunk – neutral postures were reported much more frequently in the non-flying pilot role (see Bridger et al. 2001, for further details).

Ergonomics
Pilots were presented with a list of ergonomic elements and indicated which, if any, contributed to their back pain. The most commonly selected elements for 98 pilots were: posture needed to operate the controls (89%), shape of seat (86%), shape of back cushion (85%), padding of seat (80%) and angle of seat (79%).

Vibration discomfort
Depending on the type of helicopter, vibration exposure in the pilot's seat during forward flight at constant speed could be categorised as being "*fairly uncomfortable*" to "*very uncomfortable*". Results show that the co-pilot would generally be exposed to *higher* vibration magnitudes compared to the pilot; however, there are exceptions for specific flight operations. The vibration discomfort likely to be experienced by both the pilot and the co-pilot would be influenced by other parameters specific to the aircraft (e.g. radar configuration, carrying load). Some specific operations could be expected to cause extreme discomfort (for example, automatic transition from forward flight at 90 knots at an altitude of 200 feet to hover at sea level); however, it is understood that these operations are unlikely to be sustained for long periods.

Health effects of vibration
Vibration dose values, VDVs, were calculated from the acceleration measurements in the pilot's and the co-pilot's seats. Using the guidance in BS 6841 (1987), durations were calculated that would be required for exposure to vibration to reach a tentative action level of 15 ms$^{-1.75}$. The data show that a pilot's exposure to vibration during forward flight could exceed a VDV of 15 ms$^{-1.75}$ after a period ranging from 1 hour 30 minutes to exceeding 8 hours. This large variation in duration depended on the type of helicopter and operation carried out. The corresponding exposure duration for the co-pilot could range from about 1 hour 10 minutes to 8 hours. There are some specific operations that showed large variations in exposure periods; for example exposure to vibration during hover could be expected to exceed the tentative action level after a minimum of about 2 hours. There are some operations that showed high VDVs but it is understood that these operations would not be carried out for long periods (e.g. co-pilot's seat during an operation involving transition up from hover at sea level to 90 knots at an altitude of 200 ft).

Seat characteristics
Seat isolation efficiencies, that is SEAT values, for the pilot's and the co-pilot's seats have shown values significantly greater than 100%. SEAT values in the pilot's seat during hover ranged from 118% to 197% depending on the type of helicopter. The SEAT values in the two seats for the helicopters in forward flight ranged from 99% to 244%. Similar values have been measured for most of the different operations investigated in the four helicopters. Seat transfer functions between vertical vibration on the floor and on top of the seat indicated that the seats amplified vibration for some flight operations and at some frequencies.

Discussion

In general, the findings confirm that helicopter pilots have a high prevalence of occupational low back pain and suggest that pain severity is associated with task demands. The pain seems to be most prevalent when the demands on the musculoskeletal system are likely to be at their highest - during instrument flying when the position of the head is fixed by the need to monitor the controls and the position of the hands and feet is fixed by the need to control the aircraft. Pilots report that instrument flying is more mentally demanding. The layout of the controls in all helicopters in this study was asymmetric – with the collective on the left and the cyclic between the legs. All pilots reported that, during training, they were taught to rest their right forearm on their right thigh. This is done to stabilise the forearm and obtain finer control of the cyclic. Presumably, this practice serves to relieve the muscles of the right shoulder girdle of the postural load by stabilising the right forearm. However, it also encourages the adoption of a non-neutral sitting posture with forward flexion of the trunk, axial rotation to the left and a tilt to the left or right depending on the height of the collective.

Lopez-Lopez et al. (2001) monitored electromyographic (EMG) activity in the low back muscles of helicopter pilots while flying. Low level, static muscle activity was found throughout the sortie, with greater EMG intensity on the right hand side of the body, possibly highlighting the importance of postural asymmetry.

Some pilots reported that moulded lumbar supports were helpful, but LBP intensity and frequency was no lower in users than in non-users indicating, perhaps, that the supports are used as a last resort and only succeed in making the pain more manageable.

Vibration exposure in the helicopters could be categorised as "uncomfortable". Continued exposure to these vibrations with the helicopters operating in certain manoeuvres could be expected to exceed a tentative action level proposed in a BS 6841 (1987). The high SEAT values indicate that the seats amplified the vertical vibration transmitted from the floor to the seat. Improvements to the seats would be beneficial in reducing the discomfort experienced by the crew during the different flight manoeuvres. Improvements could involve the selection and use of appropriate seat foams and cushions.

Conclusions

Back pain in helicopter pilots is associated with flying the helicopter, rather then being in the helicopter. The prevalence of mid to low back pain is high. The pain and the disability associated with the pain are not related to demographic or psychosocial factors (e.g. job satisfaction, mental stress, social support or fatalistic beliefs). The pain is greatest in instrument flying and least in the non-flying pilot role and is associated with the adoption of non-neutral sitting postures. That asymmetric postures are adopted is not surprising since the controls and displays are arranged asymmetrically. It is likely that ergonomic redesign using a symmetrical arrangement of controls will lower the pain prevalence to the levels experienced in the non-flying pilot role. Vibration levels, although not excessive in visual forward flying, were found to be high in particular manoeuvres. It would be unwise to expose pilots to vibration when postural stress is already high. Better seats and increased cockpit automation may help. Appropriate selection of foams for the seats could assist in reducing the exposure of aircrew to high levels of vibration.

References

Bongers, P.M., Hulshof, C.T.J., Dijkstra, L., Boshuizen, H.C. and Groenhout, H.J.M. 1990. Back pain and exposure to whole body vibration in pilots. Ergonomics, 33, 1007-1026.
Bovenzi, M. 1996. Low back pain disorders and exposure to whole body vibration in the workplace. Seminars in Perinatology, 20, 38-53.

Bridger, R.S., Groom, M.R., Jones, H., Pethybridge, R.J. and Pullinger, N.C. 2001. Back pain in RN helicopter pilots. Institute of Naval Medicine. Unpublished MOD Report.

Bridger, R.S., Groom, M.R., Jones, H., Pethybridge, R.J. and Pullinger, N.C. 2002. Back Pain in Royal Navy Helicopter Pilots: part 2, Some methodological consideration. Proceeding of the Ergonomics Society Annual Conference, Cambridge, April 2002.

British Standards Institution 1987. Measurement and evaluation of human exposure to whole-body mechanical vibration and repeated shock. BS 6841. London.

Fischer, V., de Witt, A.N., Troger, C. and Beck, A. 1980. Vibrationsbedingte wirbelsaulenschaden bei hubsschrauberpiloten. Arebeitsmedizin Sozialmedizin Praventivmedizin, 15:161-163.

Fitzgerald, J.G. and Crotty, J. 1971. The incidence of back pain among aircrew and ground crew in the Royal Air force. Unpublished MOD Report.

Froom, P., Barzilay, J., Caine, Y., Margaliot., S., Forecast., D. and Gross, M. 1986. Low back pain in pilots. Aviation, Space and Environmental Medicine, 57, 694-695.

Griffin, M.J. 1990. Handbook of human vibration. Academic Press Limited, London. ISBN 0-12-303040-4.

Kuorinka, I., Jonsson, B., Kilbom, A., Vinterberg, H., Biering-sorensen, F., Andersson, G. and Jorgensen, K. 1987. Standardised Nordic questionnaire for the analysis of musculoskeletal symptoms. Applied Ergonomics, 18:233-237.

Lopez-Lopez, J.A., Vallejo, P., Rios-Tejada, F., Jimenez, R., Sierra, I. and garcia-Mora, L. 2001. Determination of the lumbar muscle activity in helicopter pilots. Aviation, Space and Environmental Medicine, 72: 38-43.

Paddan, G.S. 2000. Assessment of whole-body vibration in a Sea King Mk2 helicopter. Institute of Naval Medicine.Unpublished MOD Report.

Pope, M.H., Wilder, D.G. and Donnermeyer, D.D. 1986. Muscle fatigue in the static and vibrational seated environments. 1986 AGARD Conference Proceedings, No 378: Backache and Back Discomfort, paper 25.

Raggatt, T.R. 2000. Back pain in helicopter aircrew: a literature review. Institute of Naval Medicine Report No. 2000.052.

Schulte-Wintrop, H.C. and Knoche, H. 1978. Backache in UH-1D helicopter aircrews. AGARD Conference Proceedings: operation Helicopter Aviation Medicine.

Shanahan, D.F., Mastroianni, G.R. and Reading, T.E. 1986. Back discomfort in US army helicopter aircrew members. AGARD Conference Proceedings, No 378: Backache and back discomfort paper, 6.

Shanahan, D.F. and Reading, T.E. 1984. Helicopter pilot back pain: a preliminary study. Aviation, Space and Environmental Medicine, 55(2), 117-121.

Sheard, S.C., Pethybridge, R.J., Wright, J.M. and McMillan, J.H.G. 1996. Back pain in aircrew – an initial survey. Aviation, Space and Environmental Medicine, 67, 474-477.

Symonds, T.L., Burton, A.K., Tillotson, K.M. and Main, C.J. 1996. Do attitudes and beliefs influence work loss due to low back trouble? Occupational Medicine, 46, 25-32.

Zimmerman, C.L., and Cook, T.M. 1997. Effects of vibration frequency and postural changes on human responses to seated whole body vibration exposure. International Archives of Occupational and Environmental Health, 69:165-179.

BACK PAIN IN ROYAL NAVY HELICOPTER PILOTS: PART 2. SOME METHODOLOGICAL CONSIDERATIONS

R. S. Bridger, M. R. Groom, H. Jones, R. J. Pethybridge & N. C. Pullinger

Institute of Naval Medicine, Alverstoke, PO12 2DL, UK,

Aspects of an investigation of back pain in helicopter pilots are reported. The intent was to capture data on task and postural factors in relation to pain intensity. A questionnaire was prepared. Items concerning back pain in both the pilot and co-pilot roles and in different flying regimes were included. A self-rating method was developed in which pilots rated their postures in three axes in each of the different flying roles/regimes. A prior test of the method compared pilots' ratings with those of two observers. For sagittal postures, the pilots' ratings were also compared with photographs, assessed by a third observer. It appears that self-rating of static sitting posture has some validity, but is crude and only of use in surveys with large samples. Problems associated with the practice of using a single instrument to measure exposure and outcome are discussed.

Introduction

The formulation of models that account for the relationship between ergonomic exposures and musculoskeletal outcomes depends on the development of inexpensive and practical ways of quantifying these exposures and outcomes. Stock (1991) concluded that the more refined the definition of the exposures, the more able researchers will be to demonstrate cause and effect, where it exists. Halpern et al. (2001) describe a risk factor questionnaire for outcome studies in low back pain, in which respondents estimate the time spent in different postures. The questionnaire has been found to have good test-retest reliability. Questionnaire surveys are an inexpensive way of reaching large numbers of people, so it is not surprising that researchers continue to explore their use in gathering self-reported data on ergonomic exposures.

Unfortunately, people are not very good at estimating their daily exposure to ergonomic factors in the workplace (Woodcock, 1988). They seem to be particularly bad at estimating the *amount of time* they spend in different postures or body positions (Burdorf and Laan, 1991). Hildebrandt and Bongers (1992) compared self-ratings of posture with observers' direct observations in homogenous groups of VDU workers, office workers, dispatch workers and assembly workers. There were large differences between self-ratings of posture and 'objective' measures, based on direct observation. This applied to ratings of bending (forward flexion of the trunk), twisting, and bending and twisting combined. There was no general trend. Office workers tended to underestimate how often they adopted different postures, whereas assembly workers overestimated the amount of bending but gave close estimates of the other postures. Baty et al. (1986) compared nurses' estimates of posture with direct observation and with assessment using an inclinometer. The ergonomic exposures included time spent standing, sitting, walking, kneeling, stooping and bending. The differences between the estimates of time spent in the postures and direct observation were as large as 100%. Rossignol and Baetz (1987), compared employee estimates of muscular effort, vibration exposure, static posture, bending and twisting with estimates obtained by direct observation of people at work. There was good agreement between employee and observer estimates for muscular efforts and exposure to vibrations, but not for static postures, bending and twisting of the trunk.

It is tempting to speculate on the reasons for these findings. For people to be able to complete questionnaire items about ergonomic exposures and outcomes, they must either be still suffering from the after-effects of the exposure or they will have to recall the information from memory. In order to recall the information from memory, it must have been stored in memory in the first place. If stored in memory, it will either have been encoded actively or passively (with or without attention). In the latter case, recall will be a process of inference, using other information. Or, as Sir Frederick Bartlett put it:

"We fill-in the lowlands of our memory from the highlands of our imagination".

Much of recall from long-term memory is a process of reconstructing "what must have happened". If posture, because of the location of its control mechanisms in the central nervous system, is not normally in conscious awareness, it will not normally be one of the dimensions along which an activity is encoded in memory and people will have no direct access to any information about it. Even for activities that *do* require conscious awareness, the extent to which they are stored directly in memory is eminently debatable as anyone who has ever forgotten to fill-in their time sheet will know. That the time spent in different posture*is* often underestimated is not surprising if people can't remember what they have been doing at work.

Events that are intrinsically alerting (such as exposure to vibration or sudden maximal exertions) *are very likely* to be stored in episodic memory and will be available for recall, given appropriate cues. Episodic memories are, by definition, encoded with respect to time, however crudely. In the absence of episodic memories of ergonomic exposures, recall will take place through reconstruction (as in, 'I spent the morning working on a report in my office, therefore I must have been sitting at my desk'). It would seem, then, that self-reports of the *time spent* in different postures will be unreliable unless they are linked to specific instances of exposure that reached conscious awareness and were stored in episodic memory or to memories of other activities, from which posture can be inferred. For example, in highly repetitive, time-based jobs, people may be able to infer how long they spend in different postures from how long they spend doing particular tasks. This would explain why assembly line workers fared better than office workers when estimating how often they adopted different postures in the Hildebrandt and Bongers study – the jobs of assembly line workers being more highly differentiated with respect to both posture and time. Estimates of whether or not particular postures are adopted*at all* would be expected to be of intermediate validity. They *may* be linked to episodic memories or be reconstructable without the need to reference them to information about time. There is some evidence for this assertion. Witkorin et al. (1990) found that people's estimates of whether or not they lifted things at work or whether they adopted bent or twisted postures were "acceptable" for research purposes.

In the present research, the aim was to investigate back pain in helicopter pilots (the main findings are in Bridger et al., 2001). It has been suggested (e.g. Pope et al. 1986) that the seated posture of helicopter pilots may be the cause of back pain even in the absence of vibration. For this reason, an index of posture had to be developed for inclusion in the questionnaire. The remainder of this paper describes the development and testing of the posture rating method and discusses some of the issues concerning its use.

Method

Helicopter pilots adopt a highly constrained, static work posture while flying the aircraft. The posture seems to be determined by the location of the controls and displays and, to some extent, by seat design and adjustment. These observations led to the hypothesis that posture was determined by the demands of different flight regimes and that back pain would be worse in the more stressful postures, with vibration and other factors held relatively constant. In order to capture data on the sitting postures of pilots, a posture rating method was developed using cartoons to depict different sitting postures with the emphasis on the posture of the spine. Pilots were asked to select the picture which best represented their own sitting posture in each flying role/regime. Three sets of pictures were developed to represent 4 levels of trunk flexion ("noticeably forward", "slightly forward", "sitting up straight" and "reclining against backrest"), 4 levels of axial rotation (Twisted >10 degrees to the left", "twisted 0-10 degrees to the left", "not twisted" and "twisted to the right") and 3 levels of lateral trunk flexion ("tilted to the left", "sitting level", "tilted to the right"). The researchers tested the method with 21 pilots in a stationary aircraft in the hanger. Pilots adjusted the seat and grasped the controls and were asked to adopt the usual postures in visual forward flying, prolonged hover, instrument flying or the non-flying pilot role. They then recorded this posture by selecting the closest matching cartoons from those provided. At the same time, two independent raters, an ergonomist and a medical doctor who was also a helicopter pilot, made their own ratings of the postures by direct observation of the pilot, but blinded to the pilots' own ratings. A sagittal photograph of the pilot was then taken. A third rater, blind to the others, assessed the postures in the photographs. Transparencies of the cartoons were laid over each photograph and the cartoon that best matched the actual posture was recorded.

Results

Table 1 presents the numbers of agreements and differences (with extent) between the pilots and raters and the photographic evidence.

Table 1. Trunk Forward Flexion: agreement between photographs and raters (n=21)

	Same	Different by 1 Category	Different by 2 Categories
Pilots	11	8	2
Rater 1	15	6	0
Rater 2	17	4	0

There was no statistically significant difference between pilots and raters relative to the photographic evidence (Cochran's Q test after assigning "1" to agree and "0" for otherwise). Pilots can give a rough indication of the degree of sagittal flexion when sitting in a helicopter. Table 2 summarises agreement between raters and pilots.

Table 2. Agreement between raters and pilots

	Same	1 different category	> 1 different category
Flexion			
Rater 1	11	6	4
Rater 2	8	12	1
Lateral Flexion			
Rater 1	7	12	2
Rater 2	9	10	2
Axial Rotation			
Rater 1	10	8	3
Rater 2	10	8	3

For forward flexion, raters agreed with pilots or chose the adjacent rating. For lateral flexion pilots and raters rarely disagreed as to the side of any lateral tilt but did not often agree on whether the pilot was tilted or sitting up straight (this may be because the degree of tilt was very small). For axial rotation, most disagreements were about whether the trunk was twisted but not the direction of twist.

Table 3 summarises the posture of 118 Sea King pilots in different flying regimes/roles and the percentage of those in each postural category (forward flexion) who suffered back pain (LBP3) in each category (see Bridger et al. 2002).

Seventy nine percent of pilots were back pain cases in instrument flying, compared with 58% and 60% in prolonged hover and visual forward flying respectively and 27% in the non-flying pilot role. Similarly, 96% of pilots reported being in a forward flexed posture in instrument flying compared with 76% and 81% of pilots in visual forward flying and prolonged hover. In the non-flying pilot role, only 25 % of pilots reported being in a forward flexed position. Similar results were found for the postural deviations in other axes (see Bridger et al., 2001).

Interestingly, in the non-flying pilot role, the prevalence of back pain did not differ across the posture categories.

**Table 3. Back pain (%) in different flying regimes/roles
by amount of forward flexion (Sea King pilots only)**

	IF N %	PH N %	VFF N %	NFP* N %
Noticeably forward	56 89[a]	6 67	5 100	0 -
Slightly forward	57 72[a]	89 65[b]	85 69[c]	29 34
Up straight	3 67	18 33[b]	22 27[c]	63 29
Reclining	1 0	2 0	2 50	24 17
Unknown	1 0	2 0	4 0	2 0
All	118 79	118 58	118 60	118 27

*VFF = Visual forward flying, IF=Instrument Flying, PH=Prolonged Hover, NFP=Non-Flying Pilot Role
(a) LBP3 prevalences differ p<0.05,(b), LBP3 prevalences differ p<0.01,(c), LBP prevalences differ, p<,0.05.

Discussion

The data indicate that pilots can give a rough indication of their sitting posture. Although there was disagreement with photographic evidence and with external raters' judgements, this was not statistically significant, nor was it biased in a particular direction. The technique would appear to be accurate enough for use in surveys with large samples, where crude descriptions of posture will suffice. Although pilots may not have direct access to "postural memories" they certainly remember posture-related information such as how they adjust their seat before flying and what they look at in different roles/regimes. These cues may enable them to reconstruct the kind of posture they adopt when carrying out the task. It seems likely that information on posture is not usually encoded in episodic memory about the day's work and therefore has to be reconstructed in this way. Since we are dealing with the static sitting postures of pilots carrying out a limited set of very well-defined tasks with different requirements, it seems reasonable to argue that this kind of information about posture can be retrieved by reconstruction in the presence of the usual environmental cues. On these grounds, the technique seems to be valid when used to compare the posture in different tasks when carried out by the same workers. A similar conclusion was reached by Hildebrandt and Bongers (1992) in their between subjects investigation. Although agreement between workers' self-ratings and observers' ratings was poor, when comparing individual workers, it was much better when comparing groups of workers doing different jobs.

However, in relation to the back pain data, several conflicting interpretations are possible, in the absence of other information. For example, it might be argued that the data are confounded by *differential misclassification*. It is known that, for long term recall (25 years), people with current back problems tend to overestimate their previous exposure to back stressors, such as heavy lifting and physically demanding tasks, to a greater degree than symptom free individuals (Koster et al., 1999). Current pain, it seems, can induce bias in the recall of previous exposure. Similarly, people whose behaviour changes over time (e.g. increasing consumption of alcohol) tend to overestimate their previous consumption. It might be argued that low-back pain induces recall bias, resulting in overestimation of the amount of postural asymmetry. It may be that pilots exaggerate their posture when they experience back pain for other reasons. In the present study back pain was found to be more severe in IF than in the NFP. One might argue that it is exposure time rather than posture that causes pain and that pilots experience more pain in IF because they spend more time in IF. The differences in posture ratings would then be a misclassification because the pain ratings were confounded by exposure time. Alternatively, increased mental workload in IF may exacerbate back pain due to increased muscle tension, again causing posture to be misclassified. However, direct observation of pilots carrying out these tasks indicates that this is not the case and that IF flying *does* demand close and continual scrutiny of the displays demanding a forward flexed sitting posture. Furthermore, since the pilots fly in pairs and have to fly all regimes to maintain their pilot status, it is the case that approximately half of their time spent airborn is in the NFP role. Thus, the findings with respect to NFP are *not* confounded by flying hours. If anything, pilots spend longer in NFP than in the other flying regimes but experience less back pain (Sea King pilots work in pairs, the other pilot is carrying out the different regimes, when the first pilot is in the NFP role).

The use of a single questionnaire to assess both exposure and outcome in a cross-sectional study is undoubtedly a high-risk strategy due to the threat of differential misclassification of outcomes. Analysis of the present data (Bridger et al, 2001) suggests that differences in posture across the flight regimes are the same in pilots with and without back pain (a small group of pilots suffered no back pain at all). However, for research in more complex environments, or when postures are not static, it is recommended that exposure and outcome measurement be made with separate instruments (e.g. posture rating might be performed by a trained observer). It seems that the posture rating method has some validity when it is carried out in the presence of environmental cues. Whether this holds true for assessments made in the absence of such cues (away from the workplace) remains to be seen.

References

Baty D Buckle PW Stubbs DA. 1986. Posture recording by direct observation, questionnaire assessment and instrumentation: a comparison based on a recent field study. In, The Ergonomics of Working Postures edited by N Corlett, J Wilson and I Manenica. Taylor and Francis, London.

Bridger, R.S., Groom, M.R., Jones, H., Pethybridge, R.J. , Pullinger, N. 2001. Back Pain in Royal Navy Helicopter Pilots, INM Report. Unpublished MOD Report.

Bridger, R.S., Groom, M.R., Jones, H., Pethybridge, R.J. , Pullinger, N. 2002. Back pain in Royal Navy helicopter pilots: general findings. Ergonomics Society Annual Conference, Homerton College, Cambridge, April 3-5th.

Burdorf, A. Laan, J. 1991. Comparison of methods for the assessment of postural load on the back. Scandinavian Journal of Work Environment and Health, 17:425-429.

Halpern, M. Hiebert, R., Nordin, M., Goldsheyer, D., Crane, M. 2001. The test-retest reliability of a new occupational risk factor questionnaire for outcome studies of low back pain. Applied Ergonomics, 32, 39-46.

Hildebrandt, V.H. and Bongers, P.M. 1992. Validity of self-reported musculoskeletal workload. Rev. Epidemiol Sante, 40, S124.

Koster, M., Alfredsson, L., Vingard, E., Kilbom, A. 1999. Retrospective versus original information on physical and psychosocial exposure at work. Scandinavian Journal of Work Environment and Health, 25:410-414.

Pope, M.H., Wilder, D.G. and Donnermeyer, D.D. 1986. Muscle fatigue in the static and vibrational seated environments. 1986 AGARD Conference Proceedings, No 378: Backache and Back Discomfort, paper 25.

Rossignol, M. Baetz, J. 1987. Task-related risk factors for spinal injury: a validation of a self-administered questionnaire on hospital employees. Ergonomics, 30, 1531-1540.

Stock, S.R. 1991. Workplace ergonomics factors and the development of musculoskeletal disorders of the upper limbs: A meta analysis. American Journal of Industrial Medicine, 19:87-107.

Witkorin, C. Karlqvist, L., Winkel, J. 1990. Validity of self-reported exposures to work postures and manual materials handling. Scandinavian Journal of Work Environment and Health, 19:208:214.

Woodcock, C. 1988. Self-reports in ergonomics: agreement between workers and ergonomist. In, Trends in Ergonomics/Human Factors V, edited by F. Aghazadeh, Elsevier Science and Publishers B.V. (North Holland).

A USER STUDY OF ERGONOMIC RISK FACTORS WHEN LOADING CHILDREN INTO CARS

Joseph Capitelli and Rachel Benedyk

Ergonomics and HCI Unit,
UCL,
26 Bedford Way,
London WC1H 0AP

An assessment was carried out of the task of placing young children into and out of various child seats in a range of vehicles. Data was gathered from users by interview, observation, user trial and focus group discussion. The objective of the study was to assess the perceived difficulty of the task, whether there was the risk of injury occurring, and factors that affected the task strategy adopted. Evidence was gathered showing that the risk of injury does exist, and several key features were identified that contribute to this. Other areas of perceived difficulty were also highlighted.

Introduction

When transporting young children by car it is usually necessary for the carer to lift the child into and out of a dedicated child seat restraint. This can entail lifting a relatively heavy, and unpredictable, load from knee height into the interior of a vehicle. Although much previous work has focused on the misuse of child seat restraints (e.g. Rainford et al, 1993), little or no work appears to have been carried out on the manual handling aspect of child seat restraint use. As a full postural analysis was not feasible for this study, it was decided to approach the use of child seat restraints mainly from the user's perspective: collecting data by interview, observation, user trial and focus group. The study was broadened to investigate any usage problems in the users' experience.

Method

Data was gathered by three methods. Firstly, observation of tasks and influencing factors, plus brief interviews, were carried out with a range of typical users. Users were asked about child seats and any usage problems or occurrence of injury. Three contrasting child seats were chosen for the study; Type A (a carry-out seat), Type B (seat secured by car seat belt) and Type C (both child and seat secured by car seat belt).

Secondly, a user trial was organised, with a range of sizes of subjects being observed loading and unloading children into three types of vehicle (5-door, 3-door and People Carrier (PC)), each with three different types of child seat. The vehicles used were chosen to give differing access and seating height. An ergonomic assessment was carried out focusing on postures associated with potential risk of injury.

Finally a focus group was arranged with five female subjects used to transporting children in cars. The discussion covered topics determined by the findings of the observation and interview work.

Results and Discussion

Preliminary Observation and Interview Work

During interview several subjects recounted experiencing back pain or suffering injury when transferring children into or out of cars. Furthermore, during observation, postures recognised as increasing the risk of back injury – namely bending and twisting of the spine whilst supporting the child - were adopted by several subjects carrying out the task (see Figure 1).

In many cases auxiliary lifting and handling tasks were also observed e.g. transferring pushchairs into and out of vehicles. The cumulative effect of two demanding procedures should not be overlooked as muscles may tire and the possibility of accidentally dropping the child may increase, as may the possibility of injury from cumulative strain.

Vehicles with increased seat height relative to standard cars, e.g. PCs, were observed to make the task less physically demanding due to the child seat pan being at a more accessible height (see Figure 2). Transferring children from or to the rear seat of 3-door cars was seen as a very difficult, physically demanding task. Type A seats were effortful because of the weight of child plus seat (see also Roca, 1997).

Figure 1. Removing child from rear seat of 3-door car.

Figure 2. Removing child from rear seat of PC.

Many users experienced difficulties with child seat buckles and/or the accompanying use of the car seat belts - concerning both correct operation and the physical demands required to reach and operate buckle and harness parts.

The lack of sufficient car park spaces suitable for those with children (i.e. wider bays) was seen to make the task more difficult.

User Trial

Independent of seat type, the difficulty of the task and risk of injury was greatest with the 3-door car and least with the PC. The subjects adopted the most extreme postures - involving bending, twisting, stretching and asymmetrically supporting the load - when using the 3-door car. This was much less pronounced with the 5-door car, and with the PC postures associated with increased risk of injury were effectively absent. Problems arise in the 3-door car when accessing the rear seat; this appears to involve a twisting manoeuvre to get round the pillar either when standing outside or entering the car. The benefit of the PC was the seating height, which put the child seat pan height and harnessing at a far more accessible height.

It was noted that with the Type A seat, the seat plus child were held in front of the carer at a distance in front of the trunk, thus increasing the moment experienced at the joints. The design of the seat and handle may be most suitably configured for use as a carrying device. The benefit of the Type A seat was that the load (in this case the child plus seat) tended to be placed on the edge of the car seat and then manoeuvred into position; whereas with the Type B seat the subjects placed the child directly into the seat pan, which is located further into the vehicle, involving stretching. Some severe static postures involving bending and twisting whilst reaching were observed whilst fitting seat harnesses on the child and seat.

The extra height the load needs to be lifted to clear the seat side lip of the Type B seat was evident when a subject of short stature used the PC. For a carer of shorter stature this may make the task far more difficult e.g. necessitate lifting the load above shoulder height.

When using the 3-door car the increased stature of the tallest subject was evident by occasional increased flexion of the spine relative to the shorter subject in a similar situation. The taller subject was clearly a 'tighter fit' in the rear of the 3-door car. Stature may not be the most relevant dimension as, when bending over in such spaces, leg length may be more critical. However in other circumstances e.g. if the child was struggling, it may be more influential.

The degree to which the car doors open appeared to have an influence on how the task was carried out, as often the subjects appeared hindered by the door behind them.

Focus Group

Two focus group subjects had experienced pain/suffered injury whilst removing children from car seats – in both cases from a child seat restraint placed in the rear of a 3-door car. One of these subjects specifically recalled a 'sharp sciatica-like pain' when bending and twisting to remove the child.

Two subjects recounted the difficulties that having a back injury had on transferring children to and from child car seats. One recalled dropping her baby from chest height into the car due to the pain experienced trying to carry out the task with a lumbar region injury, the other recounted devising a coping strategy to minimise the pain. Of interest was that both subjects tried to carry on their normal routine of taking children to activities whilst suffering an injury that, in many other situations e.g. paid manual work, would oblige rest from manual handling activities.

From the discussion it was clear that all the subjects could detect a noticeable difference between the different car/seat combinations with regards the ease of transferring children in and out of cars. All subjects agreed that the PC was by far the easiest vehicle to get a child into and out of – independent of child seat type. All subjects appeared to clearly appreciate the reduced physical strain when placing and removing a child from a seat situated at an elevated height relative to that in a standard car. The small 3-door car was seen as the most difficult to use by those using the Type B seat fixed onto the rear seat, due to difficulty in accessing the seat whilst holding the child.

It was considered, by all subjects, difficult and time consuming to secure the car seat belt harnesses with all three types of seating - both to thread the car seat belt through the necessary guides and to fasten the seat belt clip. With Type B seats problems develop with harnessing becoming twisted, which can cause injury to the child in a crash if uncorrected (Rainford et al, 1993), and the groin strap of the 5-point harness becoming trapped under the child, which can be difficult to retrieve when the child is in the seat.

The difficulty and length of time it takes to operate the harnessing has an impact in two main ways: the carer may be in an uncomfortable and potentially damaging position for longer (comments such as 'leaning', 'stretching', 'really lean forward' were made); and whilst securing the harnessing the carer cannot give full attention to any other children they may be responsible for e.g. those outside the car.

All subjects had experience of making journeys with children unwittingly not properly secured in the car - either with children secured into seats that were not secured into the car or with children not secured properly into seats. A major concern was that in general there was no indication that the seat or child was not secured properly without specifically checking for this – something that most of the subjects did not to do on a regular basis.

Two subjects recounted using Type B seats without them being secured to the car e.g. after transferring the seat between cars. All subjects had experienced children wriggling their arms free of the harnesses whilst in a Type B or C seat. Several subjects had tried increasing the harness tension to prevent this but without success. When using a Type C seat the children will often quickly learn to release the seat belt buckle without assistance, and may do so inappropriately and fail to reattach it (on their own seat or an adjacent seat).

Car parks were cited as being areas of difficulty for loading children into and out of cars. The lack of room afforded by a standard car park space coupled with the lack of sufficient wider parent and child bays being seen as the cause of the problem. These may necessitate moving the car partially from the bay, and may mean leaving the child outside the car whilst doing this.

Conclusions

Several subjects recounted experiencing back pain or suffering injury when transferring children in to and out of cars, showing that injury can occur from this task, especially from the rear of 3-door cars.

Postures recognised as increasing the risk of back injury (e.g. bending and twisting whilst supporting a load) were adopted by subjects carrying out the task. Vehicle type has an effect on the postures adopted (as will the stature of the subject): vehicles with elevated seating in general being preferable and 3-door cars being the least satisfactory.

Although redesign of the car and child seat could improve the situation, getting the relevant information to users concerning the risk of injury and ways to reduce it could also be of benefit, e.g. techniques and strategies that reduce the chance of injury, getting the child to help as much as possible, the importance of trying out equipment before buying it and how different types of equipment may influence the risk of injury. This may be particularly important for those with established back complaints.

Carers should be made aware of the ease and consequence of travelling unwittingly with the child and/or seat not fully secured. One way to present this message would be on the child seat itself. Both child seat manufacturers and car manufacturers should also address the problem of children travelling unsecured - making it more difficult for this to occur and, if it does so, making it immediately detectable.

Improvements could be made to the design of the harnessing and car seat belt guides in children's car seats to make the overall system less difficult and avoid the need for the user to assume potentially harmful and uncomfortable postures. Harnessing buckles were seen as unnecessarily complex and would benefit from the use of affordances or colour coding to help users clip them together in the correct manner. When using the car seat belt with child seat restraints it may make the task less demanding if the seat belt clip fastened on the door side of the car interior.

Most subjects appear to agree that wider car parking spaces were greatly beneficial to carrying out the task but that their numbers were not sufficient.

Finally, although this is a pilot study and it is not claimed that the results are representative of a larger user group, it is evident that risks exist in loading children into cars, and several topics worthy of further ergonomic investigation have been identified.

References

Rainford, J.A., Page, M. and Porter, J.M. 1993, How and Why Child Safety Restraints in Cars are Misused. In E.J. Lovesey (ed.) *Contemporary Ergonomics 1993* (Taylor and Francis, London), 190-195.

Roca, L.L. 1997, This Infant Carrier is too Heavy; An Ergonomic Redesign of Infant Carriers. *Proceedings of the Silicon Valley Ergonomics Conference & Exposition, Ergocon'97*, 142-148.

ASSOCIATION BETWEEN BACK PAIN AND POSTURE IN SCHOOLCHILDREN

Sam Murphy and Peter Buckle

Robens Centre for Health Ergonomics
University of Surrey
Guildford GU2 7TE
s.murphy@surrey.ac.uk

Contrary to common belief, back pain amongst young people is a frequent phenomenon. Epidemiological studies have found high prevalence rates of back pain. The study reported here aims to identify the extent of back pain experienced by 11 to 14 year old schoolchildren, and establish the intensity, duration and frequency of exposure to physical risk factors present in schools. This paper considers the sitting postures of schoolchildren in the classroom. The sitting postures of 66 children were recorded in normal lessons using the Portable Ergonomic Observation Method (PEO). Associations were found between high body mass index and back and neck pain, flexed trunk postures and low back pain and static postures and upper back pain. The implications of the findings are discussed.

Introduction

Back pain is a significant burden on industrialised countries. If the symptoms and causes of back pain could be identified at an early stage the opportunity for remedial action would be improved. It has been shown that a strong predictor of having future back pain is a previous history of such symptoms (Troup et al., 1987). A large portion of adult sufferers report a first onset of back pain in their early teenage years or in their twenties (Papageorgiou et al., 1996). It is commonly perceived that back pain amongst young people is uncommon, Turner et al. (1989) suggest that back pain accounts for around 2% of referrals in the under 15 age group. However epidemiological studies have found high prevalence rates of back pain, Brattberg and Wickman (1992), report 29% of children had back pain often and Troussier et al. (1994) report 51% cumulative prevalence of back pain amongst children. Mandal (1994) suggests that a seated person has a hip joint flexion of about 60° and the pelvis has a sloping axis, so that the lumbar region then exhibits a convexity, or kyphosis. This is supported by Schoberth (1962) who found from x-ray examinations of 25 people sitting upright, an average 60° hip flexion and 30° lumbar flexion. Aagaard-Hensen and Storr-Paulsen (1994) found that in one school, children remained seated between 19 and 90 minutes during a 90 minute double lesson, with older children sitting for longer periods of time and most of the children sitting on average for more than 60 minutes. Of the time spent seated, 57% was spent leaning forward (e.g. writing or painting) with 43% spent doing backward leaning activities (e.g. looking at blackboard or reading). Parcells et al. (1999) found that less than 20% of a total of 74 children could find acceptable chair/desk combinations when the anthropometric dimensions of the children were considered. Marschall et al. (1995) found less neck flexion and larger hip angles for children sitting on ergonomically designed furniture than standard school furniture. However, Troussier et al. (1999) found that although children preferred ergonomically designed furniture this did not lead to a reduction in back pain prevalence. School chairs and desks are designed for children to sit and work with a 90° flexion of the hip joint and a preserved lumbar lordosis, as recommended by Snorrason (1968). It seems that children do not use school furniture in this way.

The Robens Centre for Health Ergonomics at the University of Surrey is conducting a three-year study in conjunction with the Arthritis and Rheumatism Council's Epidemiology Research Unit at the University of Manchester.

The overall aims of the study are to identify the extent of back pain experienced by schoolchildren, aged 11 to 14, establish any physical risk factors, which may be present in school and to provide advice to prevent problems arising in the future. The aims of this paper are to monitor the sitting posture of schoolchildren in the classroom and investigate associations with self-reported health data.

Direct observation of children in the classroom was considered the most suitable method to use in schools to record posture (Murphy and Buckle, 2000). The Portable Ergonomic Observation method PEO (Fransson-Hall et al., 1992), was used in the study as the method records the posture of the children in real-time in the classroom. Time-sampled observations such as Posturegram (Priel, 1974) and OWAS (Karhu et al., 1977) only provide an estimate of this information whereas real-time observations provide information about the intensity, duration and frequency of the posture. This paper describes the results of observations using PEO in the classroom to observe the sitting posture of schoolchildren and associated self-reported health data.

Ethics
Permission was granted from the Director of Education for Surrey, the Ethics committee of the University of Surrey and Head Teachers of the schools involved. The parents and children were each sent a consent letter informing them of the study with the option to withdraw at any stage.

Methods

The subjects were 66 children (32 female and 34 male) with a mean age 12.72 years (S.D. 0.88) selected from 12 Surrey schools. Measurements of sitting posture were recorded using the Portable Ergonomic Observation Method PEO (Fransson-Hall et. al., 1995). Observations of body postures were made in real time directly in the classroom using a Viglen Dossier 486 laptop computer. 66 children were recorded for 30 minutes each. The categories included in the PEO system were selected according to risk factors in the literature (Fransson-Hall et al., 1995, pp 97). All postures were recorded in relation to an upright sitting posture, i.e. trunk flexion > 20° was activated when the subject's trunk was at an angle of 20° or more from vertical. Recording started around ten minutes after the lesson began to allow the children to settle down and become accustomed to the presence of the researcher. Measurements included trunk flexion and rotation, neck flexion and rotation, working at the desk and trunk unsupported.

A detailed health and lifestyle questionnaire was distributed and measurements of height and weight were recorded. The questionnaire was distributed during a separate lesson, and included a modified version of the Nordic Musculoskeletal Disorders Questionnaire (Kuorinka et al., 1987). Also included were questions on activities in school time and after school.

Results

As a result of the low number of subjects and the high variability of the postures observed the data were collapsed into high and low exposure categories at the 50th percentile. There were very high rates of self-reported pain, 44% of children reported experiencing low back pain in the last month and 26% in the last week. Thirty-five percent of the children had experienced upper back pain in the last month and 21% in the last week (Table 1). More than half (52%) had experienced neck pain in the last month and 24% in the last week. There were very few children who sought medical attention or had taken time off school for their pain. Only 3% had sought treatment for low back pain in the last month and 1.5% had sought treatment for upper back pain in the last month with 4.5% seeking treatment for neck pain in the last month. Three percent had taken time off school due to low back pain in the last month and 3% had taken time off school due to neck pain in the last month (Table 2).

Table 1. Back and Neck pain

Self-reported pain	Number	Percentage
Low back pain last month	29	44%
Low back pain last week	17	26%
Upper back pain last month	23	35%
Upper back pain last week	14	21%
Neck pain last month	34	52%
Neck pain last week	16	24%

Table 2. Treatment and absence

Treatment and absence	Number	Percentage
Treatment low back pain last month	2	3%
Treatment upper back pain last month	1	1.5%
Treatment neck pain last month	3	4.5%
Absence low back pain last month	2	3%
Absence neck pain last month	2	3%

Low back pain and associated factors
Long lesson length was significantly associated with low back pain in the last month as was high body mass index. High body mass index was also significantly associated with low back pain in the last week. High percentage of trunk flexion > 20° and high percentage of trunk flexion total (trunk flexion > 20° + Trunk flexion > 45°) were both significantly associated with low back pain in the last week, as was high percentage of neck flexion >20°. High level of working at the desk was also significantly associated with low back pain in the last week (Table 3).

Table 3. Low back pain and associated factors

Low back pain	Factor	P-value
Last month	Lesson length (high)	0.039*
Last month	BMI (high)	0.015*
Last week	BMI (high)	0.027*
Last week	TF 20 (% high)	0.012*
Last week	TF total (% high)	0.014*
Last week	NF (% high)	0.035*
Last week	Work at desk (% high)	0.005**

$*P<0.05; **P<0.01$ Mann-Whitney U

Upper back pain and associated factors
Low levels of trunk flexion > 20° (n and %) were both significantly associated with upper back pain in the last month as was a low number of trunk flexion total. High body mass index was also significantly associated with upper back pain in the last week. A low number of trunk flexion > 20° and a low number of trunk flexion total were both significantly associated with upper back pain in the last week. A low percentage of time spent unsupported at the desk was also significantly associated with upper back pain in the last week (Table 4).

Table 4. Upper back pain and associated factors

Upper back pain	Factor	P-value
Last month	TF20 (n low)	0.004**
Last month	TF20 (% low)	0.046*
Last month	TF total (n low)	0.006**
Last week	BMI (high)	0.014*
Last week	TF 20 (n low)	0.025*
Last week	TF total (n low)	0.033*
Last week	Unsupported (% low)	0.024*

$*P<0.05; **P<0.01$ Mann-Whitney U

Neck pain and associated factors
High body mass index was significantly associated with neck pain in the last month. Low number of trunk flexion > 45° and low number of trunk flexion total were both significantly associated with neck pain in the last week (Table 5).

Table 5. Neck pain and associated factors

Neck pain	Factor	P-value
Last month	BMI (high)	0.009**
Last week	TF45 (n low)	0.048*
Last week	TF total (n low)	0.047*

$*P<0.05; **P<0.01$ Mann-Whitney U

Discussion

There were very high levels of neck and back pain with back pain rates higher than those reported by Brattberg and Wickman (1992) but lower back pain rates than those reported by Troussier et al. (1994). This could be due to the different definitions used in different studies. There were very low rates of children seeking treatment for these complaints the same number as suggested by Turner et al. (1989). This suggests that although the reporting of such pain is common more serious pain that requires medical treatment is not common amongst this age group. Also, it maybe important to report pain experienced at different sites as upper back pain and neck pain were also common complaints in this age group. Around 25% of the time was spent with trunk flexion of more than 20° although there was considerable individual variation. Trunk and neck flexion categories were active when the children were working at their desks. The rotation categories were active when the children were talking to friends beside or behind them, and unsupported was usually active when the children were retrieving things from their bag on the floor. The lumbar support was rarely used by any of the children. There was also large variation in the amount of movement, some children moved continuously while others had very static postures. The association between long lesson length and low back pain last month suggests that lessons over one hour are too long. Also children with high body mass index are more likely to suffer from low back pain in the last week and the last month. High levels of trunk flexion more than 20° also increased the likelihood of reporting low back pain in the last week. When children were working at the desk they were in flexed postures. High levels of the flexion categories seem to increase short-term low back pain whereas sitting for long periods during lessons leads to an increase of low back pain in the last month. Upper back pain showed a different trend. Those children who did not move as much during the lesson showed increased levels of upper back pain in the last month and last week. Again high body mass index was associated with upper back pain in the last week. High body mass index was also associated with increased reporting of neck pain in the last month, while a low number of trunk flexion registrations were associated with neck pain in the last week.

Conclusions

- High body mass index is associated with increased reporting of back and neck pain
- Longer lessons may increase likelihood of low back pain
- Static postures i.e. less movements during lessons may increase likelihood of upper back and neck pain
- Trunk flexed postures may increase short term low back pain

- This study has implications for both the length and structure of lessons and the design of school furniture
- It has implications for the future workforce with many young adults entering the workplace with MSDs already present
- Further research required examining the association between sitting posture and pain

References

Aagaard-Hansen, J. and Storr-Paulsen, A. 1995. A comparative study of three different kinds of school furniture. Ergonomics, 38, 5, 1025-1035.

Brattberg, G. and Wickman, V. 1992. Prevalence of back pain and headache in Swedish school children: A questionnaire survey. The Pain Clinic, 5, 211-220.

Fransson-Hall, C., Gloria, R., Kilbom, Å., Winkel, J., Karlqvist, L. and Wiktorin, C. 1995. A portable ergonomic observation method (PEO) for computerized on-line recording of postures and manual handling. Applied Ergonomics, 26, 93-100.

Fredriksson, K. 1999. PEOflex: Portable Ergonomic Observation Method, User's Manual. National Institute for Working Life. S-171 84 Solna Sweden.

Karhu, O., Kansi, P. and Kuorinka, I. 1977. Correcting working postures in industry: a practical method for analysis. Applied Ergonomics, 8, 199-201.

Kuorinka, I., Jonsson, B., Kilbom, A., Vinterberg, H., Sorensen, F.B., Andersson, G. and Jorgensen, K. 1987. Standardized Nordic questionnaires for the analysis of musculoskeletal symptoms. Applied Ergonomics, 18, 3, 233-237.

Mandal, A. C. 1994. The prevention of back pain in school children. In: Lueder, R., Noro, K. (Eds.), Hard Facts About Soft Machines: The Ergonomics of Seating. Taylor and Francis, pp. 269-277.

Marras, W.S., Fathallah, F.A., Miller, R.J., Davis, S.W. and Mirka, G.A. 1992. Accuracy of a three-dimensional lumbar motion monitor for recording dynamic trunk motion characteristics. International Journal of Industrial Ergonomics, 9, 75-87.

Marschall, A., Harrington, A.C. and Steele, J.R. 1995. Effect of workstation design on sitting posture in young children. Ergonomics, 38, 9, 1932-1940.

Murphy, S.D. and Buckle, P. 2000. The occurrence of back pain in schoolchildren and the risk factors in schools: Can they be measured? The triennial Congress of the International Ergonomics Association and 44th meeting of the Human Factors and Ergonomic Society, July 29-August 4 San Diego, California, 5-549-552.

Papageorgiou, A.C., Croft, P.R., Thomas, E., Ferry, S., Jayson, M.I.V. and Silman, A.J. 1996. Influence of previous pain experience on the episode incidence of low back pain: results from the South Manchester Back Pain Study. Pain, 66, 181-185.

Parcells, R.N., Stommel, M. and Hubbard, R.P. 1999. Mismatch of Classroom Furniture and Student Body Dimensions: Empirical Findings and Health Implications. Journal of Adolescent Health, 24, 265-273.

Priel, V. Z. 1974. A numerical definition of posture. Human Factors, 16, 576-584.

Schoberth, H. 1962. Sitzschaden, Sitzmöbel, (Berlin: Springer-Verlag).

Snorrason, E. 1968. Tidsskrift for Danske Sygehuse, Copenhagen.

Storr-Paulsen, A. and Aagaard-Hensen, J. 1994. The Working Positions of School children. Applied Ergonomics, 25 (1), 63-64.

Troup, J.D.G., Foreman, T.K., Baxter, C.E. and Brown, D. 1987. The perception of back pain and the role of psychophysical tests of lifting capacity. Spine, 12, 645-657.

Troussier, B., Davione, P., deGaudemaris, R., Fauconnier, J. and Phelip, X. 1994. Back pain in school children - A study among 1178 pupils. Scandinavian Journal of Rehabilitation Medicine, 26, 143-146.

Troussier, B., Tesniere, C., Fauconnier, J., Grison, J., Juvin, R. and Phelip, X. 1999. Comparative study of two different kinds of school furniture among children. Ergonomics, 42, 3, 516-526.

Turner, P.G., Green, J.H. and Galasko, C.S.B. 1989.Back pain in Childhood. Spine, 14, 812-814.

ORGANISATIONAL OBSTACLES TO RECOVERY ('BLACK FLAGS') FROM MUSCULOSKELETAL DISORDERS

Serena Bartys[1,2], Kim Burton[1], Ian Wright[3], Colin Mackay[4], Paul Watson[2], Chris Main[2]

[1]*Spinal Research Unit, University of Huddersfield, 30 Queen Street, Huddersfield, HD1 2SP*
[2]*Dept of Behavioural Medicine, Hope Hospital, Salford*
[3]*GlaxoSmithKline, Brentford*
[4]*Health and Safety Executive, Bootle*

It is now recognised that psychosocial factors ('yellow' and 'blue' flags) play an important role in delayed recovery from musculoskeletal disorders, and that early identification of these obstacles to recovery is recommended. Thus, an occupational psychosocial intervention (with availability of modified work) was developed which was intended to be delivered within the first few days of absence or complaint. During a 12-month period, targets for early intervention were reached at one site (mean 3 days), but not at another (mean 15 days). Early contact with appropriate intervention was shown to significantly reduce presenting absence, compared with late contact and no intervention. These findings highlighted organisational policies ('black flags') as an obstacle to recovery, in that they can negate the benefits of early intervention.

Introduction

Musculoskeletal disorders (MSDs) remain the predominant occupational health problem in most industrialized countries, accounting for about 20-30% of all worker's compensation claims and up to 50% of all direct compensation cost (CSAG, 1994). In occupational settings, interventions to reduce MSDs and the resulting work loss have been based primarily on biomedical or ergonomic principles, but investigations into predictors of outcome have shown that the specific influence of psychosocial factors may be more important (Burton and Main, 2000).

Considerable effort has already been directed at primary prevention, but, given the prevalence of MSDs in the population, and the group of individuals who take short recurrent absences, a more realistic target is secondary prevention. In order to prevent MSDs becoming disabling or recurrent, particular effort on tackling potential psychosocial obstacles to recovery is indicated (Burton and Main, 2000). Recent research has indicated that, in addition to the established clinical psychosocial factors (yellow flags) (Kendall *et al*, 1997), occupational psychosocial factors (blue flags) can act as obstacles to recovery, thus arguing for an early workplace intervention (Bartys *et al*, 2001).

The Faculty of Occupational Medicine's recent guidelines for occupational health management of back pain (Carter and Birrell, 2000) proposed an active approach with early work return (even if the worker still has some pain). Analgesic medication is recommended, along with appropriate written information. Whilst assessment of 'yellow flags' was advocated, it is not clear how these specifically are to be addressed. It was also recognised that occupational rehabilitation differs from clinical treatment in that it places a greater emphasis on the workplace and on specific interventions directed at workplace factors assumed to inhibit return to work (e.g. temporary availability of modified work).

In recognition of the need for more effective management of workloss due to MSDs, a Health and Safety Executive funded project was initiated, with the pharmaceutical company GlaxoSmithKline as the industrial partner. The first phase was a workforce survey of all GlaxoSmithKline employees in the UK. Using the data from that survey (results of which were presented to the Society last year), and guided by the yellow flags document (Kendall *et al*, 1997), an early psychosocial assessment and intervention program was developed. It incorporated availability of modified work, and liaison with team leaders. It was designed to be suitable both for absent workers and for those still able to work with their musculoskeletal symptoms, and is intended to be implemented by the occupational health advisor (OHA) within the first week of absence (or report of symptoms).

A prospective, controlled trial was initiated to explore the effectiveness of the program. It was hypothesised that this early intervention, addressing both yellow and blue flags, would reduce the length of presenting absence.

Methods

Following several preparatory meetings with the occupational health staff at GlaxoSmithKline to seek closely matched sites for the trial, it was decided to use five manufacturing sites. The decision to use manufacturing sites was reached using the following criteria:

- all sites would be closely matched for job type and demographic data,
- all sites had similarly high absence rates due to MSDs compared to the other sites of GlaxoSmithKline,
- at all sites, ostensibly, the OHAs would be notified at the start of periods of absence.

The experimental sites were then split into 'intervention' sites and 'control' sites, matched for population size and absence rates. Two of the sites would deliver the intervention (n=1,500; 185 cases of absence due to MSDs in the preceding year), and the remaining three sites would act as controls (n=1,500; 194 cases of absence due to MSDs).

Psychosocial intervention program – training and content

The OHAs at the intervention sites were trained, in two stages over two days, to deliver the psychosocial assessment and intervention. The first stage of training incorporated the theory and research underlying psychosocial intervention principles. They were also taught how to identify both clinical (yellow) and occupational (blue) psychosocial factors that can act as obstacles to recovery from MSDs. These factors would be addressed using a structured 'counselling' technique that tackled beliefs and attitudes and reinforced evidence-based messages and advice (e.g. importance of keeping active and early return-to-work).

The second stage of the training consisted of ensuring that all the nurses followed the procedures consistently. A manual was devised in order that the nurses may follow each step required, and record the correct information at the correct point of the intervention. The recorded information for each participant was transferred to a custom database, which was monitored on a regular basis.

The program required each absent worker to be contacted by the OHA within the first days of absence, at which time they were reassured and invited to attend for further intervention. Those who took up the offer were assessed and counselled as appropriate for a maximum of 4 weeks. Temporary modified work was offered as required, along with educational booklets for back pain or upper limb disorders. In addition, contact was made with the team leader and GP.

GlaxoSmithKline complies with the Management of Health and Safety at Work regulations 1992; risk assessments are conducted, and foreseeable risks for musculoskeletal injury appropriately controlled. Each participant was fully briefed about the nature of the study and signed a consent form before being enrolled into the program. Ethical approval was obtained.

Results

This paper gives preliminary results after implementing the program for one year, and is limited to considering just the presenting spell of absence. Over the 12-month period, 212 workers were contacted for the intervention. Of these, 120 had taken absence due to MSDs. The proportion of those workers experiencing low back pain was 68% (n=143), whilst the remaining 23% (n=49) presented with an upper limb disorder.

The mean times taken to contact absent workers at the two experimental sites show that targets for early intervention (<1 week) were reached at Site 1, but not at Site 2; this difference was statistically significant (t=7.33, p<.001) (Table 1). Furthermore, the mean duration of presenting absence at Site 1 was shorter than for Site 2. Whilst this latter difference between the two sites was substantial, it was not statistically significant.

Table 1. Contact times and mean duration of presenting absence

	Site 1 (n=71)	Site 2 (n=46)
Mean time taken to contact absent worker	3 days	15 days
Mean duration of presenting absence	5 days	9 days

Since the unexpected time-to-contact differential could impact on the delivery of the intervention package, it was necessary to examine further the effects of the timing of the intervention on the duration of presenting absence. The two sites were combined, and the mean duration of presenting absence was calculated for each group, depending on whether they were contacted early or late (< or > one week). The results in Table 2 show that early contact was associated with shorter absence (t=-2.12, p<.05). In addition, it was clear that fewer workers took up the intervention if they were contacted later than one week from the start of absence.

Table 2. Presenting absence and workers taking up the intervention package

	Early contact (n=66)	Late contact (n=51)
Mean duration of presenting absence	5 days	10 days
Workers (n) taking up intervention	47	19

It is possible that these effects were due simply to an effect of 'contact', irrespective of the actual intervention. Thus, the data were split into groups of workers who had early and late contact only, compared to groups that had early and late contact *plus* receipt of the intervention package. The mean duration of presenting absence of the three controls sites was also calculated for comparison. The results are given in Figure 1.

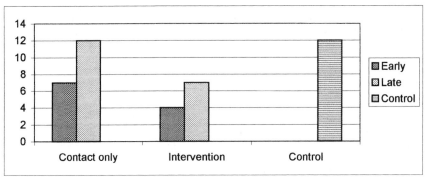

Figure 1. Mean duration of presenting absence for those workers who did and did not take up the intervention and controls

Workers contacted late who did not take up the intervention had a mean duration of absence equivalent to that of the control sites (12 days). By contrast, those contacted late who took up the intervention had a mean duration of absence equivalent to those contacted early without the intervention (7 days). The shortest mean duration of absence was found in those who were contacted early and took up the intervention (4 days).

Discussion

Addressing the issue of early return-to-work from a biopsychosocial perspective, specifically in terms of obstacles to recovery, is felt to be advantageous (Bartys *et al*, 2000). The results presented here offer preliminary support for that view, but the relatively low numbers involved do not permit firm conclusions. Nevertheless, the data do suggest that a psychosocial intervention program can reduce the duration of absence for the presenting spell of MSDs. In addition, a simple early contact (reassurance) of absent workers was somewhat effective in reducing duration of presenting absence due to MSDs, and also encouraged participation in an intervention program. Late reassuring contact without delivery of the intervention was no better than no contact (control sites) in terms of duration of presenting absence.

During the course of the trial, it became apparent that early intervention targets were not being reached at Site 2. Whilst the study sites were carefully chosen to match all appropriate variables, it transpired that Site 2 had a different organisational policy for sickness reporting. The OHAs at Site 2 were not being notified of absence immediately by the worker, and medical certificates went first to another department rather than directly to the OHA. In contrast, an early self-reporting culture was in place at Site 1, and medical certificates were delivered to the OHA in the first instance. Thus, the organisational policies at Site 2 meant that, generally, early contact and intervention was precluded, thus limiting any beneficial effects from the intervention program. Organisational policies have been proposed as obstacles to recovery, and have been termed 'black flags' (Main and Burton, 2000). So far as we are aware, this is the first empirical (albeit preliminary) evidence of the detrimental effects that black flags can have on otherwise useful rehabilitation programs. Data from the program will be collected for a further 12 months, at which time adequate numbers for a robust analysis will be available.

Many practical challenges accompany the implementation of an integrated approach to secondary prevention in occupational settings, and it would appear necessary that not only are "all the players onside" (Frank *et al*, 1998), but that fundamental procedures within the organisation are in place to help optimise such an intervention.

References

Bartys, S., Main, C.J. and Burton, A.K. 2000, Pain management in occupational settings. In C.J. Main and C.C. Spanswick (eds.) *Pain management. An interdisciplinary approach*, (Churchill Livingstone, Edinburgh), 403-418

Bartys, S., Tillotson, M., Burton, K., Main, C., Watson, P., Wright, I. and Mackay, C. 2001, Are occupational psychosocial factors related to back pain and sickness absence? In M. Hanson (ed.) *Contemporary Ergonomics*, (Taylor & Francis, London), 23-28

Burton, A.K. and Main, C.J. 2000, Obstacles to recovery from work-related musculoskeletal disorders. In W. Karwowski (ed.) *International Encylopedia of Ergonomics and Human Factors,* (Taylor & Francis, London), 1542-1544

Clinical Standards Advisory Group. 1994, *Epidemiology Review: the epidemiology and cost of back pain*, (Her Majesty's Stationery Office, London)

Carter, J.T. and Birrell, L.N. 2000, *Occupational health guidelines for the management of low back pain at work - principal recommendations*, (Faculty of Occupational Medicine, London)

Frank, J., Sinclair, S., Hogg-Johnson, S., Shannon, H., Bombardier, C., Beaton, D. and Cole, D. 1998, Preventing disability from work-related low-back pain. New evidence gives new hope - if we can just get all the players onside, *Canadian Medical Association Journal*, **158**, 1625-1631

Kendall, N.A.S., Linton, S.J. and Main, C.J. 1997, *Guide to assessing psychosocial yellow flags in acute low back pain: Risk factors for long-term disability and work loss*, (Accident Rehabilitation & Compensation Insurance Corporation of New Zealand and the National Health Committee, Wellington NZ)

Main, C.J. and Burton, A.K. 2000, Economic and occupational influences on pain and disability. In C.J. Main and C.C. Spanswick (eds.) *Pain management. An interdisciplinary approach*, (Churchill Livingstone, Edinburgh), 63-87

ASSESSMENT OF GRIP FORCES AND CUTTING MOMENTS ASSOCIATED WITH RED MEAT PACKING

Raymond W. McGorry[1], Patrick G. Dempsey[1] and Peter C. Dowd[2]

[1]*Liberty Mutual Research Center for Safety & Helath*
Hopkinton, Massachusetts, USA
[2]*AgResearch Ltd, Hamilton, New Zealand*

The meat packing industry has a high incidence of musculoskeletal disorders of the upper extremities. This may be due in part to exposure to the risk factors of non-neutral postures, the repetitive nature of cutting tasks, and high force applications. The force exposure associated with meat cutting operations has not been well documented. For this investigation, knife handles and blades were mounted to a core handle instrumented with strain gauges. Direct measurements of grip forces and cutting moments were obtained from fifteen meat cutters during lamb shoulder boning, beef rib trimming and beef loin trimming operations in two meat packing plants. The meat cutting operations were found to be high force, high repetition tasks requiring subjects to perform at a large proportion of their maximum grip force and cutting capacity.

Introduction

Forceful exertions, high repetition rate, and awkward postures are risk factors for musculoskeletal disorders of the upper extremities (MSDUEs) (Armstrong, 1983). In industries such as meat packing, where hand tools (knives) are used extensively, the incidence of MSDUEs can be much greater than the average (e.g. US Department of Labor, 1996). Though a dose response relationship for MSDUEs has not been definitively established, an intervention based on reducing exposure to known risk factors is a reasonable ergonomic approach. Knowledge of the applied moments and grip forces associated with knife use can be used to study the relationship between force exposure and injury, identify high-risk techniques, and validate task redesign. Occupation Safety and Health Association guidelines for meatpacking plants suggest that work methods analyses "....should be supplemented by addressing the force levels and the hand and arm postures involved. The tasks should be altered to reduce these and the other stresses identified with CTDs." (US Department of Labor, 1993). While task frequency and upper limb posture can be determined with conventional methods, quantification of exposure to high forces can be difficult. The purpose of this field study was to quantify exposure to cutting moments and grip forces in several meat cutting operations performed by professionals in meatpacking plants.

Methods

An instrumented knife was fabricated replicating the 15cm boning knives most typically used in the two plants. The instrumented knife has strain gauges mounted on the handle core, which allow measurement of the grip force through a 700N working range, blade cutting moments in excess of 20Nm, and 10Nm in the plane perpendicular (lateral) to the cutting plane. Blades were cut from the actual knives and welded to a fitting allowing for attachment to the handle core. Replicas of the handle were cast in polyurethane, and cut and mounted to the handle core. Photographs of an original and instrumented version are presented in Figure 1. Further description of the instrumentation and the fabrication process are presented elsewhere (McGorry, 2001). The signals from six strain gauges in the knife handle, and the two strain gauges mounted near the attachment of the blade (one detecting cutting moments, the second detecting lateral plane moments) were amplified and displayed in real-time, sampled at 100 Hz and stored in memory of the laptop computer.

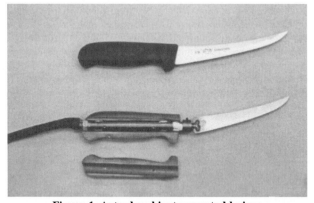

Figure 1. Actual and instrumented knives

One female and 14 male New Zealand meat cutters with a mean age of 36.1 yr. (s.d. = 11.9 yr.), and mean experience of 9.8 yr. (s.d. = 9.3 yr.) gave written informed consent for participation. A maximum voluntary contraction (MVC) for power grip and for the cutting motion (ulnar deviation) was obtained with the experimental knife handle using the Caldwell protocol (Caldwell et al, 1974). In a lamb processing plant data was collected while six subjects performed a shoulder boning operation where shoulders were removed from the rib cage and trimmed of excess fat. At a second plant, five subjects performed a beef "rib trimming" operation removing as much muscle as possible from between the ribs. A beef loin trim was performed by four subjects. Excess fat and tissue are removed from the tenderloin so that it meets market specifications. Six cycles of the operation were performed by the 15 subjects. These three operations were selected for evaluation because they represent a diversity of technique and perceived difficulty. The shoulder boning and rib trimming operation were both considered physically demanding operations, yet involved differences in technique. The loin trim provided a good contrast. Though considered less strenuous, it requires skillful cuts to produce a high quality product.

The data from the six gauges in the handle were averaged to produce a single mean grip force value for each data sample. The cutting moment was calculated by vector summation of the cutting and lateral blade moments. The data were then visually examined, and the beginning and end of the cutting cycle were marked. The period between these two points was called Cycle time. From examination of the data it was determined that a 0.5 Nm resultant moment on the blade was satisfactory as a threshold for cutting activity. Each data point was evaluated, and if the cutting moment exceeded threshold, that grip force and cutting moment data point was assigned to the cutting period. Data points below threshold were classified as quiet or non-cutting periods. The amount of time the threshold was exceeded was referred to as Cutting time. The peak and mean cutting moment and grip force during cutting were determined for each cycle. A numerical integration of the cutting moment and grip force was performed. The average of the mean and peak values for the six cycles was calculated.

Results

Videotape of the data collection was reviewed to estimate the average number of cuts required for each operation. This information combined with typical daily production rates reported by plant foremen were used to estimate the number of cutting motions per day. An estimate of the daily cutting moment "exposure" for the three tasks was defined as the mean integrated cutting moment per operation x number of operations per 8 hr day. A similar calculation was made for the grip force exposure. Of the three tasks evaluated, the lamb shoulder operation had the highest peak grip forces, and peak and mean cutting moments. The beef rib trim produced the highest mean grip force, and the highest value for the integration of both grip force and cutting moments. The number of cuts performed per section were similar for the lamb shoulder (31.8) and beef rib (31.0) operations, but the Cycle time per operation was shorter, and the number of sections processed per hour was greater for the shoulder section. The beef loin trim had the lowest values for all the measured variables. A results summary is presented in Table 1.

The MVCs for power grip were calculated for all 15 subjects and the group mean was 152.9 N (s.d.= 39.8 N). The MVCs for cutting (ulnar deviation) were calculated for only 13 subjects, as results for two subjects were rejected because of procedural errors. The group mean for the cutting MVC was 12.1 Nm (s.d. = 3.0 Nm). The relationship of each subject's peak and mean grip force to their grip MVC was calculated and expressed as a percentage, for the 15 subjects. The same calculation was made for the relationship of cutting moment to ulnar deviation MVC for the 13 subjects. The results of this analysis are included in Table 2.

Table 1. Moments, grip force and exposure for the three operations

Operation	Lamb Shoulder Mean (s.d.)	Beef Rib Mean (s.d.)	Beef Loin Mean (s.d.)
Cutting moment			
Peak (Nm)	17.2 (3.4)	12.9 (2.2)	10.6 (2.6)
Mean (Nm)	4.69 (1.11)	3.49 (.96)	2.33 (.44)
Integration (Nm-sec)	102 (47.9)	161 (48.5)	31.9 (15)
Grip force			
Peak (N)	135.9 (40.6)	97.9 (19.4)	75.1 (21.5)
Mean (N)	41.6 (10.8)	46.2 (13.0)	31.2 (3.9)
Integration (N-sec)	885 (405)	2,074 (469)	409 (136)
Cuts per operation	31.8 (2.4)	31.0 (10.7)	28.1 (6.5)
Cuts per day	12,733	9,200	9,000
Cycle time (sec)	53.6 (13.2)	77.9 (20.8)	43.4 (7.8)
Cutting time (sec)	20.7 (6.0)	46.5 (10.5)	13.0 (4.0)
Exposure			
Cutting moment/shift (Nm-sec)	40,800	51,520	10,208
Grip force/shift (N-sec)	354,000	663,680	130,880

Table 2. MVC, and cutting moments and grip forces expressed as % of MVC

	Lamb Shoulder Mean (s.d.)	Beef Rib Mean (s.d.)	Beef Loin Mean (s.d.)	Overall Mean (s.d.)
MVC				
Grip force (N)				152.9 (39.8)
Cutting moment (Nm)				12.1 (3.0)
Cutting moment				
Peak (% of MVC)	130.7 (32.0)	109.8 (23.9)	104.1 (17.7)	117.7 (28.7)
Mean (% of MVC)	35.9 (10.9)	28.3 (8.9)	23.1 (3.9)	30.2 (10.3)
Grip Force				
Peak (% of MVC)	85.8 (32.0)	66.8 (20.9)	60.3 (20.3)	72.6 (27.6)
Mean (% of MVC)	27.2 (11.9)	32.0 (12.7)	24.8 (4.4)	28.2 (10.9)

Conclusions

Lamb shoulder boning required the highest peak and mean cutting moments and peak grip force. Mean grip force was highest for the rib trimming operation. Rib trimming, which required nearly 45% greater Cycle time per section than the shoulder boning operation, also required nearly 125% more cutting time. The integration of both the grip force and cutting moment are also much higher for the rib trimming operation. When comparing the daily exposure to cutting moment and grip force, rib trimming exceeded shoulder boning. For both variables the daily exposure from rib trimming was five times greater than for loin trimming, which had the lowest values for all the measured variables. Though less physically demanding, this operation may represent a trade-off of strength for precision, as skilled trimming yields a higher quality and quantity of this more expensive product.

When the grip forces and cutting moments exerted were expressed as a proportion of the MVC, as a group, the 15 meat cutters were found to be working at a sustained (mean) level at 28% and 32% of their maximum grip force and cutting moment, respectively. The group mean for peak cutting exertions were found to be 118% of MVC. One possible reason for this observation is that the peak cutting moments were forceful but generally of brief duration, whereas the MVC protocol requires a sustained isometric contraction. Also, during the MVC protocol arm and body posture are constrained, whereas during the cutting tasks, postures more favorable for moment generation might be used. A weakness of the experimental protocol was that by necessity, MVCs for each subject were taken near in time to the their experimental trial, which occurred several hours into their workday. MVCs could be affected to some degree by fatigue. It could be argued, however, that an MVC taken at the time of data acquisition might more closely reflect the actual capacity at that instant.

In conclusion, the goal of quantifying the magnitude of the grip forces and cutting moments produced in several red meat packing operations was achieved in this field investigation. The instrumentation developed for and used in this investigation should have utility in future investigations of meat packing operations. Quantification of additional operations would contribute to the knowledge base of the exertions involved in this industry. These measures should also prove useful in the evaluation of changes in work practices such as training programs, work station design, job rotation and sharpening protocols.

References

Armstrong, T. J., 1983. An ergonomic guide to carpal tunnel syndrome. In: *Ergonomic Guides, American Industrial Hygiene Association*, Akron, OH.

Caldwell, L.S., Chaffin, D.B., Dukes-dobos, F.N., Kroemer, K.H.E., Laubach, L.L., Snook, S.H., Wasserman, D.,E., 1974. A proposed standard procedure for static muscle strength testing. AIHAJ 35 (4), 201-206.

McGorry, R., 2001. A system for the measurement of grip forces and applied moments during hand tool use. Appl. Ergon. 32 (3), 271-279.

US Department of Labor, Bureau of Labor Statistics, 1996. *Occupational Injuries and Illnesses: Counts, Rates and Characteristics, 1993.* Washington, DC: US Government Printing Office, Bulletin 2478, p7.

U.S. Department of Labor, 1993. Occupational Safety and Health Administration, OSHA 3123, Ergonomics Program Management Guidelines for Meatpacking Plants, p10.

FOREARM DISCOMFORT PROPERTIES FOR SIMULTANEOUS FOREARM TORQUES AND HORIZONTAL FORCES

L.W. O'Sullivan and T.J. Gallwey

Ergonomics Research Center, University of Limerick,
Plassy Technological Park, Limerick, Ireland

This study investigated forearm torque strength and forearm discomfort for right arm torque exertions, while simultaneously applying horizontal forces. The first part of the study examined maximum forearm torque in both the supination and pronation directions at three forearm joint rotation angles. The second part of the study involved subjects exerting intermittent isometric torques at 10% and 20% of MVC in both torque directions, at three forearm angles. These conditions were each tested while simultaneously applying a pushing, pulling and no pushing or pulling horizontal force. The results show that supination torque was strongest and that maximum torque in both directions were significantly affected by forearm joint angle. Discomfort scores from the intermittent exercises revealed significant main effects for force level, forearm angle and the application of a horizontal force simultaneous to applying torques, while direction of torque was only significant in interactions. Regression equations were developed that predict these relationships.

Introduction

Forearm pronation and supination and increased muscular activity in the wrist extensors are speculated to be linked to forearm and elbow work related injuries, but there is growing concern about the lack of epidemiological studies in this area. Based on a review of over 600 Work-related Musculo Skeletal Disorder (WMSD) studies of the neck, upper limb and back, Bernard (1997) indicated that there are fewer epidemiological studies addressing workplace risk factors of the elbow than for other WMSDs.

Studies on the work relatedness of forearm and elbow WMSDs indicate that epicondylitis, the most common elbow and forearm injury, is evident throughout all aspects of industry from keyboarding to meat processing to foundry work. At present the aetiology of many WMSDs is poorly understood (Ljung et al. 1999) and therefore no risk models have been developed in this area. As with most WMSDs, the risks appear to be multi-factorial, but with some risk factors showing predominance. The strongest of these is forearm torque that has a very strong association with lateral epicondylitis (Bernard, 1997). Much of the data on the relationships between risk factors and injuries are only based on industrial descriptions of tasks with high injury rates, but their specific contribution to injuries remains unclear due to a lack of controlled laboratory experiments in this area.

Discomfort measures are advantageous as they integrate the effects of the factors giving rise to injury and measured discomfort levels need to be incorporated into forearm/elbow WMSD prediction models. O'Sullivan and Gallwey (2001) developed discomfort equations for intermittent torque exertions, based on forearm angle, but the equations do not include parameters for any other injury risk factors. In this study, discomfort scores were collected for the combined effects of forearm torque direction, torque level, and forearm angle while simultaneously applying different horizontal forces, as applied in real tasks.

Method
Experimental design
Thirty six right-handed University students participated in a study that lasted 5.5 hours. Part 1 involved the measurement of maximum torque strength in both the supination and pronation directions at three forearm rotation joint angles, i.e. 75% supine Range Of Motion (ROM), neutral and 75% prone ROM. In the second part of the experiment subjects performed a total of 36 intermittent isometric trials of duration five minutes. The experimental condition combinations were determined using a full factorial design and presented to the subjects using Latin square ordering. The independent variables comprised torque direction (supination and pronation directions), forearm angle (60% supine ROM, neutral, 60% prone ROM), force level (10% and 20% MVC) and simultaneous horizontal force (pushing 100N, pulling 100N, No Pushing or Pulling). In each trial, subjects exerted torques according to the combination of conditions as specified by the software, for one-second duration every five seconds i.e. 10 repetitions per minute. Forearm discomfort was recorded on a 100 mm visual analog scale (0–10) in the software interface. The analysis of the discomfort data included pronation torque endurance time at 50% MVC as a covariate in an attempt to control for differences between individuals' perceptions of discomfort rating and pain threshold as detailed by O'Sullivan (1999).

Apparatus
A Penny and Giles Biometrics electo-goniometer model Z180 was used to measure forearm ROM. Signals from the goniometer were amplified and interfaced with the PC (333 MHz) using a National Instruments data acquisition and A/D converter board (model PCI-MIO-16XE-50) with a BNC adapter board (model BNC2090). The forearm torque meter consisted a purpose built T-bar mounted with strain gauges. Voltage signals from the strain gauges were also interfaced with the PC using the BNC adapter board. Horizontal forces (push/pull) applied to the torque rig during the experiment were registered using a Biometrics pinch meter model P100 positioned in the load cell of the torque rig. Force readings were passed to the PC via the serial port. Virtual Instruments (VIs) were written using G code in LabVIEW V5.0 to control the experiment. A series of separate VIs were coded for each part of the experiment and loaded dynamically into memory. The electro-goniometer and torque signals were configured within LabVIEW and displayed the readings in real time on the VDU for the VIs.

Procedure
For the duration of the experiment, subjects were positioned in front of the VDU at an adjustable height table set such that the included elbow flexion angle was 90^0. The experiment commenced with the measurement of maximum prone and supine ROM followed by the measurement of maximum forearm torque in the specified postures. This was followed by the testing of pronation torque endurance time at 50% MVC for use as a covariate in the statistical analysis of the discomfort data. For the main part of the experiment, subjects were presented with a VI containing a six second cycle analogue clock and a torque force meter that measured the subjects' exertion as a percentage of their maximum values in that direction. Subjects were requested to maintain an exertion of 20% in the direction specified by the software for one second every six-second cycle. Each exercise was five minutes in duration followed by a one-minute rest before testing in the

next posture. The discomfort scores were standardised for each subject using the following procedure.

$$\text{Standardised Discomfort Score (SDS)} = \frac{(\text{raw data} - \text{minimum data})}{(\text{maximum data} - \text{minimum data})} \times 10$$

Results

Repeated measures ANOVA was used to test if forearm joint angle (%ROM) and torque direction affected maximum torque exertion. The ANOVA results (Table 1) indicate that both Forearm Angle and the Forearm Angle * Direction factors were each significant at $p<0.001$ while the Direction main effect was significant at $p<0.05$. The plot of the mean maximum torque strength values (Figure 1) indicates that supination torque was stronger than pronation with the overall mean values 12.5 and 9.6 Nm respectively.

Repeated measures ANCOVA was used to test the null hypothesis that torque Direction, Level of MVC, % forearm ROM and horizontal force condition did not affect forearm discomfort for the intermittent exertions. The results of the ANCOVA (Table 2) indicate that torque direction did not affect forearm discomfort ($p<0.07$). However, percentage MVC exerted had a highly significant effect on the SDS values ($p<0.01$), a finding supported by the mean SDS ratings for both MVC (10% MVC 4.15, 20% MVC 4.86). Percentage forearm ROM had a significant effect on forearm discomfort in the ANCOVA ($p<0.05$). Horizontal force condition had the most significant effect of all factors in the ANCOVA ($p<0.001$). The SDS values indicate that the means for Push (5.83) and Pull (5.39) were more than double that for NPP (2.29). The Direction*Horizontal force two-way interaction and the Direction*% MVC* ROM three way interaction were also significant ($p<0.01$, and $p<0.05$ respectively).

Table 1 ANOVA results for effects of ROM and direction on maximum torque values

	SS	df	MS	F	Sig.
Forearm Angle	199	2	99	15	$p<0.001$
Direction	77	1	77	5.7	$p<0.05$
Forearm Angle*Direction	352	2	176	21.9	$p<0.001$
(Error) Forearm angle	*456*	*70*	*6*	*456.9*	
(Error) Direction	*472*	*35*	*13*	*472.7*	
*(Error) Angle*Direction*	*562*	*70*	*8*	*562.5*	

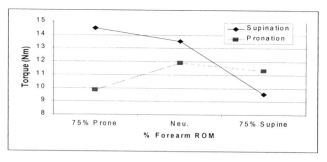

Figure 1 Mean torque strength for both directions and forearm joint angles

Table 2 ANcOVA results for effects of Direction, Force, ROM and horizontal force condition on discomfort*

	Factor	SS	df	MS	f stat	sig
Main effects	Direction	32.2	1	32.2	3.4	0.07
	%MVC	45.0	1	45.0	8.5	0.01**
	% ROM	37.1	2	18.6	3.3	0.04*
	Hor. Force	770	2	385	49.1	0.001***
	Direction * %MVC	1.3	1	1.3	0.4	0.56
	Direction * % ROM	1.5	2	0.7	0.1	0.89
	Direction * Hor. force	38.7	2	19.4	4.9	0.01**
	%MVC * % ROM	10.7	2	5.3	1.2	0.32
	%MVC * Hor. force	5.2	2	2.6	0.6	0.54
	% ROM * Hor. force	20.2	4	5.1	1.1	0.34
	Direction * %MVC * % ROM	24.2	2	12.1	4.4	0.02*
	Direction * %MVC * Hor. force	5.3	2	2.7	1.0	0.36
	Direction * % ROM * Hor. force	29.9	4	7.5	2.2	0.08
	%MVC * % ROM * Hor. force	7.9	4	2.0	0.6	0.68
	Direction * %MVC * % ROM * Hor. force	0.3	4	0.1	0.1	0.95

* data not shown for covariates due to large table size

The covariate, pronation endurance time at 50% MVC, was a significant factor combined with Direction ($p < 0.05$). It was also significant for the Direction*MVC*% Rom and Direction*% ROM*Horizontal force three way interaction, ($p < 0.05$ and $p < 0.001$ respectively).

An equation was developed to predict forearm discomfort based on the SDS data for each subject as the dependent measure and Direction, MVC, Force and Horizontal force condition as the predictor variables. The equation predicted the SDS scores with high accuracy ($R^2 = 0.92$) and the overall model was highly significant ($p < 0.001$). This equation is given by:

$$\text{Discomfort} = 6.05 + \text{Direction} + \text{MVC} + \text{ROM} + \text{Force Direction}$$

where,

Direction	Supination	= -0.037
	Pronation	= 0
MVC	10% MVC	= -0.69
	20% MVC	= 0
ROM	60% Prone	= -0.22
	Neutral	= -0.62
	60% Supine	= 0
Force Direction	Push	= 0.41
	NPP	= -3.1
	Pull	= 0

Discussion

The mean torque strengths reported in this study are similar to the values collected on twenty four males in a previous study by O'Sullivan and Gallwey (2001) using the same torque meter. The results show that forearm torque was strongest in the supination direction, but when the forearm was in 75% supine ROM, the strength in both directions decreased with a considerable reduction for supination torque. This suggests that at minimum, neutral forearm joint angles are preferable for tasks, especially those requiring forceful exertions such as screwdriver work and valve twisting operations.

The ANCOVA did not identify a significant difference between the discomfort for intermittent pronation and supination torques across all conditions as a main effect ($p>0.05$), and initially this appears to contradict the significant ($p<0.05$) main effect for Direction reported by O'Sullivan and Gallwey (2001). Further analysis of the results shows that the Direction*Horizontal force two-way interaction was highly significant ($p<0.01$). The data indicated that for the NPP condition, which was similar to that in the study by O'Sullivan and Gallwey (2001), there was a difference between both directions with the discomfort for pronation torques higher than for supination torques.

A significant main effect was identified for % ROM ($p<0.05$) and this also supports the results of O'Sullivan and Gallwey (2001). Further analysis of the discomfort scores shows that for the NPP condition the % ROM effect was more evident, but the effects of simultaneous push or pull force was a lot greater than the effect of ROM. This would indicate that the presence of horizontal force is a greater precursor to injury than forearm angle. This supports the view of Bernard (1997) who suggested that force level had the strongest association with epicondylitis while forearm angle did not. But caution is needed in this respect as the three way interaction for Direction*%MVC*ROM indicate that when the forearm is supine for high supination torques there is a much greater risk of injury.

The application of horizontal forces simultaneously with forearm torque had a highly significant effect on forearm discomfort ($p<0.001$) and thus was the factor that had the greatest effect on discomfort. This has specific consequences for the design of many tasks and tools that involve similar combinations of forces, especially in the light of the two-way interaction that was already discussed. A higher risk of WMSDs for these combinations of forces is supported by Silverstein et al. (1998) who specifically commented on high rates of injury for industries that involved pushing and pulling with twisting. Likewise Hughes et al. (1997) found a significant ($p<0.05$ min) relationship between tasks that involved forearm twisting and pushing/pulling forces.

The ANCOVA identified pronation endurance time as a significant covariate for controlling for individual differences in pain tolerance in the analysis of the discomfort scores. The presence of only three significant findings is initially surprising in the light of the successful previous use of the covariate by the authors. It is suggested this may be due to the use of the standardising routine to control for inter-individual differences in discomfort rating.

The discomfort equations from this study accurately model the significant findings from the ANCOVA and can be used to predict upper limb injury. These equations complement other discomfort models developed by O'Sullivan and Gallwey (2001) for forearm torques and can eventually be integrated with other discomfort equations for the upper limb to produce a complete injury prediction model for the upper limb. The discomfort data reported in this study add considerably to the quantification of the dose-response relationship between occupational risk factors and injury that is presently lacking in ergonomics research.

Conclusions

1. Torque direction did not affect the discomfort scores but the Direction*Horizontal force two-way interaction did ($p<0.01$).

2. Force level (%MVC) had a highly significant effect on the forearm discomfort ($p<0.01$) and this supports an association between strenuous tasks and forearm and elbow injuries.

3. Forearm angle affected discomfort scores ($p<0.05$) and this was most evident for the NPP condition. In addition, a significant three-way interaction for Direction*% MVC*ROM ($p<0.05$) indicates that, when the forearm is supine for high supination torques, there is a much greater discomfort score and risk of injury.

4. The application of horizontal force simultaneously with forearm torque had a highly significant effect on forearm discomfort ($p<0.001$) and this helps to explain injury rates for industries that involved pushing and pulling with forearm twisting.

5. An equation was developed that models the discomfort scores for the intermittent exertions with good accuracy ($R^2 = 0.98$). This can be used to predict the risk of elbow and forearm injury for tasks involving forearm torques which are known to result in considerable strain on the upper limb, as identified in the EMG data.

References

Bernard, B.P., 1997, Musculoskeletal disorders and workplace factors, NIOSH, Cincinnati, Ohio.

Hughes, R.E. Silverstein, B.A. and Evanoff, B.A., 1997, Risk factors for work-related musculoskeletal disorders in an aluminium smelter, American Journal of Industrial Medicine, 32, 66-75.

Ljung, B.O., Fridén, R. and Lieber, R.L., 1999, Sarcomere length varies with wrist ulnar deviation but not forearm pronation in the extensor carpi radialis brevis muscle, Journal of Biomechanics, 32, 199-202.

O'Sullivan, L.W. and Gallwey. T.J., 2001, Forearm discomfort for repeated isometric torque exertions in pronation and supination, Proceeding of the Ergonomics Society Annual Conference, 4-6 April, Lincolnshire, UK, ed. by Hanson., M.A., Lovesey, E.J. and Roberston., S.A., Taylor and Francis, London.

O'Sullivan, L.W. and Gallwey. T.J., 1999, Individual and gender differences in range of motion and discomfort at the wrist, In Contemporary Ergonomics, Proceeding of the Ergonomics Society Annual Conference, 7-9 April, Leicester, UK, ed. by Hanson., M.A., Lovesey, E.J. and Roberston., S.A., Taylor and Francis, London.

Silverstein, B.A., 1998, Claims incidence of work related disorders of the upper extremities: Washington State 1987-1995, Public Health, 88, 1827-1833.

PROPOSED EU PHYSICAL AGENTS DIRECTIVES ON NOISE AND VIBRATION

Neil J. Mansfield

*Department of Human Sciences,
Loughborough University,
Loughborough,
Leicestershire,
LE11 3TU*

The proposed European Union Physical Agents (Noise) and Physical Agents (Vibration) Directives are likely to be adopted in 2002. Member states will be required to introduce domestic regulation within three years of adoption. The Physical Agents (Noise) Directive is a development of the 1986 Noise Directive which is implemented in the UK as the Noise at Work Regulations. The new Directive proposes to reduce limits on personal noise exposure to a limit value 85 dB(A) and an action value of 80 dB(A), representing a 5 dB reduction when compared to existing legislation. The Physical Agents (Vibration) Directive will, for the first time in the UK, place limits on worker's exposure to whole body and hand-transmitted vibration.

Introduction

European Directives usually lead to Regulations which must be complied with. For example, the UK Manual Handling Operations Regulations (HMSO, 1992) have been derived from European Directive 90/269/EEC (European Commission, 1990). It is the purpose of Regulations to reduce risk factors sufficiently to protect workers' health.

In April 1993 the European Commission proposed a new Directive on physical agents. This directive, once amended by the European parliament in 1994, included annexes with proposed limits on exposure to 'noise, mechanical vibration, optical radiation and magnetic fields and waves' with the intention of future extension to temperature and atmospheric pressure. The details of this Directive were deemed unsatisfactory across many of the associated sub-disciplines. Therefore, it was not developed until 1999 when the German presidency proposed to limit the scope of the directive to human vibration only, with the intention of subsequent introduction of directives in the other areas. Currently, the Physical Agents (Noise) and Physical Agents (Vibration) Directives are in a mature stage of development and are likely to complete their progression through the European Parliament in 2002.

The Physical Agents Directives will approach occupational health from the perspective of the end-user. Therefore, the operator's exposure is assessed, rather than the emission of the machine itself, although these two quantities are related. There is, however, existing legislation for machinery manufactures. The Machinery Directive (Council of the European Union, 1998) requires noise and vibration emission values to be declared and for reduction of the risk factors. These are required for CE marking. For noise, the Directive states that the instruction manual must contain:

- equivalent continuous A-weighted sound pressure level at workstations, where this exceeds 70 dB(A); where this level does not exceed 70 dB(A), this must be indicated,
- peak C-weighted instantaneous sound pressure value at workstations, where this exceeds 63 Pa (130 dB in relation to 20 µPa),
- sound power level emitted by the machinery where the equivalent continuous A-weighted sound pressure level at workstations exceeds 85 dB(A).

For vibration, the instructions must also contain:

- the weighted root mean square acceleration value to which the arms are subjected, if it exceeds 2.5 ms^{-2} as determined by the appropriate test code. Where the acceleration does not exceed 2.5 ms^{-2}, this must be mentioned.
- the weighted root mean square acceleration value to which the body (feet or posterior) is subjected, if it exceeds 0.5 ms^{-2}. Should it not exceed 0.5 ms^{-2}, this must be mentioned.

Declared vibration emission data are collated for some tools and vehicles on the world wide web at 'http://umetech.niwl.se/Vibration/'.

Noise exposure legislation

Current noise legislation

The UK Noise at Work Regulations (HMSO, 1989) came into force on 1 January 1990 and were the implementation of the 1986 Noise Directive (European Commission, 1986). The basis of the regulations are the 'first action level', 'second action level' and 'peak action level'. The first and second action levels are defined as a personal noise exposure of 85 and 90 dB(A) respectively. The peak action level is defined as a level of peak sound pressure of 200 Pa. The regulations state that:

- every employer shall ensure that a noise assessment is made when any employee is likely to be exposed at or above any action level,
- records must be kept of noise assessments,
- employers shall reduce risk of hearing damage to the lowest level practicable,
- noise exposure must be reduced other than by provision of hearing protection if the second or peak action levels are reached,
- ensure that hearing protection is maintained and used,
- provide information to employees if any action level is exceeded.

In addition to these general requirements, action must be taken at each action level:

- First action level: Employees have the right to demand suitable hearing protectors,
- Second action level: Employees must be provided with suitable hearing protection which reduce risk below exposures at the second action level. Clearly marked ear protection zones must be implemented into which no employees must enter unless wearing hearing protection.

- Peak action level: Employees must be provided with suitable hearing protection which reduce risk below exposures at the peak action level. Clearly marked ear protection zones must be implemented into which no employees must enter unless wearing hearing protection.

Proposed noise legislation

The current draft of the proposed Physical Agents (Noise) Directive (as at December 2001) builds on the 1986 Noise Directive and follows a similar pattern. The main changes from the Noise Directive are that there are three categories of exposure criteria:

- Exposure limit value   = 87 dB(A)   and peak pressure = 200 Pa
- Upper exposure action value   = 85 dB(A)   and peak pressure = 200 Pa
- Lower exposure action value   = 80 dB(A)   and peak pressure = 112 Pa

For the exposure limit value, the Physical Agents (Noise) Directive takes account of the attenuation provided by hearing protection. The action values do not take the attenuation of hearing protection into account. The exposure limit value must not be exceeded. A further requirement of the general duties is that health surveillance must be introduced if a risk to health has been indicated by a noise assessment.

The Directive is being introduced through the co-decision procedure (Borchardt, 2000). Common Position was reached on 29 October 2001 and it is likely to go before the European Parliament for a second reading early in 2002. The Directive could be adopted by the end of 2002, after which Member States have three years to bring into force the laws, regulations and administrative positions required.

Vibration exposure legislation and guidance

Current vibration guidance

There is no legal requirement in the UK to limit exposure to human vibration, apart from general health and safety regulations. However, British and International Standards give guidance on human exposure to vibration and the Health and Safety Executive provide specific limits on exposure to hand-transmitted vibration which is taken as best practise (HSE, 1994).

The Health and Safety Executive (HSE) guidelines for hand-transmitted vibration recommend that preventative measures and health surveillance should be carried out if vibration exceeds a W_h frequency weighted 8-hour equivalent level (A(8)) of 2.8 ms^{-2}. This figure is based on assessment of the 'worst-axis' of vibration at either hand. Even at 2.8 ms^{-2}, 10% of exposed persons would be expected to show symptoms of vibration white finger after eight years (BS6842, 1987b).

For whole-body vibration there are two standards currently applicable: BS6841 (1987a) and ISO2631 (1997). Although these standards can be used in a way that is compatible there are important differences between them (Griffin, 1998, Table 1). It is therefore essential that the methods used are clearly specified in any report, as stating '... measured according to ISO2631...' does not imply a 'standard' method. According to BS6841, a vibration dose value (VDV) of 15 $ms^{-1.75}$ '...*will usually cause severe discomfort...accompanied by increased risk of injury*'. ISO2631 defines a 'health guidance caution zone' with a VDV from 8.5 to 17 $ms^{-1.75}$ and states that '*for exposures below the zone, health effects have not been clearly documented...in the zone, caution with respect to potential health risks is indicated and above the zone health risks are likely*'. The HSE have not produced documents for whole-body vibration aligned with their hand-transmitted vibration guidance.

Table 1. Summary of assessment techniques for whole-body defined in BS6841, ISO2651 and the proposed Physical Agents (Vibration) Directive

	British Standard BS6841 (1987)	International Standard ISO2631 (1997)	Physical Agents (Vibration) Directive
Frequency weighting	Fore-aft: Wd Lateral: Wd Vertical: Wb	Fore-aft: Wd Lateral: Wd Vertical: Wk	Fore-aft: Wd Lateral: Wd Vertical: Wk
Axis multipliers for health	Fore-aft: 1.0 Lateral: 1.0 Vertical: 1.0	Fore-aft: 1.4 Lateral: 1.4 Vertical: 1.0	Fore-aft: 1.4 Lateral: 1.4 Vertical: 1.0
Assessment method	Sum of VDVs	VDV or Σ axes r.m.s. or } or MTVV } worst axis	Worst axis VDV or Worst axis r.m.s.
Action values (VDV criteria only)	15 ms$^{-1.75}$ 'action level'	8.5-17 ms$^{-1.75}$ 'health guidance caution zone'	8.5 ms$^{-1.75}$ 'action value' 14.6 ms$^{-1.75}$ 'limit value'

Proposed vibration legislation

The core of the proposed Physical Agents (Vibration) Directive (as at December 2001) is based around 'limit values' and 'action values' for 8-hour exposures. The action and limit values are set at 2.5 and 5.0 ms^{-2} for hand-transmitted vibration and at 0.5 and 0.8 ms^{-2} for whole-body vibration. For whole-body vibration, the values are also defined in terms of VDV (Table 1). The directive states that:

- employers shall assess the levels of mechanical vibration to which workers are exposed,
- records must be kept of assessments,
- risks arising from exposure to mechanical vibration shall be reduced at source to a minimum,
- workers shall not be exposed above the exposure limit value
- if the limit value is exceeded, employers must take immediate action and identify reasons for the over-exposure
- information and training must be provided to those exposed to a risk
- health surveillance must be implemented

The proposed Directive includes annexes which described how measurements of vibration should be taken. For hand-transmitted vibration, measurements should be made in accordance with ISO 5349-1 (2001) with the evaluation based on the root sum of the squares of vibration in the three orthogonal axes. For whole-body vibration, measurements must be made in accordance with ISO2631, but restricted to only using r.m.s. or VDV with the worst axis taken as representative.

Common Position was reached on 25 June 2001 and the second reading took place on 23 October 2001. The limit and action values for whole-body vibration were reduced between reaching Common Position and being referred for a second reading. At the time of writing, the Directive is passing through the conciliation procedure and is likely to be adopted at a third reading in 2002, after which member states have three years to implement domestic regulations. Lobbying by industrial representatives is applying pressure to increase the values for whole-body vibration. Derogations from limit values have been applied to agriculture and forestry as a result of lobbying, so it is possible that important changes could still be made to the text before its third reading.

Discussion

The ultimate goal of Regulation is to avoid damaging workers' health. Unfortunately, inter-subject variation in susceptibility and imperfect agent monitoring mean that the holy grail of a perfect predictor of injury is impossible to achieve. Therefore it might be tempting to err on the side of caution when setting limits. However, if the limits are set too low then some tasks with a low risk might be prohibited and hence industry would bear financial penalties with little improvement in health. Conversely, if the limits are set too high then action that could protect health might not be taken by employers.

A dose-effect relationship has been standardised for noise and for hand-transmitted vibration exposure and so percentages of persons with adversely affected health can be estimated from measures of noise and vibration. For whole-body vibration, no such relationship is established and so setting limits is more difficult. One reason for this is that vibration is just one of a range of factors that might lead to back pain, whereas hearing damage and vibration white finger are specific to the injurious physical agent.

Models of noise exposure predict that after 20 years exposure at the current Second Action Level (90 dB L_{Aeq}), a 40 year old would have a 4% chance of a mean hearing loss of 30dB at 1, 2 and 3 kHz (BS5330, 1976). Reducing exposure to the proposed Limit Value (85 dB L_{Aeq}) would reduce the probability of the 30dB mean hearing loss to 1%. Therefore, a fourfold decrease in reports of noise induced hearing loss would be expected if current compliance with the Noise at Work Regulations is maintained.

The limit and action values for the proposed Physical Agents (Vibration) Directive cannot be directly compared with the HSE hand-transmitted vibration guidance, as the HSE specify 'worst axis' for assessment and the Directive specifies combined axes. However, a multiplier of 1.4 can be used to estimate triaxial exposures from previous single axis measurements (Nelson, 1997). Therefore, the limit value in the proposed Directive is higher than that currently suggested in the UK. Models of blanching for multi-axis stimuli predict that 10% of those exposed at the limit value would experience blanching after 6 years (ISO5349, 2001) compared to 8 years for the HSE guidelines.

For whole-body vibration, it is difficult to predict the extent of improvement in workers' health. Reducing exposure to one of the many risk factors for back pain must be welcomed, although it is possible that the limits are set too low resulting in an unnecessary burden on industry to comply. The Directive will also be helpful in clarifying a standardized procedure for application of ISO2631 at least across Europe.

It seems illogical that the hand-transmitted vibration part of the Directive specifies assessment in three axes whereas the whole-body vibration part of the Directive specifies assessment considering the worst axis only. The represents a step forwards for hand-transmitted vibration but a step backwards for whole-body vibration.

Conclusions

The Physical Agents (Noise) and Physical Agents (Vibration) Directives will limit exposures to noise, hand-transmitted and whole-body vibration. These Directives are likely to complete their progression through the EU legislative process in 2002 and must be implemented in member states within three years of adoption.

Noise exposure will be reduced when compared to current allowable limits. Limits for hand-transmitted and whole-body vibration will be implemented for the first time.

The reader should take note that the Directives discussed in this paper are currently in a draft form. It is likely that minor amendments will be made to the text of the Directives prior to adoption and it is feasible that limit values will be changed. An unlikely, but possible, outcome is that either, or both, of the Directives fail.

References

Borchardt, K.D. 2000, *The ABC of community law*, Luxembourg, Office for Official Publications of the European Communities, ISBN 9282878031

British Standards Institution 1976, BS5330. Method of test for estimating the risk of hearing handicap due to noise exposure

British Standards Institution 1987a, BS6841. Measurement and evaluation of human exposure to whole-body mechanical vibration and repeated shock

British Standards Institution 1987b, BS6842. Measurement and evaluation of human exposure vibration transmitted to the hand

Council of the European Union 1998, Council Directive 98/37/EC on the approximation of the laws of the Member States relating to machinery

European Commission 1986, Council Directive on the protection of workers from the risks related to exposure to noise at work (86/188/EEC)

European Commission 1990, Council Directive on the minimum health and safety requirements for the manual handling of loads where there is a risk particularly of back injury to workers.

European Commission 2001a, Common Position on the minimum health and safety requirements regarding the exposure of workers to the risks arising from physical agents (vibration).

European Commission 2001b, Common Position on the minimum health and safety requirements regarding the exposure of workers to the risks arising from physical agents (noise).

European Parliament 2001, Recommendation for a second reading on the Common Position on the minimum health and safety requirements regarding the exposure of workers to the risks arising from physical agents (vibration). A5-320/2001

Griffin, M.J. 1990, *Handbook of human vibration*, Academic Press, London

Health and Safety Executive 1994, Hand-arm vibration. HS(G)88. HSE Books. ISBN 0717607437

HMSO 1989, *The noise at work regulations*. Statutory Instrument 1989/1790

HMSO1992, *The manual handling operations regulations*. Statutory Instrument 1992/2793

International Organization for Standardization 1997, ISO 2631-1. Mechanical vibration and shock – evaluation of human response to whole-body vibration. Part 1

International Organization for Standardization 2001, ISO 5349-1. Mechanical vibration – measurement and assessment of human exposure to hand-transmitted vibration

Nelson, C.M. 1997, Hand transmitted vibration assessment – a comparison of results using single axis and triaxial methods, *presented at the United Kingdom Group Meeting on Human Response to Vibration held at ISVR, University of Southampton*

IMPORTANCE OF INFORMATION RECORDING AND MANAGEMENT

Teresa Ibarra and Joanne O. Crawford

*Institute of Occupational Health,
The University of Birmingham,
Birmingham, B15 2TT*

When carrying out an ergonomics assessment within industry, it can be a temptation to use readily available data from companies to help identify the risk factors for injuries at work. However, it is not always possible to obtain relevant ergonomics information from databases set up for other needs such as medical or engineering information. The following paper highlights the problems that can occur when using databases in the industrial setting for passive surveillance of Work Related Upper Limb Disorders. Firstly the issues of using databases set up for other data recording are discusses followed by the problems that this causes including incompatibility of different databases which do not interact with each other. To solve these data management problems, it is vital that those involved in information management identify the information required by all stakeholders involved. Previous research highlights the needs of the Ergonomist and how this can be built into computational databases. Essential information to be included within databases and ways to manage it are discussed including the issues of confidentiality and data protection.

Introduction

When an ergonomic assessment needs to be carried out within an industry, it is very tempting for the Ergonomist to use databases, files or any other data that is readily available across different departments within a company. However, at times, the way in which data is collected makes it difficult to obtain relevant information in order to perform the assessments. Tanaka (1996) recommended that surveillance activities should provide the backbone of any ergonomics program. Often reported injuries are an indication of poor design of the workplace or the work tasks. Musculoskeletal injuries such as upper limbs disorders, and back injuries are often called ergonomic-related disorders. Surveillance of the workplace for injuries is an important ergonomic aspect and it is common to evaluate the effectiveness of ergonomic processes by measuring injury rates and days lost through injury or ill health.

Two different types of surveillance, active and passive surveillance can be identified within the discipline of Work Related Upper Limb Disorders (WRULDs) epidemiology. Active surveillance is a term used to denote the activities of generating the data, finding cases by administering (musculoskeletal) health questionnaires and/or conducting physical examinations among workers. On the other hand, passive surveillance is used in WRULDs epidemiology to denote reviewing and analysing pre-existing records. This passive surveillance can be very tempting for the Ergonomist conducting a study since the use of information already collected such as medical records, including sickness absence and attendance at occupational health centres, accident data and quality control records; absenteeism and personnel turnover statistic, could reduce the time taken.

However, the method by which these data have been collected has usually been without the consideration of the needs of the Ergonomist and often information is lacking in essential detail. Nowadays, with the extensive use of the computers, it is more common to have computational databases. A database system can be defined as an interface between data and users with a specific purpose. Having the correct software interface could save considerable time and effort while analysing the data.

The aim of this paper is to highlight the importance of including essential information within databases across the company in order to facilitate an ergonomics analysis. The key role that the Ergonomist plays in designing or selecting the parameters to be recorded and the way, in which the data could be managed, will be also discussed.

Problem definition

Much of the time, information systems have been designed for many purposes, including collection of accident data, attendance to the medical centre, absenteeism and quality data but not necessarily ergonomics. This results in difficulties for those involved in ergonomics, as it can be difficult to trace an individual's work history. The Ergonomist has to be aware of the problems when conducting an ergonomics assessment. By examining the medical records, an Ergonomist can easily obtain information regarding the frequency of attendance to the medical centre, classification of the injuries or illness and the work tasks an individual actually carries out. However, for an ergonomic study it is important to evaluate the previous tasks that an individual has performed as the job may have changed due to injury. For example the operation where he or she is currently working may now have low risks, but previous operations that had been performed may represent a higher risk of injury. If an Ergonomist cannot trace all the tasks to which an individual has been assigned, the study is incomplete.

Another example is the use of absenteeism, sickness absence and personnel turnover statistics as predictors of ergonomic and psychosocial factors in the shop floor. Although there is concern about the number of hours or days lost per month and whether or not they are within tolerance levels. At times not enough resource is invested to analyse the causal factors involved in absenteeism, sickness absence and turnover data. This results in an inability to identify particular groups or working areas with major problems as data is often unspecific and analysed as means.

Recording and Management

The sophistication level, regarding databases installed in companies, varies according with their size, requirements, and resources invested. In larger companies there often exists different databases for different departments, for example, Human Resources, Medical Centre, and Engineering Design and Quality. However, a major issue can be incompatibility between different systems.

The integration of databases could save resources as well as facilitate ergonomics research, while allowing integrated data collection, not only the use of injury statistics. Developing an ergonomics module for extracting relevant information from other databases could allow Ergonomists to make a more complete assessment of the workplace. Injury statistics, workers' compensation records, medical records and accident records can be considered as sources of primary injury surveillance data. However, there are many other components within a company that can give valuable information to assess ergonomic implementation. There are also issues of confidentiality and data protection involved with the use of individualised data.

Accordingly, Stuart-Buttle (1999), recommended that the data that can be gathered apart from injuries include:

- Turnover rates
- Absenteeism rates
- Productivity rates and quotas
- Quality-related figures
- Workers' compensation costs
- Job profiles or descriptions
- Employees' jobs record or work history
- Discomfort survey results
- Ergonomics audit results
- Employee suggestions or reports of problems

These data should allow the Ergonomist and other managers in the company to obtain information as department; process and task where there are issues of absenteeism, sickness absence, turn over, compensation, quality and productivity. In linking this information, problem identification and analysis of the cause would be easier. When the causes are recognised, it can be identified if there is a link between these problem and ergonomics issues in the workplace.

As the Ergonomists will obtain the direct benefits of information systems, they play an important role in the interface design. Working in a multifunctional team integrated by the Ergonomic Department, Human Resources Department, Medical Personnel, Safety and Engineering Departments where everybody needs to work collectively in order to get the most of the system. The information system chosen, no matter if it will be developed or purchased, must integrate with the company's, vision, needs and resources.

The correct implementation, recording procedures and management of information systems will have a direct impact on the company's productivity as problems could be attacked directly. Woodcoock (1989), mentions the relationship between ergonomics and accidents, while Knave (1991) agrees with the fact that high incidence of illness absenteeism and occupational injuries can be regarded as an indicator of a poor work environment. The direct costs of accidents and long-term absenteeism can be calculated easily, however the impact on production may be hidden. If an employee is absent for several days, fellow-workers are forced to work harder in order to maintain production standards even if temporary staff is employed. Stress is likely to increase among them and there is the possibility that due to increase workload and increased exposure to risk, there may be an increase in symptoms associated with musculoskeletal disorders. Weyman (2001) mentions the evidence on literature of the potentially relevant influence between low levels of job-satisfaction, high job demands, and musculoskeletal symptoms. Having a well-designed information system pays off since the problems can be traced to the root and sorted out in shorter time.

In order to obtain the most of an injury surveillance systems, it must help to identify the who, what, where, when of employee exposure and use this information to correct and prevent further incidents. As mentioned in Stuart-Buttle (1999), a simple way to categorise most of this information is by using three groups: People, Places and Things as shown in Figure 1.

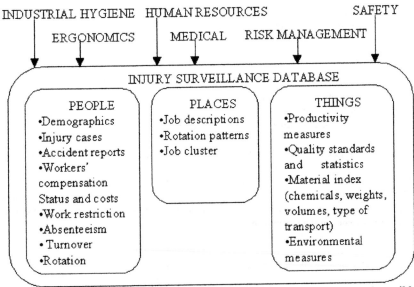

Figure 1 Fundamental model of an injury surveillance database system and possible user groups. (Modified from Stuart-Buttle, 1999).

One of the principal characteristics of the system is that it should be able to trace the different tasks an employee has performed over time. This feature will allow the Ergonomist to trace all the risks to which the employee has been exposed. If an employee experienced pain or discomfort while performing a task, the cause identification and an immediate action would be possible.

Confidentiality

Confidentiality concerns appear when an information system is shared between different users of different disciplines. Medical data is particularly sensitive in that respect. If the system has been designed using different modules for different departments or users, a password can be used for each module granting or preventing access to particular information. Thus, anonymised data could be created to identify particular workplaces with higher levels of risk or reported symptoms. Creating an extract of each module could solve the fact that the Ergonomist needs information from multiple departments. In this way the Ergonomist would be able to obtain the information relevant for the ergonomic assessment without breaching confidentiality or data protection issues. Using password-protected databases, a system can be developed that is both secure but accessible.

For the particular case of medical data, it is necessary for the Ergonomist to have access to certain anonymised information such as accident records, injuries, days lost. However, there is no need to allow access to other types of illness unrelated to workplace design or assessment. It is apparent however, that when carrying out a workplace assessment, individual workers may be identified as part of the workplace evaluation dependent on the types of ill health symptoms they have reported. In this case, interviewing by the Ergonomist to ascertain which particular processes result in pain or discomfort may give more information than any other type of record.

Conclusions

The early and accurate identification of problems in the workplace is vital as it represents savings in effort, time and money to companies. The integration of information systems can be helpful in order to trace productivity within different departments and operations, more specifically. The information system chosen should match with the company necessities and a multi-departmental involvement achieved.

Each department should identify the essential information that needs to be integrated in its module. Sometimes one specific department would need to have access to information from different sources, for example, the Ergonomics Department. The Ergonomist would need information from various departments in other to realise a complete assessment of the workplace. Access to these data should be facilitated by the use of a password protected database system. However, the development team must discuss the topic of confidentiality at early stages in order to design a reliable system.

References

Buckle, P. 1988, Health Data Bases and Their Applications. In A.S.Adam, R.R.Hall, B.J.McPhee and M.S.Oxenburgh (eds.) *Proceedings of the 10th Congress of the International Ergonomics Association, Sydney Australia*, (Taylor & Francis, London) 313-315

Knave, B., Paulson, H., Floderus, B., Gronkvist, L., Haggstrom, T., Jungeteg, G, Nilsson, H, Voss, M. and Wennberg, A. 1991, Incidence of Work-Related Disorders and Absenteeism as Tools in the Implementation of Work Environment Improvements: The Sweden Post Strategy. *Ergonomics, 34,* 6, 841-848

Stuart-Buttle, C. 1999 Injury Surveillance Database Systems. In W. Karwowski and W.S.Marras (eds.) *The Occupational Ergonomics Handbook,* (CPC Press, Boca Raton, Florida, USA), 1189-1203.

Tanaka, S. 1996 Record-Based ('Passive') Surveillance for Cumulative Trauma Disorders. In A. Bhattacharya and J.D.McGlothlin (eds.) *Occupational Ergonomics: Theory and Applications*, (Marcel Dekker, New York), 477-488.

Weyman, A., Boocock, M. 2001, Psycho-Social Influences on Reporting of Work Related Musculoskeletal Disorders – the Need for a Grounded Theory Approach. In M.A.Hanson (ed.) *Contemporary Ergonomics 2001*, (Taylor & Francis, London), 17-22.

Woodcock, K, 1989, Accidents and Injuries and Ergonomics: A Review of Theory and Practice, *Proceedings of the Human Factors Association of Canada 22nd Annual Conference*, (Toronto, Ontario, November 26-29), 1-10.

DISPLAY SCREEN EQUIPMENT

AN ERGONOMIC STUDY OF AN ORGANISED TELEWORK PROGRAMME

Colin Roscoe

3, Park View Road
Berkhamsted
Herts HP4 3EY

This paper presents the results of a survey of over 1000 employees at a large telecommunications company. The survey used an online self-assessment questionnaire to assess the employees DSE set-ups with regard to current legislation. The participants were separated into two groups Teleworkers and Office Workers. The Teleworkers were better off. Significant differences were found in 11 of 34 aspects including; health and well being, chair set-up and the general working environment. A summary of related publications indicated that this was at variance to previous research in this area. The key difference between this study and previous ones was that the company provided its Teleworkers with home office set-ups.

Introduction

Since the Health and Safety (Display Screen Equipment) Regulations 1992 were introduced employers have been obliged to assess every employee's workstation regardless of location. It is difficult to include Teleworkers in any assessment scheme, two of the main difficulties being that Ergonomists cannot access their workstations to make an assessment and that the quality of the workstations is rarely uniform.

The aim of this study was to determine whether the measures taken to address these problems were having a positive effect, by determining if the Teleworking group was being exposed to better or worse conditions than the regular Office Workers.

Office Workers generally have access to office furniture of both reasonable and uniform ergonomic value but Teleworkers have traditionally had to improvise their workstations. One solution to this problem is for the company employing the Teleworker to provide them with a workstation and I.T. set-up. This approach may be characterised as "If the worker won't come to the office then bring the office to the worker". Companies who implement this method gain a degree of influence over the working environment of the users. The approach does not reduce the need for companies to assess Teleworkers workstations but it does for ensure that there are fewer workers who need serious intervention in their workstation set-ups.

A literature review revealed that studies into the problems of Telework have investigated the entire work dynamic of Telework but were of limited sample size due to stringent matching of samples or low response rates. The studies suggested that Teleworkers suffered not only from physiological difficulties due to poor workspace design but also from psychological issues.

Due to technological advance such as ISDN, Teleworking is becoming a more attractive proposition to both employers and employees. Teleworking is increasingly important and attendant risks to health and safety are worthy of further investigation.

Industrial Context

The study focussed on a leading telecommunications company with a large Teleworking programme, employing more than 8000 Teleworkers worldwide. The company provides its Teleworkers a variety of home offices layouts.

The company provided two levels of service. A "Silver" (2-3 days of Telework a week) user of the Telework system was provided with an ISDN connection to the Internet. A "Gold" user (who gave up their office desk) was given a home office set-up with a choice of workstation, a standard chair and the ISDN line. This gave the company a measure of control over the workstation and working conditions.

Employees wishing to become Teleworkers were provided with a self-assessment questionnaire that assesses their psychological suitability to Telework. The Teleworkers were also provided with a specific Telework policy and Health & Safety Guidelines that make specific reference to Ergonomic issues.

Method

The Questionnaire
The principal method of assessing the participants in this study was an online self-assessment form known as "PC-Comfortable". The form consists of an initial section recording general information about the employee. The remainder of the questionnaire consists of 34 questions that could be answered yes, no or left blank. Throughout the questionnaire the questions are designed so that "No" is the response that indicates that there is a problem or issue to be addressed. The questions covered 7 sections: Health and Well-being, Chair Set-up, Keyboard and Mouse, Display Screen, Work Area, Task Structure, General Environment.

Data collection
The data was collected from the questionnaire over a 15-month period (December 1999 to March 2001). After data cleansing there were there were 1003 completed questionnaires. Some respondents worked both in offices and at home, for the purpose of the study they were treated as Teleworkers. There were 731 Office Workers and 272 Teleworkers.

Results

Table 1 gives the percentage of "No" responses to each of the questions that was significant at 0.05 level. The Chi Square value for each question is given giving an indication of the level to which the question achieved significance. The critical value of Chi Square was 3.84 for the 0.05 significance level.

Table 1. Chi-Square values for selected questions

Question	% "No" Response		Chi Square (3dp)
	Office Workers	Teleworkers	
During or after a days work are you free from aches and pains in the legs, feet and back?	26	20	8.997
Can you sit comfortably in the position described above?	15	7	9.648
Is the chair stable?	5	1	9.138
Do you have a foot rest of you cannot place your feet on the floor?	35	29	6.554
Have you adjusted the armrests so that they do not get in the way?	21	9	19.123
Is the screen free from glare?	14	9	3.849
Are adjustable blinds fitted to the windows that can be used to eliminate glare?	8	27	65.525
Can you vary your working posture throughout the day?	8	4	4.200
Do you find the noise level of your working environment acceptable?	9	4	8.562
Do you find the temperature of your working environment acceptable?	16	6	19.056
Do you find the lighting levels of your working environment acceptable?	8	4	5.955

Discussion

Table 1 shows that there were 11 questions with a significant difference between Office Workers and Teleworkers, they are discussed below.

Reported Discomfort
The Teleworkers reported fewer incidences of aches and pains in the legs, feet and back than the Office Workers. This suggests that the environments that the Teleworkers are working in place less stress on these areas of the body.

Seating Issues
Teleworkers reported problems with their ability to sit comfortably, chair stability and their armrests less frequently. Office Workers have a choice of four different adjustable chairs to suit their individual needs where as the Teleworkers that receive a furniture option have no alternative chairs. These results may appear surprising but can be explained. In an office it is very easy to "put up and shut up" rather than take action on any pain and discomfort that the user feels. Office based worker are far more likely to

receive a chair is either old or inherited from a previous user. Teleworker are always supplied with a new chair when they request a furniture option. Teleworkers can replace the chair that they are issued with, with a chair from their own house that they personally find comfortable. It is possible that Teleworkers by being exposed to specific ergonomic guidance in the form of the Health and Safety Guidelines issued to them when they become Teleworkers are more likely to make alterations to their workstations. Teleworkers may feel more empowered to make changes to their workstation, as they feel necessary.

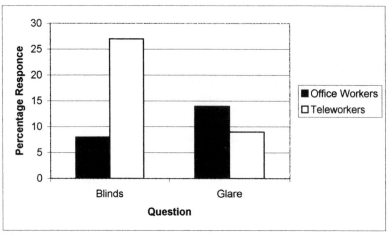

Figure 1 Graph illustrating discrepancies in "No" answers between Office Workers and Teleworkers

Incidence of Glare
Figure 1 shows that Teleworkers have blinds less frequently but also that they have fewer problems with glare. Teleworkers may not have windows in the room of their home office set-ups in the first place or have no need for blinds due to the window position relative to the monitor. In a modern open plan office it is hard for those users that do experience glare to eliminate it from their screens if there are other employees who benefit from or enjoy the light from windows. This peer pressure may stop Office Workers from eliminating the glare they experience.

Workspace Issues
Office Workers reported that they had more problems with adjusting their posture. Teleworkers as employees who are "unseen" may benefit from the ability to get up move about and otherwise vary their posture but Office Workers may feel unwilling to do so due to managerial supervision. The Office Workers in a modern office must contend with space limitations in terms of workspace and storage space together with all the other employees in their area. The Teleworker may be able to devote an entire room in their house to their job and thus have sufficient storage space. Teleworkers do not have to compromise their workspace to take into account fellow employees.

Environmental Issues

In a Teleworking environment the user has a small enclosed space to manage environmentally. In this respect the Teleworker may have more direct control over his or her environment than the regular employee who must rely on the facilities department of the office to alter the environment to suit them. If the Teleworker is too hot they can merely turn off the central heating or open a window. In an air-conditioned office an employee has little individual control over the environment they personally cannot open a window or turn the air conditioning off. In all three of the environmental issues covered by the questionnaire Noise, Temperature and Lighting the Teleworkers faired better than the Office Workers.

Overview

From the results it could be taken that Teleworking in an organisation that has a well thought out Teleworking program could provide several benefits over working in an office for the same company. Teleworking presents the employee with both benefits and difficulties. In a well-planned solution to the problems of Telework many of the difficulties that have been reported in Telework can be overcome.

The sample size has been much larger than in previous studies, but the questionnaire does not account for age, sex or amount of experience within the two sample populations.

In some cases Teleworkers perform jobs that have been designed to work well with Telework and in some cases Teleworkers work from home because it is convenient for the worker despite the job not being designed specifically for Telework.

The study also serves to identify "multi-base" employees i.e. those employees who work at several different locations either in the course of a week or in the course of a day. The problems posed by the multi-based employee are even greater than those posed by Teleworkers. They will often make use of "Hot Desks" (workstations that have no set user) for visiting employees. A "multi based" worker requiring a specific alteration to be made to a workstation would present any Ergonomist with a formidable challenge.

Teleworkers reported less stress and fatigue whilst at work although the questionnaire does not cover in depth the mental well being and attitudes of the Teleworkers. This area should perhaps be investigated further to see if the Telework programme also provides adequate psychological support for Teleworkers.

Conclusions

There have been many problems associated with Teleworking including the poor design of workstations. One solution to this problem is to provide Teleworkers with a home office with furniture of a reasonable ergonomic standard. This study analysed the responses of 1003 workers to a simple on line self-assessment questionnaire. 272 of the respondents were Teleworkers and 731 were Office Workers. The Teleworkers had the option of a home office set-up if they were a full time Teleworker. 11 of the 34 aspects examined by the questionnaire the difference were found to be significantly in favour of the Teleworkers.

References

Kerrin, M., Hone, K., Cox, T. (1998) Teleworking: Assessing The Risks. *Contemporary Ergonomics, 1998,* 174-178

Kaminsky, B.B., Schulze, L.J.H., Jimenez, L.Q. (1998) Ergonomic Evaluation of Home Computer Workstations. *Advances in Occupational Ergonomics and Safety, 2,* 533-536

Campion, S., Clarke, A. (1998) Evaluating Teleworking – a Case Study *Contemporary Ergonomics,1998,* 179-183

Budworth, N.(1998) Teleworking – Out of "site", out of mind? Institution of Occupational Safety and Health, (Technical Data Sheet)

McClay, C.J. (1998) The Development of Work at Home Safety Programs, *Professional Safety*, 43, **1**, 39-41

CALL CENTRE EMPLOYEES USING DISPLAY SCREEN EQUIPMENT (DSE): WHAT DO WE KNOW ABOUT WHAT THEY KNOW ABOUT IT?

Christine A. Sprigg & Phoebe R. Smith

Health & Safety Laboratory (HSL), Broad Lane, Sheffield, S3 7HQ

In the media, call centres have been referred to as "telephone sweatshops", whilst call-handlers are described as "battery hens" because of the suggested intensive nature of the call-handling task. Since 1998, psychologists from HSL have conducted two studies into call centre working practices. For these studies we have conducted interviews with call centre employees, and we have compiled a questionnaire specifically for this work context. Our questionnaire included job design measures and asked employees about their physical health and psychological well-being. Also included were questions about display screen equipment (DSE), performance monitoring, and the physical work environment. This paper reports findings on DSE working practices utilising questionnaire data from call handlers (N=874). We discuss hot-desking, and consider some of the challenges of the call centre context for both ergonomists and psychologists.

Introduction

The UK call centre industry has experienced huge growth in the last five years with recent estimates suggesting that it has over 420,000 employees (IDS, 2000). The call centre industry has a poor image regularly attracting negative comment in the media. The intensive nature of the call-handling task is exemplified by journalists using the terms "telephone sweatshops" to describe call centres and "battery hens" to describe those that work in them.

In 1998, psychologists from HSL were approached by the Local Authority Unit (LAU) of the Health and Safety Executive (HSE) and asked to conduct an exploratory study into call centre working practices. Health and safety stakeholders had started to express concern about the industry. These stakeholders included LA health and safety enforcement officers (EHOs), unions, employees and employers.

Thus, we embarked upon an exploratory study, the aims of which were to examine call centre working practices, highlight examples of good and poor working practices, and identify issues warranting further research. To fulfil these aims we discussed call centre working practices with EHOs, and officers from seven unions. Some unions had previously conducted their own research on call centres.

We conducted semi-structured interviews with employees (N=22) who held a variety of roles (eg., call-handlers, team leaders, occupational health nurses, managers) in six call centres. Throughout the exploratory study we continued to discuss call centre working practices with the unions, industry professional bodies, e.g., The Call Centre Association (CCA), and other technical specialists within HSL and HSE e.g. ergonomists, noise experts.

The product of our exploratory study was a Local Authority Circular (LAC) 'Initial Advice Regarding Call Centre Working Practices'. The LAC (94/1), issued in November 1999, which describes call centre working practices and the potential risks to health and safety.

Although the exploratory study and LAC were welcomed by the call centre industry and EHOs the study had several limitations. For example, the study was conducted in just six call centres (three of which were from the financial sector), and the interviewees were selected by employers prior to our arrival at a site. Acknowledging these limitations, LAU considered that more data were required. Then a decision would be made as to whether new regulations were required for the call centre industry. HSL were commissioned to gather this further data.

As psychologists, the focus of our research has been the collection of data on psychosocial risk factors and on the self-reported health outcomes that may be related to them. However, in both of our studies we have also collected information about Display Screen Equipment (DSE) working practices and hot-desking, since the Health and Safety (Display Screen Equipment) Regulations 1992 are the most pertinent sets of health and safety regulations for call centres.

This paper reports some preliminary findings on call-handlers self-reported knowledge of DSE good practice, and also employees views of the organisational practice of hot-desking. The information presented here is taken from the questionnaire we developed for our second study.

Method

Questionnaire Development
We compiled a self-report questionnaire tailored to the call centre context. In addition to questions on job design (i.e autonomy, task variety etc.) we asked questions about DSE. We included self-reported affective reaction measures of job satisfaction, mental health, job-related well-being, and also included measures of self-reported vocal, optical, auditory and musculoskeletal health.

We adapted pre-existing core job characteristic measures, e.g., the measures of job control developed by Jackson et al (1993). We developed new questionnaire items on Electronic Performance Monitoring (EPM) based on research literature, and we have used, where possible, other measures that are considered psychometrically robust, e.g. the General Health Questionnaire (GHQ-12) (Goldberg, 1972).

In collaboration with EHOs and HSE experts, we developed additional questions e.g., on auditory health and employees' understanding of headset maintenance.

Procedure
Employers participated in the study on a voluntary basis, and we requested that employees were instructed that questionnaire completion was also entirely voluntary. We negotiated access to employees in each call centre and asked that employees were given the opportunity to complete the questionnaire during work time. We composed a detailed covering letter for the questionnaire and provided a reply paid envelope for ease of return.

Results

One thousand, one hundred and thirty completed questionnaires were returned, representing an overall response rate of 38%. The sample includes employees from 20 organisations in various locations across England, Scotland, Wales and Northern Ireland. This paper uses data from call handlers only (n=874).

The quantitative data presented here are from simple frequency analysis of questionnaire data. Qualitative data take the form of written comments made by participants when they were prompted on specific aspects e.g., hot-desking, EPM etc.

Table 1. DSE Training

Questionnaire Item	% Agreement
1. I have suffient training to know how to set up my workstation e.g., chair, desk & computer	69
2. I have sufficient training to know how to adjust my workstation e.g., chair, desk & computer	68
3. I have sufficient training to know how to adjust my VDU screen contrast	56
4. I have sufficient training to know how to adjust my VDU screen brightness	57
5. I have been trained to set up my workstation in such a way as to minimise the risks to my health & safety	44

Table 1 shows the percentage of call-handlers in agreement with each questionnaire item. Here, 'Agreement' is the sum of the percentages for those that have ticked the 'Agree' and 'Strongly agree' boxes. Thus, the remaining percent is those that have ticked either 'Neither agree nor disagree', 'Disagree' and 'Strongly disagree'. It is of note that 35% of call-handlers disagreed ('Disagree' plus 'Strongly disagree') with the item 'I have been trained to set up my workstation in a such a way as to minimise the risks to my health and safety'.

Table 2. DSE Awareness & Hot-desking

Questionnaire Item	% YES
1. Are you aware of the Health & Safety Executive (HSE) regulations and guidance for DSE users?	44
2. Are you aware that, as a DSE user, you can ask your employer to arrange an eye test for you?	80
3. Has your employer trained you to set up your workstation?	51
4. Does someone else usually conduct a DSE assessment on you?	25
5. How often do you have a DSE assessment whilst sitting at your workstation? A) never had a DSE assessment	65
6. Do you hot-desk? (i.e you don't always work at the same workstn)	53
7. Do you hot-desk across the whole call-handling area?	51

Table 2 illustrates our findings on DSE awareness and hot-desking. Here, the remaining percent are those that ticked 'No'. Of note are the 80% of call-handlers who 'are aware that, as a DSE user, you can ask your employer to arrange an eye test for you'. Also of note are the 65% of call handlers who indicated that they had 'never had a DSE assessment'.

Qualitative Data on Hot-desking

Fifty-three percent of the sample reported that they 'hot-desk'. Thirty-four percent of call handlers wrote comments when prompted by the question 'Do you have any particular issues with 'hot-desking' that you would like to tell us about?'. We conducted a thematic analysis of the comments on hot-desking. The majority of comments were negative with views clustered around a number of themes. These themes were:

DSE: e.g., *"you are not even given time to make adjustments to sit at new desk stations"; "adjusting seat every time etc. is not always easy"; "unable to maintain set up according to DSE"*

Stress: e.g., *"causes extra stress before shift"; "It's often difficult to find anywhere to sit, we are argued at by our supervisors/managers for not immediately taking calls"; "I find this aspect of the job extremely difficult in that I cannot always guarantee a seat".*

Manual handling: e.g., *"moving drawers around from floor to floor is pointless and dangerous"; moving all my stuff around (team info etc) can be time consuming and heavy"*

Storage: e.g., *"I like to have all my information to hand and when moving desks something may be forgotten""The only real issues are carting your belongings around and not having any secure storage space"*

Preferences for own desk: e.g., *"would particularly like own desk/area to work at each day, stations can be untidy and also have to alter chairs etc""I don't agree with hot-desking it is better to stay at your own desk and then you have your own work items at hand"*

Hygiene: e.g., *"some keyboards and phone turrets not very clean are told wipes are available, but if busy no time to come off phone to clean them""Hygiene - equipment not cleaned regularly enough, keyboards and monitors grubby"*

Discussion and Conclusions

In the context of call centres DSE good practice is vital. Call-handlers are DSE users (in our sample, 94% of call handlers use DSE more than 75% of their working shift) who are engaged in a highly repetitive task with limited opportunities to move away from the desks and phones.

From preliminary data analysis (see Tables 1 and Tables 2) it is evident there is a level of awareness about DSE good practice, but that a greater level of knowledge is still needed. We must be cautious in interpreting these results as it is possible that call handlers might not be aware that the training they have received on DSE is connected to HSE regulation (see Item 1, Table 2) and that making adjustments to their workstations are about minimising the risks to their own health and safety (see Item 5, Table 1).

Increasingly, ergonomists and others are considering the contribution of psychosocial and work organization factors in the aetiology of stress and musculoskeletal symptoms. Indeed, there are a few papers on the contribution of psychosocial and organizational variables in predicting and mediating well-being in call centres (e.g., Fenety et al, 1999; Ferreira et al, 1997; Hook & Matta, 1997; Most, 1999; Sznelwar et al, 1999). One such organisational work practice that may contribute to call handler stress is hot-desking itself (see 'Qualitative Data'). Yet, previous research conducted for HSE (Simpson 2000) found no significant differences between a hot-desking group and a control group on stress and arousal scores. We hope to examine the hot-desking issue in more detail in our further statistical analyses.

In conclusion, call centres present a number of challenges to ergonomists and psychologists alike. We need to encourage employers to give employees time to make adjustments to DSE, especially in hot-desking contexts. We have to consider what are adequate breaks for call handlers as opposed to other DSE users. Moreover, we have to encourage call handlers, and others, to appreciate the link between making workstation adjustments, taking breaks, having changes of activity, and their own physical and mental well-being.

As a result of research we have produced two UK Local Authority Circulars (November 1999 and December 2001 [Rev]). These documents give practical advice on minimising the health and safety risks to call centre employees. As 'open' documents they are available for use by employers, unions and other industry stakeholders.

During 2002, we will produce a HSE Contract Research Report (CRR) which will detail our statistical findings. In the CRR, for example, we will report on the relationships between the job design variables (e.g., autonomy, task variety etc.) and the outcome measures (e.g., mental health, job-related well-being and musculoskeletal health etc.). This statistical analysis will aid our understanding of the psychosocial risk factors in call centres, and other modern office contexts using similar working practices.

References

Fenety, A., Putnam, C. and Loppie, C. 1999, Self-reported health determinants in female call centre tele-operators: A qualitative analysis. In G. Lee (ed.) *Advances in Occupational Ergonomics and Safety*, (IOS Press), 219-224.

Ferreira, M., Conceicao, G.M. and Saldiva., P. H. 1997. Work organization is significantly associated with upper extremities musculoskeletal disorders among employees engaged in interactive computer-telephone tasks of an international bank subsidiary in Sao Paulo, Brazil. *American Journal of Industrial Medicine, 31*, 468-473.

Goldberg, D.P. 1972, *The detection of psychiatric illness by questionnaire* (Maudsley Monograph No.21). Oxford: Oxford University Press.

HELA. 2001, Advice regarding call centre working practices. *LAC,* **94/1 (rev).**

Hook, K. and Matta, L. 1997, Organisational variables in call centres: Mediator relationships. In S. A. Robertson (ed.) *Contemporary Ergonomics 1997,* (Taylor and Francis, London), 295-300.

IDS (Incomes Data Services) 2000, *Pay and conditions in call centres 2000.* London: IDS.

Jackson, P. R., Wall, T.D., Martin, R., & Davids, K. 1993, New measures of job control, cognitive demand, and production responsibility. *Journal of Applied Psychology, 78*, 753-762.

Most, I. G. 1999, Psychosocial elements in the work environment of a large call center operation. *Occupational Medicine: State of the Art Reviews, 14*, 135-147.

Simpson, N. 2000, The effects of new ways of working on employees' stress levels. HSE Contract Research Report **259/2000.** (HSE Books).

Sznelwar, L. I., Mascia, F. L., Zilbovicius, M. and Arbix, G. 1999, Ergonomics and work organization: The relationship between Tayloristic design and workers' health in banks and credit card companies. *International Journal of Occupational Safety and Ergonomics, 5*, 291-301.

APPLIED PHYSIOLOGY

AN OPTIMISATION APPROACH TO ERGONOMIC EVALUATION AND MOTION ANALYSIS

M.A. Williams and A.J. Medland

University of Bath

Abstract

This research is aimed at providing a design tool in the form of a computer manikin to evaluate and optimise devices to aid stability. A case study to model the intermediate postures employed when rising from a chair has resulted in good comparisons with both published literature and empirical studies.

Introduction

Many people, such as the elderly and those with restricted physical abilities experience, difficulties, and sometimes injury, when undertaking daily tasks. This can often be due to instability during transitionary movements. The focus of this work has thus been centred on the creation of a manikin representation of a potential user that can be placed into various postures and simultaneously evaluated for stability. The manikin is based upon existing anthropomorphic data provided by the Technical University of Delft (TUD). It is manipulated using the constraint modelling techniques created and employed at the University of Bath. This novel approach enables the designer or ergonomist to quickly evaluate the manikin's ability to see, reach and fit into a predefined space without having to redefine the body posture of the manikin whenever a design modification is made. A study to animate and validate the movement strategies when using a chair is presented.

The Manipulation of the Manikin using Constraint Modelling

Constraint based modelling has been predominantly used at the University of Bath to redesign machinery for the packaging industry. This approach involves the manipulation of geometric entities through the use of rules that are written in the form of mathematical or geometric relationships. The rotation, translation and scaling of the geometric entities about their individual axes are considered as variables that can be invoked and resolved through the use of rules, as described by Medland *et al* (1995).

The development of the computer manikin has used a similar approach. The skeletal body segments are individually represented by rigid links, which are pivoted about each other and restricted to rotating about their individual local axes. The external body is represented by a wire frame and solid geometric model, as described in Williams and Medland (2001). The wire frame model is created from a database developed from thirty years of research undertaken at the Technical University of Delft of various body sizes (ADAPS TUD website).

The movements of the computer manikin are manipulated by invoking the rules and choosing the variables, i.e. body segments and the axes that they are required to move about or along. When designing, for example, a chair the designer is able to create a geometric representative of the conceptual design and employ the rules to sit the manikin onto the chair and evaluate the posture of the manikin, as shown in Figure 1a. If further design iterations are required the designer or ergonomist can modify the geometric representatives accordingly and invoke the rules. The manikin will automatically obey the rules which, in this case, will constrain the trunk, upper and lower legs to the backrest, the seat and the leg rest of the chair respectively, irrespective of the modifications of the geometric entities representing the chair, as shown in figure 1b. If the evaluation of a different posture is required the rules can be modified accordingly to place the feet, for example, on the ground, as shown in Figure 1c.

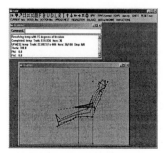

Figure 1a Evaluation of posture in a reclined chair

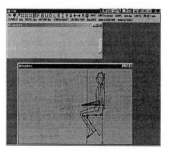

Figure 1b Evaluation of posture with different chair configuration

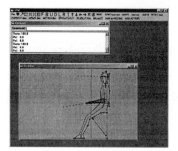

Figure 1c Evaluation of posture with different rules and chair configuration

The evaluation of Stability
The option of analysing the stability of a given posture is available through the use of a display menu. This is calculated by determining the position of the total body Centre of Mass (CoM) relative to the base of support, formed by any part of the body in contact

with the ground, for example, the feet. If the CoM is maintained within the base of support a 'balanced' statement will be returned. If both balance and the posture cannot be achieved the designer is able to invoke the rules to allow the stability of the manikin and/or the support aid, for example, to be re-positioned until a satisfactory state is found.

Advantages of this approach

This design tool also provides the options of predefining animations through motion analysis of the potential user. Whenever a modification to the design is made the animation has to be predefined. Another option provided for the user is to individually manipulate each individual body segment until the desired posture is found. The advantage, however, of using the constraint rules to manipulate the manikin occurs when a desired posture requires the manipulation of multi segments and may involve the iterative process of evaluation. The constraint rules will automatically change the posture of the manikin without further manipulation of body segments or the involvement of subjects, which can be time consuming.

The manipulation of intermediate postures

Once the initial posture of the manikin is determined the designer is able to evaluate the intermediate postures of the manikin when undertaking a given task, for example rising from a chair. This is done initially by understanding the movements of the body segments and how they interface with the external environment. The published authors, Jeug *et al* (1990), Schenkman *et al* (1990), Butler *et al* (1991) and Ikeda *et al* (1991), all describe the sit to stand movement, for example, as being divided into phases, as shown in Figures 2a, 2b, 2c and 2d respectively. When an able-bodied person initially rises from a chair, the feet remain in contact with the floor and the trunk sways forwards towards maximum flexion, as shown in Figure 2b. The buttocks then lift off from the seat of the chair allowing the total body CoM to move within the base of support formed by the feet, as shown in Figure 2c. The final movement phase is the achievement of maximum trunk extension to an upright standing posture, as shown in Figure 2d.

Comparisons to literature and empirical studies

An empirical study was carried out to validate the intermediate postures resulting from the use of constraint rules and to further understand the movement strategies employed by people when rising from a chair. This was undertaken by six able-bodied people aged between 24 – 53 years and three people with arthritis of the knee and hip joints, aged between 37 and 65 years. For comparative purposes of this study only the results from the able-bodied subjects will be discussed. The subjects were requested refrain from moving their feet from a predefined position on the floor and were instructed to look at a ball, which was placed at a given height and distance from the chair, during the whole of the movement. The aim of this study was to observe the gross total body movements without the use of the arms for support or as an aid to stability. The subjects were therefore also requested to keep their arms as straight as possible and move them in sequence with their trunk.

 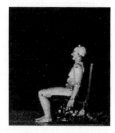

Figure 2a The sitting posture modelled by the manikin and a human

Figure 2b The forward trunk sway modelled by the manikin and a human

Figure 2c The lift off from the seat modelled by the manikin and a human

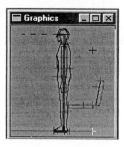 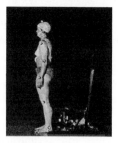

Figure 2d The erect stance modelled by the manikin and a human

Conclusions and Further Work

The observational comparisons of the intermediate postures when rising from a chair observed during the empirical study and those described in published literature showed good agreement with the postures created using the constraint rules. This shows that the constraint process can be used to manipulate the manikin into the intermediate postures that a person would commonly employ, and allow the evaluation of stability to be simultaneously carried out.

A more in-depth study employing the body sizes of each subject to represent the manikin is currently being carried out, to compare the resulting maximum trunk, hip and knee flexion of each subject observed with the manikin. This is being undertaken with and without the use of armrests. These results will be further compared to literature and published by the authors.

This constraint modelling approach has also been used to design a chair that can aid stability and increase the ease of rising from a sitting posture. The prototype of this chair is currently being employed to undertake a comparative study of the intermediate postures predicted using constraint rules, and the actual postures used by humans. This prototype is also being subjectively evaluated for comfort. The results of this study will be published by the authors.

References

ADAPS, ergonomics department, University of Delft website: www.io.tudelft.nl/research/ergonomics/research/adaps/index.html

Butler P.B., Nene A.V. and Major R.E 1991, Biomechanics of Transfer from the sitting to the standing posture in some Neuromuscular Diseases, *Physiotherapy,* **77,** 521-525

Jeug S.F., Schenkman M., Riley P. and Lin S.L 1990,Reliability of a Clinical Kinimatic Assessment of the Sit-to-Stand Movement. *Physical Therapy*, 70, 511/56-520/65.

Ikeda E.R., Schenkman M.L, Riley P and Hodge W.A. 1991, Influence of Age on Dynamics of Rising from a Chair. *Physical Therapy*, **71**, 473-481.

Medland, A.J., Mullineux, G., Rentoul A.H and Twyman B.R, 1995. MATADOR '95, Manchester, 1995, 449-454

Schenkman M., Berger R.A., Riley P., Mann R.W. and Hodge W.A, Whole Body Movements During Rising from sitting 1990, *Physical Therapy*, **70**, 638/51-651/64

Williams M.A. and Medland A.J, The creation of techniques for the design of machines compatible with human posture. ICED 2001, 347-353

THE EFFICACY OF DEEP-WATER RUNNING

T. Reilly, N. T. Cable and C. N. Dowzer

Research Institute for Sport and Exercise Sciences
Liverpool John Moores University
Henry Cotton Campus
15-21 Webster Street
Liverpool, L3 2ET

Abstract

Deep-water running has been advocated as a mode of physical training for injury rehabilitation, injury prevention and promoting recovery from strenuous exercise. The present aim was to examine the efficacy of deep-water running in a series of studies of physical training. First, the effects of a 6-week water-running programme on aerobic, anaerobic and muscle strength measures were investigated. Deep-water running was compared with treadmill running and a combination of water-running and treadmill running. The greatest improvements in $VO_{2 \, max}$ were noted in the mode of exercise practised. All three experimental groups improved anaerobic power output; muscle strength remained stable in the group performing 'combined' training, whilst increasing in the other two. Improvements in shoulder strength were evident only with deep-water running. Second, the alleviation of muscle soreness following 'plyometric' exercise by use of deep-water running was investigated. Deep-water running proved to be superior to the other methods of reducing muscle soreness and restoring muscle strength. The benefits of deep-water running were most apparent in the subjective responses to training activity. These observations indicate the effectiveness of this exercise mode in enhancing the training of individuals seeking improved fitness.

Introduction

Deep-water running (DWR) is promoted in physical training programmes, particularly during rehabilitation from injury (Dowzer and Reilly, 1998). This mode of exercise is employed also for preventive purposes, since the impact with the ground is avoided thereby reducing the load on skeletal structures. Deep-water running is also utilised by athletes as a form of recovery training in between strenuous competitive engagements.

In deep-water running the exercise is conducted in the deep end of a swimming pool with the body kept up by means of a buoyancy belt. This system of exercise can allow the individual to operate at an intensity of about 75% of the maximal oxygen uptake ($VO_{2 \, max}$). It may therefore be effective in maintaining aerobic fitness (Dowzer et al., 1999).

Due to the resistance provided by the water when moving the limbs, there may also be benefits for strength training purposes. There have been no longitudinal studies of deep-water running regimens in which the various modalities of fitness have been monitored together.

Deep-water running is used as a form of supplementary training, especially to reduce impact stresses associated with exercising on hard surfaces. Dowzer and co-workers (1998) showed that deep-water running decreases the compressive load on the spine compared with training in shallow water and with treadmill running. In these latter two cases the runner's feet contact the surface on each successive stride, whereas such contact does not occur in deep-water running.

Regular training is often interrupted by muscle soreness following competitive matches or training sessions which incorporate physical contact (Reilly, 1998). Delayed-onset muscle soreness refers to the discomfort felt for some days following a particular form of exercise known as plyometrics. Plyometric exercise refers to stretch-shortening cycles of muscle actions which are incorporated into training drills for individuals attempting to improve their abilities to generate high power output. Such drills include repetitive hopping, bounding, drop-jumping (Boocock et al., 1990) and 'pendulum swing' exercises (Fowler et al., 1997). During the periods in which delayed-onset muscle soreness is experienced, deep-water running provides a means by which training may be continued.

The utility of DWR was examined in a couple of studies entailing longitudinal and quasi-longitudinal research designs. In the first study the aim was to compare the effects of DWR, treadmill running and combined deep-water and treadmill running on aerobic, anaerobic and muscle strength measures. The aim in the subsequent study was to examine the effects of DWR on the prevention and recovery from an exercise regimen designed to induce delayed-onset muscle soreness.

Training study

Eighteen previously untrained males were divided into three groups and randomly assigned for training using DWR, treadmill running (TrR) and combined DWR and treadmill (W & TrR) modes. Training was performed three times a week for 6 weeks, increasing progressively to 45 min and 80% of mode-specific VO_{2peak}. Assessments were made of VO_{2peak} for DWR and TrR, and blood lactate minimum (La_{min}) was measured according to Tegtbue et al. (1993). Concentric muscle strength was recorded for knee flexion and extension at 1.05, 3.14 and 5.2 $rad.s^{-1}$ and shoulder flexion and extension (1.05 $rad.s^{-1}$) using an isokinetic dynamometer (Lido Active, Davis). Muscular endurance for knee flexion and extension was measured at 2.05 $rad.s^{-1}$ over a 2-min period. The Wingate cycle ergometric test was used to measure anaerobic performance over 30 s (Bar-Or, 1987).

All groups improved in VO_{2max}/VO_{2peak} irrespective of training mode (see Table 1). No changes were observed in the velocity corresponding to lactate minimum values for any of the groups. There was no significant change in maximal heart rate. The greatest improvement in VO_{2max} was observed in the group training on the treadmill. The improvement in VO_{2peak} measured in water running was observed with DWR. The group using combined training was intermediate in magnitude of improvement on mode-specific VO_2 responses.

Significant increases were noted in mean power and peak power (1 s) in all three groups (P< 0.01). No differences were evident between groups ($F_{2,15}$ = 0.87; P = 0.44). There was a significant improvement in shoulder flexion torque (mean increase 11.2 Nm) evident only in DWR (P<0.05). Muscle endurance of knee flexion and extension as expressed by a change in fatigue index, improved in the DWR and TrR groups by 8.7 and 7.1% respectively (P<0.01), but not in the group using combined training.

Table 1. Changes in highest values for VO₂, VE and heart rate (HR) over the 6 weeks of training in the 3 groups. Data for anaerobic performance are also incorporated.

	Test mode	DWR training		Combined training		TrR training	
		Before	After	Before	After	Before	After
VO₂ peak (ml.kg⁻¹.min⁻¹)	TrR	38.4± 4.4	40.2 ± 6.0	38.9 ± 2.7	43.4 ± 7.5	41.0 ± 8.0	46.7 ± 6.2
	DWR	25.1 ± 5.7	29.6 ± 6.4	24.7 ± 6.0	27.8 ± 5.3		
VE peak (l.min⁻¹)	TrR	116.2 ± 20.2	129.2 ± 22.7	117.6 ± 25.7	131.5 ± 11.9	119.0 ± 34.1	124.3 ± 20.9
	DWR	91.2 ± 19.6	105.2 ± 14.4	86.0 ± 19.0	101.6 ± 17.2		
HR max (beats.min⁻¹)	TrR	182 ± 5	181 ± 5	181 ± 21	181 ± 20	187 ± 9	185 ± 5
	DWR	170 ± 14	169 ± 6	172 ± 20	167 ± 16		
Peak power (W)	Cycle	1018 ± 236	1269 ± 279	981 ± 154	1179 ± 189	813 ± 340	1007 ± 265
Mean power (W)	Cycle	802 ± 140	1086 ± 245	844 ± 161	994 ± 244	612 ± 314	856 ± 215

Delayed-onset muscle soreness

Fifteen male and 15 females (age range 20-41 years) participated in the study. They were divided into 5 groups, differentiated according to the warm-down procedure and the type of active recovery from exercise over the subsequent 3 days. The exercise session consisted of drop-jumps every 7 s until volitional exhaustion. Exercise on the three subsequent days consisted of running for 30 min at 70-80% of heart rate reserve (HR max minus resting heart rate). The recovery exercise on each day was i) rest on all days; ii) rest on Day 1, DWR on remaining days; iii) rest on day 1, treadmill running on later days; iv) treadmill run on all days; v) DWR on all days.

Measures on each day pre-exercise included range of motion at hip and ankle, leg strength, standing broad jump, muscle soreness rated on a visual analogue scale (Talag, 1973) and punctuate soreness in the anterior thigh. Blood samples obtained before and after exercise on each of the four days were analysed for plasma creatine kinase (CK) concentrations using an automated spectrophotometer (Monarch Plus Chemistry System,

Instrumentation Laboratory, Lexington, MA). Data were analysed by means of ANOVA with repeated measures.

Deep-water running failed to prevent delayed-onset muscle soreness (DOMS) but appeared to speed the process of recovery for leg strength and perceived soreness. Leg strength was reduced by 20% on average after 48 hours but only by 7% with DWR. Soreness was reduced by 40% in the DWR group in the legs, but threefold for abdominal soreness. Furthermore, CK activity peaked 24 hours earlier than in groups not performing DWR. The sensation of soreness was also eliminated during DWR but DOMS returned on cessation of the exercise. No significant correlation was observed between CK levels and soreness in the quadriceps. Deep-water running helped to maintain hip range of motion when suffering DOMS.

Discussion and conclusions

The training study demonstrated that deep-water running is effective in enhancing the physiological fitness. The individuals who participated in the study were relatively unfamiliar with hard training and so the training effects noted may not be so large in already well –trained athletes. In trained individuals DWR would still have a role to play in maintaining aerobic and musculoskeletal fitness during rehabilitation or at busy times of the competitive season.

The changes in aerobic fitness measures were most evident when assessments were made in tests conducted in the water. Changes were also noted in anaerobic and muscle endurance. Of particular interest was the improvement of 22% in shoulder flexion torque observed in the group undertaking the DWR regimen. It is likely that this increase was due to the vigorous upper limb movements associated with the driving action through the water in balance with the contra-lateral lower limb.

The plyometric activity examined in the second study was successful in inducing a severe degree of muscle soreness. The elevations in circulating CK levels confirmed the subjective symptoms. The observations in this study support the use of DWR in the days subsequent to performing stretch-shortening exercise. Although the differences between conditions were not evident in CK levels, except for the decrease exhibited in the DWR group, the benefits of DWR were most evident in the subjective systems. The attenuation of soreness with DWR could be linked with the smaller decline in leg strength that occurred.

It is concluded that deep-water running is effective in enhancing physiological measures, including oxygen transport, anaerobic and muscle performance. Furthermore, this mode of training can provide temporary relief from delayed-onset muscle soreness whilst benefiting the process of recovery. Finally, DWR can offer much needed variety in a training regimen, thereby maintaining the motivation of the individual athlete.

References

Bar-Or, O. 1987, The Wingate Anaerobic Test: an update on methodology, reliability and validity, *Sports Medicine*, **4**, 381-394.

Boocock, M. G., Garbutt, G., Linge, K., Reilly, T. and Troup, J. D. G. 1990, Changes in stature following drop-jumping and post-exercise gravity inversion, Medicine and Science in Sports and Exercise, **22**, 385-390.

Dowzer, C. N. and Reilly, T. 1998, Deep-water running, *Sports Exercise and Injury*, **4**, 56-61.

Dowzer, C. N., Reilly, T. and Cable, N. T. 1998, Effects of deep and shallow water running on spinal shrinkage, *British Journal of Sports Medicine*, **32**, 44-48.

Dowzer, C. N., Reilly, T., Cable, N. T. and Nevill, A. 1999, Maximal physiological responses to deep and shallow water running, *Ergonomics*, **42**, 275-281.

Fowler, N. E., Lees, A. and Reilly, T. 1997, Changes in stature following plyometric drop-jump and pendulum exercises, *Ergonomics*, **40**, 1279-1286.

Reilly, T. 1998, Recovery from strenuous training and matches, *Sports Exercise and Injury*, **4**, 156-158.

Talag, T. 1973, Residual muscle soreness as influenced by concentric, eccentric and static contractions, *Research Quarterly*, **44**, 458-469.

Tegtbur, K., Busse, W. M. and Baumann, K. M., 1993, Estimation of and individual equilibrium between lactate production and catabolism during exercise, *Medicine and Science in Sports and Exercise*, **25**, 620-627.

CHARACTERISTIC ACTIVITIES AND INJURIES OF HILL-WALKERS

P.N. Ainslie[1], I.T. Campbell[2], D.P.M. MacLaren[1] and T. Reilly[1]

[1] *Research Institute for Sport and Exercise Sciences, Liverpool John Moores University, Liverpool L3 2ET;* [2] *University Department of Anaesthesia, University Hospitals of South Manchester, Withington Hospital, Manchester M20 2LR*

The aim of this study was to gather information from recreational hill-walkers to gauge their 'typical' activity, fluid and food intakes, as well as to investigate whether these activity characteristics may be related to injury. A questionnaire was designed to address the above issues, of which 100 were suitable for analysis. The main findings of the investigation were that recreational hill-walkers covered average walking distance of 18 – 26 km in 6 – 8 hours over 600 m in altitude, and that a reported 35% incidence of injury occurred during the walking. Knee and ankle injuries were the most common reported, with 63% occurring near the end of the walk and 83% during the descent. There were significant positive relationships between injury and age and also walking above 600 m, and negative correlations with energy intake and also frequency of walking. Some practical recommendations generated from this questionnaire analysis are considered.

Introduction

Hill walking is one of the most popular recreational pursuits, yet there is a lack of information regarding the characteristic activities of such events. The prolonged duration of a typical hill-walk places exceptional demands on the participants (Ainslie et al., 2001). The specific demands of hill-walking involve activity varying in intensity and duration, both of which are influenced by factors such as the fitness of the participant, dietary intake, backpack weight and weather conditions. The injuries occurring in the mountainous environment only receive attention in serious incidents involving the rescue services and consequently limit the recording of such incidents.

The aim of the present study was to examine characteristic patterns of hill-walkers and acquire information about the incidence of injury. Recreational hill-walkers completed a questionnaire detailing characteristic patterns of activity, injury, and typical nutritional and fluid intake for such activities. Information on the characteristic patterns such as distance, intensity, frequency, backpack weight and typical rest and lunch breaks, was collected in order to gauge the 'typical' activity for hill-walkers as well as to investigate whether these activity characteristics may be relevant to injury occurrence. Questions relating to diet and fluid intake were incorporated in to the questionnaire. Their inclusion enabled the typical energy and fluid intakes commonly used by hill-walkers to be calculated.

Methods

Questionnaire content: The questionnaire was designed to cover three main areas of interest. Initial questions were posed to establish the physical attributes of the subject group. The three main sections of the questionnaires consisted of details of (a) hill-waking activity, (b) injury / accidents, and (c) typical fluid and nutritional habits. Details of the hill-waking activity were sought in 10 questions regarding the distance, duration, intensity, frequency, back-back weight and typical rest and lunch breaks. Questions in the injury section referred to the amount and type of injury/accident sustained, in addition to the location of the accident in relation to the hill-walk. The final portion of the questionnaire contained questions to establish the typical fluid and nutritional intakes used by walkers. Calculation of energy intakes entailed a nutritional data analysis system and standard food tables (McCance et al., 1991).

Questionnaire distribution: The questionnaire was distributed by mail to more than 500 members of two hill-walking clubs in the UK, the Wayfarers and the Rucsac Clubs. Altogether 114 hill-walkers responded, from which 14 were discarded due to failure in providing adequate responses to a number of the questions posed.

Statistical Analysis: Descriptive statistics were initially carried out for a general analysis. Frequencies and chi-squared analysis served this purpose. The final statistical procedure was either a Spearman's rank order correlation coefficient, when the data did not meet with parametric assumptions, or a Pearson's correlation coefficient for the analysis of parametric data.

Results

Physical characteristics: In total, 91 male and 9 female subjects completed the whole questionnaire. In comparing the male and female hill-walkers, not many variables were significantly different and thus the data were combined for most of the analysis. The age distribution of the subjects were 44% over 60 years; 24% between 50 - 59 years; 20% between 40 – 49 years; and 12% under 40 years of age. The physical characteristics of the subjects were body mass (range) 72.8 (47.1 - 102.1) kg, height 1.76 (1.61 - 1.9) m.

Hill-waking activity details: When asked about their experience, 75% of the subjects had over 25 years of active hill-walking participation, indicating the experience of the group. Subjects spent 57 (10 - 200) days per year participating in hill walking, of which 67% was above 600 m in altitude. The average distance and duration of the walks are shown in Fig 1. On average, 90% of the subjects reported carrying a backpack weight of 14 kg or less.

Incidence of injury / accidents: The incidence of injury / accidents during hill walks was 35%. The incidence (% of subjects reporting the injury at least once) of many specific commonly occurring injuries was recorded in the questionnaire. In total 35 subjects reported injuries. There were 81 reported incidents of which 67% occurred towards the end of the walk, 25% near the middle and 8% near the beginning of the walk. There was a mix of reported weather conditions with injuries occurring as frequently in summer as in winter walking conditions. The majority of the injuries (83%) occurred during descent.

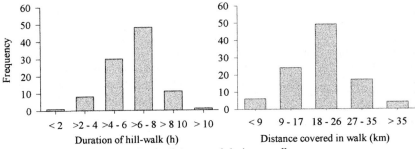

Fig. 1. Duration and distance 'typically' covered during a walk.

Typical fluid and nutritional habits: This information on fluid and nutritional intakes was divided into those used prior to the walk, during the walk and those after the walk (Table 1). Generally, the total energy intake is within guidelines for the population mean intake with respect to the age of subjects (Ralph, 2000). In total, 96% of subjects had consumed less than 1 litre prior to going out for a hill-walk, 89% and 78% reported having between 0.25 - 1.5 litres during and after the walk, respectively.

Table 1. Average energy and macronutrient intake on a hill-walking day. (mean ± SD)

Stage of walk	Energy (kJ)	Protein (g)	Fat (g)	CHO (g)
Pre-walk	2018 ± 908	16 ± 9	16 ± 10	73 ± 34
During walk	3566 ± 1402	26 ± 12	34 ± 18	118 ± 51
Post-walk	4943 ± 1482	49 ± 36	49 ± 21	142 ± 72

Correlation Analysis: The only physical attribute to correlate with any other variable was age of the subjects. Age was positively related to years of hill walking experience ($r = 0.44$, $P < 0.01$), reasons for participation, injuries sustained ($r = 0.49$, $P < 0.01$) and to the energy intake prior to walking ($r = 0.44$, $P < 0.05$). Negative correlations were found between age and the distance of walk ($r = -0.36$, $P < 0.01$), backpack weight ($r = -0.26$, $P < 0.01$) and fluid intake during the walk ($r = -0.41$, $P < 0.01$). Positive correlations were found to exist between the duration of the walk and its intensity ($r = 0.25$, $P < 0.05$), distance ($r = 0.56$, $P < 0.01$) and backpack weight carried ($r = 0.36$, $P < 0.05$).

Injuries: The analysis of data on injuries indicated positive correlations between injuries and age ($r = 0.49$, $P < 0.05$) and walking at altitudes above 600 m ($r = 0.37$, $P < 0.05$). Negative relationships were found between injuries and the frequency of hill walks ($r = -0.41$, $P < 0.05$) and of the energy intake during the walk ($r = -0.26$, $P < 0.05$). In other words there was a greater incidence of injuries occurring in the older subjects, when walking above 600 m, in the less frequent walkers, and in walkers consuming the lower energy intakes during the walk.

Discussion

The quantification of characteristic patterns of recreational hill-walkers indicated a typical distance covered of 18 – 26 km over 6 – 8 hours in duration and predominately over 600 m. Secondly, the data for injuries highlight the prevalence of knee and ankle damage, generally occurring near the end of the walk and predominately when moving downhill. Additionally, the greatest incidence of injuries occurred in older subjects, in the less frequent walkers, walking above 600 m, and in walkers consuming a low energy intake during the walk. Finally, the quantification of typical energy intakes used for such hill-walking events shows that, on average, they are not higher than the normal reference energy intakes for the different age groups.

Injuries: The high occurrence of injuries, predominantly during the downhill section is an important consideration. It is thought that injuries and pain which occur during downhill walking are caused primarily by high loads on the joints of the lower extremities (Kluster et al., 1994, 1995). These injuries are more likely to be increased by carrying an additional weight in the form of a backpack (Jacobson et al., 1997). Furthermore, the carrying of a backpack, clearly a necessity in the mountainous environment, decreases both lateral stability and balance (Jacobson et al., 1997).

An even higher incidence of injury was reported for the ankle than for the knee, including fractures and tendon damage. Excessive up-hill or downhill walking or running places increased stress upon the achilles tendon and on the ankle joint (Creagh et al., 1998; Grampp et al., 2000) . When traversing difficult, uneven terrain where the foot constantly inverts and everts is liable to risk not only a sprained ankle, but also potential micro-trauma to the achilles tendon (Mortensen, 1999). Additionally, hill-walking normally requires the use of shoes which are extremely stiff at the ball of the foot and does not allow the shoe to flex adequately, increasing the stress on the achilles tendon as the calf muscles contract more forcefully to lever the entire foot and heel off the ground.

One potential aid to combat the stress to the lower extremity joints during up-hill and downhill walking is the use of walking poles. The use of two walking poles during down-hill walking increases the maintenance of static balance and lateral stability (Jacobson et al., 1997), reduces external and internal loads on the knee and hip joints (Schwameder et al., 1999) and allows improved breathing and efficiency owing to an upright posture during up-hill walking (Jacobson et al., 1997). Any reduction in the loads incurred during downhill walking and an improved efficiency during up-hill walking should benefit hill walkers, especially walkers with lower extremity injuries.

Fluid and nutrition intake: Our recent research into the energy cost of a 12-km hill walk demonstrated a high-energy expenditure of 14.5 MJ for the walk (recorded via continuous measurement of respiratory gas exchange by means of indirect calorimetry (Ainslie et al., 2001). In this study, food and fluid were allowed *ad-libitum*; nevertheless, subjects became dehydrated and had an average energy intake from both breakfast and food consumed throughout the walk of only 5.6 MJ, indicating a marked negative energy balance of some 7 MJ as a consequence of the walk. The estimated energy intake values in the present study provides additional evidence that hill-walkers are operating at a negative energy balance. Furthermore, the typical characteristics of hill-walking suggest that the walkers, on average, cover between 18 - 26 km. This distance could elicit higher energy expenditure than that of 14.5 MJ recorded over 12 km, depending on the severity of the terrain. There was a weak but significant negative relationship between energy intake and the incidence of an injury occurring during the hill walk. A lower energy intake may lead to an increase in fatigue which could be a factor in increasing an individual's susceptibility to injury. This safety consideration would over-rule the health-benefits of such exercise when used for weight-reduction purposes. In addition, the low fluid intakes both prior to and during the walks, and the reported lower intakes in the older walkers may further increase the potential for dehydration in the mountainous environment.

The major conclusions from the present study are firstly that there is a high prevalence of lower limb injuries sustained predominately during downhill walking, nearing the end of the walk. Secondly, the typical energy intakes during a walk are probably inadequate to balance the high energy turnover of such prolonged activity. The main practical recommendations generated from this questionnaire analysis are; 1) the use of walking poles will confer some addition protection against lower limb injuries during downhill walking and may increase efficiency during up-hill sections, 2) walkers should take more foods with higher energy content to help increase energy intake and provide a measure of protection if the walk becomes unexpectedly prolonged.

This work was supported by Mars Incorporated.

References:

Ainslie PN, Campbell IT, Frayn KN, Humphreys SM, MacLaren DPM, and Reilly T. Physiological aspects of hill-walking. *Contemporary Ergonomics, Taylor and Francis,* London. pp 8 - 14, 2001.

Creagh U, Reilly T, Lees A. Kinematics of running "off-road" terrain. *Ergonomics* 41: 1029-1033, 1998.

Grampp J, Willson J, and Kernozek T. The plantar loading variations to uphill and downhill gradients during treadmill walking. *Foot Ankle Int* 21: 227-31, 2000.

Jocaobson BH, Caldwell B, and Kulling FA. Comparison of hiking stick use on lateral stability while balancing with and without a load. *Percept Motor Skills* 85: 347-50, 1997.

Kuster CA, Wood GA, Sakurai S, and Blatter G. Downhill walking: A stressful task for the anterior cruciate ligament? A biomechanical study with clinical implications. *Knee Surg Sports Traum Arthros* 2: 2-7, 1994.

Kuster CA, Sakurai S, and Wood GA. Kinematic and kinetic comparisons of downhill and level walking. *Clin Biomech* 10: 79-84, 1995.

McCance RA, Widdowson EM, and Holland B. *McCance and Widdowson's the Composition of Foods.* (5th Ed) Springer Verlag, Cambridge. 1991.

Mortensen C. Our achilles heel. *Wilderness Med (Newsletter)* 6, 1-5, 1999.

Ralph, A. Dietary reference values. *In Human Nutrition and Dietetics (10th Edition),* edited by J.S. Garrow, W.P.T. James, and A. Ralph. Harcourt Publishers Limited, Edinburgh. pp 849-861, 2000.

Schwameder H, Roithner R, Muller E, Niessen W and Raschner C. Knee joint forces during downhill walking with hiking poles. *J Sports Sci* 17: 969-978, 1999.

EFFECT OF AGE UPON HYDRATION STATUS AND PSYCHOMOTOR PERFORMANCE DURING 10-DAYS OF HIGH-INTENSITY HILL WALKING

P.N. Ainslie[1], I.T. Campbell[2], D.P.M. MacLaren[1], G. Mooney[1], T. Reilly[1] and K.R. Westerterp[3]

[1]Research Institute for Sport and Exercise Sciences, Liverpool John Moores University
[2]University Department of Anaesthesia, Withington Hospital, Manchester
[3]Department of Human Biology, Maastricht University, The Netherlands

We aimed to quantify some relevant responses that are important in the safety of hill-walkers, such as dehydration and impaired psychomotor performance, and also the possible effect that age may have on these responses. Seventeen male subjects were divided into two groups according to their age. Group 1 [younger; age mean $\pm$ SD 24 $\pm$ 3 years (n = 9)] and group 2 [older; 56 $\pm$ 3 years (n=8)]. Both groups completed 10 consecutive days of high-intensity hill-walking during the month of April in the Scottish highlands, varying between 10 – 35 km in distance and up to 1345 m in elevation above sea level. A range of gradients and terrain typical of a mountainous hill-walk was incorporated. In the morning, prior to walking, on days 1, 6 and 11, subjects provided a urine sample for the analysis of urine osmolality to assess hydration status. On these days, subjects also completed a battery of psychomotor performance tests which included choice reaction time (cognitive processing time), grip strength, flexibility and vertical jump (anaerobic power) tests. The older group demonstrated a marked increase in dehydration (381 $\pm$ 236 mosmol/kg) on day 11, relative to day 1; P <0.05) whereas the younger group remained hydrated throughout the 10 days. The progressive dehydration in the older group and the impairment of anaerobic power and cognitive functioning were highly related. Both the impaired psychomotor functioning and the marked dehydration in the older group may compromise decision-making and increase susceptibility to fatigue and injury in the mountains.

Introduction

Hill-walking is a popular leisure time activity in many of the world's developed countries. Activity tends to be light to moderate in intensity but prolonged in duration. Despite the popularity of hill-walking and the increasing problem of accidents in the mountainous environment, the ability of the safety organisations to design educational material concerning this hazard is hindered by lack of knowledge of the physiological and psychomotor responses to such events, often pursued over consecutive days.

Our recent research into the energy cost of a 12-km hill walk demonstrated a high energy expenditure of 14.5 MJ for the walk (recorded via continuous measurement of respiratory gas exchange by means of indirect calorimetry, Ainslie et al., 2002). In this study, food and fluid were allowed *ad-libitum*; nevertheless, subjects became dehydrated and subsequently lost, on average, 2 kg in body mass. Despite the high energetic cost of the walk, dehydration, and serious physiological stress, the subjects demonstrated unremarkable changes in psychomotor control during and after the walk.

Thermoregulatory and cardiovascular functions, as well as cognitive function are adversely influenced by body water deficits (Adolph, 1947; Ladell, 1955; Gopinathan et al., 1988; Sawka, 1992). For many complex tasks, both the mental decision-making and physiological functioning are closely related (Sawka, 1992). As a result, dehydration probably has more profound effects on real-life tasks than solely physiological performance measures. In hill-walking, dehydration may decrease thermoregulatory and cognitive functioning, which could impair decision making, leading to an increased susceptibility to fatigue and injury in the mountainous environment.

In this study the effect of age on the responses to hill walking was investigated. We aimed to quantify some relevant measures that are important in the safety of hill-walkers, such as the likelihood of dehydration and impaired psychomotor performance, and also the possible effect that age may have on these responses.

Method

Seventeen male subjects were divided into two groups according to their age - Group 1 [younger; age = mean ± SD 24 ± 3 years (n = 9)] and group 2 [older; 56 ± 3 years (n = 8)]. Both groups completed 10 consecutive days of high-intensity hill-walking during the month of April in the Scottish highlands. The daily walk varied between 10 – 35 km in distance and up to 1345 m in elevation above sea level. The total ascent and distance covered in the 10-day experiments were approximately 12000 m and 180 km, respectively. Intake of food and fluid was allowed *ab libitum*. The majority of the subjects were active and experienced hill-walkers.

Hydration: In the morning, prior to walking, on days 1, 6 and 11, subjects provided a urine sample for the analysis of urine osmolality in order to assess hydration status.
Urine osmolality was determined in triplicate by the use of the freezing point depression method (Advanced Micro-osmometer (model 3300), Vitech Scientific Ltd, West Sussex). Perception of thirst was assessed using a 100-mm visual analogue rating scale labelled from "not at all" to "extremely". Care was taken to ensure that both age groups interpreted the scales in a similar manner. Furthermore, measurements of nude body mass were made at the same time points.

Psychomotor performance: In the morning, prior to walking, on days 1, 6 and 11, the subjects also completed a battery of psychomotor performance tests. These included choice reaction time (cognitive processing time), grip strength, flexibility and vertical jump (anaerobic power) tests.

Statistical analysis: Data were initially tested for normality, before being analysed by repeated-measures analysis of variance (ANOVA) with age as a between-group factor. Post hoc tests were performed to isolate any significant differences. Student's paired t-tests ascertained differences between-conditions when a variable was measured once. A Pearson's correlation coefficient was used to establish any relationships between variables. Statistical significance was set at $P \leq 0.05$ for all statistical tests.

Results and Discussion

Hydration: The older group demonstrated a marked increase in dehydration on days 6 and 11, relative to day 1 ($P < 0.05$) whereas the younger group remained hydrated throughout the 10 days (Fig 1). Furthermore, the older group had lower perceptions of thirst compared with the younger group ($P < 0.05$, day 11). Potential reasons for this fluid deficit include high sweat losses, blunted thirst - especially in older people (Sawka and Montain, 2000), cold-induced diuresis, increased respiratory water losses, conscious under-drinking, and poor availability of water (Freund and Sawka, 1995; O'Brien et al., 1998). In agreement with the results from the present study, there is some evidence in the literature suggesting that older adults (>55 years) have a reduced thirst sensation, lowered ability to concentrate urine, and reduced potential to dissipate body heat (Kenny and Fowler, 1988; Mack et al., 1994; Kenny, 1995). The mechanisms for the reduction in thirst sensation, and hence potential for increased dehydration in older subjects are unclear. In spite of the increase in dehydration, body mass was relatively well maintained in both groups.

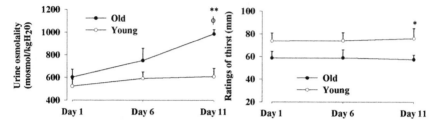

Fig 1. Change in urine osmolality and ratings of thirst. Symbol ϕ denotes significant differences ($P < 0.05$) from day 1 as a function of day in the older group. ** $P < 0.01$), * ($P < 0.05$) denotes significant between group differences.

Psychomotor responses: Table 1 presents the results for the psychomotor responses during the 10 days of walking. Generally, the younger group attained higher levels on all the psychomotor measures when compared with the older group. This finding is in agreement with the normal decline in performance associated with age (Astrand and Rodahl, 1986). Both groups showed a marked slowing of cognitive functioning after the 10 days of walking. Grip strength on day 11 remained unchanged in both groups. Flexibility did not change in the older group but showed a progressive increase in the younger group, whereas the vertical jump performance showed a progressive decrease in the older group whilst being maintained in the younger group.

Blood flow to active skeletal muscle may be an important factor associated with ageing (Martin et al., 1991; Jasperse et al., 1994). In agreement with the present study, blood flow to active muscles during and after exercise (using a small-muscle mass, such as hand-gripping), is well preserved in healthy older humans (Jasperse et al., 1994). However, during exercise utilizing a large mass of muscle, healthy older subjects had a lower whole-leg blood flow and vascular conductance compared with younger subjects during and after exercise (Wahren et al., 1974; Proctor et al., 1998). This decrease in blood flow to a large muscle mass may be one explanation for why the older subjects became compromised in their ability to perform a test such as the vertical jump which requires a large muscle mass.

Table 1. Psychomotor performance during the 10 days of high intensity walking

	Day 1		Day 6		Day 11	
	Old	Young	Old	Young	Old	Young
Grip strength	41 ± 3	52 ± 3*	42 ± 4	49 ± 2*$^{\theta}$	43 ± 3	49 ± 4
Flexibility (cm)	14 ± 6	22 ± 3*	14 ± 4	25 ± 3**$^{\theta}$	15 ± 7	27 ± 4*$^{\theta}$
Vertical jump (cm)	34 ± 4	39 ± 3*	30 ± 3	42 ± 2***	$27 \pm 2^{\phi\ \phi}$	38 ± 3**
Reaction time (ms)	717 ± 30	531 ± 29*			$793 \pm 42^{\phi}$	583 ± 47**$^{\theta}$

Values are mean ± SD based on 7 subjects in the older group and 9 in the younger group. Symbols $^{\phi}$ and $^{\theta}$ denote significant differences ($P < 0.05$) from day 1 as a function of day in the older and younger groups, respectively. *** ($P < 0.001$), ** $P < 0.01$), * ($P < 0.05$) denotes significant between group differences.

The impact of the dehydration incurred becomes apparent when the psychomotor tests are considered. In the older group, there were significant relationships between the increase in urine osmolality from day 1 to day 11 (i.e., progressive increase in dehydration) and both the slowing in reaction time and the deceased vertical jump performance (Fig 2.). No relationships existed in the younger group.

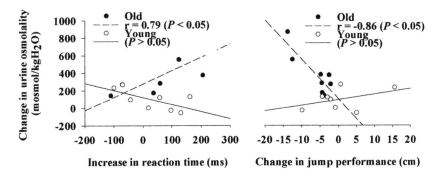

Fig 2. Relationship between change in hydration status and performance

The finding that the older group had both higher levels of dehydration and impairment in cognitive functioning and jump performance tests is an important consideration. The decrease in cognitive functioning may impair decision-making abilities (e.g., leadership and navigational decisions) and the poorer ability to employ a large muscle mass may further lead to an increased incidence of injury in the mountainous environment. Furthermore, both increasing age and dehydration lead to a decrease in thermoregulatory and cardiovascular functioning (Swaka, 1992; Swaka and Mountain, 2000). Hill-walkers can be caught unexpectedly and unprepared when rain and wind accompany outdoor activities in cool weather (Ainslie et al., 2002). Decreased thermal insulation of wet clothing presents a serious challenge to body temperature regulation, which can be compounded by fatigue associated with prolonged exercise such as hill-walking (Pugh,

1966, 1967; Thompson and Hayward, 1996). The present results suggest that the challenge to normal body temperature regulation may be increased in older participants. Taking observations collectively, due to the marked dehydration and impairment of psychomotor performance, the older age walkers may be more susceptible to fatigue, and to injury. Consequently, in adverse weather conditions the risk of potential hypothermia in the mountainous environment must be considered.

Conclusions

The impaired psychomotor functioning and the marked dehydration in the older group could compromise decision-making and increase susceptibility to fatigue and injury in the mountains. Further work and recommendations to both participants and the rescue services are clearly warranted.

The study was supported by Mars Incorporated.

References

Adolph EL. *Physiology of Man in the Desert*. New York: Interscience Publishers, Inc., 1947.

Ainslie PN, Campbell IT, Frayn KN, Humphreys SM, MacLaren DPM, and Reilly T. Physiological and metabolic aspects of a hill-walk. *J Appl Physiol, In Press, 2002*.

Åstrand P-O, and Rodahl K *Textbook of Work Physiology; Physiological Bases of Exercise*, 3rd Ed. Singapore: McGraw-Hill, 1986, pp 332-337.

Gopinathan PM, Pichan G, and Sharma VM. Role of dehydration in heat stress-induced variations in mental performance. *Arch Environ Health* 43: 15-17, 1988.

Jasperse JL, Seals, DR, and Callister R. Active forearm blood flow adjustments to handgrip exercise in younger and older healthy men. *J Physiol (Lond)* 474: 353-360, 1994.

Ladell WSS. The effects of water and salt intake upon the performance of men working in hot and humid environments. *J Physiol (Lond)* 84: 410-433, 1955.

Procter DN, Shen PH, Dietz NM, Eickhoff TJ, Lawler LA, Ebersold EJ, Loeffler DL, and Joyner MJ. Reduced leg blood flow during dynamic exercise in older endurance-trained men. *J Appl Physiol* 85: 68-75, 1998.

Pugh LGCE. Clothing insulation and accidental hypothermia in youth. *Nature (Lond)* 209: 1281-1286, 1966.

Pugh LGCE. Cold stress and muscular exercise, with special reference to accidental hypothermia. *Br Med J* 2: 333-337, 1967.

Martin WH, Ogama T, Kohrt WM, Malley E, Korte, PS, Kieffer PS, Schechtman KB. Effects of aging, gender, and physical training on peripheral vascular function. *Circulation* 84: 654-664, 1991

Sawka MN. Physiological consequences of hypohydration: exercise performance and thermoregulation. *Med Sci Sports Exerc* 24: 657-670, 1992.

Sawka MN, and Montain SJ Fluid and electrolyte supplementation for exercise heat stress. *Am J Clin Nutr* 72: 564S-72, 2000.

Thompson RL, and Hayward JS. Wet-cold exposure and hypothermia: thermal and metabolic responses to prolonged exercise in the rain. *J Appl Physiol* 81: 1128-1137, 1996.

Wahren J, Saltin B, Jorfeldt L, and Pernow B. influence of age on the local circulatory adaptation to leg exercise. *Scand J Clin Lab Invest* 33: 79-86, 1974.

EMG MEASUREMENT OF TWO SPINAL MUSCLES WHEN LOAD CARRYING WITH DIFFERENT SHOULDER STRAPS

Andrew J Walker, Robert H Campbell & Frank Bell

Department of Physiotherapy, Podiatry & Radiography,
Glasgow Caledonian University, Glasgow, G4 0BA

The effects of carrying a golf bag and the type of shoulder strap used (one shoulder strap or double shoulder strap) on the required muscle effort was quantified using electromyography (EMG). Ten healthy male subjects were tested on a treadmill using both types of strap. Muscular effort in terms of EMG activity was obtained from both left and right Latissimus Dorsi and Trapezius muscles. The results indicated that physiological differentiation did occur in the muscle groups studied during load carrying with the two different strap configurations. The single shoulder strap created high and unequal increases in muscle effort in both the loaded and non-loaded sides, while the double strap produced small but symmetrical increases in muscle effort. The study concludes that the double strap promotes symmetrical spinal posture.

Introduction

In golf many players carry their golf clubs in a bag with shoulder straps. Research has indicated that excessive levels of axial loading on the spine can cause pain and disharmony in the musculature of the back and shoulders (Holewijn, 1990).

Golf is associated with a significant number of injuries (Hosea, 1997), many of which are considered to be due to poor swing biomechanics, resulting in soft tissue injuries to the back, wrist, elbow and shoulder (McCarrol, 1996). Despite much research on swing biomechanics and related injuries (Hosea and Gatt, 1996, Metz 1999) there has been limited documented research carried out on other golfing injuries. However Batt (1993) stated the need for research on the asymmetrical loading of the spine when carrying a golf bag on one shoulder.

Load carrying creates a culmination of physiological responses to the spine and its surrounding musculature (Martin and Nelson 1986). Previous research on load carrying (Bobet and Norman, 1984; Broome and Basmajian, 1971) shows that Latissimus Dorsi and Trapezius muscles exhibit good measurable indicators of the posture of the loaded spine. Latissimus Dorsi is active during shoulder extension, adduction and internal rotation, and during depression and refraction of the shoulder girdle (Broome & Basmajian, 1971). The upper Trapezius muscle is sensitive to changes in the condition of load carriage (Bobet & Norman, 1984).

EMG of the spinal muscles has been studied since the early 1950's. It is an indirect measure of contractile activity of muscle, and has frequently been incorporated in biomechanical models to provide predictive capability (Kumar, 1998). Latissimus Dorsi and upper Trapezius could therefore be assumed to meet the criteria for a linear EMG / tension relationship during load carriage, their electrical activity can be readily isolated from that of neighbouring muscles, and their level of contraction is submaximal (Bobet and Norman, 1984).

The purpose of this study was to investigate the effects of two different types of shoulder strap configurations on selected back muscles when carrying a standard set of golf clubs. The strap configurations used were; a single strap worn over one shoulder; and double straps worn over both shoulders. The same bag was carried by each of the subjects, using both configurations. The effects of load carrying and the strap type were investigated by simultaneous electro-myogram (EMG) measurement of activity in, both left and right Latissimus Dorsi and Trapezius muscles.

Method

10 healthy male subjects who all had some experience in carrying golf bags, participated in the study. Their ages ranged from 17 to 26 years and all were within normal limits for height and weight. Two types of golf bag strap were used in this study. The single strap configuration allows the use of only one shoulder. With this type of strap the bag swings freely at approximately waist height, frequently alternating between the posterior coronal and sagittal planes. The double shoulder strap, allows the use of both shoulders simultaneously. The bag is held on the posterior coronal plane, in the hollow of the lumbar spine and is held in a secure position.   The bag used in the study was a Taylor Made[©] golf bag, which contained 10 golf clubs. The total weight of the bag was 7.2 kg. The strap configurations were adjusted to conform to each subject's size.

A Mega Me3000p8 (Mega, Electronics Ltd. Finland) EMG system was used to record the muscle effort of the Latissimus Dorsi and the Trapezius, while the subjects carried out 3 walking tasks. The EMG of the Latissimus Dorsi was recorded using a pair of bipolar surface electrodes, positioned as suggested by Paton and Brown (1995). The EMG of the Trapezius muscle was recorded by placing the electrodes as suggested by Bobet and Norman (1984). Data was collected bi-laterally to allow symmetrical analysis of muscle usage. Prior to collection of the EMG data for the 3 load carrying activities, each subject was asked to perform 3 maximum voluntary isometric contractions (MVC), for each muscle pairing (Yang and Winter 1983). The calculated MVC values allow the EMG recordings taken over the 3 walking tasks to be presented as a percentage of the subjects own unique maximum possible muscle contraction (Perry, 1992).

After the MVC's were recorded, the subjects performed 3 different walking tasks on a treadmill. The subjects walked, at a personally set comfortable walking velocity, firstly with no load, secondly with the double strap golf bag and finally with the single strap golf bag. The data was then downloaded for analysis.

Data Analysis

To calculate the final MVC value for each muscle, a 10 second span of each electromyogram graphs was selected. This contained the raw data for the maximum contraction technique and was integrated and averaged to produce 3 voluntary contraction values for each muscle. These values were then averaged, to produce a final specific MVC figure per muscle, for each subject. After the MVC values were calculated, the results for each of the 3 tasks were analysed separately. Each 8-minute walking period produced 4 electromyogram graphs that corresponded to the four muscles tested. All 4 graphs had three 15-second sections of data selected at 4, 6 and 8 minutes. The initial 4 minutes of walking were disregarded to allow subject/treadmill/golf bag familiarisation as reported by Taylor et al (1996), who concluded that movements of the lumber spine and pelvis were constant after 4 minutes of treadmill walking by normal subjects. The first 15-second section (after 4, 6 and 8 minutes) of each muscle electromyogram was then analysed. Each muscles raw data was integrated and averaged to produce a muscle effort value, related to that time. Each muscle tested therefore had 3 sets of values, which were again averaged to produce a final muscle effort value for each muscle tested.

Each walking task therefore, produced a value for the 4 observed muscles, which was calculated as a percentage of their previously calculated MVC value. This procedure was carried out for each of the 10 subjects results under the 3 test conditions, producing a table of % muscle effort for each subject. The complete results of each subject were then manipulated in the statistical software packages; Microsoft Excel v5.0 and Minitab v5.0. An analysis of variance ($p < 0.5$), 2-sample t-test was used to establish if the % effort of each muscle varied significantly under the two different strap configurations, when compared to the unloaded walking results. Taking into account the results of the ANOVA's for each of the carrying conditions, the percentage efforts for each muscle were ranked, corresponding to their contribution under each condition. When a muscle was ranked *1* it was said to have given a *very high* contribution. When a muscle was ranked *2* it was said to have given a *high* contribution to the entire activity, and a ranking of *3* equated to a *moderate* contribution. Finally, if a muscle was ranked *4* for a contribution it was said to have given a *low* contribution to the entire activity. This method of analysis corresponded to previous research by Paton and Brown (1995).

Results and Discussion

The aim of this study was to compare the muscle effort increase of the Latissimus Dorsi and the Trapezius muscles when using two different types of shoulder strap used to carry a golf bag. Examination of the results indicated that there was evidence of a significant difference in the EMG results of the Latissimus Dorsi and Trapezius, under the different carrying conditions. As previously stated the relative contributions of the muscles studied were ranked into *low, moderate, high* or *very high* and are illustrated in figure 1.

Analysis of the results of the single strap configuration showed that all subjects used a higher % muscular effort in the 4 muscles tested in order to maintain a balanced posture. The increased effort in the right Trapezius (ranked as very high) can be explained by the mechanical effects of asymmetrical loading resulting in a significant increase in Trapezius muscle activity, which is required to maintain the shoulder on a horizontal plane. (Bobert and Norman, 1984, Anderson et al, 1977). Consequently, since the right shoulder is taking the majority of the load, the left Trapezius and left Latissimus Dorsi require to increase effort, in an attempt to counteract the asymmetrical load on the spine. The increase in muscle effort by the left Trapezius was ranked as high and the left Latissimus Dorsi was also ranked as high. An unexpected increase was seen in the right Latissimus Dorsi where the effort was ranked as moderate. This may be due to the muscle being compressed from above by the weight of the bag and also working harder to counteract the forward hunching of the shoulder, while reacting to the golf bags random swaying during the gait cycle.

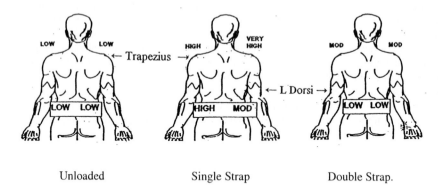

Unloaded   Single Strap   Double Strap.

Figure 1.
Relative contributions of Latissimus Dorsi and Trapezius for three load conditions.

The results of the 10 subjects for the double strap carrying method supported Tanner's (1996) theory of symmetrical loading. 70% of the subjects displayed a slight increase (Mean ≅ 2%) in required muscular effort over all 4 muscles, when using the double strap configuration and the remaining 30% of the subjects' results were equal to or slightly less than the normal (unloaded) results. This can be explained by the weight of the bag being spread across the two shoulders maintaining the even distribution of forces across the back and shoulder muscles. Although the double strap configuration still required an increase in muscle effort the increase is evenly transmitted across all 4 muscles. Thus, the weight of the bag spread over both Trapezius and Latissimus Dorsi muscles reduces the potential for high asymmetrical forces acting on the spine.

In single strap configuration the load carried was observed to be less stable and exhibit a tendency to sway considerably as the subject walked. This increased sway may be compensated for by increase in activity of Trapezius and Latissimus Dorsi in an attempt to maintain spinal posture. Whereas, the 2-strap configuration appeared more secure and less sway was observed since the golf bag was held tightly against the back in the natural curvature of the lumbar spine. Therefore, due to the reduction in movement of the bag, fewer forces may be transmitted to the shoulders and spine, causing the reduction in % muscle effort. This view is supported by Winsmann and Goldmann (1976) who reported on two backpack harness designs and the consequent differences in weight distribution. They observed that a backpack, which was held close to the body and lower down the back, spread the weight of the backpack more evenly across the shoulders.

Conclusion

The results of this study have indicated that physiological differentiation did occur within the Latissimus Dorsi and Trapezius muscles during the carrying of a golf bag with two different types of shoulder strap. The single strap configuration created unequal muscle effort over the spinal musculature studied, caused by either the direct weight of the bag forcing muscles to increase their effort, or a counteraction effect on the opposite side to the load, which may have been required to maintain posture. The double strap configuration however, created a symmetrical effort in the muscles measured. This is probably due to the weight of the bag being distributed evenly by the double strap, which produced small but equal increase in effort by Latissimus Dorsi and Trapezius. A decrease in asymmetrical muscular effort may reduce the risk of back and spinal injury.

It can therefore be concluded that the double strap device should be the preferred choice for golfers; in a bid to reduce asymmetrical loading and possible resultant back pathologies (Tanner, 1996).

References

Anderson, G.B.J., Ortengren, R. and Herberts, P.N. 1977, Quantitative EMG studies on back muscle activity related to posture and loading, *Orthop. Clin. N. Am.,* **8**, 85 – 95

Batt, M., 1993, Golfing injuries, an overview, *Sports Medicine,* **16**, 64 – 6.

Bobet, J. and Norman, R.W., 1984, Effect of load placement on back muscle activity in load carriage. *Eur. J. Appl. Physiol.,* **53**, 71 – 75

Broome, H.L. and Basmajian J.V. and (1971), Survival of the ilio-psoas muscle after Sharrard procedure. *Am. J. Phys. Med.,* **50**, 301 – 302.

Holewijn, M. 1990, Physiological strain due to load carrying. *Eur. J. Appl. Physiol.,* **16**, 237 – 245

Hosea, T.M., 1997, Golfing Injuries, The American Orthopaedic Society for Sports Medicine. {Online] http://www.sportsmed.org/d/answers. (Accessed, 28/02/2000).

Hosea, T.M. and Gatt, C.J. 1996, Back pain in golf. *Clin. Sports. Med.,* **15**, 1 – 7

Kumar, S., 1998, EMG of spinal muscles, *J.EMG and Kinesiology,* **8**(4), 195 – 196

Martin, P.E. and Nelson, R.C., 1986, The effects of carried loads on walking patterns of men and women. *Ergonomics,* **29**, 1191 - 1202

McCarrol, J.R. 1996, The frequency of golf injures, *Clin. Sports. Med.* **15**(1), 1 – 7

Metz, J.P., 1999, Managing golf injuries, *Physic. Sports. Med.,* **27**(7), 1 – 9

Paton, M.E. and Brown, J.M.M., 1995, Functional differentiation within Latissismus Dorsi. *Electomyograph. Clin. Neurophysiol.,* **35**, 301 – 309

Perry, J., 1992, *Gait Analysis: Normal and Pathological Function.* Slack, Thorofare, NJ.

Tanner, J., 1996, *Beating Back Pain,* Darling Kindersley, London

Taylor, N.F., Evans, O.M. and Goldie, P.A., 1996, Angular movements of the lumbar spine and pelvis can be reliable measured after 4 minutes of treadmill walking. *Clin. Biomechs.,* **11**(8), 484 – 490.

Winsmann, F.R. and Goldman, R.F., 1976, Methods for evaluation of load carriage systems. *Precept. Mot. Skills.,* **43**, 1211 - 1218

Yang, J.F. and Winter, D.A., 1983, EMG reliability in maximal and submaximal isometric contractions, *Arch. Phys. Med. Rehabil.,* **64**, 417 – 420.

RAIL

ALLOCATION OF ATTENTION AMONG TRAIN DRIVERS

Natasha Merat[1], Ann Mills[2], Mark Bradshaw[1], John Everatt[1] and John Groeger[1]

[1]*Department of Psychology, University of Surrey, Guildford, GU2 7XH, U.K.*
[2]*Railway Safety, Evergreen House, London, NW1 2DX, UK.*

This paper reports the results of a study initiated by Railway Safety to study train drivers' eye-movements during a familiar train journey. Results illustrate drivers' distribution of attention in a dynamic visual field, and indicate that fixation time represents some 63% of the final 15-20 seconds of an approach to a signal. Drivers' allocation of visual attention is shown to be profoundly influenced by the aspect of the approaching signal and the aspect of the signal just passed. It is anticipated that the results of this initial study and further work will provide an understanding of the factors that are important to signal sighting, and future signal design.

Introduction

A failure to effectively attend to railway signals, and problems associated with signal visibility have been responsible for a relatively high number of signals passed at danger (SPADs) within the British rail industry - 568 between April 1999 and March 2000. Indeed, according to a recent report prepared for the public inquiry into the Southall train accident of 1997, the need to divide attention between a number of objects in a dynamic scene can lead to confusions in signal detection amongst drivers. Consequently, in order to attract drivers' attention to the signal at the appropriate time, the use of auditory alerts such as the automatic warning system (AWS) are recommended by this report.

While such recommendations are doubtless beneficial, they are not necessarily supported by empirical findings, and indeed until now, a systematic investigation of train drivers' visual behaviour within a dynamic scene has not actually been accomplished. In contrast, over thirty years of eye-movement recording amongst car drivers have identified a number of visual search characteristics. For example, early studies suggest that novice drivers' eye-movements are concentrated on an area a little ahead of the bonnet of their car, while experienced drivers look further ahead (Mourant & Rockwell, 1972; but see Underwood, 1998). However, in many respects, train driving is a substantially different task to car driving. For instance, signal installation standards require that signals are visible for at least seven seconds before they are reached, while they must not be obscured by overhead cabling or other railway paraphernalia during the final four seconds of this approach. These requirements apply to signals irrespective of the number of signals visible on approach, their complexity, or the speed at which they are approached. For those of us more used to considering car drivers, a reaction time of seven, or even four seconds would seem more than adequate to allow a driver to bring his vehicle to a stop. However, even if train drivers respond to a red signal as soon as it becomes visible, the weight of trains and the speeds at which they travel would prevent them from stopping the train in adequate time. Because of this, although each signal must be responded to appropriately, railway signals also serve to indicate what the state of the next signal is likely to be (see Table 1). In addition, automatic warning systems, located on the tracks at approximately 180 metres before the signal, are intended to attract drivers' attention to the oncoming signal. Accordingly, cautionary signals (double yellow, yellow and red) are preceded by an AWS horn, and a failure to cancel this warning by drivers results in the automatic application of the train's brakes. The approach to green signals is marked by an AWS bell, and requires no response.

Table 1 - The four aspects of railway signals

SIGNAL	INSTRUCTIONS	SUBSEQUENT SIGNAL (most likely)
Green	Enter track at current speed	Green Double Yellow
Double Yellow	Proceed with caution at reduced speed	Double Yellow Single Yellow
Single Yellow	Proceed with caution and be prepared to stop	Red
Red	Do not enter next sector of track	Green Double Yellow Single Yellow Red

While signaling guidelines available to engineers within the rail industry are based on decades of knowledge and experience, it is equally important that these guidelines are supplemented by empirical studies. For instance, it is crucial to establish whether the distribution and/or capture of drivers' visual attention varies with respect to signal mountings, signal aspect, or as a result of auditory warnings. Here, we present preliminary data on some of these issues, by describing findings from a pilot field study which examined the allocation of train drivers' visual attention, using eye movement recordings.

While attention can be independent of the movement of the eyes (Posner, 1980) there is good reason to believe that they are highly interrelated in ordinary, real-world activities. For instance, in a complex activity such as driving, changes in visual inspection patterns are thought to be a strong indicator of the reallocation of spatial attention (Moray, 1990). To inspect a scene, the eyes progress in a series of saccades, and fixations, which occur when the eyes are stationary. For eye movements to be re-positioned to a new location or object, parafoveal or peripheral processes must be utilized, and a system of covert attention has been proposed for this purpose. There are several reasons to assume that this is the case. First, stimuli that attract attention, due to some salient feature or contrasting attribute (Theeuwes, 1993) also attract eye movements. The most likely explanation for such findings is that attention has been attracted to an object or location and has initiated an eye movement to that object/location. Similarly, the expectancy of an observer can determine relocations of attention which evoke corresponding movements of the eyes (see Henderson, 1993). Finally, increases in fixation duration are thought to be directly related to increased task processing load (Zingale & Kowler, 1987). Therefore, this natural link between eye-movements and the focus of visual attention provides a ready and objective means for studying how the allocation of attention relates to task performance, outside laboratory induced environments.

Method

The data reported are derived from eye position recordings of 10 experienced train drivers, who volunteered to participate in the study. The route, which was familiar to all drivers, comprised of a section of track between St. Albans and East Croydon, and involved substantial portions of low density rural, and higher density urban signaling. Drivers completed the south and north-bound stretch of the journey (each lasting about one hour), whilst wearing a custom designed face-mask, bearing a video camera to track head position, and a modified visor which continuously detected drivers' pupil position, using infra-red technology (ASL Model 5001). A subset of the data collected is reported below. Video taped sequences of the scene, which were superimposed by cross hairs designating driver eye position, were analysed on a frame by frame basis (i.e. in 40 ms steps). Coding was object rather than space based, and so on each frame, the particular object or feature in the centre of the crosshair was recorded. Subsequent analyses determined the number of fixations (i.e. > 3 frames or 120 msecs) and fixation duration for each object.

Results and Discussion

Of the signals approached, 250 showed a green aspect, 15 a double yellow, 52 a single yellow, and 29 a red aspect. Results revealed a reliable difference in the time taken to approach each signal aspect ($F(3,342)= 71.986$; $p<0.001$), with the lengthiest approach to red signals ($M = 27.45$). This mean value was found to be reliably longer than the approach time to all other aspects, which were an average of 16.30 seconds for green signals, 17.00 secs for double yellow and 18.90 seconds for single yellow signals. Therefore, drivers reduced speed slightly when approaching a double yellow signal, slowed substantially more when approaching a single yellow signal, and still more when approaching a red signal. In view of this difference in approach time, (i) total fixation time, (ii) fixation frequency and (iii) point of first fixation to signals and non-signal objects[1] were computed as proportions of the total approach length. The other variable of interest, average fixation duration, was not subject to bias because of approach length and thus was not transformed. Results are illustrated with respect to signal aspect.

---

[1] Non-signal objects include the track ahead and encompassing fields in rural surroundings, as well as platform signs and passengers in urban settings.

The proportion of the approach spent fixating signals was found to vary as a function of signal aspect ($F(3,342) = 3.180$, $p<0.05$), but the proportion of time fixating non-signal objects and locations did not ($F(3,342) = 2.086$, $p>0.1$). Drivers were found to spend a significantly less proportion of the approach time fixating green signals ($M = 0.16$) compared to cautionary signals ($M = 0.21$). In addition, when approaching green signals, non-signal objects or locations were, on average, fixated at a higher rate ($M = 0.90$) than when cautionary signals were approached ($M = 0.74$), see Figure 1. Signal aspect also governed the time of first fixation on signals ($F(3,342) = 7.284$; $p<0.001$), with later fixations on green signals ($M = 0.59$) than cautionary signals ($M = 0.77$), see Figure 2.

Results illustrate that the signal aspect approached by train drivers has a substantial impact on the allocation of their visual attention. Non-cautionary green signals are looked at for less time overall and first fixation to these signals occurs later in the approach. In addition, other objects and locations in the scene are looked at more often, while the signals themselves are inspected less frequently relative to cautionary signals. In each case, there is only a marginal difference between green aspects and double yellow, but a very reliable difference is manifest between green, single yellow and red aspects. However, fixation duration did not vary with signal aspect, lasting about half a second in each case.

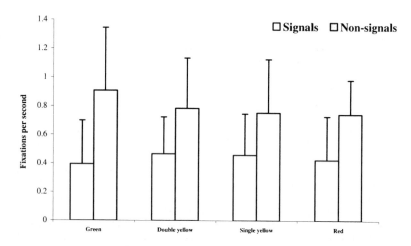

Figure 1 - Fixations per second on signals and other objects as a function of signal aspect approached

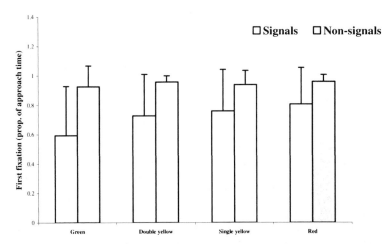

Figure 2 - Earliest look at signals and non-signals as a function of aspect approached

To an audience unfamiliar with the railway, some of these results may seem somewhat puzzling, in that the first time a green signal is fixated is much later than, say, a red signal. If a fixation to or near a signal is required in order to determine signal aspect, then how is it that this difference in early fixation arises? As outlined in the introduction, the state of an upcoming signal that cannot yet be seen is partly predicted by the aspect already passed, and for instance a red signal is almost inevitably preceded by passing a yellow aspect. Thus, in the next set of analyses, we examined the allocation of attention to an approaching signal, as a function of signal just passed.

The proportion of time fixating non-signal objects was found to vary as a function of the signal just passed ($F(3,342) = 3.66$; $p<0.01$). A lower proportion of the approach time was spent fixating on non-signal objects when drivers had just passed a yellow signal (0.37), compared to passing a green (0.49) or red signal (0.57). Analyses showed a clear strategic effect on allocation of attention in that having just passed a green signal, which is least associated with having to stop the train, drivers fixated subsequent signals later in the journey, and spent more time attending to objects other than signals. However, after passing a yellow or double yellow aspect, greater attention is dedicated to the subsequent signal. However, it is important to note that the aspect of the signal just passed simply influenced the strategy by which drivers distribute their attention towards the next signal, and was not an indication of first fixation time.

SUMMARY & CONCLUSIONS

The data reported above show that during the final seconds of an approach to a signal, train drivers fixate on some object about two-thirds of the time. Fixations on signals are substantially shorter than fixations on other objects, lasting less than half a second on average. In addition, the aspect of the signal just passed exercises considerable influence on how drivers organise and control their visual attention for upcoming signals. Passing a non-cautionary signal results in later attention to the next up-coming signal, regardless of its aspect. However, passing a cautionary signal enforces an earlier fixation on the subsequent signal. Once a signal is fixated, the drivers' distribution of attention appears to be determined by that signal aspect. Double yellow signals appear to be responded to more like non-cautionary signals, whereas single yellow or red signals cause drivers to attend more, and more often, to the signal than if it is green.

Although the current study cannot demonstrate this conclusively, these results imply that a driver who 'misreads[2]' an approaching signal as green, may attend to any aspect of the next signal later than appropriate. Misreading that subsequent signal, if a single yellow, could well lead to a SPAD, especially if the train is running at high speed. In addition to providing more precise information about how signal aspect influences drivers' viewing patterns, we regard the practical implications of later viewing of signals as highly important and hitherto unknown within this domain of application.

Acknowledgement

The authors wish to acknowledge the support of all at Railway Safety, Gary Davis, Fiona Bellaby and Andy Baker (Thameslink 2000), as well as Ian Duncan and Ian Price (Thameslink).

References

Chapman, P.R. & Underwood, G. (1998). Visual search of driving situations: Danger and experience. *Perception*, 27, 8, 951-964.

Henderson, J.M. (1993). Visual attention and saccadic eye-movements. In G. d'Ydewalle & J. van Rensbergen (Eds), *Perception and cognition, Advances in eye-movement research*. Amsterdam: North-Holland.

Moray, N. (1990). Designing for transportation safety in the light of perception, attention and mental models. *Ergonomics,* 33, 1201-1213.

Posner, M.I. (1980). Orienting of attention. *Quarterly Journal of Experimental Psychology,* 32, 3-25.

Theeuwes, J. (1993). Visual selective attention: A theoretical analysis. *Acta Psychologica,* 53, 93–154.

Underwood, G. (Ed), (1998). *Eye Guidance in Reading and Scene Perception.* Oxford: Elsevier.

Zingale, C.M. & Kowler, E. (1987). Planning sequences of saccades. *Vision Research,* 27, 1327-1341.

---

[2] The term 'misreads' includes (i) the unlikely event of a driver mistaking a cautionary signal for green (ii) the more likely event that sunlight or other objects renders the current aspect invisible, and (iii) the perhaps stronger still possibility that by switching between lines, the driver reads the wrong signal.

THE INFLUENCE OF BACKPLATE DESIGN ON RAILWAY SIGNAL CONSPICUITY

Guangyan Li[1], Stuart Rankin[1] and Clive Lovelock[2]

[1]*Human Engineering Limited*
Shore House, 68 Westbury Hill, Westbury-On-Trym, Bristol BS9 3AA, UK
[2]*Railtrack Plc Great Western Zone*
125 House, 1 Gloucester Street, Swindon SN1 1GW, UK

Experimental trials were conducted to investigate the effects of railway signal backplate design on driver responses, using an impoverished computer simulation. Measures were time taken/viewing distance to see the signal and then to identify the signal aspect. Subjective data were also collected regarding how easy/difficult it was to identify the approaching signal. The results of 43 professional train drivers showed that signal backplate size had a significant effect on driver responses tested. A signal with a backplate of at least 150% of the current standard size resulted in significantly faster response times (thus could be identified at a greater distance) when compared with the current operational signal. The majority of the 14 signal backplate shapes tested performed as well as the current standard design. The addition of a white backplate border reduced performance for signal recognition.

Introduction

In recent years, increasing public attention has been paid to railway safety. As a result, there has been an increase in research into the human factors issues associated with the sector. It is already well known that human error is a major factor associated with 'SPADs' (Signal Passed At Danger) (British Railway Board, 1995). However, there has been little research into how the physical design of railway signalling affects driver performance.

In the UK, 3-aspect colour light signals were first installed in 1923 and 4-aspect signals were then introduced in 1926 to allow trains to operate at closer headways (Hall, 1996). In order to increase the conspicuity of the railway signal, it has been suggested that single-headed fibre optic signals should be used in the future. As a result, an opportunity has arisen to re-design the signal backplate. The current standard signal backplate (with either 3 or 4-aspect signal heads) has a unique rectangular shape. It is not well understood whether a change in signal backplate size and shape, or the addition of a white border will affect signal conspicuity. Previous research in this area has largely focused on signals for road traffic where the signals are viewed over shorter distances. However, railway signals must be identified at a greater distance than those for road traffic to give the train drivers a minimum of 7 seconds to take the necessary control operation (Railtrack Group Standard GK/RT0037, 1997).

This paper describes a study by Human Engineering Limited that investigates the role of signal backplate design on driver performance. The objectives of the study were to test the effects of signal backplate size, shape, border and background conditions on driver responses, including the time taken to see the signal and then to read or identify the signal aspect, and the drivers subjective opinions on each signal design. Driver abilities to identify an offset between two signals at different viewing distances were also assessed. This study was designed to provide some preliminary baseline data to support a series of ongoing trials.

Experimental Studies

Design of the trials

The trials consisted of four tests for the effect of signal size (Test-1), shape (Test-2), border (Test-3) and offset (Test-4) on driver responses. There were six conditions in test 1, including the current standard 4-aspect signal backplate (600x1245mm) (GK/RT0031, 1996), 125%, 150%, 175% and 200% of the standard size and 'spotlight design' (600x480mm). Fourteen different shapes were tested in test 2 (see Li (2001) for detailed designs) and five types of border conditions were included in test 3, (i.e., no border, a white border of 5%, 10%, 20% of standard backplate width, and 100% white backplate). These border conditions were presented in three background lighting conditions, forming 15 conditions for Test 3.

In Test-4, two signals (spotlight design) were presented side by side, at an apparent lateral separation of 3.142m. Immediately after presentation, the two signals began to separate vertically (at 1cm/sec), until the subject could identify a difference between the signal height and press a key. The signal offset value was recorded at this point. This was repeated at equivalent viewing distances from 100m to 450m, with increments of 50m. All conditions in Tests 1 to 4 were randomly presented to the participants.

Participants

A total of 43 professional train drivers (36 male and 7 female) took part in the trials. Their average age was 39.3 years (range=23-63, SD=9.33), with driving experience ranging from 1 month to 41.5 years (average=12.0 years, SD=12.0).

Procedure

All trials were conducted on a desktop PC using a specifically developed programme. The signals were shown at full apparent size (based on visual angles), starting at an equivalent distance of 1200m and moving towards the subject at a speed of 60 mph. The letter keys on the keyboard were divided into three areas and were visually marked green, yellow and red respectively. The subject was required to press the 'space' bar as soon as he/she could see a signal (regardless of its colour) and then press a coloured key when he/she could identify the signal aspect. The computer automatically recorded the key presses and the time/distance data, while the subjective rating score was noted by the experimenter.

Of the four tests, Tests 1, 2 and 4 were repeated twice by each subject (the average results were analysed). The seven test sessions were randomised and balanced so that the two repeated tests were not completed consecutively. Each trial lasted approximately 1.5 hours, including a 10-15 minutes break in the middle. The room lighting was maintained at approximately 10 lux during the trials.

Results

Effect of backplate size on time taken to see and to read the signal

The results of 42 subjects were analysed using ANOVA. The results were highly significant for the effect of backplate size on time to see the signal [$F(5,41)=7.713$, $p\leq0.0000001$] and then to identify the signal aspect [$F(5,41)=188.905$, $p\leq0.000000001$]. Figure 1 illustrates the quantitative effects of varying signal backplate size on driver response (time taken to see and then to identify the signal aspect, starting distance 1200m), with driver opinions overlaid. The results indicated that, among the six types of signal sizes tested, the signal with 175% current operational size was the best one to read, and the Spotlight design and the current operational signal size resulted in slower responses and were considered more difficult to read.

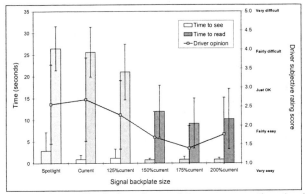

Figure 1. Effect of backplate size on driver response time (n=42)

Post hoc testing showed that there was a significant difference between Current Operational Size and Spotlight design with respect to their influence on 'time/distance to see', but there was no significant difference between the two for 'time/distance to read' the signal. The Spotlight design requires significantly more time both to see and to read than those with increased backplate sizes (125%-200% current). Statistically, there was no significant difference in response times for signals with 150%-200% current sizes.

Effect of signal backplate shape on driver responses

The ANOVA results of 41 subjects showed that the backplate shape significantly affected the time taken to see [$F(13,40)=2.970$, $p\leq0.001$] and to read the signal [$F(13,40)=6.634$, $p\leq0.0000001$]. Most shapes tested resulted in similar or better driver responses as compared with the current standard design.

Effects of backplate border size and background conditions on driver responses

The ANOVA results of 43 subjects showed that both the border size and background condition had highly significant effects on 'time/distance to read' the signal aspect ($p\leq0.0000001$); but these two factors did not have any significant effect on time/distance to see the signal. Figure 2 shows that as the border size increased, time needed to read the signal also increased, particularly against a dark background.

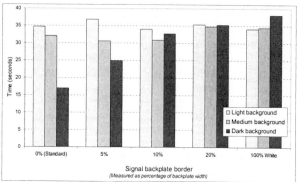

Figure 2. Effect of border size on time taken to identify the signal aspect (n=43)

Driver ability to identify a minimum signal offset at different distances

Regression analysis of signal offset vs. viewing distance with the data from 43 subjects showed an estimated linear relationship between the two variables: Y=66.8+0.22X, where: Y=signal offset value in cm; X=viewing distance in meters. [$F(1,342)$=56.637, p≤0.000001].

Discussion

This study found that signal backplate size had a highly significant effect on driver response in terms of time/distance taken to see a signal. Less predictable however, was the finding that increasing the backplate size also improved the signal conspicuity (Figure 1). Further analysis indicated that signals with backplate sizes of 150%, 175% and 200% current operational size were not significantly different from each other with respect to their influence on response times. It is thus expected that a signal with a backplate area of at least 150% current size will improve driver responses. This is possibly because the larger backplate increases the contrast between the signal and the background, an important factor for signal detection at distances greater than 100m (Janoff, 1994).

Signal backplate shapes were also found to significantly affect driver responses. Of the 14 shapes tested, most performed at least as well as the current standard design. The results suggest that signals with these different backplate shapes maintain or improve driver responses compared to the current standard. A literature review on traffic signs and signals revealed little existing information regarding driver responses to signal shapes (Robertson et al., 1999), although a variety of signal types/shapes have been used in modern railway signalling networks (Hall, 1996).

This study showed that a white border did not improve signal reading. In contrast, it took longer to read a signal aspect with a white border than for the current operational signal (Figure 2). This result contradicts the DERA study (Buxton, 2000), which showed that the white border and white signal backboard significantly improved conspicuity. However, the DERA study used a relatively short viewing distance (e.g. less than 100m) and only required signal, not aspect identification. Of the few studies that used border to improve salience of warning signs, they either showed no effect (Young, 1991) or a negative effect (Laughery and Young, 1991). The lack of consensus on this topic suggests the need for further study on this subject.

Driver abilities to identify a signal offset demonstrated a linear increase in required minimum offset value as viewing distance increased. Based on Railway Group Standard (GK/RT0037), train drivers should have an approach view of a signal for a minimum of 7s and an uninterrupted view for at least 4s. If a train travels at 60mph, for example, a signal should be recognisable from at least 190m. The study predicts that a minimum separation of 108.6cm is required at this distance so that drivers can distinguish an offset between the two signals.

Conclusion

This study indicated that driver response times can be improved if railway signal design includes a larger backplate (the size of which is at least 150% of the current operational size). Most signal backplate shapes tested in this study displayed similar or better performance as compared to the current standard design. The study suggests that these shapes can be used to improve line-signal association whilst still keeping similar signal readability for each line. A white border did not aid signal readability when viewed in the distance. A signal offset can be identified by drivers, depending on viewing distance. Therefore, it can also be used to improve signal-line association in complex railway sections.

Acknowledgements

This work was funded by Railtrack plc Great Western Zone. Thanks are due to Ian Finch at Railtrack for his assistance, and to the ergonomists at Human Engineering, especially Ian Hamilton, Karl Rich and Elizabeth Hoodless, for their technical support.

References

British Railway Board, 1995, *Red Alert*, January, Issue 1.

Buxton, A.C., 2000, Visual and Mental Acuity Study. Farnborough: Defence Evaluation and Research Agency (DERA), Report No: DERA/CHS/PPD/CR000323.

Hall, S., 1996, *Modern Signalling Handbook*. Shepperton, Surrey: Ian Allan Publishing.

Janoff, M.S., 1994, Traffic Signal Visibility: A synthesis of human factors and visual science literature with recommendations for required research. *Journal of Illuminating Engineering Society*, 23, 1, 76-88.

Laughery, K.R. and Young, S.L., 1991, Consumer product warnings: Design of pictorials, color, signal icon and border. In: *Proceedings of the 11th Congress of the International Ergonomics Association*. Toronto, CA: IEA.

Li, G., 2001, The influence of backplate design and offset on railway signal conspicuity. Technical Report (HEL/RGWZ/01543/RT2) to Great Western Zone, Railtrack plc.

Robertson, S.A., Ward, H.A. and Evans, A.W., 1999, Report to Railtrack: Review of Traffic Signs and Signals Literature in the Context of Drivers' Responses. Centre for Transport Studies, University of London.

Young, S.L., 1991, Increasing the noticeability of warnings: Effects of pictorials, color, signal icon and border. In: *Proceedings of the Human Factors Society 35th Annual Meeting*, Santa Monica, CA: Human Factors Society, pp.249-253.

THE DEVELOPMENT OF A HAND HELD TERMINAL FOR USE IN POSSESSION MANAGEMENT BY RAILWAY TRACKSIDE WORKERS

Karl J.N.C. Rich[1], Andrew M. Sutherland[1], Ben O'Flanagan[1] and Ian Lewis[2]

[1]*Human Engineering Limited*
Shore House, 68 Westbury Hill, Westbury-On-Trym, Bristol BS9 3AA, UK
E-mail: Karl@humaneng.co.uk
[2]*Bombardier Transportation Signalling*
Estover Close, Estover, Plymouth, Devon PL6 7PU, UK
E-mail: Ian.lewis@uk.transport.bombardier.com

Keywords: Task Analysis; Hand Held Terminal; Usability; User Requirements

A rail company is currently developing a pilot European Rail Traffic Management System (ERTMS) line and requires a hand held system which can be used for the protection of work and shunt zones and can be directly controlled by trackside workers. The development of these terminals required human factors support. A task analysis based on a concept of operations and structured interviews with subject matter experts and the designers was produced. A prototype interface was reviewed and a set of ergonomics requirements, which the software team would incorporate into their design, was developed. A proof of concept was reviewed by subject matter experts and recommendations were made which would enhance the design of the hand held terminal.

Introduction

Trans-European rail travel has generated the demand for a standard European Rail Traffic Management System (ERTMS), which provides inter-operability between countries. A rail company is currently developing a pilot ERTMS line, which is intended for widespread application in their own country and across Europe. It is their objective to afford the same level of safety to trackside workers as they currently do to their passengers. As part of this programme, Hand Held Terminals (HHTs) will be developed, from which the protection of work and shunt zones can be directly controlled by the trackside work team. These Terminals will allow the maintainer to take and give the control of work zones, and to manipulate the infrastructure (e.g. points) within them.

Human Engineering Limited were contracted by the system developer to provide support to the project in the form of requirements capture, human factors specification for the Graphical User Interface (GUI) and prototype testing.

The Hand Held Terminal - A Concept of Operations

Bombardier Transportation Signalling have developed a concept of operations for the HHT in collaboration with the client and the national rail regulator. This concept took the form of a series of operational scenarios which encompassed such activities as shunting and staff protection during train movements through or adjacent to areas under possession. These scenarios set out levels of supervisory control for the maintenance teams, described how teams would interact with each other and with signal control staff and described in broad terms the functionality of the software.

The software would be based on a proprietary system of signalling control centre software called "Ebiscreen" which is designed to control and supervise small to medium sized rail traffic areas. "Ebiscreen" runs on Windows NT and uses a mimic display to represent the traffic picture consisting of simplified drawings of a traffic area showing track objects (points, signals etc.) and, in the case of the HHT, areas under possession.

The concept of operations was not detailed enough to support human factors requirements capture and it was therefore necessary to undertake a Task Analysis.

The Task Analysis

A hierarchical task analysis (HTA) was undertaken using the concept of operations as a starting point but supplemented with structured interviews which were undertaken with the system developers, representatives from the rail regulator and end users who acted as Subject Matter Experts (SMEs). Since the HHT is a concept and not an existing product this required a degree of imagination and guided knowledge elicitation. It was recognised that the HTA would require updating as the project progressed, because some early thoughts and ideas would inevitably become redundant as the concept progressed to a "beta" stage.

The HTA identified a range of activities and levels of responsibility which would form the basis of a functional specification. These included the three basic modes of operation:
1.   Master. This allows the trackside supervisor to take and give control of zones, move infrastructure elements or pass control of elements to other members of the team.
2.   Restricted. These are secondary terminals used by other workers to control a single element within a work zone. The giving or taking of zones is not permitted in this mode.
3.   Slave. This is a "read only" version of the HHT, allowing workers to see the state of the work zones but not alter them.
Principal Sub-tasks included work zone allocation, decreasing or increasing the size of work areas made up of smaller work zones, controlling local infrastructure elements (points etc.) and controlling engineering train movements through the area under possession.

The HTA identified a wide range of issues that needed to be incorporated into the GUI if the design was to succeed. These included the need to maintain an overview of the work area on the HHT screen at all times, a clear representation of the transfer of responsibility from the signal controller to the on-site supervisor and, during train movements within the work zone, information regarding the track plan, position of points, track occupation, element ownership. Also, elements to be moved should be readily available at all times. Procedural control of the possession during shift handovers was recognised as a key area. The HTA, once agreed, formed the basis of the human factors requirements which would in turn be supplied to the software developers responsible for the interface design.

Human Factors Requirements

The HTA identified a range of task specific requirements, but these were insufficient in themselves. A further review of the HHT was therefore undertaken and a range of fundamental human factors issues were identified from the human computer interface literature and compiled into a requirements document. The requirements ranged from the high level (task and user characteristics) to the highly specific (colour philosophy, the use of advisory messages and prioritisation of alarms).

These requirements were reviewed and agreed by the system developer, the software specialists responsible for their implementation and the end users. At this point it was agreed that a further review of the "Ebiscreen" software would be prudent in case any of the human factors requirements that had been identified as essential, could not be incorporated.

This review was undertaken by the design team using a checklist developed by Ravden & Johnson (1989). Ravden and Johnson identified 9 goals:

1. Visual clarity (clear, well organised, unambiguous and easy to read)
2. Consistency (consistent operation, look and feel at all times)
3. Compatibility (meets user expectations and conventions)
4. Feedback (clear informative feedback at all times)
5. Explicitness (the way the system works should be clear at all times
6. Appropriate functionality (meets user needs and requirements)
7. Flexibility and control (suits the needs and requirements of all users and allows them to feel in control of the system)
8. Error prevention and correction (minimises user error, detects and handles errors and allows user correction)
9. User Guidance and support (provision of informative, easy-to-use and relevant support

"Ebiscreen" met the above requirements in most areas but its safety integrity level (SIL) and dependence on mouse driven navigation required further investigation.

The Proof of Concept

Following a period of software development a hardware platform was identified as being suitable for the presentation of the software to the end user (although not necessarily as the final platform for use in the field). It was originally intended to run a usability trial in a trackside environment but it was decided not to do this for the following reasons. It was considered vitally important that the users were happy with the GUI and felt that it could deliver the functionality they required. The software could have been overshadowed by the hardware platform (even though it was only a temporary medium for demonstrating the software) and this could have led to inappropriate rejection of the software interface as a result of contamination of the field trial by the host hardware. Instead, it was decided to run a short series of workshops with the users to gain their acceptance and suggestions regarding the GUI and its functionality.

Over a three day period trackside maintenance workers and shunters were given demonstrations of the software and, in the case of the maintenance workers, an opportunity to use the software was provided. This was followed up with guided group interviews and discussions which were very encouraging for the design team. This was because many of the comments and issues which were raised were related more to "look and feel' than software functionality. This gave considerable validity to the task analysis and also confirmed the accuracy of the operational concept. A wide range of trackside and shunting activities were discussed and it became apparent that if the HHT was to be accepted by both work groups some changes to the interface would be required. For example all verbal communications between trackside workers and signal controllers are currently recorded and it was considered essential that all HHT actions were recorded in a similar fashion. The ability to rotate the screen view through 180 degrees (orienting the "map" to the ground) was also requested. Mobile telephones were widely used and the transmission of alerts via telephone would ensure receipt. Text messaging was seen as a very useful tool.

For shunters there were concerns regarding the size of the shunt area, which in practice, can be very large - this could impact on the graphical representation of the entire work area. Using touch screens in such a dirty environment could also be problematic and the use of styluses may be desirable.

As well as user-identified issues, the ergonomist and system developer noted a number of potentially significant problems. Most notable, were concerns regarding HHT failure. Aside from the physical aspects of hardware failure i.e. provision of back-up terminals, spare batteries etc, there are fundamental safety issues associated with the provision of degraded information. A number of recommendations emerged, including some that would feed into the HHT safety case, such as the need to provide immediate warning to the operator of HHT failure, or the requirement that the system would fail to a safe mode and that an on-screen indication of "system health" was available. Work is ongoing in this area.

Discussion

This project has developed a product which will give significant control to trackside workers and will interlock with ERTMS. The software has been demonstrated to have utility and its usability is acceptable. However, the desire to "prove" the software has meant that the identification of a suitable hardware platform remains on the agenda. This is crucial since general operability, ruggedness, portability and all-weather functionality will be paramount. The design team recognises that a safety case will have to be made for both hardware and software and work continues on the assessment of a candidate hardware platform. The paradox of course is that whilst the software may be accepted, there is a risk that a suitable hardware platform will not be found as there are very few that can meet the levels of ruggedness and portability required. The use of a host platform for the proof of concept activities ran the risk of user rejection if the platform was seen as un-usable even if the software was of high utility. Fortunately, the participants were sympathetic to the constraints within which the team were operating and it is hoped that as the design matures, a suitable platform will emerge.

As with many projects of this type, it is not always possible to follow a classical human factors integration model. This may be as a result of cost constraints or merely that it is inappropriate to focus undue effort on a particular aspect of a design (in this case the hardware) if the purpose is for demonstration or proof of concept. The danger with this approach is that the good work that is put in to one aspect of the design is let down by the facet deemed irrelevant at the time (in this case the platform selected to demonstrate the software).

The development of a concept of operations proved highly valuable as it was used firstly to "sell" the idea of the HHT to the client and regulator and secondly because it formed the basis for the task analysis. Operational concepts are a vital first step in the Human Factors Integration Process and it is usually unwise to avoid this activity. The same rule applies to the task analysis. This is a relatively expensive activity, but once undertaken can be revisited time and again throughout the project lifecycle. For example, it will underpin the safety case in the form of human reliability analyses, provide operational support in terms of workload assessment and contribute to the Reliability Availability and Maintainability model.

Early user involvement was also vital to the project. This enabled the design team to approach the problems it faced with the confidence that the team had obtained the support of the end user. Conversely, early user involvement can also occasionally stifle creativity since there may be resistance to change or an inability to grasp the full technical implications of what the designers are trying to achieve. This requires careful management if there is not to be too much pull on the design from one stakeholder.

In the case of this project, early user involvement in the design and application of the concept HHT greatly facilitated the design team's understanding and appreciation of what the final product should be capable of. This created a very cooperative working environment and should ensure that the HHT becomes a successful and popular tool, offering a distinct benefit to trackside workers.

Conclusions

An integrated design team which includes user representation and early (managed) input from regulatory bodies can increase the probability of success when developing technology demonstrators or proofs of concept. The authors believe that this project demonstrates the benefits of early human factors intervention.

References

Ravden, S. & Johnson, G. 1989, *Evaluating Usability of Human Computer Interfaces: A Practical Method,* (Ellis Horwood Limited)

AIR TRAFFIC CONTROL

REAL TIME AIR TRAFFIC CONTROL SIMULATION WHAT, WHY, WHY NOT, WHITHER?

Hugh David

7 High Wickham, Hastings, East Sussex, TN35 5PB
Hughdavid4@Aol.com

Real-time simulators are widely used in Air Traffic Control. They originated as training tools, by analogy with pilot trainers, and have become more elaborate as the tools of air traffic control have become more elaborate. Simulators are generally used for training, and, to a lesser extent, as research tools. As research instruments, they suffer from some major drawbacks. They are extremely expensive, and at the same time rigid and difficult to control. Modern digital simulators are extremely difficult to program, and, although the experimental psychology paradigm is generally accepted, its application is, in practice, fraught with difficulties. It is assumed by users of real-time simulation that training and experimental results derived from simulation transfer to the real world, but there are some reasons to think that this is not always the case. Subjective and 'objective' methods are available for the measurement of simulations, and some statistical analysis is employed, but few of these methods could be defended in a court of law. Simulation can, however, be used effectively where the purpose of the simulation is well understood, for example, in the formation of teams, the rehearsal of emergency procedures or the maintenance of formal skills.

Introduction

Air Traffic Control (ATC), the ground-based element of the global air traffic system, has derived most of its traditions and practices from aviation. In many countries, Civil ATC developed from military Air Defence systems and, even in some European countries, was an Air Force responsibility until relatively recently. Much ATC technology (primary radar, Secondary Surveillance Radar (SSR)) derived from military equipment. Selection and training equally copied military methods, including strong male (even 'macho') and authoritarian biases. In spite of some pioneering work, such as that of Bisseret and Enard (1970), much ATC training involves the learning of regulations and routines, supplemented by individual and group training simulations, and completed by slow, wasteful and dangerous 'on-the-job' training. (The presence of a trainee is often mentioned as a contributory cause in incident reports.)

This paper examines real-time simulators in general, as training and research tools.

Real-time Simulators

Apart from arrivals, departures and ground manoeuvres at and around airports, (which are conducted from the 'towers' familiar to all users of civil airports), most air traffic control takes place in remote locations. Controllers work in large rooms, soundproofed and traditionally windowless, receiving information on telephones, radio links, printers and radar displays. Their contact with the 'real world' is restricted to these links, and should, in principle, be easy to simulate.

Physical Equipment.
A real-time ATC simulator is essentially a control room, with up to 40 individual working positions. Most of these will be occupied by controllers from the area being simulated, but a number will be 'ghost', 'feed' or 'dummy' controllers, who simulate the interfaces of the controllers with adjacent areas which are not being simulated. Their task is to ensure that the incoming and outgoing traffic behaves in a realistic manner.   An adjacent room contains a similar number of working positions for the 'pseudo-pilots', who simulate the pilots of the aircraft. These 'pseudo-pilots', who are not usually qualified aircraft pilots, speak 'radio' voice messages on prompting by the computer controlling the system, and input standardised messages corresponding to the instructions given by the controller to the simulated aircraft. (Attempts have been made to develop computer simulations of pilots, using automatic speech recognition systems, but these are generally useful only in training, where rigid speech formulations may be preferable and prolonged training of the speech recognition system is acceptable. Experienced controllers tend to employ more relaxed and idiomatic speech patterns than are officially specified.[1])

Method

In a typical simulation run, which lasts from 60-100 minutes, up to 200 aircraft may be active simultaneously in up to 20 sectors. A complete exercise may include up to 60 runs, three or four per day, over several weeks. Although an experimental design is usually planned beforehand, it is rarely completed exactly as planned.

In spite of the best efforts of the software specialists, unplanned system stops often require a simulation run to be halted before its planned end.

'Re-runs' of the same traffic sample may produce misleading results since the controllers will be partially familiar with the traffic.

It may become clear that one potential organisation is simply not acceptable. Controllers will not accept that it is necessary to complete a planned simulation program merely to provide a balanced design for analysis.

Revised or combined organisations may be evolved during a simulation. It is difficult to design a simulation program that does not confound learning effects.

---

[1] "Air France 513, Where the hell are you going, scheisskopf?" (Personal Observation)

Data Preparation.

Every real-time simulation requires extensive data preparation. Different types of traffic sample may be required to test revisions to an existing system, to evaluate a different working method or to evaluate a new interface.

Real traffic is frequently found to be 'unrealistic'. In the real world, major and minor incidents occur continually. In an evaluation of a new tool, these incidents introduce 'noise' into the system, and reduce the sensitivity of the simulation. When evaluating a new working method, they may be decisive, showing that it cannot cope with a particular type of emergency.

It is rarely possible to simulate even a fraction of the foreseeable emergencies. By definition, unexpected emergencies can be simulated only once, since the controllers will be expecting problems from then on.

Participant Training.

Traditionally simulations begin with a 'training period' of a few days, during which the participants familiarise themselves with the system. The 'measured exercises' begin when the participants are satisfied that they know the system.

However, electroencephalographic studies show that participants are still learning up to ten days into the simulation (Eriksen and Harvey, 1999).

Air Traffic controllers normally work in stable teams or 'watches', with the same colleagues for months or years. They are very much aware of the strengths and weaknesses of their colleagues, and develop a 'watch style' that may differ greatly from that of other watches, even in the same centre. Controllers participating in a simulation are usually volunteers from the centres concerned. They are not a complete 'watch' accustomed to working together, and have to 'learn' their colleagues at the same time as they learn the new system (Dubey, 2000).

Running Problems

Although a simulation in progress may appear to an external observer to resemble the normal environment of the controllers, there are inevitably differences.

Air Traffic Control is an 'expert' or 'overlearned' skill. A competent controller relies on very specific experience, which has reached an 'unconscious' level to carry out control. This type of learning, Rasmussen's (1986) 'skill' level is inaccessible to the conscious mind.

Expert skills are extremely specific. Generic Real-time simulators do not provide exactly the equipment that the controllers use in their daily work. This may disturb he efficiency of the controllers.

Communication problems, weather or uncooperative pilots are rarely simulated. Although regarded as exceptional events, they are in some areas the norm.

One or two simulator pseudo-pilots handle voice communications for all the aircraft in contact with a controller. In real life, controllers recognise and rely on differences in voices to identify aircraft. It is technically feasible to apply simple distortions to the communications of specific aircraft to provide different voices, although this is rarely done.

Subjective Measures

The subjective opinions of controllers are greatly respected, and may be decisive in determining the outcome of a simulation.

The project leader is traditionally a controller or ex-controller. A vital part of his skill is to extract and evaluate the opinions of the controllers in the course of de-briefings. Few controllers are trained in interviewing techniques, although several Eurocontrol project leaders have spontaneously undertaken human factors training.

Controllers often have trouble in expressing their reservations about a new method because they lack the vocabulary to express their problems. They may formulate complaints about topics they consider they are qualified to discuss, such as the samples of traffic, to express underlying unease (Dubey, 2000). Software engineers and others may spend much effort on symptoms since they are unaware of the underlying causes.

On-line Instantaneous Self-Assessment (ISA) is often used to provide a quantitative estimate of subjective workload. This method requires the controllers to assess their workload on a 1 to 5 scale, using a dedicated keyboard, at two-minute intervals. Although this interval may be sufficiently short to disrupt controllers' mental processes, the technique provides valuable immediate feedback.

Questionnaires are usually employed to obtain information on the acceptability of new or revised methods or organisations. Because the effects of the changes concerned may be very different for different working positions, it is rarely possible to combine sufficient numbers of responses to provide statistical significance on an inter-subject basis.

The NASA-TLX (Task Load Index) (Hart and Staveland, 1988) is a widely used measure of subjective effort. It measures effort on six scales, Mental, Physical and Temporal Demand, Performance, Effort and Emotional Stress, combining these to form an index. It is normally employed immediately after each simulation run, and can provide useful quantitative measures, particularly if individual scales are also analysed.

Objective Measures

Objective measurements (so called) taken from simulators are generally derived from the more easily measurable activities – such as the opening or closing of communication channels, or the displayed positions of aircraft at regular intervals corresponding to the renewal of the System Dynamic Display (SDD) image.

When an existing organisation is being modified for operational reasons, measures of overall workload – the total time spent speaking on the frequency, the number of input orders per hour, or the peak numbers during a defined period of the simulation are relevant. When the interest is in a revised display, or a new device, the amount of speech per aircraft, the number of computer inputs required per aircraft, the number (and type) of windows called up, the number of keying mistakes or cancelled orders are more important.

It is rarely possible to install special software to collect data not included in the standard analyses. Specialist measuring equipment, eye movement recorders, psychophysical measurement devices, etc. usually require calibration, which interferes with simulation running, and produce data which is not synchronised with the simulation.

Analysis takes time and requires effort from specialists, who are usually fully occupied during the exercise. The project leader will usually have formed his conclusions, so that 'objective' analyses are seen as redundant if they agree with the conclusions or wrong if they do not.

Discussion

Large real-time simulations have been the preferred method for the resolution of problems in ATC for many years. There has been tension between the desire to adopt a more 'scientific' approach, and the pragmatic approach, concerned to reach an agreed solution to the immediate problem. Simulators are widely used in training, for individuals and small groups. Simple simulators may be cost-effective in early training, to assist in implanting 'drills' in long-term memory. More elaborate and realistic simulators appear to be equally cost-effective in conversion training of, for example, airline pilots. As Dubey (2000) points out, these simulators are used to instil attitudes and form emotionally bonded working groups. Any measurement of performance is incidental.

The use of large simulators as experimental tools is less justifiable. The experimental psychology paradigm, in essence, assumes that when a few variables are changed under control, the changes in measured variables are due to these changes. There may be some uncontrolled 'nuisance' variables, which can be accounted for by proper experimental procedure, but the underlying situation does not vary systematically during the experiment. In ATC, these assumptions simply do not hold. The results obtained are rarely sufficiently well controlled to achieve statistical significance, and many 'nuisance' variables are simply unknown or ignored. (This is not to say that large real-time simulators have no role to play in the development of ATC. They are invaluable as the final validation stage, where controllers must be convinced that a system change is acceptable.)

Conclusion

Large real-time simulators should be used for development or validation, not for research.

References

Bisseret, A. and Enard, C. 1970, Le problème de la structuration de l'apprentissage d'un travail complexe. Une méthode de formation par interaction constante des unités programmées (MICUP). (The problem of structuring the learning of a complex skill. A training method using continuous interaction of programmed units.) *Bulletin de Psychologie,* **23** (11-12), 1969-70, 632-648.

Cushing, S. 1994, *Fatal Words: Communication Clashes and Aircraft Crashes*: (University of Chicago Press, Chicago, Ill.)

David, H., and Pledger, S. 1995, *Intrusiveness of On-Line Self-assessment in ATC Simulation using Keyboard and Speech Recognition.* EEC Report No. 275 (EUROCONTROL Experimental Centre: Bretigny-sur-Orge, France)

Dubey, G. 2000, *Social Factors in Air Traffic Control Simulation.* EEC Report No. 348 (EUROCONTROL Experimental Centre: Bretigny-sur-Orge, France)

Eriksen, P. and Harvey, A. 1999, *SweDen 1998 Real-Time Simulation.* EEC Report No. 335 (EUROCONTROL Experimental Centre: Bretigny-sur-Orge, France)

Hart, S.G. and Staveland, L. E. 1988, *Development of the NASA-TLX (Task Load Index): Results of Empirical and Theoretical Research* IN Hancock and Meshkati (Eds) Human Mental workload, pp 139-183

Rasmussen, J. 1986, *Information Processing and Human-Machine Interaction: An approach to Cognitive Engineering.* (North-Holland Publishing Co.: New York.)

HUMAN ERROR IN AIR TRAFFIC MANAGEMENT: DEVIATION OR DEVIANCE?

Anne R. Issac

Human Factors and Manpower Unit
Eurocontrol, Belgium

The Air Traffic Management (ATM) system in Europe is inherently safe, but the demands from the aviation industry will inevitably require changes in the way ATM operates. These changes are already impacting on the operator's environment in which procedures are changing to increase traffic efficiency. In harmony with these procedural changes are the increases in technology to 'assist' the controllers. Although it is difficult to predict how these changes will affect the controller, it is known that humans within an increasingly complex technological environment display an ever increasing number of errors, some of which are unpredictable. This paper deals with a new approach to ATM human error analysis that is being developed for the European ATM system. This approach aims to determine, not only how and why human errors are contributing to incidents, but also how to improve human reliability within the system.

Introduction

The paper discusses a model and methodology to investigate both past incidents and errors caused in a real-time simulated ATM environment. The model itself is flowchart-based, and contains a number of classification systems for determining, in some detail, the nature and causes of the error. Information with regard to air traffic contextual factors is also considered and incorporated into the analysis.

This methodology has been developed over the past four years (and will continue to be developed for another year) and was applied to a range of European (and some non-European) incidents. Most recently simulation experiments have been undertaken within Europe to investigate the erroneous events observed within a real-time ATC operational environment, and to ascertain how the controllers detect, recover and manager these events.

The Approach to Analysing Human Errors in High Hazard Industries

Human error has always been part of psychology, but in the industrial setting its beginnings are usually traced to the late fifties and early sixties, when formal methods for identifying and classifying human errors in missile development systems were developed along with hardware reliability approaches. Human error classification systems and human error databases were developed in the sixties and seventies although their main application was in the military domain and in some early nuclear power plant developments.

Given the desirability of a methodology for analysing human errors, it is useful to research the methodologies that already exist. Currently, there are no 'off-the-shelf' ATM-oriented Human Error Analysis methodologies. This is partly because ATM has been a relatively high reliability organisation - human reliability and system reliability is higher than many other industries. There has therefore been little demand for such approaches. This could mean that ATM is somewhat 'naive' compared to other 'high risk' industries (e.g. nuclear power, chemical process and offshore petro-chemical industries). These other industries have developed approaches following large-scale catastrophes and accidents such as the Three-Mile Island (TMI) and Chernobyl nuclear accidents, the Bhopal poisonous gas release, the Challenger explosion and the Piper Alpha oil platform fire. However, since human error is mainly a function of the human rather than the operational working context, this 'naiveté' may not matter. ATM can however borrow from other industry knowledge and experience and from general psychological understanding that has evolved over the past three decades concerned with industrially-related research in this area.

A review of error taxonomies, models and classifications was undertaken throughout a variety of literature and it was established that fifteen areas of human performance and fifty-four approaches within these various areas should be considered within this research. From the review of literature it was judged that the model of human information processing provided a good underlying model for a human error classification system. This model of human information processing appeared to be the most suitable model, and was adapted to make it more applicable to ATM. [A summary of the approaches and the model with full tables and flow-charts of the chosen technique can be found in EATMP a and b, in press].

The following categories of error detail and mechanisms appeared to be the most comprehensive approach and enabled an analysis of human errors within ATM incidents in a similar way to other industries:

Error / Violation – the external manifestation of the error;

Error Detail – the internal manifestation of the error within each cognitive domain (perception and vigilance/ memory/ decision making/ response selection);

Error Mechanism – the psychological or internal mechanism of the error within each cognitive domain (perception and vigilance/ memory/ decision making/ response selection)

Contextual Conditions – A set of ATM specific contextual conditions were grouped into 9 categories including the personal, team and working environment as well as issues dealing with the rules, procedures, technology, training and organisational structure.

All the details regarding each of the above categories, known as the Human Error in ATM (HERA) technique, were developed in tables and flow-charts allowing the analyst the opportunity to record all the individual cognitive problems as well as the context in which the individuals were working at the time of the incident.

Analysing Errors in ATM Incidents

Initially it was decided to attempt to identify the types of human error which are typically reported within incident investigation. This activity is not reported in any literature and typically is fraught with several problems. Firstly it is known that very few incident investigation practices within the air traffic profession are standardised (Lee, 1996). Secondly, when reassessing the involvement of the human in incident and accident reports it is known that few investigations take account of the human performance limitations, (Zotov, 1996).

A large number of incidents were randomly chosen to be re-analysed from a human performance perspective. These incidents (all non fatal occurrences) were obtained from European countries (N= 47), the United States of America (N= 5) and Australasia (N= 15). Results indicated that there were 257 errors and 21 violations. The most reported error/violation types were found in the areas of omission and situational violation. Most error types appear in the selection of action – 152 errors, and the next most reported area was associated with information transfer – 86 errors, 27 being concerned with incomplete information transmitted or sent. The following table indicates the Error Mechanism failures.

Table 1. Identification of the Error Mechanism within the Error

Cognitive failure	Number	Cognitive failure	Number
Perception and Vigilance	**52**	*Long Term Memory*	**1**
Hearback error	14	No recall of temporary information	1
Mishear	7	*Planning and Decision Making*	**119**
Late auditory recognition	1	Misprojection of a/c	22
No detection (visual)	17	Incorrect decision or plan	74
Late detection (visual)	8	No decision or plan	21
No identification	3	Late decision or plan	1
Misidentification	1	Insufficient plan	1
Misread	1	*Response Execution*	**11**
Working Memory	**15**	Selection error	3
Forget to monitor	1	Information not transmitted	1
Forget to perform action	1	Unclear information transmitted	1
Forget planned action	5	Incorrect information transmitted	5
		Omission	1
Forget previous action	2		
Forget temporary information	3		
Inaccurate recall of temporary information	3		

In terms of the Error Mechanisms, the highest number of errors is seen in Planning and Decision Making – 119 errors, more precisely the problems lie in the area of incorrect decision or plan. The second highest area of concern can be seen in Perception and Vigilance – 52 errors, and more precisely in the problems of hear-back. Although not explicit (because of confidentiality) the numbers of errors in each category are similarly reflected in the different country grouping, showing comparable trends.

In terms of the contextual conditions, the following issues were the most reported across all incidents and countries: Team Factors (22%), Traffic and Airspace (21%), Workplace design and HMI (14%) and Procedures and Documentation (13%).

These results, led to the investigation of a more accurate and detailed picture of how human errors are generated and managed in the ATM environment.

Analysing Errors in the ATM Operational Environment

Research in ecological psychology (Flach, Hancock, Caird, and Vicente 1995; Zsambok and Klein 1997; Amalberti 2001) indicate that human behaviours such as poor decision making, situation assessment and the reluctance to recover error, are in fact adaptive behaviours aimed at a compromise between the costs and benefits in complex demanding situations. Any assistance from machine systems aiming at assisting some tasks and consequently suppressing this natural behaviour could paradoxically result in an inappropriate division of attention or excessive workload, with the consequence of new and uncontrollable errors (Noizet & Amalberti, 2000). In other words, the cognitive control of situations, particularly in such environments as air traffic control, demands a continuous compromise of different issues such as available time, task priorities, and available resources.

When considering the categories of errors in dynamic systems, violations appear to be as frequent as classic routine errors and mistakes (Reason,1990). Although no evidence is available on this issue in the ATM environment it is known that violations represent 54% of the overall errors observed on the flight deck, (Helmreich *et al,* 2001).

An experimental protocol was established in order to observe controllers (particularly the radar and planning controller) in a simulated ATM environment. This was undertaken in May 2001 at the Eurocontrol experimental Centre in France. The simulation which was chosen was associated with a 'Free Routes' Airspace Protocol (FRAP). Three radar controllers, and their planning controllers were videotaped (both visual and audio recordings) for approximately 1 hour. During the 1 hour sessions two observers (an ATC expert and a human factors specialist) also took notes of all 'outstanding events' in a predefined written protocol. Following the recorded sessions the two observers compared their observations on the 'outstanding events' in preparation for an interview with the controllers who were the subject of the observation. During these 'auto-confrontation' sessions the expert observers talked with the controllers about the video recording and questioned the controllers about all the 'outstanding events'. A total of 4 1/2 hours of recording was collected over 3 simulation sessions. . [All details about the protocols used and the results can be found in EATMP c, in press].

Results from these sessions were extremely interesting, as the controllers themselves were able to explain if and when they had chosen to defer decisions with regard to the 'outstanding events' and when they had taken calculated risks with aircraft planning. Firstly a table of the main error results is shown below.

Table 2. Identification of Errors in the ATM Operational Environment

Error Category[*]	Number Recorded	Prevention[1]	Detection[2]	Recovery[3]
Perception & Vigilance				
did not identify	5	Nil	2	2
did not detect	1	Nil	1P	Nil
hearback	8	Nil	4P	4
Working Memory				
forget information	3	Nil	2	2
forget to perform action	3	Nil	2	2
Long-Term Memory				
recall wrong information	2	Nil	Nil	Nil
Judgement, Planning & Decision Making				
incorrect separation	20	Nil	20 CA[4]	11 RC/ 5 PC
Response Selection				
wrong positioning	1	Nil	Nil	Nil
wrong keying	11	Nil	1	1
wrong communication	2	Nil	2 RC	2 RC
TOTAL	**56**	**-**	**34**	**29**

As well as the identification of the errors into the HERA classification system, the observers were also able to identify whether the errors were prevented, detected and recovered. It can be seen that no errors were prevented which may lead to future classifications not assessing this ability. It was interesting to recognise, however, that the controllers were able to recover 51% of the errors.

From the auto confrontation it was interesting to note some new issues in terms of the classical definition of error. It would seem when all the interview data was compared to the expert observations there were some discrepancies with the results. These were mainly due to the erroneous classification of errors by the expert observers in some situations i.e. the controller was observed not acting on an alert, or not transferring an aircraft. When interviewed it was clear that in some situations the controllers had already determined the outcome and managed the events before the machine had alerted them, or each controller had assessed without confirmation with each other the outcome of the aircraft transfer. The

---

[*] As in HERA methodology
[1] P- Pilot, RC- Radar Controller, PC- Planning Controller
[2] P- Pilot, RC- Radar Controller, PC- Planning Controller
[3] P- Pilot, RC- Radar Controller, PC- Planning Controller
[4] CA - Short Term Conflict Alert

results suggested that whereas the expert observers had assumed the 'outstanding events' would either be a correct action or an error (including a type of violation) in reality this judgement was incorrect. Further investigation with the interview data suggested that the 'outstanding events' could, in fact, be classified into three sub-categories. Firstly correct actions – of which there were 15% - performance errors – of which there were 21% - and lastly a category to be labelled *expert judgement deviations* – of which there were 63%. Further analysis revealed that approximately 1.5% of performance errors included violations and 8.5% expert judgement deviations included violations. This finding was intuitively compatible with the knowledge of highly skilled operators in time constrained and risk critical environments, but had never been explored in an ATM operational situation.

Results also suggested that in the 4 1/2 hours of observation there were approximately 2.5 errors, that is between two and three errors in an hour, which is complimentary to the classical error rate of experts in other working environments. There were approximately 1.6 violations an hour, again similar to other similar environments.

Further work has and will be undertaken to verify and clarify the above results. Future work will also endeavour to ascertain what risk is involved with the *expert judgement deviations,* particularly concerned with new technologies and the reliance on other members of the controlling 'team' in ATM.

References

Amalberti, R. 2001, The paradoxes of almost totally safe transportation systems. *Safety Science* **37**:109-126.

EATMP a , in press, *Short Report of Models of Human Performance and Taxonomies of Human Error in Air Traffic Management.* (HRS/HSP-002-REP-02) (Brussels:EUROCONTROL).

EATMP b, in press, The Human Error in ATM (HERA) Technique (HRS/HSP-002-REP-03) Brussels:EUROCONTROL.

EATMP c, in press, The Investigation of Human Error in ATC Simulation (HRS/HSP-002-REP-05) Brussels:EUROCONTROL.

Flach, J., Hancock, P., Caird, J. and Vicente, K. 1995, *Global perspective on the ecology of human-machine systems.* (Hillsdale-New Jersey: Lawrence Erlbaum Associates).

Helmreich, R., Klinect, J.R., Wilheim, J.A., and Sexton, J.B. 2001, The Line Operations Safety Audit (LOSA). *Proceedings of the first LOSA week.* Cathy Pacific, Hong Kong,

Lee, R. 1996, Aviation psychology and safety: Implementing solutions. In B.J. Hayward and A.R. Lowe (Eds) *Applied Aviation Psychology: Achievemnet Change and Challenge.* (Aldershot, UK: Averbury Aviation).

Noizet, A. and Amalberti, R. 2000, Le controle cognitif des activities routinieres des agents de terrain en centrale nucleaire: Un double systeme de gestion des risques. *Revue d'Intelligence Artificielle.* PEC'2000, **14**, 1-2, 73-92

Reason, J. 1990, *Human Error.* (Cambridge University Press: UK)

Zotov, D. V. 1996, Reporting human factors accidents. *ISASI Forum,* 29 (3).

Zsambok, C. and Klein G., Ed, 1997, Naturalistic decision-making. (Hillsdale, New Jersey: Lawrence Erlbaum Associates).

IDENTIFICATION OF AREA CONTROLLER CONFLICT RESOLUTION EXPERTISE

Barry Kirwan

Eurocontrol Experimental Centre
Bretigny sur Orges
F-91222, France
barry.kirwan@eurocontrol.int

In the medium term future, air traffic will increase significantly, potentially leading to an increased risk of conflicting aircraft, and a need to manage the separation of aircraft more efficiently. Therefore, a tool is being developed in Eurocontrol, called Conflict Resolution Assistant (CORA), to assist the controller in conflict resolution. This tool is intended to assist, but not replace, the controller. It was decided in the CORA project to inform the design of this tool with controller expertise in conflict resolution. Accordingly, controllers in seven European states were interviewed, and each controller was asked to provide solutions for 14 conflict scenarios. This study has led to a database of expertise on conflict resolution, which is now being used to inform the developing advisory tool. The study and its results are illustrated.

Introduction

European air travel is predicted to double in capacity over the next fifteen years, placing considerable burden on the current air traffic management system, and controllers. This doubling in traffic density can lead to a more than doubling in the number of conflicts. Additionally, some of the current fixed flight route structures may become more flexible, leading to more efficient routes for aircraft. However, this freer routing leads to a less predictable traffic pattern for the controller, therefore requiring more monitoring for potential problems, particularly threatened losses of separation between aircraft (called 'conflicts'). One approach to alleviate this burden of extra and less predictable conflicts, and to reduce controller workload, is to provide automated support, in the form of computerised tools, for key tasks, such as conflict detection, and conflict resolution. Such tools, in theory, will allow the controllers to manage more aircraft, and to avoid conflicts, and at the same time to give a better service to aircraft in terms of preferred routes and minimised deviations.

Currently, controllers are masters of real-time conflict detection and resolution, and this expertise in these particular system functions is the result of rigorous selection and intensive training in air traffic control over a prolonged period. Conflict detection and resolution are indeed seen as core functions of the controller today, i.e. controllers, when asked to define their job simply, often say 'separating aircraft'. Any tools that therefore purport to support such functions have two main obstacles to overcome. The first is the development and provision of a viable alternative that is at least as good as controller expertise (and preferably better). The second is ensuring that such tools will be used by controllers, when those very tools can be seen as a threat to those same controllers. This latter aspect is poignant, since conflict resolution is seen as a core task and skill of the controller, especially given their responsibility and culpability should separation be lost.

Therefore an advisory tool for conflict resolution needs to be 'human-centred' (Billings, 1997) or at least 'human-informed' (Kirwan, 2001a). This is the intent of the CORA Project, using the specific variant of 'Human-Centred Automation' (HCA) known as a 'Cognitive Tool' (Nijhuis, 2000). This entails building a software-based system along the same lines as the controller's mental model, to enhance human-machine 'rapport'. This in turn requires understanding the conflict resolution expertise that resides in the controllers.

A major problem here is that such expertise is not well documented, and in fact it is often said that individual controllers differ greatly in their conflict resolution strategies. Such variation poses significant problems if it is true, since a computerised tool trying to emulate human thought processes will be hard to develop, if those thought processes are in essence 'unstable' or inconsistent. In order to ascertain the stability of the expertise base of controllers in the area of conflict resolution, a study was undertaken, some of the results of which are detailed below (see Kirwan and Flynn, 2001; Kirwan 2001b & c for more detail).

Eliciting controller conflict resolution expertise

Forty-five air traffic controllers from seven European countries were interviewed individually and in groups, with the same standardised set of 14 conflict scenarios. Each scenario (e.g. see Figure 1) contained at least two aircraft in conflict, in a static representation of a generic sector of airspace. Each controller was asked how (s)he would resolve the scenario. The answer, due to the minimal amount of information in the scenario representation was often 'that depends...'. The controller would then ask questions or make assumptions, until he or she was happy enough to state the principal resolution he or she would propose. The interviewer would note the order of questions/assumptions and list these as 'factors' affecting the determination of the resolution advisory. Controllers were also asked if there were any potential resolutions (i.e. theoretically possible) that would be considered poor practice, and should therefore definitely not be recommended. Such potential resolution advisories (called *No-no's*) might appear reasonable to a non-controller and might even appear mathematically optimal, but would be seen as incorrect by a controller, and would be rejected immediately by real controllers and could cause loss of trust in the tool.

The study, which took place over a six month period, resulted in significant amounts of information on controller expertise in conflict resolution. It identified the following types of expertise:

- Formal rules controllers adhere to (generally)
- Principles that controllers use to decide the best course of action
- Factors which they utilise to help prioritise the resolutions
- No-no's – negative principles, i.e. things they would not do

Examples of some of the principles controllers apply in various scenarios are given in Table 1, and the factors that controllers considered most important are given in Figure 2.

Overall, the study showed more convergence of opinions than was initially hoped for. In practice, this means that a resolution advisory system can offer a relatively small number of suggestions (e.g. 4) and satisfy most controllers. This is indeed a positive result, and suggests that this critical component of the HCA philosophy, i.e. that the machine and human can have a degree of understanding and rapport, is achievable in this area. The next phase of the work is to 'inform' the computerised tool with the information gained in this study. This will be carried out under another contract, and so it cannot be stated at this point exactly how this will be achieved, but one suggestion of a general framework is illustrated in Figure 3.

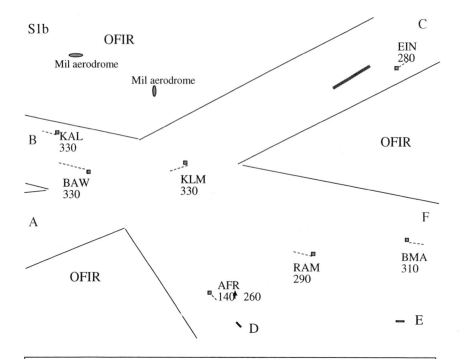

S1b

KAL 567 is an airbus A320.
EIN 536 is a 757
The BMA and AFR a/c are 737s, the RAM is an A320
Currently the BAW will pass in front of the KAL 567, with > 6-7 miles separation predicted.

Figure 1 – Example conflict scenario (catch-up between BAW and KLM)

Table 1 – Extract of Principles & Strategies used by controllers

Category	Ref.	Principle
Scenario-based		**Crossing conflicts**
	S1	Turn slower a/c behind (in order to minimise extra distance flown)
	S2	Stabilise until after crossing points
		Converging/Head-on
	S3	When there are few a/c, a temporary ODL is acceptable
	S4	Ask the pilot whether (s)he prefers a level change or a vector
	S5	Normally if vectoring, vector both a/c
	S6	In turbulence, not always good to have level solutions, since they may not maintain their levels
	S8	Solve the head-on first
	S9	Turn faster one direct to route so leaves sector before slower one on same route
	S10	Safe if locked on headings
	S12	Give a short-cut which can end the conflict
	S13	Better to put a/c behind than trying to go through the middle

Figure 2 – Main factors identified

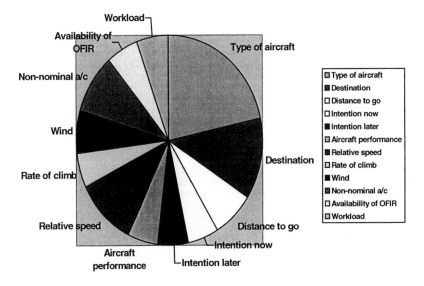

Figure 3: How Controller Information Informs a Computerised Tool

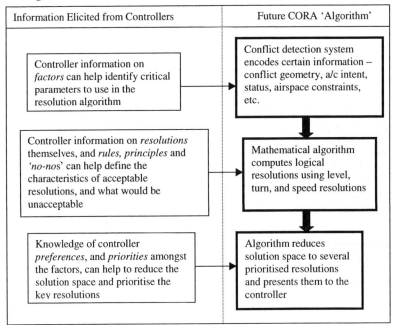

Figure 3 is based on an 'informing' philosophy. The computerised tool will contain an algorithm that calculates all the different ways to resolve the conflict, and there could easily be 20 – 30 such theoretical resolutions. The controller expertise data can then be used as an 'expert system filter' to reduce the resolution set (called the solution space) to an acceptable set of prioritised resolutions, which are optimal for the controller and the aircraft.

The above is explanative but is not prescriptive enough for a software-based expert system style of approach, as CORA may ultimately need. Below is an example attempt (and the author is neither a software engineer nor cognitive scientist) at construing the rules, principles and factors into an algorithm summarising the expertise discussed above, for a 'Head-on' conflict scenario (Table 2):

Table 2: Example synthesis of controller expertise for 'Head-on' Conflict

IF <conflict> is <pair-only> AND <category Head-On> AND <T =10+ mins>
THEN IF <air traffic control centre> is <level-favoured> AND <no turbulence at level> AND <levels above available> THEN
[OPERATION: Determine Opposite Direction Level {ODL} a/c = A] THEN
IF <A not recently taken off> AND <a/c performance category = ok/Flight Level> THEN
CLIMB A 1000 feet
ELSE IF <A recently taken off> OR <A performance inadequate> OR <turbulence> OR <no levels available above> AND <levels available below> THEN
DESCEND A 1000 feet
ELSE IF <B far from destination> AND <B climb performance ok?FL> AND <levels available above> AND <no turbulence> THEN
CLIMB B 1000 feet
ELSE IF <no OFIR boundaries nearby> AND <no context a/c left/right> AND <wind = negligible> THEN
[OPERATION: Determine non-ODL's destination short-cut = Right} THEN
TURN BOTH A/C RIGHT 10 DEGREES
ELSE IF <OFIR boundary nearby> THEN
[OPERATION: Determine boundary location: Left (East)] THEN
TURN A/C A 15 DEGREES AWAY FROM BOUNDARY
ELSE IF <wind not negligible> THEN
[OPERATION: determine significant wind direction – Eastwards] THEN
TURN A/C B RIGHT 10 DEGREES
Etc.

The above is an example of how the approach could be explicated formally, following which it could be coded to work on the output of the computer algorithm deriving the original first cut set of options.

Discussion

One aspect of HCA that has not been particularly elaborated in the literature is exactly when HCA is needed. Effectively, the type of HCA being suggested here is a team approach, the team comprising the human and the automation. A simple question that is often not asked, is what each of these 'team members' is bringing to the task, and whether the blend of these attributes and skills is a sensible one likely to improve performance. In the conflict resolution context, the relative strengths of the human and machine do in fact seem to complement each other (see Table 2).

It is also necessary to ensure that the controller does not become too reliant on the tool, and that the degree of rapport between human and machine does not lead to 'complacency'. It is intended that the controllers will retain their skills, so that they can still carry out this task, particularly if the tool should fail for any reason.

Table 2 – Allocation of Function – in a conflict resolution context

Human strengths in conflict resolution	Machine strengths in conflict resolution
Ability to recognise and categorise problems and infinite scenario permutations	Very fast and reliable processing
Ability to make decisions under uncertainty and stress	Ability to predict and project with more accuracy and range
Ability to make judgements that will satisfy most parties	Ability to optimise according to many criteria (where these can be 'parameterised')
Ability to address temporary local or special conditions	Not bound by biases, personal experience, or memory or processing limitations

Conclusion

The results of this study suggest that a human-centred approach is viable in this critical area of air traffic performance and safety. There is still much to do to produce a viable conflict resolution tool for controllers, in terms of incorporating the expertise, validating the expertise, designing the interface and ways of using the tool, and training controllers. The tool is aimed for 2007, so it is hoped that it is achievable in this timescale.

Acknowledgements

This research was sponsored by EUROCONTROL as part of the CORA project. The opinions are those of the author, and do not necessarily represent those of participating organisations. Nevertheless, the author wishes to thank all the controllers that took part, and their management for facilitating their participation.

References

Billings, C. (1996) *Aviation* automation: the search for a human-centred approach. Lawrence Erlbaum Inc., New Jersey.

Kirk, D.B., Heagy, W.S. & Yablonski, M.J. (2000). *Problem Resolution Support for Free Flight Operations*. MITRE, Center for Advanced Aviation System Development, McLean, Virginia.

Kirwan, B. (2001a) *Human Centred Automation – Walking the Talk*. Paper presented at (Man-Machine Communication in Technical Systems), NFA, 21-22 November, Hotell Olavsgaard, Lillestrom.

Kirwan, B. (2001b) *Towards a Cognitive Tool for Conflict Resolution Assistance – A literature Review*. EATMP *Report in Progress*, Eurocontrol Experimental Centre, Bretigny-sur-Orges, F-91222, France.

Kirwan, B. (2001c) *Investigating Controller Conflict Resolution Strategies*. EATMP *Report in Progess*, Eurocontrol Experimental Centre, Bretigny-sur-Orges, F-91222, France.

Kirwan, B., and Flynn, M. (2001) *Identification of Air Traffic Controller Conflict Resolution Strategies for the CORA (Conflict Resolution Assistant) Project*. ATM 2001, Santa Fe, Dec 3 – 7th.

Nijhuis, H (2000) *Role of the Human in the Evolution of ATM (RHEA)*. Final Report. NLR: Netherlands.

DRIVERS AND DRIVING

DRIVERS' RESPONSES TO CHANGEABLE MESSAGE SIGNS OF DIFFERING MESSAGE LENGTH AND TRAFFIC CONDITIONS

José H. Guerrier[1]
Jerry A. Wachtel[2]
Donald L. Budenz[3]

[1]Dept. of Psychiatry & Behavioral Sciences, University of Miami School of Medicine, 1695 NW 9th Avenue, Miami, FL 33136
[2]The Veridian Group, Inc., 567 Panoramic Way, East Cottage, Berkeley, CA 94704
[3]Bascom Palmer Eye Institute, University of Miami School of Medicine, 900 N.W. 17th Street, Miami, FL 33136

Specific guidelines have been developed in the US to inform the deployment of Changeable Message Signs (CMS). However, the legibility distance used in these guidelines is greater than that achievable by a large number of drivers due to age-related visual deficits. This paper evaluates the effect of the number of CMS phases and traffic conditions under more realistic visibility distances on driver performance. Thirty-five persons (17 young, and 18 old) were instructed to comply with instructions displayed on a simulated Changeable Message Sign (CMS) while driving an interactive driving simulator. Both display format and traffic conditions were significantly related to driver's performance. These findings have implications for CMS designs to facilitate compliance and driver safety.

Introduction and Background

Traffic congestion on US highways continues to increase. In the State of Florida, three urban areas rank among the top ten nationwide for the largest increases in vehicular travel. This growing congestion costs $3.5 billion annually in additional fuel and lost time to motorists in Florida alone, and is expected to worsen because of: a) increasing population, b) projected increase in highway travel of 35% by 2015, and c) Florida's recent implementation of legislation limiting the capacity that can be added to the State's highway system (TRIP, 2000). The Intermodal Surface Transportation Efficiency Act (ISTEA) was passed by Congress in 1991. Among its major objectives is the improvement of mobility for "elderly persons, persons with disabilities, and economically disadvantaged persons ... ". This is particularly applicable to Florida which has a large elderly population. Considerable research has been devoted to the design and operation of CMS, and several sets of guidelines have been promulgated. These guidelines address such issues as: (a) visibility and legibility (letter size, font, brightness, contrast, etc.), and (b) understandability (e.g. number of lines etc.). Adoption of such guidelines promises standardization of CMS operations nationwide. Nevertheless, sign design and operation still varies widely from one jurisdiction to another.

These circumstances pose challenges for the deployment of ITS technologies such as CMS that depend on information processing speed and/or linguistic ability. Accordingly, there is a need to evaluate the design and operation of CMS with Florida's elderly. A key concern with CMS is the attention demanded of drivers due to the number of phases required to present a complete message, and the time taken to present that message. Specific age-related deficits, among them cognitive processing ability, and perceptual skills (e.g., visual) affect driving performance (TRB Special Report 218, 1988). Further, the graying of the U.S. population reinforces the need for appropriate design and operation of such traffic control devices as CMS. Research has shown (Mace, 1988; Synthesis of Human Factors Research on Older Drivers and Highway Safety (1997)) that while the accepted visibility distance or legibility index (LI) for signs has been traditionally 6m/cm, this LI corresponds to a visual acuity of 6/8 which surpasses that of 30% of drivers 65 years old and older and exceeds the minimum acuity requirements for licensing (i.e., 6/12). Based on a review of relevant studies on visibility distance of signage by drivers, the Synthesis of Human Factors found that 85th percentile LI values reported were between 3.6m and 4.8m/cm, and thus suggested the adoption of the more conservative LI of 3.6m. This has implications for establishing letter size requirements and at a minimum in determining reasonable visibility distance expectations when older drivers are considered.

Many factors influence the visibility of CMS, including not only those common to any signage (e.g., font type, size, ambient light conditions, reflectivity, luminance) but also variables related to the display technology used (e.g., LED, flip-disc). Notwithstanding the many CMS variables that might affect drivers' performance, some guidelines have been offered for appropriate CMS performance standards. Specifically, Dudek (1991) has suggested a legibility distance of 274m, with a letter height of 45.7cm to achieve an LI=6m/cm Dudek suggests that all messages should be readable at least twice by the approaching driver. Based on Mace's work, however, one can expect a conservative legibility distance for a 45.7cm high letter to be 164.6m. Of course, any reduction in legibility distance of a CMS will reduce the driver's exposure to the displayed message. Consequently, it is important to determine whether the display time for CMS messages under the type of legibility distance likely to be expected on the road is sufficient to permit drivers in general and older drivers in particular to read and comply with the message posted. This study, funded by the National Institute on Aging (NIA), sought to investigate the impact of the number of CMS phases and their display time on the driving performance of young and old drivers under a realistic legibility distance.

Method

We used a low-cost, interactive driving simulator developed by Time Warner Interactive Simulation Products and The Veridian Group, Inc., supplemented with a flat-screen video monitor placed above the simulator's main display. While the simulator presented interactive road and traffic conditions in real time, including static signs, the supplemental monitor presented the experimental CMS. Seventeen young drivers (25-53 years old) and 18 older drivers (55 and older) drove the simulator under one-phase and two-phase CMS presentation conditions. They were asked to respond to information on road closure/detour information presented (Figures 1 & 2). The two CMS used in this study were developed in accordance with widely accepted guidelines (Dudek, 1991; MUTCD, 2000) and were reviewed for content by independent State DOT experts. Each panel had a maximum of three lines, no line exceeded twenty characters, and all characters were capitalized.

<table>
<tr>
<td>

ACCIDENT AHEAD
EXIT NEXT STREET
LEFT AT 12 ST

</td>
<td>

HEAVY CONGESTION
AT 14 & J ST.
DETOUR AT I ST

TRAFFIC TO I ST
NEXT RIGHT

</td>
</tr>
</table>

Figure 1. One Phase CMS   **Figure 2. Two Phase CMS**

Our equipment set-up did not permit us to vary the visual angle as the driver approached the sign as would be the case in the real world. Accordingly, we created the simulator sign letter height such that it would subtend the visual angle equivalent to a 45.7cm high letter at 99m site distance (half the distance suggested by the MUTCD). This resulted in a letter height of 7.8cm inch for a sign located 106.7cm from the subjects' eye. The design speed of the two scenarios was 80.5km/hr. This was accomplished by requiring all subjects to follow and keep up with a lead vehicle which was programmed to travel at this speed. However, since the driver's speed is not under experimental control, we took an additional step to ensure that the exposure duration of all drivers to the CMS would be equal and equivalent to the exposure they would encounter at 80.5km/hr. Thus, we displayed all one-phase scenarios for 8.86 seconds and all two-phase scenarios for 4.43 seconds per phase. This yielded the exposure that would be achieved at 80.5km/hr from a site distance of 198m from the sign.

Each scenario was run under two Traffic conditions: the "Low" condition featured light traffic with no passing vehicles whereas the "High" condition included more traffic and several passing vehicles. Subjects were instructed to drive as they would on the road, using appropriate caution.

Prior to starting the test scenarios, all participants were familiarized with the simulator's functions and controls and were allowed to practice driving the simulator until they felt comfortable and reported being ready to start.

<u>One Phase CMS</u>: In the scenario using a one-phase CMS, the driver was instructed to follow the red sedan traveling at 80.5km/hr and to try to keep up with it as much as possible. The participant was further told that at some point during the exercise, he/she would hear a beep indicating that important information was about to be displayed on the supplemental display monitor representing the CMS. Upon seeing the message displayed, the driver was to comply with it as quickly and as safely as possible. Once the CMS display was completed, the screen became blank. The CMS messages are shown in Figure 1.

<u>Two Phase CMS:</u> The instructions given to the driver were identical to those given under the one-phase scenario. The CMS message is shown in Figure 2. The entire message was visible for 8.86 secs (i.e., 4.43 secs for each phase)

Results and Discussion

Chi-Square analyses of the performance of old and younger drivers under the one-phase and two-phase scenarios showed that under the one-phase scenario, 27% of older drivers performed the action requested by the CMS compared to 63% of young drivers (X^2 (n=31)=4.01, p=.05). In the two-phase CMS condition, 14% of the older drivers performed the action requested compared to 41%. of younger drivers (X^2(n=31)=2.70, p=.10). The majority of older drivers had difficulties regardless of the format of the CMS (i.e., one- vs two-phase message) (X^2 (n=23) = 4.23, p=.04), see Figure 3. We also found that both younger and older drivers drove significantly slower in the two-phase than in the one-phase condition (F(1,28)= 46.99, p<.0001) (see Table 1) with older drivers driving significantly more slowly under both conditions (F(1,28)= 9.19, P=.005).

	Age	Mean (MPH)	Std. Deviation	N
Mean Speed Two-Phase	Young	33.849	3.758	16
	Old	28.234	5.810	14
	Total	31.229	5.528	30
Mean Speed One-Phase	Young	46.123	6.128	16
	Old	40.117	11.165	14
	Total	43.320	9.198	30

Table 1. Mean Speed by age group and CMS condition

As mentioned earlier, in order to examine the relationship of traffic condition to drivers' performance, each scenario was run under two traffic conditions while all other components of the scenario were kept identical. Chi square analyses were conducted to compare the actions taken under each of these traffic conditions

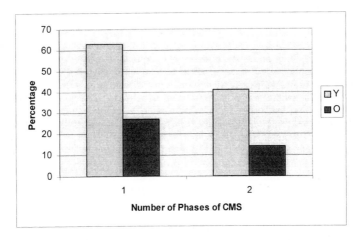

Figure 3.Younger and Older drivers who obeyed CMS messages

(i.e., heeding or not heeding messages) and the outcomes of these actions (collisions or no collisions).

In the one-phase scenario there was no significant difference in the action taken under low or high traffic condition ($\underline{X}^2$ (n=32) = 1.01, $\underline{p}$= .31). Participants were equally likely to heed messages under high or low traffic conditions. Likewise, there was no greater likelihood for participants to have a collision under low or high traffic condition ($\underline{X}^2$ (n=32) = .42, $\underline{p}$= .71). Under the two-phase scenario, although there was no significant difference in the action taken under low or high traffic condition ($\underline{X}^2$ (n=32) = 2.57, $\underline{p}$= .14), there was a significant difference in the outcome of the action under low or high traffic condition (($\underline{X}^2$ (n=32) = 5.23, $\underline{p}$=.04). Specifically, under the two-phase CMS condition, 46% of drivers under the high traffic condition had at least one collision compared to 11% of drivers under the low traffic condition. In fact, the odds of having a collision in high traffic condition are over seven times greater than in low traffic (see Table 2).

		95% Confidence Interval	
	Value	Lower	Upper
Odds Ratio for Outcome of Action (.00 / 1.00)	7.286	1.173	45.255
For cohort Traffic Condition = High	2.571	1.226	5.394
For cohort Traffic Condition = Low	.353	.103	1.204
N of Valid Cases	32		

Table 2. Odds of Crash Under Two-Phase CMS & High Traffic

Conclusion

These findings have implications for CMS design especially as it impacts elderly drivers. Of great concern are the findings that even in a one-phase CMS, fully 38% of younger drivers and 73% of older drivers failed to heed the warning. In the two-phase CMS nearly 60% of younger drivers and 86% of older drivers did not heed the warning. Further, drivers were more likely to be involved in a crash under high traffic conditions when trying to comprehend warnings on a two-phase CMS. This is particularly critical in that the study was carried out under ideal conditions. Specifically, participants were told to expect a message to be displayed and they were alerted about the message by an advanced beep. This, of course, is not the case in the real world. Further, the exposure time of the CMS displays was much longer than that currently implemented in various localities in the US. in addition to the constantly increasing traffic congestion in urban areas. Consequently, if CMS are to serve a useful function to reduce traffic congestion as part of a larger ITS implementation, they must be designed to ensure that drivers in general and older drivers in particular are allotted sufficient time to process the information displayed. Only then can we expect to improve drivers' compliance and safety. This study offers a glimpse of what may be expected if CMS format and display time are not seriously considered. Nevertheless, more research needs to be done to determine whether the findings derived from a simulated road environment with a relatively small sample can be corroborated in the field.

Acknowledgement
This research was funded by National Institute on Aging Grant No. 3P50 AG11743-05.

References

Dudek, C (1991). Guidelines on the Use of Changeable Message Signs, Publication No. FHWA-TS-90-043, Federal Highway Administration, Washington, D.C.

Mace, D.J. (1988). Sign Legibility and Conspicuity. In: Special Report 218. Transportation in an Aging Society. Improving Mobility and Safety for Older Persons. Vol. 2. Transportation Research Board, National Research Council.

MUTCD 2000 – Manual on Uniform Traffic Control Devices: Millennium Edition. December 2000.

NEXTEA: National Economic Crossroads Transportation Efficiency Act (U.S. DOT, 400 7th St., Washington, D.C. 20590)(March 12, 1997).

Special Report 218 (1988). Transportation in an Aging Society: Improving Mobility and Safety for Older Persons. Transportation Research Board, National Research Council, Vol. 2., Washington, D.C.

Synthesis of Human Factors Research on Older Drivers and Highway Safety (October 1997). Publication No. FHWA-RD-97-095.

Traffic Congestion in Florida: Trends and Solutions (October 2000). The Road Information Program (TRIP). Washington, DC 20036.

LOW COST SIMULATION AS A TOOL TO ASSESS THE DRIVING ABILITY OF PERSONS WITH COGNITIVE IMPAIRMENTS FROM BRAIN INJURY

Jerry Wachtel[1], William K. Durfee[2], Theodore J. Rosenthal[3], Elin Schold-Davis[4] and Erica B. Stern[2]

[1]The Veridian Group, Inc., [2]University of Minnesota, [3]Systems Technology, Inc., [4]Sister Kenny Institute

Brain injury can impair the cognitive skills needed for safe driving as well as the metacognitive skills needed for realistic self-assessment of those skills. Traditional on-road evaluation methods are inadequate for this purpose because they cannot test the skills needed for real-world driving challenges. Simulation has become a more viable alternative, but its use for testing brain-injured persons is quite recent and costs have been prohibitive. This paper reports on the findings of a study using a low cost, PC-based driving simulator to study the performance of brain-injured and a matched sample of non-injured persons in challenging, realistic driving tasks.

Background

Brain injury from stroke (CVA), trauma (TBI), or other cause can permanently impair the cognitive skills needed for safe driving, e.g. the ability to process simultaneous information, rapidly shift attention, limit distractibility and control impulses (Brooke, *et al*, 1992). There is no consensus about the best way to identify those who should drive after brain injury, and few such patients receive any formal driving evaluation (Fisk, *et al*, 1998). In the US, adults rely on the automobile to maintain a social life and to accomplish routine tasks (Leigh-Smith, *et al*, 1986). Because this need contributes to motivation, and because brain injury can impair the metacognitive skills necessary for realistic self-assessment (Gianutsos, 1994), brain injured persons may hold unrealistically positive views of their driving abilities.

The physician and rehabilitation team are often asked to decide if and when a person with brain injury may drive again; a decision that must balance the anticipated benefits for the individual against potential risks to society. Driving rehabilitation programs for brain-injured persons commonly use a combination of evaluation measures (Korner-Bitensky, *et al*, 1994). Chief among these is the individual's performance while actually driving a car. There are two types of such evaluations: closed course and on-road. Although reasonable as measures of vehicle control, closed-course evaluations cannot assess the decisions and responses that real-world demands make on driving (Fox, *et al*, 1998). On-road evaluations may be hampered by subjectivity, lack of standardization, unstable reliability, and questionable validity (Galski, *et al*, 1997; Sprigle, *et al*, 1995). Evaluators often conduct such assessments on residential streets in favorable weather, thus reducing the very challenges of real-world driving that on-road assessments should offer (Galski, *et al*, 2000). Lacking a realistic challenge to important cognitive skills, brain-injured individuals may pass such evaluations and drive again. To overcome these limitations, simulation is increasingly used in driving assessment (Korteling and Kapstein).

Though the term 'driving simulator' is applied to a variety of tools that place the 'driver' in an artificial environment where technology creates an impression of driving a vehicle on a road, such tools vary considerably in cost, fidelity, data capability, and degree of interaction (Aaronson and Eberhard, 1994). Most practitioners agree that interactive simulation is imperative for reasonable assessment. However, there is evidence that expensive, sophisticated technology is unnecessary for driver assessment and training, and may even be detrimental to the achievement of such goals Hays and Singer, 1989). In addition, the costs of mid- and high-end simulators are simply too great to justify their use in clinical practice (Galski, *et al*, 2000). With increasing speed and graphics capabilities of personal computers (PCs), PC-based interactive simulators that can provide sufficient fidelity have become available for less than $30,000, a cost that most clinics can consider.

Research on the use of driving simulation with brain-injured persons is quite recent (Galski, *et al*, 1992; Hirsekorn and Taylar, 1998). Galski *et al* (1992) compared on-road scores to data from a simulator and found that the percent of signaling errors and percent of attempts to steer away from hazards were strong predictors of on-road performance. In a later study, these authors found that the three strongest correlates were: anticipatory braking, defensive steering and complex attention (Galski, *et al*, 1997). Liu *et al* (1999) compared the simulator driving performance of individuals with head injury to that of a non-disabled cohort. They found four measures that discriminated between the groups: lane positioning, collision avoidance, running onto the shoulder, and compliance with STOP signs. Our study expands upon this earlier work. This paper reports some of the results of a study that compared brain injured and non-disabled drivers during a simulated driving experience.

Method

Instrumentation

We used a proven PC-based driving simulator known as STISIM Drive (Marcotte, *et al*, 1999; Risser, *et al*, 2000). An orientation drive allowed subjects to become familiar with simulator control, road geometries, and traffic control devices. Subjects continued the orientation until they felt comfortable in the simulator. No data was collected during the orientation drive.

We created a 19-mile long test scenario to represent the key elements of an existing on-road evaluation protocol, including roadway features such as horizontal and vertical curves, traffic, and traffic control devices. The drive included 2- and 4-lane roads as well as a 6- lane limited access highway. The scenario required about 40 minutes to complete. Subjects were told to maintain the speed limit and to keep right except as needed to turn, pass, or follow a command from the researcher.

The scenario was divided into a "simple" segment followed, after a short break, by a "complex" segment. In both, the subject was required to perform basic vehicle, speed, and lane keeping control while interacting with other traffic and responding to traffic control devices. Typical challenges included merging, negotiating intersections with pedestrians, and changing lanes. Whereas the simple segment was designed to reflect the demands of the existing on-road evaluation, the complex segment added realistic challenges that were absent in the simple phase: e.g. vehicles cutting off the driver or ignoring traffic signals, potential head-on collisions on two-lane rural roads, and slower traffic forcing passing decisions.

Subjects

Brain injured persons volunteered for the study. Inclusion factors were: age 25-65, mild to moderate cognitive impairment, at least 3 months post injury with physician approval for a driving evaluation, active driver at time of injury, valid driving license, expressed an interest in driving as a goal, and sufficient literacy in English to read road signs. A sample of 5 adults (3 women, 2 men) participated. Cognitive deficit was assessed using the Neurobehavioral Cognitive Status Examination (NCSE, Cognistat) (Schwamm, *et al*, 1987). Participants ranged from 29-54 years old (mean = 40.4, s.d. = 11.04). The average length of time post injury was 1.9 years (range = .5 – 4 yrs, s.d. = 1.34). A control group was matched for sex and age.

Results

We collected and evaluated data for both discrete and continuous performance. We examined three types of discrete data that, in the real world, could result in a serious accident: running off the road, crashing into a moving or fixed object, and failing to stop at a STOP sign. We also considered an error that might be categorized as "failure to follow instructions", specifically the failure to turn at intersections where it was verbally directed. In addition, we evaluated the subjects' responses to the specific challenges presented during the complex scenario. We also studied continuous (steady state) driving performance on both straight and curved roadways. These findings are reported elsewhere (Wachtel, *et al,* 2001). One brain-injured subject failed to complete the drive. Thus our results are based on data from 10 subjects for earlier events and 9 for later events.

Driving Errors

None of the five control group members committed any discrete driving error, whereas each of the brain injured subjects committed at least one: 1 failed to stop at a STOP sign on two separate occasions, 3 ran off the road (one in the presence of pedestrians), and 2 were involved in a crash with another vehicle when they failed to observe that vehicle running a red light. In addition, two of the brain-injured subjects failed to turn at one or more required locations: 1 missed a single turn; the other missed 3 turns. None of the control group members failed to execute a required turn.

Response to Specific Demands

We developed three specific cognitive challenges for the complex scenario. In one, as the subject completes a left curve on a two-lane rural road, a bus approaches from 1000 feet away. The bus obscures a small sedan. When this sedan is within 8 seconds of the subject it pulls into the subject's lane to pass the bus. When the sedan and the subject's vehicle are within 2 seconds of colliding the sedan pulls back into its lane in front of the bus, leaving a brief opportunity for the subject to act by applying the brakes and/or making an evasive steering input, or the vehicles will collide.

We took our measurements from the point where the subject's action indicated recognition of the risk (i.e. the moment they removed their foot from the throttle or initiated steering). We measured time to collision, time from throttle to brake, and lateral position change (indicating evasive steering). Our measures showed generally poorer performance by the brain injured subjects. Figure 1, for example, shows the differences in lateral position shift. (Note that one brain injured subject actually steered into the path of the oncoming car).

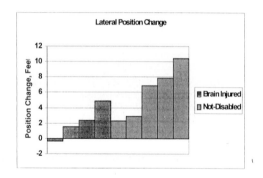

Figure 1. Lateral Position Change to Avoid Imminent Hazard

All subjects hit the brakes upon recognizing the hazard. As shown in Figure 2, however, brain injured subjects were generally slower to move from throttle to brake than were the non-disabled subjects.

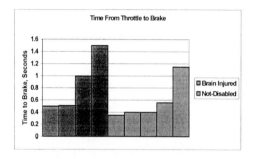

Figure 2. Time from Throttle to Brake in Response to an Imminent Hazard

A second situation involved recognition that a vehicle driving on a crossroad had failed to stop for a red signal. Thus, even though the subject was proceeding on a green light and clearly had the right-of-way, a crash would occur if the subject did not take timely evasive action. Two of the four impaired subjects crashed into this vehicle, whereas none of the five controls did so. This finding seems to indicate that brain-injured subjects might be susceptible to "improper lookout" and/or a lack of situation awareness, both deemed important to safe driving.

A third situation forced the subject to remain behind a slower vehicle for a considerable distance until the pavement markings permitted passing. This lead vehicle prevented the subject from maintaining the posted speed limit as directed. We measured the distance that elapsed before the subject executed a passing maneuver once the roadway pavement markings permitted passing. Figure 3 shows these data. All 4 brain injured individuals passed sooner than any of the 5 non-disabled individuals. The mean elapsed distance before passing was nearly twice as great for the non-disabled (143.2 ft) as it was for the brain injured (76.25 ft). These data seem to indicate a greater degree of impulsivity on the part of the impaired subjects and a general failure to attend to the potential for oncoming traffic prior to initiating their passing maneuver.

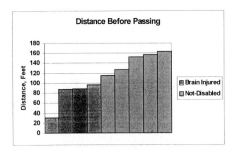

Figure 3. Distance Before Passing a Slow Moving Vehicle

Discussion

Heikkila *et al* (1999) argue that "driving should always be evaluated after stroke" (p. 349) but recognize that safety and availability of on-road tests often preclude use of those measures. Simulation, once used only in driving research, is just beginning to be used for driving assessment. Recent studies found that only 3% of TBI and 1% of CVA patients surveyed had been evaluated using a simulator (Fisk, *et al,* 1998).

Discrete driving errors are readily accessed from the simulator software, and thus can more readily support clinical evaluation and driver self-awareness. Our findings suggest that this type of data discriminates well between non-disabled subjects and those with brain injury. These findings are similar to those of Liu *et al* (1999) and Wald, *et al.*, (2000) who reported failed stops as a common error among brain injured adults during simulated driving. Further, our challenging events discriminated quite clearly between our two groups. Such events hold strong promise for the use of simulation in these settings, because they are plausible in real world driving, yet simply cannot be safely or ethically tested in on-road evaluations.

Although promising and consistent with other recent findings, our results are based on a small sample and are preliminary. Further, they may be confounded by differences between the groups unrelated to the variables of interest, because the brain injured participants had not driven, on average, for 2 years, while the non-disabled group drove regularly. It is also known that brain injured individuals have problems with fatigue (LaChapelle and Finlayson, 1998). Differences in the complex portion of the simulation could reflect differing fatigue levels during the later half of the drive rather than differing cognitive abilities.

References

· Aaronson D, & Eberhard J. An evaluation of computer-based driving systems for research, assessment, and advisement. *Behav.Rsch. Methods, Inst., & Cmptrs.* **26**(2): 195-197, 1994.
Brooke, M.M., Questad, K.A., Patterson, D.R. & Valois, T.A. (1992). Driving evaluation after traumatic brain injury. *Am J Phys Med Rehabil*, **71**(3), 177-182.
Fisk GD, Schneider JJ, Novack TA. Driving following traumatic brain injury: prevalence, exposure, advice and evaluations. *Brain Injury*. **12**(8): 683-695, 1998.
Fox GK, Bowden SC, Smith DS. On-road assessment of driving competence after brain impairment: review of current practice and recommendations for a standardized examination. *Am J Phys Med Rehabil* **79**:1288-96.1998.
Galski, T., Bruno, R.I., & Ehle, H.T. (1992). "Driving After Cerebral Damage: A Model with Implications for Evaluation." *Am J Occup Ther* **46**, 324-332.
Galski, T., Ehle, H.T., & Bruno, R.L. (1990). An Assessment of Measures to Predict the Outcome of Driving Evaluations in Patients with Cerebral Damage, *Am J Occup Ther*, **44** (8), 709-713.
Galski, T, Ehle, H.T., McDonald, M.A., Mackevich, J. Evaluating fitness to drive after

cerebral injury: basic issues and recommendations for medical and legal communities. *Journal of Head Trauma Rehabilitation.* **15**(3): 895-908, 2000.

Galski T., Ehle, H.T. & Williams, B. (1997). Off-Road Driving Evaluation for Persons with Cerebral Injury: A Factor Analytic Study of Predriver and Simulator Testing. *Am J Occup Ther*, **51**(5), 352-359.

Gianutsos R. Driving advisement with the elemental driving simulator (EDS): When less suffices. *Behavior Research Methods, Instruments & Computers* 26(2): 183-186, 1994.

Hays, R.T. & Singer, M.J. (1989*). Simulation Fidelity in Training System Design.* New York: Springer-Verlag.

Heikkila, V.M., Korpelainen, J., Turkka, J., Kallanranta, T., & Summala, H. Clinical evaluation of the driving ability in stroke patients. *Acta Neurologica Scandinavia* **99**:349-355, 1999.

Hirsekorn, L. and Taylar, S. VR technology applications in determining fitness to drive. *CyberPsychology & Behavior.* **1**(4) 1-5, 1998.

Klavora, P., Gaskovski, P., Martin, K., Forsyth, R.D., Heslegrave, R.J., Young, M., & Quinn, R.P. The effects of Dynavision rehabilitation on behind-the-wheel driving ability and selected psychomotor abilities of persons after stroke. *Am J Occup Ther* **49**(6); 534-542, 1995.

Korner-Bitensky, N., Sofer, S., Kaizer, F.,Gelinas, I. & Talbot, L. (1994). "Assessing ability to drive following an acute neurological event: are we on the right road?" *Can J Occup Ther* **(61)** 3, 141-148.

Korteling J.E., and Kapstein N.A. Neuropsychological driving fitness tests for brain-damaged subjects. A*m J Phys Med Rehabil* 77:138-146, 1996.

LaChapelle, D.L., & Finlayson, M.A. An evaluation of subjective and objective measures of fatigue in patients with brain injury and healthy controls. *Brain Injury.* 12(8):649-59, 1998.

Leigh-Smith J, Wade DT, Hewer RL. Driving after stroke*. Journal of the Royal Society of Medicine.* 1986.

Liu L, Miyazaki M, & Watson B. Norms and validity of the DriVR: A virtual reality driving assessment for persons with head injuries. *CyberPsychology & Behavior* 2(1); 53-67, 1999.

Marcotte, T.D., R.K. Heaton, et al. "Impact of HIV-related neuropsychological dysfunction on driving behavior." *J International Neuropsychological Society* 5(7): 579-592, 1999.

Risser, M. R., & Ware, J.C. "Driving Simulation with EEG Monitoring in Normal and Obstructive Sleep Apnea Patients." *Sleep* 23(3): 393-398, 2000.

Schwamm, L.H., Van Dyke, C., Kiernan, R.J., Merrin, E.L., & Mueller J. The Neurobehavioral Cognitive Status Examination: comparison with the Cognitive Capacity Screening Examination and the Mini-Mental State Examination in a neurological population. *Ann Intern Med.* **197**:486-491, 1987.

Sprigle, S., Morris, B.O., Nowachek, G., & Karg, P.E. (1995). Assessment of the Evaluation Procedures of Drivers with Disabilities. *Occup Therapy J Research* **15** (3), 147-164.

Wachtel, J., Durfee, W.K., Rosenthal, T.J., Schold-Davis, E., & Stern, E.B. (2001). "Evaluation of a Low-Cost, PC-Baed Driving Simulator to Assess Persons with Cognitive Impairments Due to Brain Injury," *Proceedings of the First International Driving Symposium on Human Factors in Driver Assessment, Training and Vehicle Design.* Iowa City, Iowa: The University of Iowa (in press).

Wald J, Liu L, Hirsekorn L, Taylar S. The use of VR with the assessment of driving performance in persons with brain injury" In J.D. Westwood, H.M. Hoffman, G.T. Mogel, R.A. Robb, and D. Stredney *Medicine Meets Virtual Reality 2000*, pp. 365 - 367. Washington, DC: IOS Press. 2000.

Wilson P, Foreman N, Stanton D. Virtual reality, disability and rehabilitation*. Disability and Rehabilitation.* **19**(6):213-220, 1997.

NAVIGATION AIDS:
A PREDICTIVE TOOL FOR INFORMATION DESIGN

Jason Duffield, Tracy Ross, Andrew May

Transport Technology Ergonomics Centre (TTEC)
Research School in Ergonomics and Human Factors.
Loughborough University
Holywell Building, Holywell Way, Loughborough,
LE11 3UZ

Landmarks can potentially enhance the guidance provided by current vehicle navigation systems. However, there is a need to ensure that if landmarks are presented, only 'good' landmarks are chosen, since poor landmarks are likely to be detrimental to the driving and navigation task. A regression model was derived based on data from a requirements study into valued landmarks, and from subjective ratings of the factors likely to explain the effectiveness of landmarks. A model significant at the 0.05 level explained a total of 34% of the variance in the dependent variable, with the most important factors being the degree to which a driver interacted with the landmark, the usefulness of its location, and its visual characteristics. Further analysis will be carried out to try to improve the predictive power of the model, before providing recommendations to industry.

Introduction

Future projections (Rowell, 1999, Zhao, 1997) imply that there will be a dramatic increase of vehicle navigation systems in the developed world. These projections are likely to mirror Japan where already 3.5 million vehicles have navigation systems installed (Rowell, 1999).

Based on a navigable database, navigation systems employ route calculation algorithms to calculate a journey to a destination, using GPS and map matching to work out the current location of the vehicle. Current systems are largely 'on-board' navigation systems, in that the navigable database is held on a CD in the car. Future systems may feature 'off-board' navigation, where telematics links transmit a 'strip map' of navigable data to the vehicle, either instead of a CD-based system, or to supplement this information.

In order to guide a driver to a destination, a navigation system will offer three basic aspects of functionality: – destination entry, turn-by-turn instructions and route overview. If an incorrect turning is taken, the system will re-route automatically (often without the driver even noticing). The integration of travel and traffic information will be the next development in navigation with systems automatically re-routing around congestion to achieve the most efficient route.

A feature of the majority of systems is that they rely heavily on distance to identify the next turn to the driver. A typical system may provide a first preview at 300m, a second preview at 150m and a final turn instruction at 50m. The use of distance is poor from a human factors perspective as drivers find it difficult to judge distances correctly (Burnett, 1998). Several authors have suggested that navigation systems could be improved by incorporating landmarks. In a survey of over 1100 drivers, Burns (1997) identified landmarks as one of the most important pieces of information drivers would like to receive (second to left/right directions with distance ranking fourth). The safety and acceptability benefits of using landmarks include reduced number of glances to the display (Burnett, 1998), less indicator errors (Bengler *et al*, 1994), increased confidence in identifying the correct turn (Alm *et al*, 1992) and higher user acceptance of the system (Green *et al*, 1993).

There is a need to ensure that appropriate landmarks are chosen to aid the navigation task; in particular to improve navigation performance and minimize any impact on the primary task of driving. This means presenting to the driver those landmarks that are optimal with regard to the driver's information requirements. These requirements include presenting landmarks when they aid the driver, presenting only the best landmarks when there are several available, and not presenting landmarks when they potentially detract from the navigation or driving task.

Presenting a 'good' landmark to a driver therefore either means selecting a potential landmark from existing information on the navigable database, or adding new information on that landmark to the database such that it may subsequently be presented to the driver during the navigation task. This issue of selecting the 'best' landmark is one where the navigable map database providers, the navigation system developers and the car manufacturers need support, and is the aim of the tool being described in this paper.

Selecting landmarks

There are various means by which suitable landmarks could be chosen for inclusion within navigation instructions. For a *particular* route, a list of landmark could be specified, based on:
(1) knowledge of that route/area, the geographical region and country;
(2) the specific objects or places on that route which that would appear to be useful landmarks;
(3) the individual characteristics of the driver (and hence the types of landmarks that they would tend to notice).

However, this 'specific' approach would only identify useful landmarks for that particular route, and generate a set that was most relevant to a particular set of drivers. Even if a generic list of landmarks was generated, based on experience of which ones were useful for a range of different routes, this would still tend to be driver, environment and/or country specific, and would not enable good landmarks to be chosen in contexts of use that fall outside of those in which the requirements were derived.

To enable generic rules for incorporation of landmarks within navigation system instructions, a theoretical basis is needed for choosing those which will aid the driver. This selection process could either apply to individual landmarks (e.g. St Mary's Church in Loughborough), or categories of landmarks (e.g. churches) if there was sufficient homogeneity within that category.

This paper describes such a basis for selecting 'good' landmarks, derived from basic perceptual, information processing and task related criteria. This is based on the derivation of a regression model that identifies the key constructs that determine whether a landmark is good or poor.

Having established a theoretical model that helps explain and predict 'good' and 'poor' landmarks, account can be taken of the operational constraints which influence how these selection landmarks might be undertaken in practice. The operational constraints will necessitate a simplified and more generic model and a cost-effective means of taking into account the most important factors that determine whether a landmark is 'good' or 'poor'.

Development of the predictive model

The aim of the current study was to develop a *predictive* model that could be used to identify the value of any object (or category of objects) for the navigation task. The intended basis for the model was a regression equation of the form:

$$V = (w1)F1 + (w2)F2 + (w3)F3 \ldots (w10)F10.$$

Where 'V' is value or effectiveness of the landmark, 'F' is an influencing factor and 'w' is the weighting that should be applied to that factor.

The intention was to develop a list of key predictive factors on which to rate individual landmarks. To generate a regression equation from this it was necessary that the landmarks to be rated had been assigned a 'Value' by some other means. This was achieved by performing a direction giving study which is reported elsewhere (Burnett et al 2001). Two groups of subjects provided navigation directions for a selection of routes. One group who were unfamiliar with the area used a video of the route ('Video' condition). The other group (who was very familiar with the area) provided directions from memory ('Cognitive Map' condition). The assumption was made that the more people who used a landmark and the greater the consensus between the two groups, the greater its value. This was represented by the equation $V = (Co + Vi) +/- (Co - Vi)/2$ where 'V' is landmark value, 'Co' and 'Vi' are the number of subjects mentioning that landmark in the Cognitive Map and Video conditions respectively, the second term always being negative.

The factors on which to rate the landmarks were developed by a process which began by identifying all possible features of a landmark that could influence its value. This was through analysis of the landmarks chosen in the direction giving study, the research literature and previous work by Burnett *et al*, (1994). The comprehensive list of features was then reduced by a grouping and sorting technique to ten main factors, shown in Table 1.

Table 1. Attributes and brief description

Attribute/Factor	Basic Description
1. Visual Effort for Scanning (VEFFSCAN)	How much effort is required to locate the object
2. Pre-Warning (PREWARN)	The appearance, before the object is actually visible
3. Familiarity (FAM)	Visual appearance of an object that would be familiar with a British driver
4. Ease of Naming (EOFNAM)	Extent to which the object can be given one unique, unambiguous name
5. Influence of Surroundings. (INOFSUR)	When close to the object, how easy is it to pick out an object from its surroundings
6. Similarity of Appearance (SIMOFAP)	Is the landmark similar to surrounding environmental structures
7. Usefulness of Location. (USEOFLOC)	In relation to its use either in (1) helping you to identify a manoeuvre, or (2) providing confirmation of your progress along a route
8. Level of Task Demand (TASKDEM)	The demand on the driver while they are looking for the object and using it for navigating. Not interested in the complexity of any manoeuvre the driver actually carries out
9. Degree of interaction. (DEGOFINT)	Degree of interaction with the object while driving
10. Visual Characteristics (VISCAR)	Visual aspects relating to the object itself

The GRADA, (Graphical Ratings Acquisition and Data Analysis) program was developed to enable participants to rate landmarks on each factor on a 0 to 10 scale. The data generated was then transferred into a SSPS program file. Five raters, experienced in navigation systems research, gave ratings for 40 landmarks (those generated by the direction-giving study) on all factors.

Results

A multiple linear regression with a stepwise method was used, with a significance level set at 0.05. The statistical package used, entered each predictor variable (factor) in a sequence and tested it against the criterion value (landmark value mentioned earlier). If the variable proved effective and contributed to the model, the statistics program kept the variable and re-tested the rest of the variables to see if they still contributed to the effectiveness of the model, until the 0.05 significance level was exceeded. The final output produced three variables that explained the greatest amount of variance in the dependent variable, noted in Table 2 below.

Table 2. Model Summary

R	R Square	Adjusted R square	Sig. F change
.594	.353	**.343**	.045

As there are a large collection of predictor variables (ten) it is wise to accept the model at the adjusted R square value. As the adjusted R square value is 0.343 it can be reported the model has accounted for a third of variance in the criterion variable.

Although the initial findings from the regression model suggest that the factors are 34% effective, the results from the ANOVA demonstrate that the regression model is significant, see Table 3 below.

Table 3. ANOVA

	Sum of squares	df	Mean square	F	Sig.
Regression	544.786	3	181.595	35.615	**.000**
Residual	999.379	196	5.099		
Total	1544.165	199			

For the regression model the Unstandardised Coefficients B will be used to calculate the weightings of each factor (see Table 4), for example:

$$V = c + (.300)F9 + (.333)F7 + (.210)F10$$

(NOTE: the low adjusted R square value suggests that this model should be used with caution and further statistical investigation is required).

Table 4. Coefficients

	Unstandardised Coefficients		Standardised coefficients		
	B	Std. Error	beta	t	Sig.
(constant)	-2.479	.775		-3.201	.002
DEGOFINT	.300	.061	.340	4.899	.000
USEOFLOC	.333	.089	.255	3.728	.000
VISCAR	.210	.104	.134	2.017	.045

Discussion and conclusions

The results to date have produced a regression model that, though statistically significant, would only be 34% effective at predicting the value of a landmark (i.e. 34% of the variance in the dependent variable was explained by the model, but it was statistically significant in explaining this level of variance). This is not unusual for a model based on subjective data, e.g. the explanation of variance for many models based on subjective rating data typically has an R squared value of this order. Further statistical investigation will be conducted to study in-depth the relative influence of each factor, and determine how a more powerful model may be derived. The unexplained variance within the model could have come from a number of sources: the most likely being: (1) a symptom of the 'Value' construct and the derivation of this from experimental data, (2) the theoretical link between the dependent and independent variables, and the likelihood of other influences on landmark effectiveness, (3) the concise definitions attached to the factors and the degree to which these are able to be reliably rated, and (4) issues to do with the validity and reliability of online rating scale techniques. It is also possible that there were individual landmarks, or individual factors that were particularly difficult to rate consistently.

Having undertaken further statistical analysis, the model will be simplified and adapted for practical use by industry.

Acknowledgements

This paper is based on work undertaken within the REGIONAL project, funded by the EPSRC. The authors wish to thank the other project partners (Jaguar Cars Ltd, Alpine Electronics of UK Ltd, Navigation Technologies, RAC Motoring Services and the Motor Industry Research Association).

References

Alm, H., Nilsson, L., Jarmark, S., Savelid, J., Hennings, U. (1992), (Swedish Prometheus, Tech. Rep. No. S/IT-4). Linköping, Sweden: VTI.

Bengler, K., Haller, R. and Zimmer, A. (1994) In First World Congress on Applications of Transport and Intelligent Vehicle Highway Systems, Vol. 4 Artech House, Paris, France, pp. 1758-1765

Burnett, G.E. (1998). "Turn right at the King's Head": Drivers' requirements for route guidance information. Unpublished PhD dissertation, Loughborough University, UK.

Burnett, G. E., A. J. May and T. Ross (1994). The Prediction of the Effectiveness of Landmarks for Use Within a Route Guidance System. CEC DRIVE II Project V2008 HARDIE Deliverable 15. HUSAT Research Institute: 45pp.

Burnett, G., D. Smith and A. May (2001). Supporting the navigation task: characteristics of 'good' landmarks. Contemporary Ergonomics 2001: Proceedings of the Annual Conference of the Ergonomics Society, Taylor & Francis.

Burns, P. (1997) In Human SciencesLoughborough University, Loughborough, UK.

Green, P., Hoekstra, E., Williams, M., Wen, C., George, K. (1993b). Examination of a videotape-based method to evaluate the usability of route guidance and traffic information systems, (Tech. Rep. No. UMTRI-93-31). Ann Arbor, MI: University of Michigan Transportation Research Institute.

Rowell, J. M. (1999) In Proceedings of seminar on Integrated Solutions for Land Vehicle NavigationMotor Industry Research Associaton (MIRA), Nuneaton, UK, pp. 1-8.

Zhao, Y. (1997) Vehicle location and navigation systems, Artech House, Boston.

ASSESSING THE 'DOH!' FACTOR: THE ROLE OF PERFORMANCE FEEDBACK IN VEHICLE NAVIGATION SYSTEMS

Andrew Bradbeer[1], Sarah Nichols[1], Gary Burnett[2]

[1]*School of Mechanical, Materials, Manufacturing Engineering and Management, University of Nottingham;* [2]*School of Computer Science and Information Technology, University of Nottingham*

This paper describes an on-road experiment investigating the effects of performance feedback on drivers' abilities to judge distances in a navigation context. Twelve participants (6 male and 6 female) aged between 20 and 40 were driven around an experimental route and asked to judge which turnings were at a specified distance (200m or 300m) by the experimenter and to give a confidence rating of their decision. Half the participants were given feedback as to which was the correct turning. There was a trend for females to perform better, although they were significantly less confident than males. Feedback did not significantly improve performance but an associated learning effect was found, in which feedback gave a steady reduction in errors, but with no feedback this improvement was absent. The results and their potential consequences for the design of in-vehicle navigation systems are discussed.

Introduction

Feedback has been defined as the 'sending back to the user information about what action has been done [and] what result has been accomplished' (Norman, 1988, p.27), and is a strongly advocated human factors principle (Sanders and McCormick, 1992). In the context of guided training, feedback about errors has been shown in many application domains to be a valuable tool for improving task performance (Wickens and Hollands, 2000).

This paper addresses the role of performance feedback within GPS-based vehicle navigation systems. Such technology, as an example of ubiquitous computing, is increasingly available to the everyday driver to support him/her in wayfinding and navigation tasks. With respect to the Human-Machine Interfaces (HMIs) for current systems two aspects are of relevance here:

- The emphasis placed on absolute distance information as a primary means for drivers to identify an oncoming manoeuvre (e.g. the use of voice messages such as 'left turn in 300 metres'). Absolute distance judgement, particularly in a dynamic environment, is a difficult cognitive task, and previous empirical work has quantified this level of demand in the driving and navigating context (Burnett, 2000).

♦   The lack of explicit feedback provided to drivers on navigational performance (correct/wrong turnings). Typically, with current systems, if a driver makes a wrong turning a new route is calculated automatically (often in less than a few seconds). As a result, error feedback is only offered in an *implicit* fashion, via a brief on-screen message (e.g. 'route re-calculating') and/or the context of the error and the new route instruction (e.g. an up-dated voice message 'left turn in half a mile').

This paper describes a road-based study which investigated a range of issues relevant to the use of distances within the HMI for vehicle navigation systems. For the purposes of this paper, the focus will be on the impact of explicit performance feedback on drivers' abilities and confidence in judging navigational distances in a moving vehicle.

Method

Twelve participants, six male and six female, (age range 20-40) completed the experiment. All participants had been driving with a full driving licence for at least two years, and had driven an average of 14,900 miles in the previous twelve months (range 1,000-62,000).

A post-experiment questionnaire was administered to obtain information about participants' individual characteristics and driving experience. In particular, the questions asked about any strategies used to judge distances during the experiment and preferred units of distance (imperial or metric).

The independent variables were performance feedback (feedback or no feedback) and gender (both between subjects), and the dependent variables were accuracy of distance judgement and confidence in distance judgement. In addition to the IVs of performance feedback and gender, the impact of road type, distance to be estimated and direction of turning were also examined, but results from these variables will not be reported in this paper.

Participants were seated in the front passenger seat of the experimental car, which was driven by an assistant. The purpose of the experiment was explained to the participant, and they were allowed to practice the task of estimating a distance, to familiarise themselves with the question and response format. They were then driven to the start of the selected routes. Two routes were used during the trials, one of which consisted primarily of a straight road, the other being a curved road. Both routes were deliberately chosen so that a number of potential side turnings (left and right side) were present along them.

All participants completed both routes, and the order of presentation of the two routes was balanced. Participants were then issued a judgement instruction, which would be in the typical form of "left turn, 200 metres from now". Participants indicated which turning they thought was at the requested distance (as that turning was passed), and then rated their level of confidence on an interval scale of 0 to 10, where 0 indicated no confidence and 10 indicated total confidence. In the case of the group receiving feedback, the participants were told if their decision was correct, and if not which was the correct turning. This was repeated for all sixteen judgements. A constant speed of 30 mph (± 3mph) was maintained by the driver while the judgement was being made. If during any judgement the driver was required to slow down excessively or stop, that particular judgement was repeated.

Analysis and results

Table 1 shows the overall results for performance (number of correct distance judgements) and confidence ratings grouped according to gender and the presence/absence of feedback. The results show some differences related to gender, but no differences with feedback. The table also shows the wide variability present in the data as measured by standard deviations.

Table 1. Mean performance scores and confidence ratings (with standard deviations in brackets) related to gender and feedback condition

Independent Variable	IV levels	Mean performance (SD) (Maximum possible score = 16)	Mean confidence (SD) (0 = no confidence, 10 = total confidence)
Gender	Male	7.8 (3.66)	7.3 (1.15)
	Female	11.0 (1.79)	5.6 (1.09)
Feedback condition	Feedback	9.5 (3.27)	7.0 (1.49)
	No feedback	9.3 (3.44)	5.9 (1.13)

To investigate whether a learning effect was present, a plot was made of the total number of errors made by participants over subsequent exposures (i.e. for each of the 16 distance judgements made in turn). Figure 1 provides the results of this analysis and reveals evidence of some limited learning when feedback was provided.

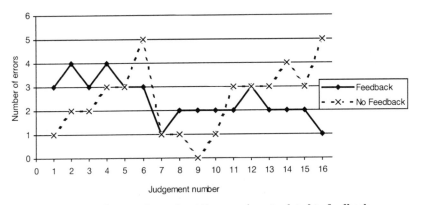

Figure 1. Number of errors throughout the experiment related to feedback

A between subjects 2*2 ANOVA was performed for the IVs gender and feedback and the DV of performance. Although the descriptive data suggests that females performed better than males, no significant main effects or interactions were obtained (Gender: $F=3.059$; $df=1,8$; $p>0.05$; Feedback: $F=0.008$; $df=1,8$; $p>0.05$; Interaction: $F=0.415$; $df=1,8$; $p>0.05$).

This lack of a significant result could be attributed to the large variability in the data and the relatively small number of participants.

The same analysis was performed for the DV of confidence. Females were found to be significantly less confident than males ($F=8.847$; $df=1,8$; $p<0.05$). The overall effect of feedback on confidence was not significant ($F=3.858$; $df=1,8$; $p=0.085$) and there was no significant interaction ($F=0.050$; $df=1,8$; $p>0.05$).

The strategies reported by the participants were categorised by type of distance measurement. The vast majority of participants used some form of relative measurement of distance related to a mental picture of a distance. In some cases this mental picture was related to a definite object (e.g. a swimming pool or a running track) and in some cases it was related to an estimate of a distance 'chunk' found on a route (e.g. residential plots or dimensions of concrete paving slabs). Two participants used absolute strategies that were time-based (counting seconds).

Discussion and conclusions

Contrary to expectations, the results showed that there was no *overall* difference in performance between those who received feedback and those who did not. Nevertheless, Figure 1 shows that performance over repeated exposures did tend to increase with feedback whereas there was no evidence for this effect without feedback. Such a result indicates the need for further research in which participants make distance judgements either over a longer exposure period or following a longer training period.

There was considerable variability in the data and several individual effects may have accounted for the lack of any strong differences due to feedback - two particular effects are discussed here. A study with a greater number of participants could help to understand these individual effects better.

Firstly, it was clear that *gender* had a significant impact on task performance and confidence ratings. Such findings are consistent with the literature which suggests that males and females react differently to feedback, with males showing overconfidence and a reluctance to accept information (Mannering et al., 1995) and females making use of clear, unambiguous feedback (Lenney, 1977).

Secondly, it was evident from the self-reporting of *strategy* that people responded in varying ways to feedback. Of the six people who received feedback, three stated that they either changed or revised their strategy because of the feedback received. As reported above, the most common strategy was distance estimation in relation to a fixed mental image. Significant improvements in performance occurred when participants switched from this strategy to a time-based strategy, using feedback to determine the time equivalent for the distance to be judged.

As noted in the introduction to this paper, it is commonly assumed that the inclusion of feedback in human-machine systems will have a positive effect on performance. However, Wickens and Hollands (2000) highlight the fact that decision makers often suffer from biases when presented with feedback. These biases include *misleading feedback*, where the user interprets positive feedback as a sign that the decision rule (such as a time-based distance estimation strategy) employed is generically applicable because it resulted in a correct outcome on one occasion, and *selective perception of feedback*, where decision makers tend to attend to positive feedback rather than negative. A further factor that could have had an impact on the results was the nature of the feedback itself. In this study, primarily for practical reasons, human verbal feedback was provided. Zakay (1992) found that people (especially naïve users) respond more favourably to computerised feedback. Zakay hypothesises that this occurs because people generally see computers as infallible and so do not question their judgement. A further factor which could have had an impact on the results was the nature of the feedback itself. In this study, primarily for practical reasons, human verbal feedback was provided. A study by Zakay (1992) found that people (especially naïve users) respond more favourably to computerised feedback. Zakay hypothesises that this occurs because people generally see computers as infallible and so do not question their judgement.

As a final point, the fact that so many people used relative distance strategies supports the use of distance countdown bars on a navigation display. However, instead of abstract 'bars' it may be more beneficial to relate distances to commonly understood distance models (e.g. a graphic of a swimming pool or running track).

In conclusion, this study, although limited in size, was effective in identifying many points of interest. The use of explicit feedback in the HMI for vehicle navigation systems shows some potential, but further research is required to investigate in depth the various individual differences present and the design of the feedback information.

Acknowledgements

This authors would like to thank the Motor Industry Research Association (MIRA) for the use of the vehicle, in particular Mark Fowkes and Zaheer Osman.

References

Burnett, G. 2000, "Turn Right at the Traffic Lights": The Requirement for Landmarks in Vehicle Navigation Systems, *Journal of Navigation,* **53(3)**, 499 - 510

Lenny, E. 1977, Women's Self-Confidence in Achievement Settings, *Psychological Bulletin,* **84(1)**, 1 – 13.

Mannering, F., Kim, S-G., Ng, L. and Barfield, W. 1995, Travelers' preferences for in-vehicle information systems: an exploratory analysis, *Transportation Research – Part C,* **3(6)**, 339 – 351.

Norman, D.A. 1988, *The Psychology of Everyday Things,* (Basic books, New York)

Sanders, M.S. and Mccormick, E.J. 1992, *Human Factors in Engineering and Design,* 7th Edition, (McGraw-Hill, New York)

Wickens, C.D. and Hollands, J.G. 2000, *Engineering Psychology and Human Performance,* 3rd Edition, (Prentice-Hall Inc., New Jersey)

Zakay, D. 1992, The influence of computerized feedback on confidence in knowledge, *Behaviour and Information Technology,* **11(6)**, 329 – 333.

INFORMATION REQUIREMENTS FOR FUTURE NAVIGATION SYSTEMS

Steve Bayer, Andrew May, Tracy Ross

Transport Technology Ergonomics Centre (TTEC)
Loughborough University, Holywell Building,
Holywell Way, Loughborough,
Leicestershire, LE11 3UZ

An experimental route description study was undertaken to identify drivers' requirements for navigation information within a complex urban environment. A factorial design was employed, with participants writing down navigation instructions based on either watching a video of urban routes, *or* using extensive local knowledge of those routes. Data was analysed in terms of the information used, its role within the navigation task, and the importance of that information. Results showed that the most popular information category used for navigation was landmarks, followed by junction description; these were both predominantly used as primary (as opposed to redundant information). Information was used to both help identify manoeuvres *and* confirm progress along the route.

Introduction

Navigating (i.e. finding your way) in unfamiliar road environments is a common and demanding cognitive activity for drivers (Burnett, 2000). Research has long highlighted the problems that drivers have in planning and following efficient routes to destinations (e.g. Streeter and Vitello, 1986; Wierwille *et al*, 1989). If efficient routes cannot be planned and followed, the consequences are stress, frustration and delays for the driver, potentially unsafe road behaviour (e.g. late lane changes) and inappropriate traffic management such as traffic diversions through small villages. Vehicle navigation systems offer a technological solution to these problems enabling drivers to navigate unfamiliar routes, visit Points of Interests on route and arrive at a predefined destination. Due to reductions in their cost, vehicle navigation systems are now commonly offered as options on medium 'family sized' cars and are appearing at the compact end of the market.

In order to provide the driver with navigation instructions, most systems currently provide 'turn-by-turn' instructions to the driver: for each turn or manoeuvre the driver is required to make, the navigation system will present a symbol to indicate the type and direction of the turn that is required, and simultaneous voice guidance. A visual indication of the distance to the next manoeuvre may be given via a count down bar. The turn-by-turn instructions are supplemented with map overlays of either the whole route, or individual junctions, or both. In addition, some systems also incorporate basic road-sign information for motorways, and Points of Interest such as hotels, petrol stations and car dealerships.

The factors determining *what* information is presented to the driver by navigation systems are a combination of historical (this is how it has been done in the past), technological (this is technically feasible at this price point) and pragmatic (this is actually do-able). Human factors criteria have not necessarily been the key drivers.

By identifying the information actually used by drivers for navigation (and enabling the design of enhanced navigation systems), there is the potential to: (1) enable navigation systems to be more effective in aiding navigation decisions; (2) reduce the cognitive effort and distraction imposed by these systems; (3) design systems which are more accepted by the user.

A limited number of previous studies *have* either empirically tested the provision of varying forms of navigation information to the driver used survey or route description techniques to identify preferred information elements. For examples, see Burns (1997); Burnett, (1998); Akamatsu *et al*, (1994); Alm, (1990).

The aims of this study were (1) to identify the key navigation information categories drivers use in a complex, unfamiliar environment, (2) identify how these information categories were used to support the navigation task, and (3) identify the relative importance of these information categories at a task level.

Method

The study was based on a factorial design with 36 subjects in total; 18 of whom used a recorded video of a series of complex urban routes to write down directions that a driver unfamiliar with those routes could use to navigate them successfully. These subjects had no prior knowledge of the routes provided and based their directions purely on the video. The other 18 participants had extensive local knowledge of the test routes, and provided directions based on recall of the salient features of the routes. These subjects used a schematic representation of those routes, where all road names and other map features were removed. Only the start and end points were labelled on these schematic maps. This dual approach attempted to address some of the methodological concerns with these requirements capture type studies, and captured both the visual and cognitive aspects of information employed within a navigating task. It also makes common sense: where information items are used to help with way-finding, they are likely to be those which people can see easily, and those which people can remember and assess as important.

An initial pilot study with 32 subjects was undertaken to refine the experimental procedure; results are not presented here.

Subject, Route & Questionnaire Design

A quasi-random sampling technique was used to identify a subject sample reasonably representative of the driving population (as opposed to the current users of navigation systems). Subjects were balanced for gender and age as far as possible; age varied between 20 and 65.

Three routes were used within the experiment: an urban route skirting a town centre and including a complex ring-road, a route through a town centre, and a out-of-town route comprising mostly dual-carriageway. As suggested in the literature (e.g. Burnett, 1998), complex routes were chosen that made the experimental task demanding. The UK city of Coventry was used, as it provided the diversity of driving environments, and for pragmatic reasons of subject selection. Across all three routes, there were 41 main navigation decision points. This is where the driver would expect to receive some form of navigation instruction.

A detailed pre-trial questionnaire was used. This questionnaire was largely based on earlier work by Burnett (1998) and Lawton (1994). The questionnaire covered user demographic details, subjects driving experience and the different types of strategies subjects used when navigating in unfamiliar places. Due to space limitations, the questionnaire results are not presented in this paper, but are available from the authors on request. A post trial field dependency test was carried out; results are not reported here.

Procedure

On arrival, subjects received standardised written and verbal instruction on the experimental procedure. A consent form was signed, and the questionnaire administered. Subjects were asked to imagine that they providing written directions to another driver wishing to navigate a series of routes, this driver being totally unfamiliar with the routes in question. Subjects were asked to use written textual descriptions only, and to refrain from using diagrams. No further limitations were placed on the subjects.

Subjects then wrote down directions based on either using the schematic maps, or the video, without any time limit being imposed. The video subjects were allowed to rewind and fast-forward the tape as many times as they liked until they were happy with the directions they had written down. At the end of the trial, a field dependency test and debrief were carried out.

Data coding

Each information element contained within the written instructions was coded according to the following criteria: (1) what the information element was, categorised according to the schema shown in table 1 below; (2) whether that information was used at a manoeuvre or between manoeuvres (progress point); (3) whether the information was used to preview, identify or confirm the manoeuvre or progress point; and (4) whether the information was used as primary or secondary information at each manoeuvre or progress point. Primary information was defined as that *needed* to identify any manoeuvre or progress point; if this information was not present, the driver would either be unable to complete a manoeuvre, or identify a progress point on the route, or would have high levels of uncertainty.

Results

The following results show *what* navigation information elements were used within the scenario and *how* these information types are used by the driver. The data was categorised according to the main information type as shown in table 1. Other categories with low frequency of use are not shown here. Some main information types were further subdivided e.g. landmarks into subcategories like traffic lights, petrol stations etc, however these detailed results are not shown in this paper.

Table 1. Definitions of main information categories

Information type	Examples
DIST Distance	e.g. turn left soon, turn left in 200m
DSN Direction sign	e.g. follow signs to ....
EN Environment	e.g. Residential area, park, industrial area, commercial,
GN Geometry of node	e.g. sharp, gentle, (incl. Large small e.g. for roundabout)

GP Geometry of path	e.g. Bendy, straight, long, narrow, splits into, hilly
JN Junction	e.g. cross roads, T-junction, roundabout, Y-junction
LC Lane change	e.g. get in left hand lane
LM Landmark	e.g. Pub, Petrol Station, Traffic Lights,
RT Type of road	Visual appearance, e.g. Ring road, dual carriageway etc.
SNN Street name/ no	e.g. turn onto Holyhead rd, turn onto the A6

Graph 1.

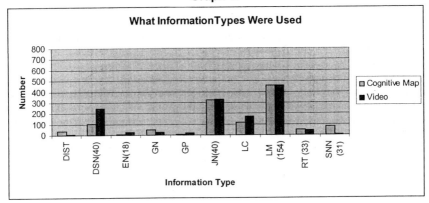

Graph 2.

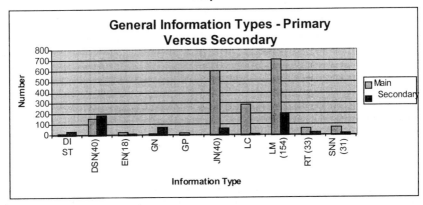

Graph 1 shows how many times in total subjects referenced a particular category of information e.g. lane change. This graph represents the top ten information types used by subjects. The numbers in the brackets on the x-axis represent the *actual number* of individual objects in that category that were mentioned by the subjects e.g. for the LM category, Town Hall, Holyhead pub would be counted as 2 objects). This number does not represent the *total number* of instances there were on the route. On some information categories there are no numbers in brackets this is because these information types have an infinite amount of times they could be used (they are not physical objects).

Graph 2 represents how the information given by the subjects was used differently. This relates to primary versus secondary information. The graph also shows the differences in use according to information type. As before the graph represents the top ten information types even though other categories were considered.

Discussion

The findings indicate that landmarks are the predominant information category (Graph 1). They are the most widely used information category across both the cognitive map and the video condition. This relates to the way people naturally give directions to drivers unfamiliar with an area. Overall there were 154 different landmarks mentioned and over 900 references to these in total. There is an equal split between the cognitive map and video conditions in relation to the number identified. This finding implies that landmarks are both visually noticeable and easy to store in the long-term memory.

Junction information is the next most used category. It is useful to identify to the driver what type of junction they are approaching prior to them reaching it. This increases driver situational awareness and decreases the opportunity for driver error. In relation to the different conditions there is a fairly equal split in the identification of each junction. In the video condition all subjects identified all of the junctions. However in the cognitive map condition some junctions were omitted. This could be because, from memory two junctions may be remembered as one or a junction has recently been built.

A directional sign was another used information type. It is a navigational aid that is already used by drivers to help them reach their desired destinations. It was found to be more predominant in the video condition as their visibility causes them to be used. In the cognitive map condition it is unlikely that subjects will recall all available signs especially if, as locals, they rarely use them. Another freqent information type was lane change. This information is important to be presented prior to the junction so that driver workload can be minimised and also any risks of accidents reduced. It was identified more frequently in the video condition. This can probably be explained in the same way as direction signs.

Distance information is currently used by navigation systems as the main way of identifying the location of a manoeuvre. However it was only the seventh most used information type. Also it was never identified in the video condition. This can be explained by the difficulty subjects had in judging a distance from watching a video plus the predominance of other more visually identifiable objects. Even so its usefulness as an information type is still very low for one that is currently being adopted by industry.

Graph 2 shows that landmarks and junctions are the main information categories identified by the subjects (as shown in Graph 1). Also when these information types have been identified they are being used as primary or secondary information. A reason for landmarks being used frequently, as primary information is that they are valued as information items by drivers. In a survey of 1100 drivers they were rated the second most popular information type (after left-right directions) for aiding navigation (Burns 1997). Lane change is another category that is used as primary information. This is because making an incorrect lane change when approaching a manoeuvre will add to the complexity of carrying out that manoeuvre and increase both the cognitive load of the driver and the potential for traffic conflicts. As a result it is important to present this information type as primary information rather than secondary.

The findings showed that distance is mainly used as secondary information. This is because people find it difficult to estimate distances with any precision thus the efficiency and reliability of this information type when compared to landmarks and junctions is low.

Conclusion

The findings associated with this investigation enable us to determine the primary information types that driver's use when navigating. The results indicate that distance, as an information category is not sufficient for effective navigation. The inclusion of a number of other information types into the navigation databases is likely to increase driver comfort, traffic safety and navigation performance. The candidate information types are landmarks, junction type and direction signs. Some navigation systems already make use of the latter types of information. However this data is not recorded comprehensively on navigable databases. Landmarks are a category rarely incorporated in current databases except as Points of Interest (POI's) i.e. a place to go to rather than navigate by. Further work in this research area will concentrate on how different information types are combined with navigation instructions and the features of a manoeuvre that cause particular information types to be used.

Acknowledgements

This paper is based on work undertaken within the REGIONAL project, funded by the EPSRC. The authors wish to thank the other project partners (Jaguar Cars Ltd, Alpine Electronics of UK Ltd, Navigation Technologies, RAC Motoring Services and the Motor Industry Research Association).

References

Akamatsu, M., Yoshioka, Imacho, N., Kawashimo, H. (1994). Driving with a car navigation system - analysis by the thinkin aloud method. International Ergonomics Assocciation (IEA), Human Factors Association of Canada

Alm, H. (1990) In Laboratory and field studies on route representation and drivers' cognitive models of routes (DRIVE II V1041 GIDS, Deliverable GIDS/NAV2)(Eds, Winsum, W. v., Alm, H., Schraggen, J. M. and Rothengatter, J. A.) Groningen, The Netherlands: University of Groningen, Traffic Research Centre, pp. 35-48.

Burnett, G.E. (1998). "Turn right at the King's Head": Drivers' requirements for route guidance information. Unpub. PhD dissertation, Loughborough University, UK.

Burnett, G.E. (2000) ""Turn right at the traffic lights" The requirement for landmarks in vehicle navigation systems", The Journal of Navigation, **53**(3), 2000, pp.499-510

Burns, P. (1997) In Human SciencesLoughborough University, Loughborough, UK.

Lawton, C.A., (1994) Gender differences in way-finding strategies: Relationship to spatial ability and spatial anxiety. Sex Roles, **30**, (11/12), 765-779.

Streeter, L. A., Vitello, D. (1986) Human factors, **28**, 223-239.

Wierwille, W. W., Antin, J. F., Dingus, T. A. and Hulse, M. C. (1989) In Vision in vehicles II(Ed, Gale, A. G.) London: Elsevier Science,, pp. 307-316.

IN-CAR ASR:
SPEECH AS A SECONDARY WORKLOAD FACTOR

Alex W. Stedmon[1], John Richardson[2] and Steven H. Bayer[3]

[1]Virtual Reality Applications Research Team
School of MMMEM, University of Nottingham, Nottingham NG7 2RD
[2]Transport Technology Ergonomics Centre
Loughborough University, Loughborough, Leicester LE11 3UZ
[3]Department of Human Sciences
Loughborough University, Loughborough, Leicester LE11 3TU

This paper presents two complementary experiments that investigated the effects of workload on driving and on the use of a speech interface whilst driving in a simulator. The experiments were designed so that comparisons could be made between the two situations, to investigate differences between identical conditions in each, and also simple speech effects by comparing matched conditions with a speech interface present or not. The results illustrated how different workload factors impact on the safe use of a speech interface whilst driving. In addition, whilst participants did not consider the *use* of the speech interface itself to be demanding, *using* it contributed to poor performance when other workload factors were present. Guidelines are presented that indicate the need to understand how dynamic workload is and the importance of integrating in-car systems so that they support the user in a variety of driving environments. Only by developing an understanding of how driver's may want to *use* a speech interface and the impact of *using* it has on their driving abilities will it be possible to design an interface that truly meets the expectations of the user and supports safe driving.

Speech Recognition, Driving Behaviour & Workload

The successful application of Automatic Speech Recognition (ASR) in the driving domain relies upon the careful design of the interface to match the expectations, preferences and abilities of various user groups (Stedmon & Bayer, 2001). What might be a potential aid could just as easily prove to be hazardous if it distracted drivers from safe control of their vehicles (Stein, Parseghian, & Wade Allen, 1987). As standard in-car systems generally require drivers to operate visual displays and manual controls whilst driving, ASR may potentially improve the usability and safety of in-car systems including voice-dialling of mobile phones, operating entertainment systems and Intelligent Transportation Systems (ITS) such as route guidance or travel/traffic information services (Graham & Carter, 2000).

The use of ASR to control in-car systems is fundamentally different from a number of other ASR applications as the task of using ASR is secondary to the primary task of safe driving. The driver's ability to attain an acceptable level of performance in terms of the ASR process (commands, menus, vocabulary, etc) may be constrained, as the car can be considered a 'hostile environment' for ASR (Baber & Noyes, 1996). Such environments are typically characterised by high levels of noise, stress and workload affecting the way people talk and how their speech is recognised by the interface.

Maintaining driver workload within acceptable boundaries (i.e. preventing underload or overload) is a primary objective when new in-car systems are integrated into a vehicle. Workload may arise from various sources (Graham *et al*, 1998):

- the driving task itself (e.g. lane-keeping, speed choice, keeping a safe headway and distance from other vehicles);
- the driving environment (e.g. traffic density, poor weather, road geometry, etc.);
- the use of in-vehicle systems (e.g. the presentation, amount and pacing of information to be assimilated and remembered).

However, workload is more than merely "doing a task, it encompasses an individual's perception of its complexity in relation to their ability to perform it" (Stedmon *et al*, 2001). In order to assess the effects of workload, it is useful to combine measures that take account of an individual's perception of their workload as well as their ability to perform a task.

A number of studies have sought to correlate different measures of workload, suggesting that psycho-physiological measures such as heart rate and heart rate variability serve as an index of mental workload (Aasman *et al*, 1987). Objective workload may be measured as a function of the task difficulty/environment. In the driving domain workload may be manipulated through driving conditions (traffic, weather, road surface conditions, time of day etc). Measures for assessing driving performance include: accidents, speed, lane deviation, speeding tickets, and traffic light violations (Stein *et al*, 1987). Accidents are a clear measure of traffic safety and can be caused by lapses in attention, excess speed, poor speed control or poor lane keeping. Excess speed may cause a driver to lose control of their vehicle (due to road geometry or hitting obstacles) whilst driving faster or slower than other road users (poor speed control) may increase the likelihood of an accident. Lane deviation is another indicator of driving performance. If a driver's ability to maintain lane position is impaired, then the probability of exceeding the lane boundaries and hitting another object/vehicle also increases (Stein *et al*, 1987). Speeding offences and traffic light violations may be taken as indicators of driver Situational Awareness (SA). These parameters provide some indication of driver vigilance over the speedometer (inside the vehicle), and road signs, traffic lights and the behaviour of other traffic (outside the vehicle).

By reviewing previous research, it is clear that driving behaviour may benefit from a transfer of loading from the over-burdened visual/manual modality to the auditory modality (Graham & Carter, 2000). However, the underlying assumption that speech exists as an untapped resource is a contentious issue, for speech may already be an active or semi-active mechanism (Linde & Shively, 1988). By using multiple measures of workload, therefore, it is possible to assess relationships between actual task difficulty (objective measures), perceptions of task difficulty (subjective measures) and how individuals react to their perception of task difficulty (psycho-physiological measures).

Baselining Behaviour - Driver Workload & Speech

Two complementary experiments investigated the effects of workload on the use of a speech interface and on driving behaviour in a simulator. Each experiment is covered in more detail in Stedmon *et al*, (2001) and Stedmon & Bayer (2001).

Experiment 1:   Baselining Behaviour - sought to define a range of baseline driver workload factors so that future comparisons could be made for workload effects on the use of in-car ASR.   Traffic behaviour (density, flow, speed changes, etc) and road layout/conditions (geometry, speed restrictions, fog, etc) were manipulated to assess the validity of different workload levels (Stedmon *et al*, 2001).

Experiment 2:   Driver Workload & Speech Trials - investigated a range of driver workload factors and the use of an in-car ASR interface.   Within a series of scenarios traffic behaviour and road conditions were manipulated to assess the impact of different workload levels (Stedmon & Bayer, 2001).

The design adopted made it possible to compare matched conditions between the two experiments. It was thus possible to consider the impact of the workload factors with and without use of the speech interface.

Method
A summary of the common method for both experiments is presented below.   Participants were required to drive a number of scenarios with varying levels of workload.   In Experiment 1 no speech interface was used, in Experiment 2 a smaller number of conditions were used and all but one combined the use of a speech interface for various driving related tasks.

Participants - all held full UK driving licences, drove at least 2-3 times per week and 6,000 miles a year.   All had normal, or corrected to normal, vision, did not wear pacemakers and were not taking any prescribed medication.

Apparatus - driving scenarios were generated and displayed using a full-size, interactive, driving simulator running STI-Sim simulator software.   Heart rate data were collected on ADI-Instruments MacLab/8 & Bio-Amp hardware and Chart v3.5 software. 2020Speech Mediator 6 software was used on an Aurix speech recogniser with a head mounted microphone and Press-To-Talk (PTT) switch on the dashboard.   NASA-TLX and Bedford-Harper subjective workload questionnaires were administered.

Design Repeated measures, within-subjects, designs were used.   The independent variable (workload) was manipulated across time, traffic and fog in both experiments, and also for speech in Experiment 2.   To minimise any carry-over effects, conditions were counterbalanced across subjects using a Latin Square.   Dependent variable measures were collected for psycho-physiological workload (heart rate & heart rate variability), subjective workload (Bedford-Harper & NASA-RTLX scores), and objective workload (accidents, lane deviation, speed, speeding & traffic light violations).

Baseline comparisons between Experiments 1 & 2

Mean data scores for the same condition in each experiment were obtained and analysed using T-tests. Few significant effects were observed.   Heart Rate was higher in Experiment 1 [$t(19) = 2.704$, $p<0.05$] whilst Vehicle Speed was slower [$t(19) = 2.537$, $p<0.05$].   No other significant effects were observed ($p>0.05$).

These results illustrated a degree of consistency between the experiments.   The reason for the lower Heart Rate in Experiment 2 is not fully understood.   Taken with the increased Vehicle Speed, this finding might highlight an aspect of cognitive dissonance. The condition in Experiment 2 was the only one without speech, whereas in Experiment 1 the whole of the experiment was conducted without speech.   Participants may have considered, therefore, the 'no-speech' condition to have been different in some way to the rest of Experiment 2, perhaps not taking it as seriously or relaxing knowing that they did not have to attend to the interface and speeding up as a result.

Driver workload & speech

Mean data scores for the conditions compared in each experiment were obtained and analysed using T-tests.

Heart Rate - significant effects were observed between Experiments 1 & 2. Heart Rate was lower in Experiment 2 across all four conditions:
- Control: $t(19) = 3.265$, $p<0.001$
- Traffic: $t(19) = 2.123$, $p<0.05$
- Time & Traffic: $t(19) = 3.993$, $p<0.01$
- Traffic & Fog: $t(19) = 3.430$, $p<0.01$

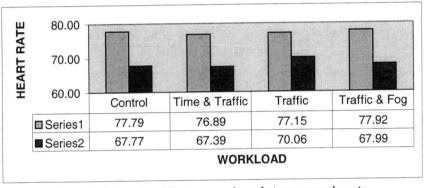

	Control	Time & Traffic	Traffic	Traffic & Fog
Series1	77.79	76.89	77.15	77.92
Series2	67.77	67.39	70.06	67.99

Figure 1: Mean Heart Rate comparisons between experiments

Road Traffic Accidents - only one significant effect was observed between Experiments 1 & 2. More accidents occurred in Experiment 2 than 1, but only under high time pressure and with increased traffic density: $t(19) = -3.115$, $p<0.05$.

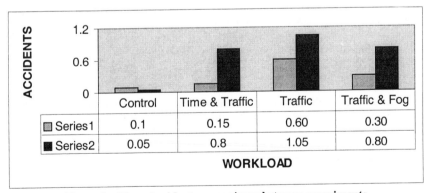

	Control	Time & Traffic	Traffic	Traffic & Fog
Series1	0.1	0.15	0.60	0.30
Series2	0.05	0.8	1.05	0.80

Figure 2: Mean Accident comparisons between experiments

Lane Deviation - only one significant effect was observed between Experiments 1 & 2. Drivers showed an increased tendency for Lane Deviation in Experiment 2 than 1, but only in fog and with increased traffic density: $t(19) = -2.268$, $p<0.05$.

Vehicle Speed - significant effects were observed between Experiments 1 & 2. Vehicle Speed increased in Experiment 2 when traffic density was low and decreased in when traffic density was high:

- Control: $t(19) = 2.588$, $p<0.05$ • Traffic: $t(19) = 2.198$, $p<0.05$

Traffic Light Violations - only one significant effect was observed between Experiments 1 & 2. Drivers showed an increased TLV tendency in Experiment 1, but only under increased time pressure: $t(19) = 2.565$, $p<0.05$.

No significant effects were observed between Experiments 1 & 2 for *Heart Rate Variability, NASA-RTLX* or *Bedford-Harper Scores, or speeding offences* ($p>0.05$).

Discussion
It was anticipated that significant differences between the same conditions in Experiments 1 & 2 would highlight differences due to the use of the speech interface.

Across a number of measures (NASA-RTLX, Bedford-Harper, and Heart Rate Variability) no differences were found. As such it can be argued that participants neither perceived the use of the speech interface to contribute to workload, nor did the use of the speech interface evoke a physiological response as reflected in the heart rate measures.

The results for accidents, lane deviation, speeding and traffic light violations, illustrate that when the speech interface was used, more accidents occurred, lane deviation increased, speed increased and traffic light violations increased. However, when the speech interface was used, accidents only increased when both traffic density and time pressure were high; lane deviation increased only when both traffic density was high and fog was present; and traffic light violations only occurred when both traffic density was high and time pressure were high. As such, the impact of using the speech interface for the tasks used in the experiment would only appear critical under particular conditions where other workload factors are present. As such speech would appear to be a secondary workload factor in this situation.

The results illustrate how different workload factors impact on the safe use of a speech interface in the driving domain and are consistent with widely accepted models of driver workload which are generally additive. What is not understood from the data is how the interface would naturally have been used in a driving environment where workload is high. It would be reasonable to propose that drivers would stop using the interface in such situations, however, in the current research they had been instructed to use the interface throughout the experiment.

These findings suggest provisional guidelines for the use of ASR interfaces while driving. Admittedly these results are only directly related to the simulator trials and any guidelines would need to be evaluated further in a more realistic environment, but they offer a provisional basis for developing guidelines for automotive speech interface use:

- *Speech interfaces should not be used when driver workload is already elevated because of the co-occurrence of at least two other factors. These factors may include traffic density, time pressure and the presence of fog*

Using a speech interface under these conditions would appear to lead to more accidents even though the user may not perceive the use of the speech interface to impact on their overall workload. Lane deviation increases due to the use of the speech interface but only when visibility is poor and there is more traffic on the road.

These guidelines would appear to reinforce common sense and even anecdotal evidence for speech interface use. Even in normal driving, drivers may stop conversing with passengers when the driving task requires more attention and so this could also be the case for using a speech interface. What is apparent from the data, however, is that whilst participants did not consider the *use* of the speech interface itself to be demanding, *using* it contributed to poor performance when other workload factors were present.

The results of both experiments indicate the need to understand how dynamic workload can be and the importance of integrating in-car systems so that they support the driver in a variety of driving environments. Only by developing an understanding of how driver's may want to *use* a speech interface and the impact of *using* it has on their driving abilities will it be possible to design an interface that truly meets the expectations of the user and supports safe driving.

Acknowledgements

This research was carried under the UK Government LINK Inland Surface Transport (IST) programme, funded jointly by the Economic and Social Research Council (ESRC) and the Department of the Environment, Transport and the Regions (DETR).

References

Aasman, J, Mulder, G., & Mulder, L.J.M., 1987. Operator Effort and the Measurement of Heart Rate Variability. *Human Factors* 29, 161-170.

Baber, C., & Noyes, J., 1996. Automatic Speech Recognition in Adverse Environments. *Human Factors* 38(1), 142-155.

Graham, R., Aldridge, L., Carter, C., & Lansdown, T., 1998. The Design of In-Car Speech Recognition Interfaces for Usability and User Acceptance. In, D. Harris (ed). *Engineering Psychology & Cognitive Ergonomics - Vol.4*. Ashgate.

Graham, R., & Carter, C., 2000. Comparison of Speech Input and Manual Control of In-Car Devices whilst on the Move. *Personal Technologies* 4, 155-164.

Linde, C., & Shively, R., 1988. Field study of communication and workload in police helicopters: implications for cockpit design". In, *Proceedings of the Human Factors Society 32nd Annual Meeting*. Human Factors Society, Santa Monica, CA. 237-241.

Stedmon, A.W., Carter, C., & Bayer, S.H., 2001. Baselining Behaviour: Driving Towards More Realistic Simulations. In, D. Harris (ed). *Engineering Psychology & Cognitive Ergonomics*. Ashgate.

Stedmon, A.W., & Bayer, S.H., 2001. Thinking of something to say: workload, driving behaviour & speech. In, M. Hanson (ed). *Contemporary Ergonomics 2001. Proceedings of The Ergonomics Society Annual Conference, Cirencester. 2001*. Taylor & Francis Ltd. London.

Stein, A.C., Parseghian, Z., & Wade Allen, R., 1987. A Simulator Study of the Safety Implications of Cellular Mobile Phone Use. *Proceedings of 31st Annual Meeting of the American Association for Automotive Medicine*. New Orleans, USA, 28-30 September, 1987.

THE USE OF VISUALLY IMPAIRED PARTICIPANTS TO EXPLORE THE DESIGN OF HAPTIC AUTOMOTIVE CONTROLS

Steve J. Summerskill[1], J. Mark Porter[2], Gary E. Burnett[3],

[1]*Department of Design and Technology, Loughborough University*
E-mail: S.J.Summerskill2@lboro.ac.uk
[2]*Department of Design and Technology, Loughborough University*
E-mail: J.M.Porter@lboro.ac.uk
[3]*School of Computer Science and Information Technology*
University of Nottingham
E-mail: Gary.Burnett@cs.nott.ac.uk

With the advent of satellite navigation systems and computing based technology the driving environment is becoming more complex in the 21st century. The interfaces of these devices currently rely on vision to provide feedback to the user. Drivers are resource limited with respect to their vision when attempting secondary tasks. People with visual impairments interact with consumer products, and develop strategies that allow them to 'map' the controls and functions of electronic devices using their sense of touch. In this way, people with visual impairments are focused on the tactile cues of controls that allow them to distinguish function. The aim of the study was to explore the tactile cues that visually impaired people find useful when interacting with unfamiliar consumer electronic products. The results are being used in the design of non-visual controls for the driving environment.

Introduction

A consortium of organisations (Including Honda Research & Development, Visteon, Loughborough University and the University of Nottingham) has been formed under the title of BIONIC (Blind Operation of In-car Controls) in order to explore the issues surrounding the design of haptic control interfaces in the automotive environment. The BIONIC project aims to design an interface that will allow the control of secondary and ancillary functions of a car, with minimal visual distraction for the driver.

An analogy can be drawn between drivers and people with severe visual impairments. The driver's primary task is the control of the vehicle using visual cues from the exterior of the car to determine the correct speed and direction of travel. Secondary tasks such as the use of radios and navigation systems share the visual attention of the driver. In terms of the secondary task, drivers can be considered to be resource limited with respect to their visual sense.

It is generally recognised that people with visual impairments do not have enhanced tactile perception skills when compared to non-disabled people, indeed diabetes, which can result in visual impairment, can also reduce tactile sensitivity (Caruso G. 2001). People with visual impairments do however interact with consumer products, and develop strategies that allow them to 'map' the controls and functions of electronic devices using their sense of touch. The presumption is that people with visual impairments are focused on the tactile cues of controls that allow them to distinguish function.

This paper reports the results from a baseline study. The study was performed using twelve visually impaired participants. The aim of the study was to explore the tactile cues that visually impaired people use to facilitate the initial exploration of an unfamiliar piece of audio equipment.

Experimental design

The initial step in designing the experiment was to explore a methodology that would allow visually impaired people to express the skills that are used when interacting with controls. A technique was developed which involved a list of tasks that the participant would be requested to perform with a piece of audio equipment e.g. play the third track on the CD. The technique required the participant to verbalise the process of exploring controls in order to understand user preconceptions of the type of controls that would be associated with a certain function. Therefore at each stage in the assumed task process the participant was asked to perform a certain task, and at an appropriate stage, they were asked to verbalise what they were looking for, and subsequently what allowed them to find it. The task selected for initial experimentation was using a piece of audio equipment, typically comprising of a radio tuner, cassette tape deck, and CD player. The experimentation was video taped.

Although it may have been useful to be able to quantify performance in terms of time to complete a certain task, it was considered more important to explore the preconceptions of control design for a certain control type, and the tactile cues that allowed the participants to determine control types. As the conversation between the researcher and participant would artificially extend the time for a certain task, the time for completion was not recorded.

It was decided that information on control usage by people with visual impairments would be gathered in two ways. Initially the participants were asked to interact with a piece of audio equipment that was unfamiliar to them, thus allowing them to recreate the 'first contact' situation. The participants were also asked to discuss and show the electronic devices that they own, with reference to the reasons that they had selected a particular model, and also any control designs that they have found particularly useful.

Selection of the experimental equipment

The selection of the audio equipment was a crucial stage in the experimental design. The device selected contained the following attributes; CD player, Radio and Cassette player. Therefore the participant would be required to perform mode switching operations via sliding switches. The device contained similar large rotary knobs for volume and radio frequency selection. The difference between the two was a small raised arrow on the volume control. A non standard array of controls formed in a circle containing the CD play, stop, and programme functions. The selected product was intended to be representative of such equipment in some respects (i.e. cassette controls) but to also contain novel, maybe poor, design features in order to allow the examination of coping strategies and problem solving. As this array did not conform to the assumed stereotypes it was considered important to explore how participants would react to it. The cassette player controls had indented symbols and were of the mechanical type. This allowed the exploration of the use of the sense of touch to identify small indented symbology. The device also contained a power on/off button that protruded further from the mounting surface than the other controls.

The sample frame

It was initially assumed that people with an acquired visual impairment, who had driven before the onset of symptoms, might be useful when exploring the issues surrounding the design of the BIONIC device. In order to explore the issues of participants with some level of vision, (as would be expected for the majority of people with acquired visual impairment), a pilot study using two participants was designed. One participant had complete visual impairment; the second participant had a level of impairment central to the OPCS (1991) scale for vision, i.e. "Cannot see well enough to recognise a friend who is an arm's length away" (the OPCS severity scale is based upon a survey of 14000 disabled people in the UK).

In this way the methodologies of two different groups of visually impaired participants could be examined and compared. The two participant pilot was performed using participants from the Royal National Institute for the Blind (R.N.I.B) College based in Loughborough. After analysing the video recording of the two trials a difference was noted between the methodologies that were used by the two participants. The participant with some level of vision used a methodology to explore the device that included the use of the limited vision available to her. The participant spent much longer than her counterpart with no vision exploring the device, and tried to read the labels associated with controls (without success), and to determine the overall layout using a combination of tactile and visual skills. The participant with no vision explored the device with a more methodical approach, and spent approximately half the time on the initial examination of the device, with no apparent detriment to performance in the tasks that followed. In order to explore the difference between participants with no vision, and those with some level of vision, it was decided to include both groups in the sample. Seven females and five males took part in the experimentation. The age of the sample ranged from 17 to 42 (mean = 24). Many participants were unable to name the medical term for their visual impairment. Therefore a severity scale taken from the OPCS survey was used to rate the severity of visual impairment. The sample included three participants with no vision, and nine participants with some level of vision. The participants with the best level of vision fell into the OPCS category of "Cannot see well enough to recognise a friend who is an arm's length away" and could not read the button labels on the device.

Summary of findings from the user trials

The user trials provided results from the experimentation with the device that was unfamiliar to participants, and also from the examination of the devices that the participants owned.

Findings from the audio device used in experimentation

The initial exploration of the device allowed the majority of the participants to identify the major interaction areas and control groups. The initial inspection of the device was performed using different methodologies depending on the level of visual impairment. Participants with little or no vision performed a very fast, but methodical examination of the device, tracing the outlines of controls to determine their shape, and identifying control centers. This was illustrated by the speed with which the volume control was found by the participants when the volume was initially set to a high level. An accidental benefit of the users exploring the device, with the power off, was that generally they would turn and push various controls in order to gauge feedback. On a number of occasions this left the volume control set to a high level. When the CD player, cassette player or radio were used by the participants, the initial reaction of the participant was to silence or reduce the volume as rapidly as possible. The speed with which the participants identified the correct control and reduced the volume illustrated that they had identified the possible location of the volume control during the initial examination. This point was strengthened by the preconceptions that the majority of participants had about the volume control, i.e. a large rotating knob.

Participants with some level of vision performed the initial inspection using a combination of vision and touch. In general these participants took longer to perform an inspection of the device as they seemed to rely on their vision. The visually impaired pilot study has shown that the desire to use even a little vision is usually very compelling for people with visual impairments. However, the study has shown that the use of this vision was not necessarily of any benefit when performing the tasks.

The dichotomy between viewing being of benefit for conceptualisation, and of detriment when driving, as it may be a form of distraction, needs to be addressed. The solution may be as simple as coding the display layout so that it represents the layout of the 'non-visible' BIONIC system. This would allow for the initial examination of the device using touch with a visual frame of reference.

When exploring the device over half of the participants mistook the microphone for an active area due to the textured surface. This shows that the participants were looking for a textured surface to indicate an interaction point. It also indicates that 'tactile noise' should be avoided in any device which is to be controlled using touch alone.

The participants indicated that different levels of protrusion of controls from the surface of the device gave clues as to the function, where there were two protrusion levels adjacent to each other. The power switch was identified using this cue by the majority of participants.

None of the participants could identify the shape of the indented symbols on the cassette player buttons. Most participants stated that protruding symbols are better due to the fact that one could trace around the shape improving identification.

Findings from the participant's own equipment

Some participants used hand-control reference points (H-CRP) with their own equipment. An example of this behaviour was the use of a bevel in the surface of a speaker grill. The bevel allowed the user to locate the hand near to the controls. The bevel then allowed the user to select various controls on the basis of whether they were above or below the bevel, effectively improving the location coding of the controls.

Several found that adaptations to devices allowed for easier use. For example one participant had removed the majority of buttons from her HIFI remote control. She stated that all of the buttons were the same, so she had found it hard to find the button that she wanted. She therefore had over three quarters of the buttons removed, leaving controls for playing, stopping and reviewing the tape, along with function and power buttons. Three of the participants owned audio equipment similar to that used in the experimentation (i.e. portable audio systems, containing radio, tape and CD functions). The main control panel of the devices was in the same location as those of the test device, but the buttons were shaped to match the standard symbols for play etc. The large buttons also protruded by approximately 2mm from the mounting surface. Two of the participants stated that the button shapes and protrusion level were useful when they first owned the device, but that they subsequently found the buttons as they knew their location in reference to the side of the mounting panel. The other participant used the shape coding of the buttons on a regular basis.

Summary of the coding techniques to be explored in further research

Location coding

Location coding of the controls should be considered in relation to a specific and obvious H-CRP. Control location coding should be performed on the basis of the dynamics of hand position and hand anthropometry when interacting with the H-CRP. The H-CRP may take the form of a hand rest which allows the hand to be steadied when interacting with panels, or it may be a joystick type device that allows the hand to wrap around the controls.

Shape coding

The shape coding of raised symbols for use on tactile buttons should be explored to determine if symbols can be recognised and associated with a specific function by touch alone. Also the height of such raised symbols should be explored. This requires the design of a set of icons that aim to represent the various functions available.

The shape coding of the actual buttons should also be considered. The combination of a shape coded button, and an embossed symbol on that button, provides redundant coding which it is presumed will aid button discrimination. The interaction with visually impaired participants has raised the issue of using haptic interfaces to provide active tactile feedback regarding system state information.

Size and protrusion coding

The visually impaired trials have indicated that the size of buttons which have a shape around which the user will trace a finger is crucial. If symbols are too small, visually impaired people find it difficult to determine the shape. This links to protrusion coding. Not mentioned as a coding technique in the standard ergonomics literature, protrusion coding has been shown to be a valuable technique for visually impaired users.

The amount that a button protrudes from the mounting surface has the possibility of allowing critical buttons to be found more quickly, and may help in the avoidance of inadvertent operations.

Texture coding

Texture coding may help in the location of H-CRP's when not looking at the hand fascia interaction. If the target of the hand is of a low coefficient of friction, and the surrounding area of a higher coefficient of friction, the hand can be guided to the H-CRP.

Conclusions

The use of visually impaired participants has highlighted areas for further work which would not have been identified without their unique perspective regarding the interaction with controls without vision. The analysis of the first contact situation has proved to be valuable, and further work in this area is planned, with larger participant groups. The design of tactile symbols to be used in conjunction with a haptic display will also be explored using visually impaired participants, as the subtleties of tactile icon design can be more thoroughly examined with people who are more focussed on the tactile sense.

References

Office of Population Censuses and Survey 1991. OPCS surveys of disability in Great Britain. The prevalence of disability among adults. HMSO. London.

Caruso G., Nolano M., Crisci C., Lanzillo B., Di Lorenzo N., Lullo F., D'Addio G. 2001. Electrical and tactile sensory conduction velocity in peripheral nerves of diabetic patients. Department of Neurological Sciences, University of Naples Federico II, Naples, Italy.

LEARNING TO USE MINI-ROUNDABOUTS

Tay Wilson and F. A. Clube

Psychology Department
Laurentian University, Ramsey Lake Road
Sudbury, Ontario, Canada   P3E 2C6
tel (705) 675-1151   fax (705) 675-4889

It is contended that early behaviour observations of new traffic engineering interventions could form the basis of locale specific driver improvement programmes to assist drivers otherwise left to their own devices to learn and develop driving norms. Here, as exemplar, are some findings from observations of a mini-roundabout in Bedford which appeared likely to assist drivers to become more skilful and sociable. Expect relatively few mistakes on Monday, many on Tuesday but relatively few intimidations; Friday is the big day for intimidations. Expect relatively more piggybacking and cutting up by vehicles from so called "main" entrances. When traffic volume is moderate watch for entering vehicles looking left only. When traffic volume is high watch for piggybacking and cutting up. Cyclist and "failing to see the roundabout" incidents tend to occur at one specific entrance.

Introduction

All too often traffic engineering interventions ranging from new motorways to speed bumps are introduced with apparently insufficient attention devoted to proper "instructions for the user". Users are left severally and collectively to learn how to use new interventions and further to develop "cultural" norms of behaviour at them which may prove to be inappropriate and dangerous. With the aim of looking for such inappropriate behaviour road users were studied at a recently introduced mini-roundabout in Bedford at a time when only a couple such entities were present.

Method

Road-users were observed using the then recently converted four entrance roundabout defined by a .125 m raised wooden circle but still cluttered by disabled traffic lights at the Bromham Road/Greyfriars-Union junction in Bedford on 10 consecutive weekday afternoons between 2:30 and 4:30 starting Monday 16th of July, 1979. Bromham was the "main and important" heavy volume road with road signs directing traffic from Cambridge and the A1 on the east to Northampton and the M1 to the west. Signs for the roundabout were on the entrance give-way lines apparently causing considerable surprise to drivers on a wide road. Westward travelling Bromham road users encounter a Zebra crossing about 90 m after the roundabout in front of which a lamppost obscures waiting pedestrians resulting in frequent emergency stops which had been recently highlighted in the local press. Eastward travelling Bromham road users were relatively slowed by the zebra crossing before encountering the roundabout. The city bus station and city centre was down Union street.

Observations were preceded by a some weeks of "learning how to look" during which time the observer settled upon watching each entrance for 30 minutes in randomised order, writing down descriptions of passages that were in any way eventful and subsequently sorted them into about one hundred incident categories. Traffic counts were taken for three minutes at each entrance in randomised order before and after the observation period.

Results

The estimated traffic flows for five hours of observation at each of the four roundabout exits Bromham Road West, Union Street, Bromham Road East and Greyfriars respectively were 3165, 1698, 2510 and 3410 for a total flow of 10783 of which 75% were cars, 13% lorries and buses and 12% bicycles and mopeds. Over ten days 790 incidents — about one in fourteen drivers — were described and classified as various types of intimidation (425), mistakes (177), timidity/receiving aggression (104), engaging in communication (17), involving cyclists (59), or as miscellaneous (8).

Under intimidation were piggybacking involving either loss of turn (44), near miss (5), cutting up another vehicle (73), enough room (4) or not looking (9); squeezing in involving large vehicle before small vehicle (24), small before large (22), same sized vehicles (73), or one vehicle before several vehicles (16); cutting up involving large vehicle before small (22), small before large (20), same sized vehicles (69), one vehicle before several (10) or out of turn causing stoppage (1); driving fast forcing another vehicle to wait (5) or being forced to stop by another driver (1); blocking roundabout when own exit is blocked (5); usurping other vehicles by approaching roundabout apparently not intending to stop (6) or being forced to slow by other vehicles (1); pretending not to see usurping other vehicles (4); edging out continuously (8); and driving inside another vehicle at same entrance when other driver is driving abnormally (3).

Under mistakes were apparently ignorant of roundabout right of way rules - looking both ways then giving any vehicle at any entrance right of way (8); looking left only (25), blocking roundabout by stopping over give-way line then continuing on (1); entering before right entrance driver who is looking right (14) or has stopped (6); stopping on roundabout when realizing other driver is being cut-up; reasoning failure — entering roundabout without looking right upon seeing left driver signalling right apparently thinking right lane will thus be blocked (4); looking right only not noticing entering driver from opposite entrance who is cut-up (5); failing to notice driver signal and missing turn when apparently mesmerized by traffic stream to right (1); causing opposite entering driver to wait through failure to see signal (4); stopping for vehicle simultaneously entering opposite then exiting right (2); blocking right entry by wrongly entering roundabout simultaneously with opposite vehicle turning left (3); driver in entering stream failing to signal turn and being stopped by vehicles entering opposite (19) or causing such entering vehicle to stop (2); entering just as opposite entering vehicle is exiting and slowing other drivers (51), entering out of turn when opposite entering vehicle almost exiting (3) causing others to miss turn (3), cutting up another vehicle (10), being cut-up (1) or having to slow down (1); stalling at roundabout (2); and apparently failing to look and/or see the roundabout (12).

Under timidity/receiving aggression were flummoxed driver waits for roundabout to clear (3); fails to take first good opportunity (7); fails to take second (5); takes second opportunity (3); causes traffic to slow (4); cut up by larger vehicle (7), smaller (13), same sized (24) or more than one vehicle (3); squeezed out by larger vehicle (6), small (2), same sized (18); waits for second vehicle entering opposite just as first vehicle entering opposite completes roundabout (2); going around edged-out-roundabout-blocking vehicle on left (2); eventually given right away by another driver (1); looks back to check after correctly manoeuvring roundabout; cut-up by opposite entering vehicle signalling right (2); and missing a turn because driver at right failed to signal (2).

Under communication were waving on another driver (3); special communication making others wait (4); guessing that a not signalling vehicle will turn (1); communication ambiguity (3); honks at vehicle ahead for not entering with opposite vehicle (3); using horn (1); and other drivers help stuck driver (2).

Under cyclists/moped were failing to give way to cyclist then stopping (10) or making cyclist stop (8); failing to see and cutting-up cyclist when driving normally (2) or when breaking highway code (8); cyclist apparently failing to look (18) and causing other driver to slow (2) or cutting-up other vehicle (1); cyclist weaving through vehicles (1) or alongside vehicle (6); and cyclist acting as pedestrian (3).

Under miscellaneous were giving right of way out of turn to foreign driver (2), to learner driver (1) or to police car (1); taking advantage of foreign driver (2) and using unique strategy for negotiating roundabout (2).

Many of the above incidents are self-explanatory but some may benefit from clarification. "Piggybacking" refers to a second vehicle following a first vehicle onto the roundabout without stopping or checking that the way is clear. "Squeezing" involves entering the roundabout in front of a vehicle to the right to whom priority should have been given but not causing such driver to slow down. "Cutting up" refers to the same event when such driver is forced to stop or slow down to avoid collision. Under mistakes note that it is okay for two vehicles entering opposite to traverse the roundabout together but timing/right of way problems can occur when vehicles do not exit straight across particularly if not signalling.

Using chi square analysis data were analyzed by incident type, roundabout entrance and week day. Only chi square results significant at $p < 0.001$ are reported in this paper. Intimidation, mistake, timidity and cyclist incidents were overall significant. Among intimidations "squeezing in and cutting up - same sized vehicle" and "piggybacking causing someone to lose a turn" or "cutting up someone" were relatively most frequent. Among mistakes "slowing traffic by entering roundabout just as opposite vehicle is about to exit" and "looking left only" were relatively most frequent. Among timidity incidents "being squeezed out or cut up by a same sized vehicle" were relatively most frequent. Among cyclist incidents "apparently failing to look" and "vehicle failing to give way to cyclist then stopping" were relatively most frequent.

Significant daily differences in total incidents and intimidation incidents alone relative to traffic flow were noted with a smaller relative frequency on Tuesdays and a larger on Fridays. Significant differences in mistakes were also noted but the pattern showed relatively more mistakes on Tuesdays and fewer on Mondays.

Although no significant differences in total or intimidation incidents by roundabout entrance relative to traffic flow were noted, significant entrance by individual intimidation incidents were noted. "Piggybacking causing turn loss" or "cutting up a vehicle" was relatively more frequent on both Bromham entrances and relatively less frequent at the Union Street entrance, while "squeezing in" occurred relatively more frequently at Greyfriars and Union entrances and less frequently at Bromham entrances.

Overall significantly fewer mistake incidents occurred at Greyfriars. Significant individual mistake incidents by entrance were also noted. Relatively more "ignorant of who has right of way" and "looking left only" mistakes occurred at Bromham East while relatively fewer "looking left" occurred at Greyfriars. "Entering roundabout out of turn when seeing driver on right looking right" occurred relatively more frequently at Greyfriars and less at the Bromham entrances. "Entering with an opposite stream and causing a not — signalling member of stream to wait to turn off" and "causing vehicles to slow down by entering roundabout just as opposite vehicle has almost left roundabout" occurred relatively more frequently at Bromham West. "Apparently failing to see the roundabout" occurred relatively more frequently at Bromham East.

Although no significant overall entrance by timidity incidents were noted, significant individual timidity incidents by entrance were found. Being "cut up by a smaller or same sized vehicle" was relatively more frequent at Bromham West. Being "squeezed" was relatively more frequent at Greyfriars.

Significant individual cyclist by entrance incidents were found with relatively more (all of them) "failing to see cyclist when driving incorrectly according to the Highway Code" incidents occurred at Bromham East and relatively more "cyclist apparently failing to look" incidents occurred at Bromham East.

In order to investigate relative incident rate by level of traffic flow, car flows per half hour at each entrance were divided into three flow rates (1 - 152, 153 - 238, and 239 -325). Lorry flows were similarly divided into three flow rates (0 - 24, 25 - 44, and 45 - 65). No significant differences were noted for the nine combinations of lorry car flow rates by general incident type.

Significant individual intimidation incidents by traffic flow were noted. "Piggybacking causing turn loss" was relatively more frequent when both cars and lorries had high flow or cars had medium and lorries high flow. "Squeezing in large vehicle in front of small" was relatively more frequent when both cars and lorries had high flow. Both "cutting up large vehicle before small" and "cutting up small before large" were relatively more frequent when cars had high flow and lorries had medium flow.

Significant individual mistake incidents by traffic flow were also noted. "Looking left only" was relatively more frequent when both cars and lorries had medium flow while "causing vehicles to slow down by entering roundabout just as opposite entering vehicle has almost exited" were relatively more frequent when both cars had medium flow and lorries had high flow. Overall significant timidity incidents by traffic flow were noted with relatively more timidity incidents when car and lorry flow were both high or when car flow was high and lorry flow was medium.

Incident frequency was also analyzed for cars and lorries separately. For cars significant differences in type of intimidation incident were noted with "piggybacking, squeezing in - same size vehicle" and "cutting up - same sized vehicle" occurring relatively more frequently. Overall relative to car flow significantly fewer car mistakes were made when car flow was high while individually "causing vehicles to slow down by entering roundabout just as opposite entering vehicle has almost exited" and "looking left only" were relatively more frequent. Finally, for timidity incidents being "cut-up by same sized vehicle" was significantly relatively more frequent.

For lorries, significant differences in intimidation incidents were noted with "piggy backing resulting in cutting-up a vehicle or causing turn loss", "large vehicle squeezing in and cutting up" occurring relatively more frequently. Significant mistake types for lorries were noted with "causing vehicle to slow down by entering roundabout just as opposite vehicle has almost left roundabout" being relatively more frequent. Significant timidity type incidents for lorries were noted with "being cut up by a smaller vehicle" occurring relatively more frequently.

Discussion

It is contended that findings such as these could form the basis of locale specific improvement interventions. For example, the following skeletal driver improvement programme could be developed from the above data. Watch out for an overall the high frequency of squeezing in, cutting up and piggy backing causing sudden braking. Be aware that vehicles opposite an entering vehicle are likely to "hop-in" slowing traffic suddenly when that entering vehicle is just about to exit. Note that many drivers (1/400 here) look left only and might not see you on the right. Day of week appears to make a difference. Expect relatively few mistakes on Monday, many on Tuesday but relatively few intimidations. Friday is the big day for intimidation. Individual entrances also appears to make a difference. Expect more piggybacking and cutting up which is likely to cause sudden braking by vehicles from so-called "main road" entrances (Bromham East and West). At the Bromham road East entrance watch out for drivers looking left only, those apparently ignorant of right-of-way rules, those apparently failing to see roundabout and those apparently failing to see a cyclist while they are contravening the highway code. At the Bromham west entrance watch out for being cut up by drivers and for cyclists failing to look. Traffic flow also appears to make a difference. Watch out for piggybacking when car flow is medium and lorry flow is high and cutting up when car flow is high while lorry flow is medium. Watch for entering vehicles looking left only when both cars and lorries have medium flow. Watch for instances of timidity (e. g. turn missing) when car flow is high or when both car and lorry flow is high. A repeat study of mini-roundabouts now that drivers are much more used to them could prove interesting.

References

Wilson, Tay, 1991. Locale Driving Assessment — A Neglected Base of Driver Improvement Interventions. In *Contemporary Ergonomics*, Praeger, London, pp 388-393.

Wilson, Tay and Godin, Marie, 1993. A study of co-operation extended to trapped merging drivers. In *Contemporary Ergonomics*, E. J. Lovesey, Editor. Praeger, London, pp 387-391.

Wilson, Tay and Chisel Christine., 1997. On-Campus pedestrian crossings: Opportunity for locale based driver improvement. In <u>Contemporary Ergonomics</u>, Praeger, London, pp 86 -91.

MOTORCYCLE ERGONOMICS

MOTORCYCLING AND CONGESTION: DEFINITION OF BEHAVIOURS

Sandy Robertson

Centre for Transport Studies
University College London, Gower St. London WC1E 6BT

It has been suggested that increased use of motorcycles could result in lower congestion, reduced fuel consumption and thus lower emission of pollutants. DTLR wished to explore this proposition and commissioned a study. This paper reports on part of that study in which a set of behaviours of TWMV users were identified and defined to assist in determining the impact of TWMVs on congestion funded by DTLR. The behaviours identified in this paper are: Moving to head of queue. Filtering (traffic stationary). Filtering (traffic moving). Lane changing behaviour. Inaction. Balking, (TWMV progress is impeded). Wriggling, (where the rider moves around stationary vehicles/objects). The views expressed in this paper are those of the author alone.

Introduction

The Government encourages alternative modes of travel that could help to reduce overall levels of congestion and pollution. Various groups have suggested that the increased use of TWMVs (two wheel motor vehicles) could result in lower congestion, reduced fuel consumption and thus lower emission of pollutants. The Department of Transport Local Government and the Regions wished to explore this proposition. The project was awarded to Halcrow Group with UCL providing input for the behavioural study.

Within the overall research programme, one objective that was being investigated was to understand the behavioural characteristics of TWMV users and their association with the journey time savings that TWMV users may experience in differing situations. To this end, journey time surveys were undertaken and these were linked with observations of the behaviour of TWMVs at the sites where the journey time surveys were undertaken. This paper describes the behaviours used in this study together with the process by which they were identified.

Method

Behaviours were identified through an iterative process using 1) a theoretical assessment of what behaviours were possible, 2) video recordings of the road system using roadside and in vehicle recording, and 3) informal discussions with TWMV users. A final list of behaviours and their definitions were then developed for use in the later study. Figure 1 shows schematically the process for the development of the list of behaviours.

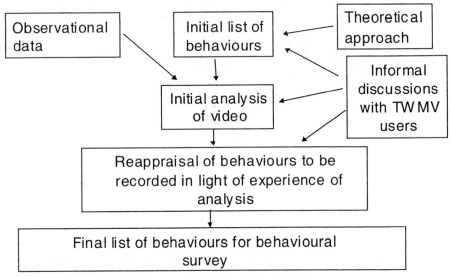

Figure 1 The process for identifying behaviours.

The TWMV in the context of the road system.

The road system may be considered to be a semi-constrained system, in which the majority of vehicles can generally only change their speed (or direction of travel within a lane) or move laterally from one lane to another. In some cases the road width is such that single track vehicles (cycles and TWMVs) may move between lanes of traffic in various ways. The types of behaviour of particular interest were those that could be described as 'congestion busting'. In other words, the types of behaviour which would facilitate the rider of the TWMV making progress through traffic. Other types of behaviour of interest were those which would impede progress, for example being balked by other vehicles or remaining within a queue even if there was an opportunity to make progress. The classification of motorcycle behaviours had also to relate to the behaviour of other road users and road conditions.

Results

Figure 2 shows the behaviours initially identified from the theoretical assessment. This was followed up with the observational survey in which several of the behaviours were grouped together and frequencies of these behaviours were recorded. These behaviours are shown in Table 1 for an in-vehicle survey (duration 3hrs.) covering a range of road types and for a roadside survey in the centre of London (duration 1hr.). As can be seen the relative frequencies of the behaviours were different.

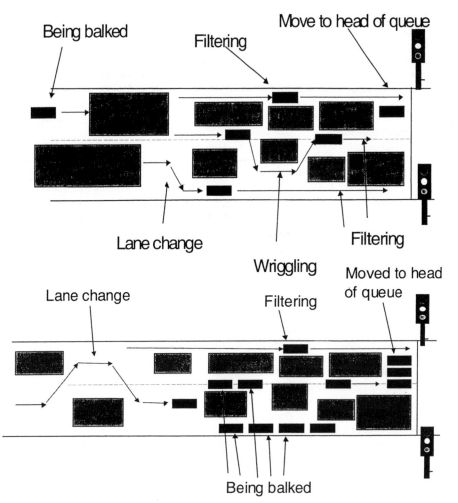

Figure 2 Types of manoeuvre

Table 1 Frequency of behaviours observed in initial observational studies

Behaviour	In vehicle 3hr	Roadside 1hr.
Go to head of queue	15	27
Filter (centre of traffic)	4	116
Filter (Nearside)	8	2
Filter (Offside)	29	1
Wriggle (Traffic)	3	4
Wriggle (Object)	0	0
Lane change	4	16
Inaction	1	11
Balk (roadside only)		16

After further discussion with TWMV users, the following behaviours were selected and the definitions developed prior to and during a period of training of a researcher for extraction of the data for a study reported later. The training materials were for the urban situation where the subsequent study was planned. The training consisted of 3 extractions of a sample video by one researcher and one extraction by another researcher with the definitions being refined until there was general consistency in the extraction of the data. At the end of the training period, the discrepancy between the number of behaviours recorded was between 2 and 6 percent depending on the behaviour recorded.

The behaviours were:-

1) Moving to head (or near the head, for example 2 vehicles back) of queue of non-TWMVs. The TWMV passes to the head or to within two vehicles of the head of a queue at a junction or roundabout. A TWMV that arrives at the head of the queue unimpeded should not be counted for this behaviour. The TWMV should have exhibited some form of behaviour whereby queuing is avoided to be included in this category. For example filtering or moving from the lane which would otherwise have been blocked (for example to a turning lane or an empty lane).

2) Filtering (nearside, offside, and centre) where traffic is stationary. The key characteristic of filtering is that the TWMV makes use of road space other than the conventionally marked lanes. It is therefore making use of road space that could not be used by other vehicles. This behaviour is where a TWMV passes other traffic that is stationary. The passing may be on the nearside or the offside of road, or it may be in the middle of two streams of stationary traffic. If there was a stream of stationary traffic with a full width road lane available to the offside (or nearside) and a TWMV was moving down that lane, it would not be categorised as filtering. If there was a single stream of stationary traffic on a single lane road and a TWMV passed down the offside of stationary traffic than this behaviour would be counted.

3) Filtering (nearside, offside, and centre) where traffic is moving. The criteria for filtering when traffic is moving is essentially the same as for filtering when traffic is stationary, except that at least one of the adjacent streams of (usually non-TWMV) traffic are moving.

4) Lane changing behaviour is defined as when the TWMV moves between streams of traffic that are travelling at different speeds so as to make progress through the traffic. This is different from filtering in that the TWMV is moving from one stream of traffic to another. This type of behaviour may also be seen in conjunction with filtering, for instance when the TWMV moves from one stream into another to allow access to road space where filtering is possible.

5) Inaction, failing to make use of an opportunity to use a congestion busting behaviour. This may be considered to be a proxy to identify TWMVs whose drivers are less inclined to undertake congestion busting behaviours.This category is relatively subjective. It should be identified when it is clear that a TWMV is remaining within stream of traffic when there is a clear opportunity to engage in filtering or other behaviours. This may be most easily identified when other TWMVs are filtering while the one in question remains in position. There is a thin dividing line between this category and that of balking (in which there is no opportunity to undertake a specified behaviour). In the case of doubt, a behaviour should be placed in the balking category.

6) Balking, where progress is impeded by another vehicle or where there is no opportunity to undertake a behaviour due to the positioning of vehicles on the road. This behaviour is again relatively subjective. It is defined as another vehicle (TWMV or other) either actively or passively placing itself into a position so as to prevent a TWMV engaging in a congestion

busting behaviour such as filtering.

7) Wriggling, where the rider moves around stationary vehicles/objects that are balking progress. This behaviour occurs specifically in stationary traffic. It is where the TWMV moves around another stationary vehicle or object to allow the TWMV an opportunity to make progress. This could include moving around the vehicle to allow access to an open section of road suitable for filtering. Wriggling is often characterised by very slow speeds (often walking speed or less) and the TWMV being at least 45 degrees and often nearly at right angles to the normal flow of traffic. Once traffic is moving this type of behaviour becomes categorised as lane changing and the TWMV is more likely to be parallel to the flow of traffic.

8) No identifiable behaviour (No congestion). None of the other specified behaviours are observed. This can occur anywhere, but typically this would be where there was no congestion.

Discussion

A set of behaviours of TWMV users were identified and defined for use within a study of the impact of TWMVs on congestion. It was clear that there were a range of behaviours specific to two wheeled vehicles (as exemplified by TWMVs) that could be identified in the context of the urban road system. These behaviours were associated with reducing the impact of delay imposed by the presence of other road users. From the initial observations and discussions with road users, it appeared that these behaviours might not impose additional delay on other road users. This was due to the behaviours often being associated with use of parts of the road that were not usable by other types of vehicle. From this early work it appears that TWMV users adopt an active approach to their use of the road system, and that a range of behaviours are used to ameliorate delay associated with queuing by other road users. From the initial observations, it appeared that a high proportion of TWMV users were using these behaviours in some locations but not at others. The behaviours observed generally gave the impression of not being associated with high speeds, though the instantaneous speeds were not quantified. Initial indications were that there a higher proportion of TWMVs than was initially anticipated were observed. This may reflect suggestions that TWMVs are becoming an increasingly popular form of transport. The frequency of observed behaviours appeared to be different on different types of road though from the data obtained at this stage of the study it is not possible to quantify these differences. The next phase of the study (which quantifies the behaviours) is reported later in these proceedings Robertson (2002). It is not clear how behaviours may be affected by different proportions of different vehicles within the road system and further investigation of the impact of high proportions of TWMVs in the UK context would be useful..

References

Martin B, Phull S. and Robertson S (2001) Motorcycles and Congestion. Proceedings of the European Transport Conference 10-13 September 2001 Homerton College Cambridge. Proceedings published as a CD, PTRC, London

Robertson S.A. (2002) Motorcycling and Congestion: Quantification of Behaviours. In P.T. McCabe (ed.) *Contemporary Ergonomics 2002,* (Taylor and Francis, London),

MOTORCYCLING AND CONGESTION:
QUANTIFICATION OF BEHAVIOURS

Sandy Robertson

Centre for Transport Studies
University College London, Gower St. London WC1E 6BT

It has been suggested that increased use of motorcycles could result in lower congestion, reduced fuel consumption and thus lower emission of pollutants. DTLR wished to explore this proposition and commissioned a study. This paper reports on part of that study in which a set of previously defined behaviours of TWMVs (two wheel motor vehicles) are quantified in congested conditions. This work was funded by DTLR as part of a larger study into TWMVs and congestion. TWMV users appear to make use of strategies to reduce delay which appear not to impede other traffic. Up to 90% of TWMVs were observed to filter in traffic. The views expressed in this paper are those of the author alone.

Introduction

The Government encourages alternative modes of travel that could help to reduce overall levels of congestion and pollution. Various groups have suggested that the increased use of TWMVs (two wheel motor vehicles) could result in lower congestion, reduced fuel consumption and thus lower emission of pollutants. The Department of Transport Local Government and the Regions wished to explore this proposition. The project was awarded to Halcrow Group with UCL providing input for the behavioural study. The background to this study is described in Robertson (2002) which paper describes the behaviours used in this study together with the process by which they were identified. The aims of this part of the study were to quantify the behaviours that might contribute to reducing delay or congestion used by TWMV users.

Method

Behaviours were recorded by video cameras at five London sites. Four sites were signal controlled junctions and one was a roundabout. These are shown in Table 1. A journey time survey and traffic counts were undertaken at the same time as the video recordings were made as part of the DTLR project (see Martin, Phull, and Robertson 2001). Video recordings were undertaken in the morning peak period (7am to 10am). At each site, a single video camera was used, usually pointing upstream so as to include the junction. The frequency of behaviours as described in Robertson (2002) were recorded for the 3 hour period for every TWMV observed. The TWMVs were also categorised as being a motorcycle or a scooter. This level of disagregation was the finest that could be undertaken given the constraints of video location and quality

Table1. The sample

Site	N (TWMV)	% TWMV	Weather	Junction type	Lanes
Farringdon St	280	7.8	rain	Signal	2 two way
Commercial St	308	2.1	dry	Signal	1 two way
Lwr. Thames St	952	15.5	dry	Signal	2 dual
Upr. Thames St	498	9.7	rain	Signal	2 dual
Monmouth St	107	6.7	rain	R'dabout	1 one way

Results

The results are shown in Tables 2 and 3. Table 3 indicates differences between behaviours observed for motorcycles and scooters and statistical tests on these are shown in Table 4. The results for each behaviour are discussed in turn.

Move to head of queue.

The proportion of TWMV users who exhibited this behaviour varied from none at Monmouth St to over half the riders at Commercial St. The data suggests that at signal controlled junctions in London, at least one in five TWMVs reach the head of the queue. There were no differences in the proportion of behaviours exhibited by different TWMV types. The implications of this are that, at typical signal controlled junctions in London, the users of between one quarter and one half of TWMVs will not contribute any delay at traffic signals (based on estimates quoted by Powell 2000). This may, however be a questionable assumption. This reduction to delay in the road network might not have much impact where there are only a small percentage of TWMVs in the traffic. Where higher proportions of TWMVs use the road system, there is potential for reducing congestion.

Table 2. Proportion of TWMVs users exhibiting behaviours.

Site	Behaviours							
	Move to head of queue	Filtering: stationary	Filtering: moving	Lane changing	Inaction	Balking	Wriggling	No observed behaviour
Farringdon St	0.19	0.19	0.20	0.56	0.05	0.23	0.01	0.26
Commercial St	0.56	0.69	0.36	0.50	0.08	0.35	0.08	0.06
Lower Thames St	0.20	0.46	0.92	0.23	0.02	0.56	0.00	0.04
Upper Thames St	0.18	0.32	0.57	0.27	0.08	0.22	0.02	0.24
Monmouth St.	0.00	0.14	0.22	0.30	0.10	0.25	0.03	0.50

Table 3. Proportion of TWMVs of each vehicle type, Motorcycle (M) or Scooter (S), exhibiting behaviours.

Site	Behaviours								
	Move to head of queue	Filtering: stationary	Filtering: moving	Lane changing	Inaction	Balking	Wriggling	No congestion	Type
Farringdon St	0.19	0.17	0.23	0.51	0.05	0.23	0.02	0.29	M
	0.20	0.24	0.15	0.68	0.04	0.23	0.01	0.21	S
Commercial St	0.57	0.70	0.37	0.50	0.06	0.35	0.07	0.08	M
	0.52	0.66	0.31	0.50	0.17	0.36	0.09	0.02	S
Lower Thames St	0.19	0.45	0.92	0.233	0.02	0.54	0.00	0.04	M
	0.23	0.49	0.92	0.226	0.03	0.61	0.01	0.03	S
Upper Thames St	0.17	0.32	0.55	0.30	0.08	0.21	0.02	0.26	M
	0.21	0.34	0.61	0.21	0.09	0.25	0.01	0.21	S
Monmouth St.	0.00	0.16	0.24	0.37	0.14	0.26	0.04	0.42	M
	0.00	0.10	0.19	0.13	0.00	0.23	0.00	0.71	S

Table 4. Chi squared test on the proportion of motorcycles and scooters exhibiting behaviours, df 2. (only values significant at p>0.05 are shown.)

Site	Behaviours							
	Move to head of queue	Filtering: stationary	Filtering: moving	Lane changing	Inaction	Balking	Wriggling	No congestion
Farringdon St				7.51				
Commercial St					8.91			
Lower Thames St		4.214	5.7	6.3		24.3		29
Upper Thames St				4.65				
Monmouth St.				6.02	5			7.3

Filtering in stationary traffic

There were a range of proportions of TWMVs observed to filter in stationary traffic ranging from 14 to 69 per cent. This behaviour appeared to be associated with the length of queue on the link (see Figure 1). Possible reasons for this are: a) This behaviour is associated with the opportunity to do so, and longer queues mean that there is more opportunity and b) Riders may be more likely to undertake such behaviours where there are long queues leading to perceived delay for them. These findings indicate that a high proportion (4 out of 5) TWMV riders are filtering at some sites and that not less than one in six riders were filtering at the sites in this study. Filtering is a behaviour that appears to be a classic 'congestion busting' behaviour which is associated with the presence of congestion as indicated by the presence of queues of traffic. Filtering by TWMV may be seen as a method of making use of the small size and manoeuvreability of the TWMV to make effective use of parts of the roadway that could not be used by larger vehicles. In terms of time saving and queue length, there is some relationship between queue length and time saving estimated as 38 seconds per 100m of queue (See Martin, Phull and Robertson 2001).

Filtering in moving traffic

This was more prevalent than filtering in stationary traffic. The site with the highest proportion of TWMVs filtering was also the site with the highest flows.

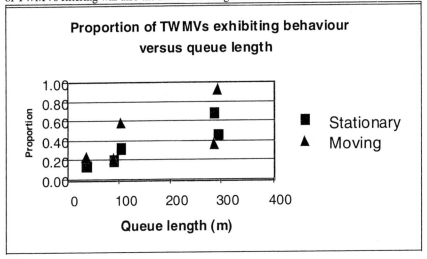

Figure 1 Proportion of TWMVs filtering in traffic vs. queue length.

A plot of average queue length versus the proportion of TWMVs filtering (Figure 1) shows a similar broad pattern to that observed. While traffic is moving it cannot, by definition, be queued, but there may be traffic moving in one lane but not the other. The small vehicle width of a TWMV means that it is possible to filter in streams of traffic that are moving.

One may describe filtering behaviour in terms of 'virtual lanes' or additional streams of traffic which may allow additional capacity to be obtained from the existing network. Observations indicated that the TWMVs would sometimes position themselves such that they were in a position between the conventional lanes, even if they were not actively making progress. Such positioning of the TWMVs may provide additional viewing distance and possibly braking distance. The latter would of course only be if the existing streams of traffic did not change their lateral position on the road under braking.

Lane changing

There were a range of proportions of TWMVs exhibiting this behaviour. For the purposes of this study, the lanes were defined as streams of traffic. There did not appear to be a relationship between the queue length and the proportion of TWMVs that were observed to change lanes. Lane changing behaviour was, therefore, observed equally within and outside queues.It is possible that this behaviour allows TWMVs to select the fastest moving traffic stream at any given time or get access to a 'virtual lane' as a precursor to filtering behaviours (though this has not yet been quantified). A higher proportion of motorcycles (compared to scooters) appeared to change lanes at 3 of the 4 sites that had statistically significant differences. These differences were, however, small in magnitude.

Inaction
Few TWMVs were observed not to act when an opportunity presented itself. It may be concluded that TWMV users generally take up an opportunity to reduce delay where it was presented. These observations also support the suggestion the TWMV users take an active approach to their driving and make use of opportunities to make progress where possible.

Balking
The proportion of TWMV users being balked was between 0.22 and 0.56. Balking would appear to occur frequently, but as it was often a transitory phenomenon, it is not yet clear to what extent it might reduce any time savings made by TWMV users.

Wriggling
This behaviour was rarely seen at most sites but Commercial St was an exception. It appears that wriggling is a site specific activity relating to the need of the TWMVs to change traffic streams. Wriggling may have occurred at Commercial St due to a slight pinch in the road some distance back from the signals such that filtering became impossible in the stream originally chosen and that the TWMVs had to move to a different stream to make progress.

Discussion
As this study only covers 5 sites caution should be exercised in generalising these findings. Some behaviours appear to be more prevalent in different road types. For example, filtering in moving traffic was more prevalent at the Upper and Lower Thames St. sites. These are dual carriageway sites which were conducive to filtering in the middle. At the other sites, which were not grade separated, there was a smaller proportion of TWMVs that filtered. This pattern was not, however, repeated for filtering in stationary traffic.

TWMV users appear to make efficient use of the road space when queuing. Observations indicate that a great number of TWMVs may get close to the stopline by a) forming a line across the road at the stopline and b) by queuing back between cars waiting. While the numbers have yet to be quantified the initial observations indicate that 3-4 TWMVs may queue across a single lane and that 2 or more TWMVs may queue in the space between 2 cars. It must be emphasized that these are rough estimates and that a count will need to be undertaken to confirm these figures.

References
Powell. M (2000) A model to represent motorcycle behaviour at a signalised intersections incorporating an amended first order macroscopic approach Transportation Research 34A 497-514, 2000

Martin B, Phull S. and Robertson S (2001) Motorcycles and Congestion. Proceedings of the European Transport Conference   10-13 September 2001   Homerton College Cambridge. Proceedings published as a CD, PTRC, London

Robertson S.A. (2002) Motorcycling and Congestion: Behavioural Identification. In P.T. McCabe (ed.) *Contemporary Ergonomics 2002,* (Taylor and Francis, London).

AN ON-ROAD COMPARISON OF FEEDBACK AND COGNITIVE PROCESSING IN MOTORCYCLISTS AND CAR DRIVERS

Guy H. Walker, Neville A. Stanton & Mark S. Young

Department of Design
Brunel University
Runnymede Campus
Egham
Surrey [UK] TW20 0JZ
United Kingdom

This exploratory on-road study compared motorcyclists with car drivers, and detected key differences in rider versus driver cognition. The main findings are that motorcyclists possess greater situational awareness due to the close interaction they have with their machine and the road environment, but curiously also seem to exhibit very low mental workload.

INTRODUCTION

Even 'normal' drivers exhibit, "*a very high differential sensitivity to changes of* [vehicle] *response time, and reasonably good ability to detect changes of steering ratio and stability factor*" (Hoffman & Joubert, pg 263), and can perceive differences in vehicle handling of a magnitude equivalent to "...*the difference in feel of a medium-size saloon car with and without a fairly heavy passenger in the rear seat*" (Joy and Hartley, 1953, pg 119). This 'feel' is an expression of the information that the driver or rider perceives through their senses as the vehicle responds to the demands made upon it by the driver or rider, and those imposed upon it by the environment Vehicle feedback (f/b) represents the vehicle's input into the driver's cognitive information processing chain.

Motorcycle riding, as an alternative type of driving task provides a further level of feedback that can be examined experimentally with that of car driving. The motorbike rider is more intimately involved with both the machine's dynamics and the external environment, rather than being cocooned or isolated from it. The human factors of motorcycles and riders is not a well studied domain. This is particularly concerning considering the very prominent safety issues surrounding motorcycling, coupled with the potential human factors insights that could help to design motorcycles and cars (and new vehicle technology thereof) for safer, more efficient, and more enjoyable human use.

This exploratory on-road study represents an initial step into examining the issue of vehicle feedback, providing further research questions that are currently being examined in more detail, and under more experimentally controlled circumstances in the Brunel University Driving Simulator. Motorcycles represent a new and interesting vehicle feedback manipulation, as well as a further and much needed area of human factors application.

METHOD

Design
This exploratory study was comprised of three groups, motorcyclists, high feedback cars, and low feedback cars. The feedback status of the vehicle is operationalised with reference to automotive engineering design criteria and specifications. (Anecdotally, high feedback cars are 'driver's cars' whereas low feedback cars are 'average cars'). The test protocol and route were fixed, and dependent measures were questionnaire responses and concurrent verbal reports on what drivers or riders were perceiving, comprehending, and doing. A Content Analysis paradigm was employed for the analysis of verbal data.

Participants
There were fifteen male riders/drivers who took part in the present study, all of whom used their own vehicles. Six of these were in the high feedback car category, six in the low feedback car category, and three in the motorcycle category.

Procedure and Materials
The participant initially completed a Driving Style Questionnaire (DSQ) and MDIE (Locus of Control) pre run questionnaires. For the car driving participants, a miniature video camera and microphone assembly was installed in their vehicle. If the participant was a motorcyclist then a microphone device was installed in their crash helmet and a mini disc recorder secured on them. The vehicles that participated are shown in table 1 below.

Table 1. Participating Cars and Motorcycles

Motorcycles	High F/B Cars	Low F/B Cars
Triumph Daytona 900	Audi TT Quattro	VW Golf TDi
Suzuki TL1000R	Masarati 3200GT	Toyota Tercel
BMW R1100GS	Holden HSV GTS	Mitsubishi Space Runner
	Morgan 4/4	Renault 18 GTL Estate
	BMW 325i Sport	VW Golf CL
	Toyota MR2	Peugeot 309 GLD

Standardised instructions, emphasising the desired form and content of the Concurrent Verbal Protocol (CVP) were then issued. Participants drove over a specially designed 14 mile test route on public roads, which comprised of six definable road types and ten junctions. They provided a running commentary (i.e. CVP) as they drove. The experimenter rode as a passenger in the cars. As a control for this with the motorcyclists group, the experimenter adopted an Institute of Advanced Motoring paradigm of following the participant, ensuring that he was positioned in the lead rider's mirrors so that directional indications could be seen and responded to. Upon completion of the course, the SART (self report situational awareness (SA)) and NASA TLX (mental workload) post run questionnaires were administered and completed.

Data Reduction
The CVP was subject to a comprehensive content analysis. Three encoding groups were defined: behaviour, cognitive processes, and feedback. The behaviour group was

comprised of own behaviour (OB), behaviour of vehicle (BC), behaviour of the road environment (RE), and behaviour of other traffic (OT). The cognitive processes group was sub divided into perception (PC), comprehension (CM), projection (PR), and action execution (AC). The feedback category offered an opportunity for vehicle feedback to be further categorised according to whether it referred to system or control dynamics (SD or CD), or instruments (IN).

RESULTS

Analysis of Pre Run Questionnaires
No significant differences were detected between the three groups on the Driving Style Questionnaire's dimensions of speed, calmness, focus, social resistance, planning, and deviance.

The MDIE questionnaire provides scores on two dimensions of locus of control, internality and externality. No significant differences were detected at the 5% level between drivers and riders on the internality dimension. Differences were detected at the 10% level (F=3.17, df=14, p<0.1). Scheffe post hoc tests, however, did not discern a strong effect between any of the pairwise comparisons of vehicle type, although a weak trend was detected between high and low feedback car drivers (at the 11% level), suggesting very tentatively that low feedback car drivers *might* tend to rate themselves higher on the internality dimension. A much stronger effect was detected for the externality dimension (F=8.73, df=14, p<0.01). Scheffe post hoc tests confirm that motorcyclists measure significantly higher than both groups of car drivers (figure 1).

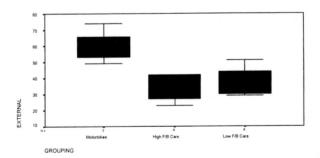

GROUPING

Figure 1, Boxplots of Externality Locus of Control Dimension versus F/B

Analysis of Concurrent Verbal Protocol Data.

Analysis of Total Verbalisations
A grand total of 12732 encoding points were derived from the content analysis of riders/drivers verbalisations. Motorcyclists supplied a mean of 1183.67 encoding points per run compared to high f/b cars (mean of 853.83 encodings per run) and low f/b cars (mean of 676.33 per run). A one way ANOVA revealed significant main effects of total encodings vs vehicle type (F=3.87, N=15, p<0.05).

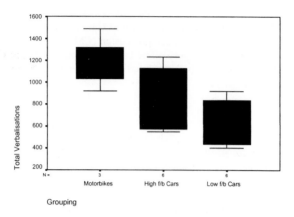

Figure 2. Boxplots of Mean Encodings Per Run Supplied by Each Vehicle Group

Scheffe post hoc tests reveal that the total number of encodings supplied by High feedback car drivers did not differ significantly from low feedback car drivers or motorcyclists, but motorcyclists had significantly more encodings per run compared to low feedback car drivers (figure 2). The implication at this stage is that greater vehicle feedback gives rise to a greater number of verbalisations.

Analysis of Encoding Categories
Across the eleven content analysis encoding categories, inter-rater reliability (IRR) was established with the use of two independent raters (IRR 1: Rho=0.7, n=11, p<0.05 and IRR 2: Rho=0.9, n=11, p<0.01). This demonstrates the reliability of the encoding scheme.

The method detected a number of key differences in the structure of encoding between motorcycles and cars. In other words, not only are motorcyclists talking more, but according to the content analysis they are talking about qualitatively different things.

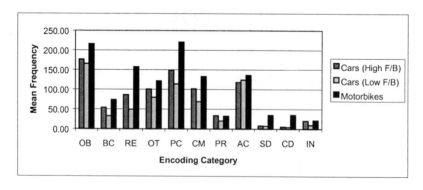

Figure 3. Mean Frequency of Encodings per Content Analysis Category

Significant differences were detected in the number of encodings between car drivers and motorcyclists. Motorcyclists made more verbalisations concerned with the Road Environment (F=21.64, df=14, p<0.01), Perception (F=5.24, df=14, p<0.05), Comprehension (F=4.22, df=14, p<0.05), System Dynamics (F=7.37, df=14, p<0.01) and Control Dynamics (F=34.86, df=14, p<0.01). Figure 3 is supported by Sheffe post hoc tests to illustrate that motorcyclists talk more than car drivers about the Road Environment, System, and Control Dynamics. Motorcyclists are not significantly different from high feedback car drivers, but *are* significantly different from Low feedback car drivers in terms of Perception and Comprehension. In summary, this analysis would suggest that motorcyclists are more aware of the road environment, talk more about levels 1 and 2 SA, and furthermore, report more about the feel and interaction of their machine when compared to groups of car drivers.

Analysis of Post Run Questionnaires

In terms of self reported SA (measured through the SART questionnaire), no significant differences were detected between the three vehicle groups on the dimensions of demand on cognitive resources, supply of cognitive resources, understanding of the situation, and overall SA.

Mean mental workload, as measured through the NASA TLX questionnaire, suggests that low feedback car drivers experience the *highest* mental workload (mean score of 43.2), compared to high feedback car drivers (mean score 32.22). Motorcyclists have the *lowest* workload out of the group, scoring a mean of just 12.05. These results differ significantly (F=17.66, df=14, p<0.01). Scheffe post hoc tests demonstrate that all groups are significantly different from each other in the manner that is depicted in figure 4 below. Based on this evidence there is quite a clear inverse relationship between feedback and MWL.

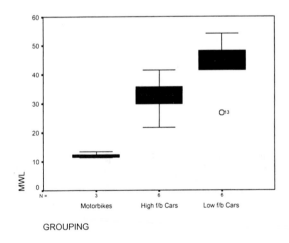

Figure 4. Boxplots Illustrating the Relationship Between Vehicle Feedback and Mental Workload

DISCUSSION AND CONCLUSIONS

The key findings of this exploratory study can be summarised as follows:

1. In terms of locus of control, motorcyclists rate highly on dimensions of externality. This could be reflective of a slightly more fatalistic approach that riders might tend to adopt towards crashing in general.

2. The driving style questionnaire showed no significant difference between any of the groups of riders or drivers. This offers a degree of reassurance that during the content analysis we are not merely measuring artefacts of driving style, instead it is the effect of the driver/vehicle interaction that is being observed.

3. There appears to be a direct relationship between level of vehicle feedback and the total number of encoding points (and by implication, verbalisations). Feedback provides more information for the driver or rider to process, and therefore to talk about.

4. Motorcyclists talk more about the road environment (in particular the condition of the road surface) and the way the vehicle and its controls feel and respond. This greater perception of vehicle feedback by motorcyclists is borne out in terms of cognitive processing by increased levels of SA.

5. SART scores show no significant difference between groups in terms of self reported SA, compared to significant differences detected in the content analysis. In terms of SA, ignorance does indeed appear to be bliss (Endsley, 1995). Self awareness of SA, or any shortfall in perceived SA is poor.

6. Feedback appears to have a strong inverse relationship with mental workload. The most obvious explanation is concerned with the quality of feedback and the gains in efficiency in terms of riders/drivers cognitive processing. A further subtle explanation might be related to the idea that feedback could be causing a growth in the attentional resources supplied to the task. So in other words, not only may feedback be making the driving task easier but it may be causing greater attentional resources to be supplied to it anyway.

REFERENCES

Endsley, M. R. 1995, Measurement of situation awareness in dynamic systems, *Human Factors,* **37**, (1), 65-84

Hoffman, E. R., & Joubert, P. N. 1968, Just noticeable differences in some vehicle handling variables, *Human Factors,* **10**, (3), 263-272.

Joy, T. J. P., & Hartley, D. C. 1953-54, Tyre characteristics as applicable to vehicle stability problems, *Proc. Inst. Mech. Eng., (Auto. Div.),* **6**, 113-133

DEVELOPING AN ERGONOMIC TOOL TO ASSESS OCCUPATIONAL USE OF MOTORCYCLES

Philip D. Bust

Consultant Ergonomist COPE
www.cope-ergo.com

With an increase in the take up of motorcycle use there is a need to appreciate the effect that regular riding of motorcycles has on individuals whether used as part of work, for commuting or leisure purposes. When asked to carry out an ergonomic assessment of an office worker's workplace and motorcycle, a tool was developed from vehicle assessment documents, a search through ergonomic papers and motorcycle publication articles. The tool was used to gather sufficient information in a short time to enable considerations of motorcycle use to be evaluated for a report and give recommendations with regard to musculoskeletal loading.

Introduction

A workplace assessment was required for an office worker who had been referred to a Physiotherapist by the company doctor with tendonitis in his elbow. He worked as an IT specialist and commuted to work on a Honda CB400N motorbike.

Four months prior to the assessment he had had a bike accident from which he received bruised ribs and a stretched tendon around his thumb. The week after his accident he was on holiday then for two weeks while his Honda was being repaired he rode his main bike, Yamaha 900, to work and for the four weeks after that he had no problem. Then he began to experience discomfort in his left elbow. Adjustments were made to his handlebars following discussions with a Physiotherapist and the discomfort then became apparent in both elbows.

An ergonomic assessment of his workplace and vehicle (motorbike) were requested to see if any improvements could be made. This is normally carried out with a combination of observation, checklist and questionnaire guided interview techniques. The questionnaire information being contained in the checklist.

Method

In order to carry out the assessment it was necessary to replace the vehicle assessment sheets of the checklist usually used which was based on the use of a car with a new assessment sheet that would be appropriate to assessing the use of a motorbike. As there was not an appropriate one available it was necessary to put one together. In order to do this an existing vehicle assessment sheet was used as a base and information was obtained from a literature search of ergonomic research and Internet search of motorcycle information.

Motor Vehicle Assessment – Checklists

Existing previously used motor vehicle assessment sheets were used as a base to put together a checklist for the motorcycle assessment sheet.
These concentrated on:-
1. Adjustments available in the seat
2. Clearance in seat and through door
3. Use of controls (especially steering wheel and pedals)
4. Schedule of daily/weekly use
5. Manual handling of equipment in the boot and from the back seat

With the motorcycle assessment sheet the following areas were considered:-
1. Adjustments available in handlebars and foot rests
2. Positions of handlebars and foot rests
3. Use of controls (by hands and feet)
4. Schedule of daily/weekly use
5. Clothing

Literature Search of Ergonomics of Motorcycle use

From a search through the Ergonomics Abstracts and vaults of the Nottingham University work by Robertson and Minter, 1996, anthropometrics of motorcyclists in the UK and knee position on static motorcycle rig and Robertson and Porter, 1987, preferred riding position were found.

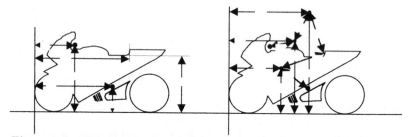

Figure 1 showing Motor Cycle Template

Motorcycle Publications

An Internet search was carried out which uncovered a number of magazine articles usually in the form of reviews of particular motorcycles. From these it was possible to draw a template for use in the assessment sheets to obtain relevant dimensions of the riding position and motorbike. References to Paul Kuhn and ergonomic articles in the Motorcycle Consumer News on Ergonomics of Motorcycles provided useful background information.

Tool Development

Using the information obtained for the various searches and checklist requirements a draft tool was developed in the following format:-

Sheet One
General Information: Essential information (name, job,…) together with basic physical details (height and weight) and specific information (Bi-deltoid width) to assist with consideration of use of handlebars. Travel Details: To assess usage of bike, discomfort experienced and bike type. Details specific to vehicle: Adjustments on vehicle, controls and instrumentation. Dimensions of Vehicle: To be used in conjunction with diagrams to position fixed points on vehicle (handlebars, footrest, seat) and joint locations on rider.

Sheet Two
Side elevations of bike and rider with ground level and front most point of bike as axes. Diagrams loosely based on modern sports bike but essential locations pretty much the same unless chopper type bike being assessed.

Sheet Three
Additional information: This consisted of a table with 'Area', 'Information' and 'Comments' columns. Prior to assessment the areas included were leisure use as opposed to work or commuting use (including whether used 'two-up'); modifications to vehicle and especially anything since accident; PPE worn; most discomfort in use of bike; and what would you change that you are unable to.

Assessment

For the Work Assessment, the worker's day (7 ½ hours) was almost entirely spent at a dedicated workstation. Tasks included writing documents, manipulating data, entering data and dealing with e-mails. There was little to no telephone and no paper work. There were no time pressures and breaks could be taken as required. The worker had stopped using his right hand, due to discomfort, for mouse work and was now using left hand and experienced pain with use of the keyboard.

At the workstation the office desk had a curved front profile and there was a five-wheel office chair. Use included operation of cursor by mouse to the left of the keyboard, telephone to right of computer and footrest provided and used. Files for reference to the left-hand side of the desk and filing tray to the right hand side.

With the Vehicle Assessment a Honda CB400N was used to commute to work each day. This comprised journeys of 30 minutes twice a day accounting for 120 miles travel each week. With use of bike he experienced additional discomfort to elbows directly after each journey also some discomfort to bottom, 'numb bum'.

The motor bike was assessed in an underground garage of the worker's office. It was positioned on a level floor area in front of a vertical wall in order that dimensions could be taken of the bike and rider relative to the vertical and horizontal planes.

The worker was observed riding the bike around the garage and all questions relating to the bike and its use were asked with the bike available for further inspection.

During the assessment it was noted that the tool had only included for dimensions in side elevation. Dimensions to indicate width of handlebars, distance between foot rests and rider's elbows were therefore included at the time. While these dimensions were being taken it was found that the handlebars were not aligned symmetrically thus twisting the riders position on the bike.

Recommendations

From the assessment it was recommended that a trial of a contoured keyboard with integrated touch pad be carried out to help reduce the loading from mouse work. The screen height needed to be lowered so that the top of the screen was at eye level. A trial of a document holder was recommended to be carried out in order to reduce twisting of the neck while working on the computer and advice was given on site with regard to chair adjustments, key board legs and taking regular breaks. With the vehicle assessment it was recommended that the handlebars should be checked for alignment and a recommended riding position with alterations of the handlebars was given in sketch form. The preferred posture was determined using REBA, Hignett and McAtamney 1999, postural analysis scores.

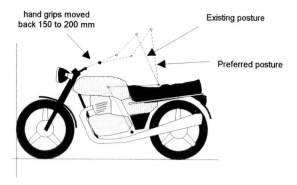

Figure 2 Showing recommended position for rider

Follow up

The office worker was contacted about 5 months after the assessment to see how he was coping with the changes. He reported that the touch pad, which had replaced his mouse as input device, had made a huge difference. He was using his right arm again for input work and was able to use the keyboard for typing without any pain. The effectiveness of the touch pad was underlined when he said that it had broken and while it was being fixed he returned to using the mouse and the pain in his elbow returned within a couple of days. One drawback was that he found this to be less productive.

With moving the screen he was experiencing neck pain when turning away from the screen so reminded him about taking breaks and if he found this difficult to use software reminders.

With regard to the bike the handlebars had been re-aligned so that they were symmetrical and while this was being carried out the position recommended to reduce musculoskeletal loading was looked at. From this appraisal it transpired that the replacements forks from repairs after his accident were too short thus moving his riding position forward.

Discussion

In the assessment the individual concerned had loaded his elbows from both work activities (extensive mouse use) and vehicle use (almost locked elbows in riding position). The work activity intervention (keypad) appears to have reduced the loading sufficiently enough to remove the previously experienced discomfort. Although the handlebars were re-aligned they were not extended so it is difficult to say that any benefits were obtained.

Self selection has resulted in motor cyclists being larger than average population, Robertson and Minter, 1996. The rider in the assessment was shorter than average and had an 8th percentile shoulder to grip resulting in lack of flexion in his elbows, therefore decreasing their ability to act as natural shock absorbers. Motor cycles do not provide the adjustments to ride position that cars offer and motor cyclists make adjustments, often with changes to handlebar positions, to improve the comfort in riding, Robertson and Porter, 1987.

Considered ergonomic assessments of individuals, their workplaces, vehicles, environment and work and leisure activities can lead to successful ergonomic interventions. However, the time is not always available to do this. In order that assessments can be made quickly while still capturing all the relevant information required for successful interventions, tools are needed which are based on sound ergonomic principles and research information. In this assessment the use of the tool enabled an assessment to be carried out on the work, workstation and vehicle at approximately the same time as previously spent on assessments where cars had been assessed. Unfortunately, with the time required to put the assessment sheets together and process the information obtained the exercise was considerably longer. Although we can draw on the ergonomic principles in creating tools to carry out ergonomic assessment of motorcycle use more research is needed to fill holes in our knowledge of motorcycle ergonomics.

Reference list

Robertson, S. and Porter J. M. 1987, Motorcycle Ergonomics: An Exploratory Study. In Megaw, E. D. (ed) *Contemporary Ergonomics 1987*, (Taylor and Francis, London), 173-178

Robertson, S. and Minter, A. 1996, A Study of some anthropometric characteristics of motorcycle riders, *Applied Ergonomics*, **27**, 223-229

Stanton, N. and Young, M. 1997, Is utility in the mind of the beholder? A study of ergonomic methods, *Applied Ergonomics*, **29**, 41-54

Hignett, S. and McAtamney, L. 1999, Rapid Entire Body Assessment(REBA), *Applied Ergonomics*, **31**, 201-205

DESIGN

POLICE RADIO HANDSET DESIGN: A USER-CENTRED APPROACH FOR THE NEXT GENERATION

Alex W. Stedmon[1] and John Richardson[2]

[1] Virtual Reality Applications Research Team (VIRART)
School of MMMEM, University of Nottingham
University Park, Nottingham NG7 2RD
[2] Transport Technology Ergonomics Centre
Loughborough University, Loughborough
Leicester LE11 3UZ

Simoco Digital Systems design and build complete TErrestrial Trunked RAdio (TETRA) systems and has developed a TETRA handset for use by Police forces throughout the UK. In order that design specifications could be confirmed prior to the initiation of new tooling for production and an initial request was made for user centred evaluations of a prototype handset. An initial review of the handset was made by experts in mobile communications, user interface design and 'context of use' analysis. This was followed by more specific focus group exercises with a number of Police authorities from around the UK. This paper summarises the findings of both activities which are then discussed from a methodological standpoint in terms of the points that were generated by the different groups, and in relation to the scope of the project.

Design Confirmation & Expert Evaluations

Simoco Digital Systems design and build complete TErrestrial Trunked RAdio (TETRA) systems with expertise in public safety handsets. TETRA is the international standard for digital mobile radio communications and provides digital Private Mobile Radio (PMR), with duplex telephony, extensive text messaging, advanced mobile data capability and secure encryption. Simoco have used this technology to develop a handset and now supplies over 60% of TETRA handsets to UK Police Forces.

Prior to the final tooling for production, it was important to confirm the design through expert evaluations. With only a non functioning prototype and a very short timescale, it was decided that the most effective method of data capture would be a number of focus group exercises. So that there would be no delay an initial review of the handset was made by Human Factors experts. At the same time, contact was made with a number of Police authorities around the UK so more specific user centred focus groups could be scheduled. From both these activities a final report was delivered within 3 weeks of the initial review and has since been used to support the development of the SRP2000 handset.

Figure 1. The Simoco SRP2000 Handset

From Analogue to Digital: The Generation Gap

The SRP2000 TETRA handset is designed to replace the previous generation (G-1) analogue handset. The G-1 is still in use but is now considered to be rather dated in it's technology and design. The G-1 is heavy (680g) and bulky when compared to the SRP2000 handset which weighs only 220g and is roughly half the size.

Due to it's size and weight the G-1 handset is most conveniently accommodated on an equipment belt along with other critical items such as: baton, handcuffs, and CS-spray (Edmonds & Lawson, 2001). In addition, First Aid kits, pen knives, torches and personal mobile phones are often carried. This can give rise to problems of equipment being caught on obstructions; discomfort when driving; back pain; hip pain; and poor equipment accessibility (Lomas & Haslegrave, 1998; Edmonds & Lawson, 2001).

With the advent of digital technology, the SRP2000 handset has developed into a hybrid between the previous G-1 and modern GSM mobile phones. As such, this not only increases the functionality of the handset but should also make it easier to carry and use, and may assist in re-distributing the equipment load of Officers.

A Review of the SRP2000 Handset

An initial review was carried out at the HUSAT Research Institute involving 6 staff members with expertise in mobile communications, user interface design and 'context of use' analysis. Comparisons were made between the G-1 and SRP2000 handsets and focused primarily on physical design and general usability issues. Participants were given a brief presentation on the SRP2000's features and the purpose of the review exercise; and an opportunity to handle both handsets and ask questions regarding their physical and operational characteristics.

From this, more specific focus group exercises were conducted with representatives from Police Forces around the UK and from a range of operational environments (metropolitan, urban and rural). The Police focus groups were organised depending on staff availability and, as such, ranged from 4 to 10 members and varied according to their ranks and duties. Due to the short timescale and staff availability no female Officers were available, however consideration was still given to different user group needs.

The Police focus groups followed a similar design throughout with a short presentation on the aims of the exercise and introduction to the new handset. The Police all had their own personal G-1 handsets and some had GSM mobile phones for comparative purposes. They were instructed to speak freely about their opinions and experiences and that none of their responses would be identified by name.

Due to commercial sensitivity, the information presented here is limited to first impressions, overall physical characteristics of the handset and usability issues. All quotations are taken from the various Police focus group exercises.

First impressions – "a big improvement on the old brick"

In the initial review the SRP2000 handset was considered to be a more portable, stylish and modern looking hybrid between the G-1 style handset and standard GSM mobile phones.

From the Police focus groups, the SRP2000 handset was generally very well received and regarded as a big improvement on previous generation/network handsets.

Shape & size - "nice size and shape with no sharp edges"

Within the initial review the shape and size of the SRP2000 handset was considered to be a distinct advantage over the G-1 handset. The SRP2000 handset is considerably smaller, making it easier to manipulate and carry in normal use, but also possibly easier to drop, lose from clothing or 'forget'.

The SRP2000 handset is designed to be held in much the same way as a GSM handset. This allows the thumb freedom to operate most of the keys on the keypad whilst only the index finger is needed to operate the Press-To-Talk (PTT) button. Under extended use this could prove tiring, especially if the PTT button is not directly in line with the index finger reach. Other issues related to the functional design of the handset, such as the handset not being free-standing and whether left- or right-handed uses would be able to use the handset equally well. Without a fully functioning handset these points were raised but not evaluated.

Within the Police focus groups some Officers expressed concern over making the handset any smaller in the future "there is a point when things get too small to hold easily or give the impression of being robust, it [the SRP2000] seems the smallest you could go without losing some of the functionality".

The SRP2000 handset features an ancillary port on the right hand side of the body. The location of the port (when not in use) was considered a comfortable thumb rest for users and could, therefore, be uncomfortable with ancillaries connected.

Although the issue of the SRP2000 handset not being free-standing was raised by the initial review, in the Police focus groups this was not considered to be a disadvantage unless the radio signal suffers. The feedback from the police focus groups showed less concern with both left- and right-handed Officers believing the handset could be used without any problem.

Weight – "we don't tend to use the radio for self defence anymore"

Participants in the initial review considered that, in extreme conditions, the bulkier G-1 handset could be used in self defence or to break a window, etc. The SRP2000 device, being lighter, would not be as effective in these situations although it would be less of a threat if used against an Officer.

The Police focus groups did not consider the use of the handset as a tool for self-defence, or for gaining entry/opening windows, to be a high concern and overall the weight of the SRP2000 handset was regarded as a big improvement. It was still considered "heavy enough to know when you don't have it with you" and generally considered easier to carry, conceal and hold than the G-1 handset.

Usability Issues

A number of general usability issues were discussed in both the initial review and the Police focus groups.

Carriage and Handling – "might be more acceptable to female Officers"
The initial review regarded the SRP2000 handset small enough to be mounted on the chest and therefore free from most of the problems associated with belt carriage. Chest mounting would probably prove more convenient in the majority of situations provided that a suitable method of attachment is available and clothing is suitable for the task (Health & Safety Executive, 1992). However, issues associated with chest mounting need careful consideration:

- A secure attachment is required to prevent loss when running.
- A chest mounting may require the handset to be removed or re-oriented if the screen needs to be viewed.
- It will be particularly important that the primary controls are accessible and convenient to operate when the device is mounted.
- Although use of the SRP2000 handset should be more convenient when in vehicles there may be a problem with seat belts catching on the handset.
- Left-handed operation, or personal preference, may require mounting points on the left and right side of the chest.
- In an assault or vehicle accident a chest mounting may lead to injuries.

The Police focus groups considered that the reduced size and weight of the SRP2000 handset will radically change the way it is carried and handled. This, in turn, may lead to changes in the way it is used particularly with respect to ancillary devices. For example, the size and weight of G-1 handsets necessitates the use of a belt mounting and the consequent use of a Remote Speaker Microphone (RSM), whilst the SRP2000 handset could be carried on the chest without the need for an RSM.

Officers could easily wear the radio in covert mode, in riot gear, or with a front harness. This might be more acceptable to female Officers although it would be necessary to identify a number of potential carriage options, along with ancillaries to support them, so that Officers have some choice about wearing the radio. In general, "Officers should not wear more than three items on their belts so wearing the radio somewhere else would be a big benefit", although it was generally considered that wearing the SRP2000 on a belt would not be a problem as it is lighter than current kit. However, Officers will need to look at the screen more than they look at their current radios so if they opt to wear it on the belt this could be a problem, especially under a coat or stab-vest.

Training- "should be fairly straight forward if it mimics GSM style & functionality"
The initial review considered that an extensive training programme would be required to support the introduction of TETRA based communications. Concerns were expressed if

user expectations are greater than the technical ability to deliver them, resulting in users becoming frustrated with the technology.

The Police focus groups considered that training on the SRP2000 handset would not be a major concern as most Officers will be familiar with conventional communication procedures and will also have experience of GSM based communications from their own use of mobile phones.

Operational Issues – "TETRA is clearer and potentially offers more extensive coverage"
The initial review identified issues concerning the type of ancillary devices that may accompany the handset. It became apparent that whilst the handset is well designed and suitable for a number of roles, specialist use (covert operations, etc) will be dependent on the successful integration of peripheral devices.

The Police focus groups considered that TETRA based communications should be accepted as a major improvement over current communications technologies and handsets will allow for much more flexibility in communications traffic between Officers. TETRA is seen as a distinct advantage over traditional analogue transmission as it is clearer and offers more secure and extensive coverage. It is anticipated that senior Officers will use TETRA for command and control (C^2) functions and that TETRA will allow for more communications within and between call groups.

Safety – "personal safety is paramount"
The Police focus groups agreed that whilst Officers are not instructed to wear radios in vehicles, they sometimes do. As such a chest mounted radio may have serious implications for personal safety if the Officer is involved in an accident. Another safety concern raised by the Police was that Officers may feel vulnerable wearing a radio on their chest rather than an RSM as with the G-1 handset. If handsets are snatched and taken/broken, then Officers could lose contact with their colleagues and station.

Discussion

A principal limitation on the work was that only a non-functional prototype was available for the initial review and the Police focus groups. Whilst this was, in itself, a very useful tool for demonstrating the physical design of the new handset and every effort was made to take account of the anticipated functionality and operational characteristics, it was not possible for such issues to be addressed directly.

The expertise of the Human Factors researchers highlighted physical ergonomics issues and more general issues concerning the context of use, whilst the Police focus groups highlighted practical usability issues that were drawn from first-hand experience in the field. Without specific operational scenarios or a fully functional handset the findings are somewhat generalised but serve to illustrate the importance of a user centred approach and, more specifically, the use of focus groups.

Focus groups are a useful method of gaining a number of views from a target audience and can be employed at any stage of the design process. They allow participants to discuss particular issues in an informal manner and because the method is loosely structured participants have the opportunity to raise issues that the investigator might not have anticipated (Jordan, 1993). As such, with the Police focus groups, the issue of personal safety was particularly significant and participants discussed various points based on their experiences. The size of a focus group is also important, with the optimum generally between 8-12 participants (Jordan, 1998). Care must be taken to acquire a representative sample of participants and facilitate an inclusive discussion by the careful use of questions and session structure. Participants for the Police focus groups were based on their availability and it was unfortunate that no female Officers were present. Another factor that can affect the success of focus groups is the mix of participants. Even though Officers of different ranks were present in the Police focus groups they all seemed to be comfortable expressing their views in front of their colleagues and superiors.

From the initial inquiry to the final report, the work was carried out in three weeks over the Christmas period (which is one of the busiest times for the Police), with a strict deadline for the report to be back to the designers before they could authorise the tooling for production. The SRP2000 was launched in 2001 and subsequently won "Best New TETRA Product" award.

Conclusion

By taking an integrated approach, combining the specialist knowledge of Human Factors experts and the practical experience of Police Officers, a more balanced perspective was generated that focused not only on the physical ergonomics of the handset, but also more subtle issues derived from the context of use and probable usability of the handset by various types of user.

Acknowledgements

This work was funded and carried out for Simoco International Limited. The authors would like to thank Mr Steve Barber & Mr Roger Lapthorn for their assistance throughout.

References

Edmonds, J., & Lawson, G., 2001. The design of methods for carrying police equipment. In, M. Hanson (ed). *Contemporary Ergonomics 2001*. Taylor & Francis Ltd. London.

Health & Safety Executive, 1992. *Personal protective equipment at work regulations.* HMSO

Jordan, P.W., 1993. Methods for user interface performance measurement. In, E.J. Lovesey (ed). *Contemporary Ergonomics 1993*. Taylor & Francis Ltd. London.

Jordan, P.W., 1998. *Introduction to usability.* Taylor & Francis Ltd. London.

Lomas, S.M., & Haslegrave, C.M., 1998. The ergonomics implications of conventional saloon car cabin on police drivers. In, M. Hanson (ed). *Contemporary Ergonomics 1998*. Taylor & Francis Ltd. London.

PRODUCT HANDLING AND VISUAL PRODUCT EVALUATION TO SUPPORT NEW PRODUCT DEVELOPMENT

Anne Bruseberg[1] and Deana McDonagh[2]

[1]*Department of Computer Science
University of Bath, Bath BA2 7AY, UK*
[2]*Department of Design and Technology
Loughborough University, Loughborough
Leicester LE11 3TU, UK*

User experience, knowledge and understanding offer product developers (industrial designers and ergonomists) a valuable resource. Eliciting user information requires a range of techniques and approaches. Product evaluation techniques are vital to retrieve users' perceptions on products' functionality and appearance – both to identify new requirements and evaluate product concepts. This paper discusses two distinctive, yet complementary, techniques – *product handling* and *visual product evaluation*. They have proven to be effective when integrated within focus group sessions and are suitable for a wide range of product types. The paper will highlight research findings from a recent EPSRC-funded[1] project and offer recommendations for using the techniques.

Introduction

Customers/consumers often make purchasing decisions based on whether a product 'looks' durable and functional. Likewise, the initial visual impact is a major factor when making purchasing decisions. There is little opportunity to test product functions at purchase point. The evaluation is often carried out within a short space of time. Hence purchasing decisions are often made with a significant lack of information. This is the case for buying items from product showrooms (e.g. Currys), but even more so when purchasing through catalogues or the Internet (e.g. Argos, Index). Here, two-dimensional images are one of the main resources for making purchasing decisions, besides lists of product functions and the position of the product within a price range. For many mainstream products (e.g. small domestic appliances), the choice is made more complex by the availability of vast range of comparable products. The customers/consumers needs to 'guestimate' the product's appropriateness and suitability whilst comparing it against other products competing for attention.

A previous study underlined the importance of the social element of retail activity/purchasing is important. Male users were, on average, more influenced by price, while female users tended to consider the product within a wider context – for example, its

---

[1] Engineering and Physical Sciences Research Council (Grant number: M98654) *Supporting New Product Development by Applying a User-Centred Design Approach.*

impact upon others within the home environment, or the question whether it would fit in with the existing décor (McDonagh-Philp and Denton, 1999).

Through focus group discussions, the authors have revealed that, on average, users anticipate a product costing between £30-£40 (e.g. kettle, toaster) will only 'last' for 2-3 years. Users have reported a degree of cynicism regarding how quickly product styling becomes 'out-of-date', and products are constantly up-graded by new looks and features. A trend is emerging where the customer/consumer is beginning to question the value of gross consumerism. Though they are in the minority, users are becoming more in tune with sustainability/ecological issues and socially concerned with the (perceived) limited product life expectation of the majority of consumer products. It is often the users' product expectations, needs and aspirations that may change, rather than its usefulness, thus rendering products *unwanted*, resulting in then being discarded. Users have expressed the desire for such products to last longer (durability and emotionally) and they claim to be prepared to invest more financially in such purchases.

This can clearly be seen in the higher retail price end (e.g. Siemens Porsche kettle, Dualit toaster). These products have successfully met user aspirations and needs, though both products are not actually functionally superior to other less expensive mainstream products. This highlights the need for product developers to tap into the emotional domain of user-product relationships to enhance understanding, knowledge and awareness of the 'real' user experience.

Wilson *et al* (2001) specifies the main factors that lead to a decision to purchase an item in preference to another as availability, affordability, desire and need, which in turn are influenced by other factors such as technology, geography, income, status and culture. The factors influencing product choice were grouped into six categories: aesthetic, cultural and traditional, social-psychological, economic and political, managerial, and physical.

Industrial designers/new product developers need to be aware of user needs and aspirations. They have to continually learn from user feedback to identify the factors that may distinguish future products from others and increase consumer satisfaction. The information gained from studying the benefits and drawbacks of existing products is vital to establish new design specifications. It is particularly important to identify those criteria that potential customers/consumers recognise as significant when evaluating products without being able to actually use them (e.g. not filling a kettle with water, not placing bread in a toaster).

The authors have tailored several studies to explore the ways in which users evaluated and investigated consumer products under such conditions. Product handling exercises limit users to the evaluation criteria available in a retail showroom scenario. Visual product evaluation only allows the study of two-dimensional images, as found in mail order/web based purchasing scenarios. This paper demonstrates how the techniques may be applied, and explores how the data may be analysed and used.

The application of the techniques

During the studies, both visual product evaluation and product handling exercises have been integrated as part of focus group session. They are relatively short exercises that provide a different type of activity in contrast to group discussions, thus keeping participants interested and focused.

An effective way of capturing feedback is through the use of questionnaires (refer to Figure 2 for an example). Although questionnaires limit the depth of data that can be captured and reduce the scope of responses by the questions that have been prepared in advance, the feedback tends to be short and precise, focusing on the most important aspects. They also enable the use of rating scales to quantify the feedback. The questionnaires used a rating scale ranging from 1 (very poor) to 5 (very good) (Likert, 1932). Whilst questionnaires are suitable to retrieve a standard set of data that can be used for direct comparison between participants, it is important to note that the small number of participants cannot supply statistically secure data. The most useful information stems from the combination of the ratings with the comments made regarding the reasons for the selection (Dumas, 1998).

When using the exercises as part of a focus group session, it is important to ensure that participants conduct them in a self-contained manner – i.e. without conversing or sharing their opinions with others. Otherwise strong positive or negative emotional reactions of some individuals may influence the judgement of others. The issues identified by the users during filling in the forms may be explored in more depth during a subsequent group discussion.

Visual product evaluation

The data relied upon during visual evaluation includes aspects such as the product's shape, form, the use of materials, colour and weight. Participants are provided only with an image of the product for visual evaluation, similar to a situation where users evaluate products under the restricted conditions of a catalogue (i.e. mail order, television, internet purchasing), where only two-dimensional visual information is available. The assumptions that users make when perceiving aspects such as shape or style help to elicit insights regarding product semantics (e.g. a product might 'look' heavy and may therefore be rejected).

The emphasis is on the initial impact; hence the time given for evaluation is limited (e.g. 5 minutes per product). Users are given an image of the product only (e.g. slide projection, photograph or rendering). Ideally, the images should represent the artefacts as realistically as possible. When planning to compare the impact of several products, then the images should have the same background colour and should have been photographed from the same angle. Visual product evaluation is suitable for both evaluating existing products as well as product concepts (see Figure 1).

**Figure 1: Evaluating concepts visually based on renderings
(feedback capture through questionnaires).**

The data captured in a questionnaire may retrieve ratings regarding the perceived shape and styling of the product. Space for comments need to be included. It is useful to provide separate boxes for making comments about liking and disliking of attributes. The questionnaire should retrieve information as to whether the participant has any previous knowledge of the product (as this may influence the visual evaluation). It is useful to ask questions regarding purchasing intentions and estimated retail price.

Product handling

The product handling questionnaire (see Figure 2) has been designed to capture immediate feedback regarding the perceived functionality of the product samples. User evaluation was based on a retail showroom scenario to extract immediate 'gut' reactions to the products. Participants were required to assess the suitability of the product for the different activities involved in its use, based on a simplified task analysis of the product operation. Additional verbal user feedback was retrieved and later analysed (e.g. visual appearance, perception of quality and durability).

Product handling exercises, when carried out as part of a focus group session, do not provide a 'natural' context for observation, such as the home of the user. Hence, product operation cannot be studied in relation to associated tasks or physical settings. Likewise, some merits or problems of product operation may not be identified because the product functions cannot be actually tested. It may be useful to back up the study with user trials or home observations. However, in the authors' experience, the information from product handling offers rich design data – just by drawing on the personal experiences of users.

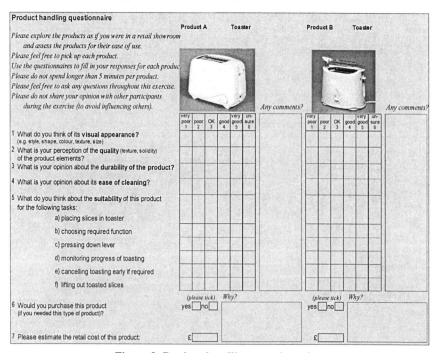

Figure 2: Product handling questionnaire.

Data analysis

Quantitative data (i.e. ratings, *yes/no* choices, price estimates) from both visual questionnaires and product handling questionnaires can be averaged for each product, to begin to understand users' preferences and priorities (e.g. Table 1). The results need to be interpreted in conjunction with the participants' comments, and through comparison of product attributes. In this case, the participants preferred simple lines and traditional shapes. Visual appearance, low cost, and small size appear to be some of the most important factors influencing overall preference (e.g. Toaster D has 'lost' against Toaster C in terms of purchasing intentions in spite of its highly rated functionality).

Table 1: The results of the product evaluation questionnaires
(both visual and product handling) for 5 toasters (sample size: 11 participants);
the average ratings are situated between 1 (poor) and 5 (very good); the products are
sorted after purchasing intentions (number of answers saying 'yes'); average overall
ratings are shown in italics; bold figures indicate highest ratings.

Toasters:	C	A	D	E	B
Visual evaluation results					
shape and styling	**3.73**	3.55	2.73	3.27	1.73
colour	**4.00**	3.45	3.82	3.64	3.18
materials used	3.82	3.40	3.82	3.27	2.91
size and weight	**3.45**	3.18	2.73	3.00	2.60
average rating (visual)	*3.75*	*3.40*	*3.27*	*3.30*	*2.60*
suit kitchen – answer: 'yes'	3	**6**	4	2	0
buy? (visual) – answer: 'yes'	**4**	3	2	2	0
estimated retail cost (visual)	28.18	18.55	52.00	35.73	26.27
Product handling results					
estimated retail cost (handling)	33.27	18.82	57.00	46.36	29.82
visual appearance	**4.09**	3.91	2.82	3.55	3.27
perception of the quality	**4.18**	3.09	**4.18**	4.09	3.27
durability	4.18	3.00	**4.36**	4.09	3.09
ease of cleaning	3.18	3.67	2.82	**3.73**	3.36
placing slices in toaster	4.00	3.36	**4.09**	3.60	3.18
choosing function	3.18	**3.64**	3.50	3.00	3.36
pressing down lever	3.36	3.55	**3.89**	3.36	3.40
monitoring progress	2.80	3.00	**4.00**	3.00	3.00
cancelling toasting	3.36	**3.73**	3.56	3.11	3.73
lifting out slices	3.36	3.27	**4.20**	3.70	3.09
average rating (handling)	*3.57*	*3.42*	*3.74*	*3.52*	*3.28*
buy? (handling) – answer: 'yes'	**5**	5	3	2	2

Preferably, the data should be evaluated through qualitative data analysis approaches. This may include:

a)   The extraction of adjectives to express product attributes, and grouping them into those communicating positive meanings (e.g. unfussy, clean-looking, simple) and negative meanings (e.g. flimsy, clunky, pretentious) for each product (see Bruseberg and McDonagh-Philp, 2002).

b)   Summarising the comments from the questionnaires in conjunction with the focus group results through 'category analysis' – by grouping similar ideas and examples into suitable clusters and estimating relative importance through the frequency of mentioning them (see Bruseberg and McDonagh-Philp, 2001).

Conclusions

Capturing users' immediate feedback to the visual appearance of products (existing or conceptual) enables the design team to gain further insight into values that are being assigned and attributes that are recognised as valuable. Though these techniques clearly have limitations, they can offer the product developer, designer and ergonomist, flexible 'tools' that can be incorporated into product research to retrieve both quantitative and qualitative data. The authors advocate employing the techniques discussed within this paper during user-centred design research. They will benefit the product developer throughout the designing process, but in particular during the pre-concept generation stage.

References

Bruseberg, A. and McDonagh-Philp, D. 2001, New Product Development by Eliciting User Experience and Aspirations. *International Journal of Human-Computer Studies,* **55** (4), 435-452

Bruseberg, A. and McDonagh-Philp, D. 2002, forthcoming. Organising and Conducting a Focus Group: The Logistics. In Langford, J. and McDonagh-Philp, D. (eds.) *Focus Groups: Supporting Effective Product Development*, (Taylor and Francis, London), in press

Dumas, J. 1998, Usability Testing Methods: Subjective Measures; Part II - Measuring Attitudes and Opinions. *Common Ground (Newsletter of the Usability Professionals' Association)* (October 1998), **8** (4), available from: http://www.upassoc.org/html/usability_testing_methods.html

Likert, R. 1932, A Technique for the Measurement of Attitude. *Archives of Psychology,* **140**, 1-55

McDonagh-Philp, D. and Denton, H. 1999, Using Focus Groups to Support the Designer in the Evaluation of Existing Products: A Case Study. *The Design Journal,* **2** (2), 20-31

Wilson, J., Benson L., Bruce M., Hogg, M. K. and Oulton, D. 2001, Predicting the Future: an Overview of the Colour Forecasting Industry. *The Design Journal,* **4** (1), 15-31

EMBRACING USER-CENTRED DESIGN: THE *REAL* EXPERIENCE

Ian Storer and Deana McDonagh

Department of Design and Technology,
Loughborough University,
Loughborough,
LE11 3TU

This paper highlights the views of a small sample of practising industrial designers in relation to User-Centred Design (UCD). These views were considered important in order to compliment the literature in the area and inform the integration of UCD within undergraduate design training at Loughborough. UCD appears to offer benefits in relationship to enhancing the designing process, particularly in terms of sensitivity to ergonomics.

A small scale survey of designers from diverse product areas shared their understanding and experience of User-Centred Design via an electronic questionnaire. This indicated that in industry these approaches are not always fully utilised. The discussion explores the limitations and perceived barriers to integrating such user centred methodology.

Introduction

User-Centred Design has developed from its roots in market and political research into mainstream design philosophy over the last ten years. UCD, in Industrial and Product Design, may be defined as taking the users' needs into consideration and securing their views as part of the designing process (McDonagh-Philp *et al* 1999). UCD seeks to bring the designer closer to the user, often reducing the step function of market research which tended to act as a barrier between the designer and the user. It offers a flexible approach which may incorporate a range of specific methodologies e.g. focus groups, generative methods. These methods may be used at any stage within a design process and not simply at the beginning. It is an approach that brings the designer closer to the user, to increase understanding, awareness and empathy. But is this approach accepted, adopted and employed within the industrial context?

A small scale questionnaire was conducted which **aimed** *to sample the understanding, practice, and attitudes of a group of practising designers* from a range of product areas. The results are intended to feed into a larger follow-up survey.

Method

The authors recognised the difficulties of gaining feedback in surveys and therefore decided to use an Email questionnaire and limited it to nine questions. It was anticipated that this would improve the return rate. The final question was an open-ended opportunity for respondents to add any points they felt important and not covered by the previous questions. The questions were generated by mind mapping around the central aim. For brevity they are reported below together with a snap shot of responses.

The sample of 12 was selected to include designers from a range of product areas from consumer, medical and automotive. All were male, university educated and ages from 24 to 44. They ranged from recent graduates to experienced designers. The designers were mainly from the UK with one from the Netherlands.

The majority of the participants required additional contact via the telephone to remind them to complete the questionnaire. One consultancy contacted refused to comment/respond considering UCD as a *secret weapon* of their own and that explaining their activities further would erode any competitive advantage currently held.

Results

The first question (*What is your understanding of User-Centred Design?*) provided the opportunity for the designers to communicate their working definition of the term. This helped to place their responses within a context, to assist interpretation. All the respondents indicated that they were familiar with the term and provided their own working definition which placed the user as a focal point within the designing process.

The designers highlight the importance for stakeholders to immerse themselves within the user experience to provide *"insights, visions and embodiments"*. The general feel, was that the users should become an integral component to the designing process prior to concept generation, to assist in defining the actual design brief. This shift in thinking moves away from employing the user in a purely evaluating role. Nevertheless, it also shows that these designers perceived UCD as a 'front end' process. The did not see it as employable during a design process, for example, to check the initial concepts.

The second question (*Within the commercial environment, do you actually employ any User-Centred Design approaches?*) aimed at confirming if the designers actually employed UCD in their day-to-day work. The majority claimed to have employed UCD with varying degrees of interpretation. UCD was generally considered to contribute to a products commercial success.

Question three requested the methods, techniques or approaches actually used. This question aimed to find out how aware the designers are of methods and techniques available to them. This would provide the opportunity to highlight novel approaches that the authors may not be aware of. All the responses indicate a general awareness of standard approaches, without raising any unfamiliar techniques. The range of methods reported included verbal communication, sketches, three-dimensional models, experiential profiling and creating collages.

The fourth question, *(What are the barriers to User-Centred Design?)* focused on what prevents the designers utilising UCD. Limited resources were reported as one of the main obstacles to UCD. Companies perceived designers as universal experts on products and user needs.

A second barrier reported was the question of a shared language between users, designers and stakeholders would assist the flow of communication. Industrialist use the 'language' of commerce, designers primarily use visual language, while the users may understand neither. The designers report that understanding and interpreting the data collected is a challenge, an alien activity. From the designers perspective, how can you turn market research data (typically quantitative data in the form of graphs and statistics) into a desirable kettle?

The respondents perceived users as conservative and lacking creativity, and their input may not lead to innovation solutions. Another designer was anxious to avoid the *"Perception that the designer is not confident if they cannot make design decisions without having to refer to Joe-Public"*.

The fifth question explored the benefits of employing UCD (*What are the benefits to User-Centred Design?*). This question aimed to elicit the advantages to employing UCD. Generally the designers accepted that there were benefits as it *"adds rigour to a fuzzy process of creation"*. It also discourages *"the ego driven designer concept way of working"*. This is a crucial element to evidence based design decision making, which supports the designer in the design process by enabling a path of action, to proceed with confidence and reduce the risk of commercial failure (e.g. *'Ford Scorpio'*).

The sixth question (*In an ideal world, if anything were possible, what sort of data/information would you like to extract from users?*) attempted to relieve the designer of all constraints. This gained a range of responses from the emotional, *"Dreams. True feelings"* to the simple and direct, *"Comfort/are they comfortable? Can they see out? Should the product reflect or look forward"*. A respondent commented that there was *"Nothing that cannot be extracted already,"* and that information received may not be suitable, *"if you have 10 people in 10 groups and you come away with 100 ideas, what have you gained? It becomes too personal, too subjective"*. Whereas *"identifying true users responses through brain wave patterns or something that removes confusing feedback"* was desired.

Question seven attempted to explore the area of design training (*In your experience, what sort of methods, techniques and approaches should be taught at undergraduate level to prepare the students for the commercial environment?*). Unfortunately the responses were too general and the question should have been more focused on UCD. However, respondents did identify *"The basics in observation techniques and interviewing skill; basic understanding of people, society, cultures"*.

Question eight asked respondents to reflect upon their own experience (*In your experience does the user hinder or enhance the designing process?*). This produced some conflicting responses; *"Neither – the user is simply **part** of the design process – it is their requirements the designer must try and meet (whether physical, intellectual or emotional) if you have no user you have no need/but they are not the designer – no-one has a monopoly on good ideas – but creativity is required to interpret users wants and observations"*. User data contributes to the designing process like ingredients in a recipe, it is the designer that creates the final outcome, it is not user-driven but user-informed.

The questionnaire finished with the opportunity for comment. Not all responded but the general essence was that there needs to be a more collective creativity. A greater emphasis on problem formulation from the perspective of the user, rather than form giving. A greater emphasis on methods that probe and generate user creativity. Designers should have a command of visual based generative research methods as these are likely to be more user-friendly. They should also be competent in the traditional evaluative methods.

Discussion

The responses highlighted the sensitivity of the designers to the dichotomy of working for a client company and specific product users. This was most apparent in the comments regarding cost (money, resources and time) of UCD. Respondents gave the impression that companies (with the one exception that saw UCD as its *secret weapon*) that they worked for did not yet see the value of UCD out weighing the costs and were reluctant to inject the money and time necessary. The feel from the feedback was that companies expected the designers to be universal experts on products and user needs.

The literature on UCD frequently refers to benefits (Wengraf 1990) in terms of the product. The authors also consider the potential benefit to the designer in terms of strengthening theirs abilities to empathise and design for people beyond their socio-economic profile.

As many of the benefits of UCD are intangible and incapable of being incorporated in a conventional cost/benefit analysis there is a grave danger that projects remain more client-centred than user-centred. Clearly further research is needed in this area, however, it must be recognised that many benefits will never be easily measurable. There are so many variables involved within product development, it is not possible to compare similar design projects to each other due to their unique nature. The market place is ever changing so are the needs and aspirations of users, and to a degree the skill of the designer in employing UCD approaches.

In contrast, UCD should not be assumed to be a straight forward approach to adopt. Sensitivity and experience is required (a) to enable users to express their feelings openly and (b) to interpret such qualitative and highly individual data into design relevant information. The specific methods (including focus groups, task analysis, product handling) all have their advantages and limitations, of which the designer must be aware in order to support users in reaching and expressing their feelings. None of the methods or techniques are prescriptive. Any one context will require a range of methods in order to gain the most effective feedback. It must be recognised that such research does cost money and the designer has a responsibility to use what resources they have available to the maximum effect. The authors have not found specific work which describes the effectiveness of one individual method compared to another, if indeed that is possible. A designer, therefore, needs experience, understanding and skill in selecting and employing methods in any given unique context. In interpreting UCD data, it must also be appreciated that external and internal factors can be highly influential and transient. For example, the impact of specific events reported in the media may have a considerable impact on the users response without them actually referring to it (e.g. USA 11 September 2001). This example is of course an extreme, however, there may be many other events which are far more subtle and yet are transient. UCD data, like any other data, is a snap shot of opinions at a point in time.

Designers also need to be sensitive to the benefits of UCD. It does of course mean that they will be partly removed from the actual practice of design which some designer may feel is not efficient use of their skills and time. Indeed, the questionnaire showed the designers had had little training in design research methods at undergraduate level. This is also supported by Garner and Duckworth (2000).

Conclusion

The survey indicates that in practice designers do not tend to have the resources to use UCD methods iteratively within the design process. In the authors experience as practising Industrial Designers and observing colleagues, design solutions are often based upon minimal testing and product analysis work, using a 'best guess' and past experience to drive the process. However, manufacturers need to understand and appreciate the value of design research and UCD. Designers are not typical users (Norman 1988) they cannot rely or expect themselves to have *intimate* knowledge of the user. For the practitioner, it is crucial that approaches, techniques and user design resources are easily accessible and in an appealing format. Designers are visual and the prospect of having to wade through tables and text would be a barrier. With the emerging design research culture there is a shift towards designers, designing methods and approaches for designers. Which in itself is user-centred!

User-centred design clearly has its benefits and drawbacks. The 'real' experience of UCD of practitioners indicates that it is fundamentally important. However, in reality, it becomes a luxury due to limited resources. Data collection is a relatively simple task, but it is the actual conversion/translation of that data in to design relevant information. Turning the data in to a form, with colour, texture and personality. This becomes the ultimate challenge to the designer.

References
Emmison M. and Smith P. 2000, *Researching the Visual*, (London: Sage)
Garner S and Duckworth A (1999) Identifying key competencies of industrial design and technology graduates in small and medium sized enterprises. In Roberts P H and Norman EWL (Eds.), the Proceedings of the *International Conference on Design and Technology Educational Research and Curriculum Development*. (Department of Design and Technology, Loughborough University, Loughborough) pp. 88-96.
McDonagh-Philp D, Lebbon C and Torrens G E (1999) An Evidence Based Design Method Within a User-Centred Approach. In the Proceedings of *the 4th Asian Design Conference International Symposium on Design Science*, Japan, October. Program committee of the 4th Asian Design Conference: Japan. (CD Rom)
Norman D. A. 1988, *The Psychology of Everyday Things*. (Cambridge MA: MIT Press)
Riggins S. H. 1994, *Fieldwork in the living room*. In Riggins S. H. (ed.) The Socialness of Things. (New York: Mouton de Gruyter)
Wengraf, T. (1990) Documenting *Domestic Culture by Ethnographic Interview* in (Putnam and Newton, Household Choices, Futures Publications)

OPTIMIZING THE PRODUCT SHAPE FOR ERGONOMICS GOODNESS INDEX. PART I: CONCEPTUAL SOLUTION

Niels C.C.M. Moes and Imre Horváth

Delft University of Technology, Dept OCP/DE
Section Integrated Concept Advancement
Jaffalaan 9, 2628 BX Delft, The Netherlands
C.C.M.Moes@IO.TUDelft.nl
I.Horvath@IO.TUDelft.nl

Abstract: An optimization procedure was developed to search for the optimum product shape. The criterion for optimization is an objective functional, the Ergonomics Goodness Index, which is based on medical, physiological and ergonomics criteria. The input configuration to be optimized is a Finite Elements Model of the human body. The independent variables for optimization were the pressure values in the contact area. The actual optimization algorithms were defined according to the gradient search procedure. The measurements and calculations were developed for a sitting support without a backrest or arm rests. The initial load was a flat, hard and horizontal support. This paper presents the construction of the Finite Elements Model, and the first results of the analysis for a simplified assembly of the upper leg and the buttock area. The model includes the skin, the bony parts and in between a matrix of soft tissue. The validity testing of the model is based on predictive formulations of the sitting pressure distribution.

Keywords: ergonomics, Finite Elements Model, shape conceptualization, pressure distribution, ischial tuberosities, optimization.

Introduction

Current CAD programs for the shape of a product are not able to incorporate design concepts such as ergonomics, manufacturability and aesthetics, to balance them in design decisions and design optimization, and to generate a a real prototype for user trials or production. The aim of the Integrated Concept Advancement (ICA) project (Horváth, 1998) is the development of an Intelligent Balanced Comprehension Engine to assemble and synthesize the knowledge of the relevant design concepts into actual product design proposals for shape, physical properties etc. This subproject explores the possibilities and the feasibility of the implementation of ergonomics knowledge, rules and guidelines in such a system. Since a vast amount of valuable ergonomics knowledge was created during the last century, A general comprehension is virtually impossible. Thus a convergence of the focus is necessary. We concentrated on the ergonomics aspects that contribute to the conceptualization of the shape of the contact area of physically handled products. In particular the transmission of the interactive force between a seat and the body is considered, which is naturally explained by the pressure distribution.

Such a system of procedures to automatically generate an instance for the product shape

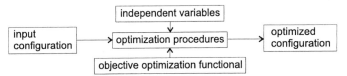

Figure 1.   The main aspects of the shape optimization.

based on ergonomics rules would be considered a great benefit for the conceptual design of several product classes, especially those where the shape of the contact area and the physiologic functioning of the skin and its underlying tissues are strongly related. Examples are seats, footwear, beds and hand tools. Since such 'analytical' solutions have not yet been developed we have to apply optimization techniques.

Methods

Optimization

A typical general optimization procedure is shown in figure 1. It consists of (i) an object to be modified, which is called the *input configuration*, (ii) a variable to control the optimization, (iii) a (set of) object-quantity(ies) that will actually be varied, called the *independent variables*,and (iv) the resulting modified object. The object to be modified is the part of the human body that is involved in sitting. The control variable to be optimized, which is usually called the *objective optimization functional* is the *Ergonomics Goodness Index* ($\mathcal{E}$). The independent variables is the pressure distribution in the contact area. The modified configuration is the body with the modified properties.

Reasoning model

To optimize an initial shape, an unloaded deformable model of the upper leg and the buttock areas of one half (right) of the body is loaded by a flat, undeformable support. The loading force equals half of the weight of the upper body. After applying this load the contact area becomes a flat surface. The characteristics of the resulting pressure distribution in the contact area are the maximum pressure over the ischial tuberosity area, the pressure gradient, the magnitude of the contact area, the location of the maximum pressure points,and the average pressure (Moes, 2002b). A change of the pressure distribution *within* the continuum of the body introduces an internal pressure gradient. If it is sufficiently large, it may result in problems, such as decubitus (Hobson, 1988; Staarink, 1995).

The optimization is controlled by the objective functional that contains information of (i) the current pressure and (ii) the maximum pressure values that are allowed from the physiological point of view. After applying the flat surface load the actual optimization starts. First the direction of change of the contact pressure is calculated for increased $\mathcal{E}$. The calculation of this direction is based on the gradient search method (Bevington, 1969). Then the pressure modification is iteratively applied until termination. The termination criteria control (i) running out of CPU-time, (ii) the convergency of the iterations, and (iii) the number of iterations. When the optimization has finished the final, deformed Finite Elements Model (FEM) is analyzed for the location of the nodes of the contact area. These locations are needed for the geometry of the actual physical prototype.

User trials form an integral part of the system. If they give satisfactory results the shape can be transferred to the next design stages. If not, then the definition of $\mathcal{E}$ must be adjusted and another optimization cycle must be performed.

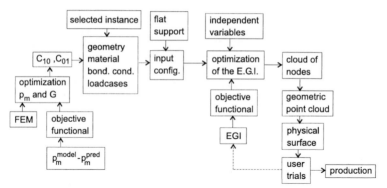

Figure 2.   Overview of the procedures to conceptualize the ergonomics optimized shape.

Implementation I: Finite Elements Model

The geometry part of the 3D FE model describes the shape of the elements, the localization of the nodes, and the boundaries. To enable easy manipulation of the model the nodes and the elements are stored in sets. The model properties were assigned to the sets. The basis for the geometry is the measured surface models of the involved body parts. A closed surface model is generated from this partial model as input for the solid mesher.

Instance generation of the surface geometry
Since human characteristics are always vague, or statistically uncertain, all calculations for geometry, material; properties, contact area, etc. should include vagueness. In (Moes et al., 2001) a methodology is presented to generate geometric instances from a generic geometric model that is based on the theory of reference vectors, metric occurrences and the location index (Rusák et al., 2000). Ideally the FEM, that is derived from such geometric model should also be defined in vague terms. Unfortunately such FE analysis engines do not yet exist. The escape is to analyze a stratified sample that is representative for an intended user group. The parts I and II of this paper present the implementation of one crisp instance. Figure 2 shows the complete setup of the involved procedures.

Geometric aspects of the Finite Elements Model
During the construction of the FEM, assumptions were done to simplify the model. The current FEM reflects only three components of the body: the skin, the bony tissue, and a matrix of soft tissue. Further assumptions included the following. The bony tissue is undeformable. Between the skin and the bony tissue exists only isotropic, non-viscous, non-creep soft tissue. The hip joint and the sacro-iliac joint have no freedom of movement.

Material properties
(Manschot and Brakkee, 1987b; Manschot and Brakkee, 1987a) showed that the elastic and viscous behaviour of the human skin depend on age, environmental conditions, physical constitution and even seasonal variation. Comparable data on other tissues was not found. Despite such complexity it was decided to apply a most simple mechanical model for the material properties.

The expected deformations are large so that elasticity must be defined as non-linear. Human tissues contain mainly water, so that incompressibility was assumed. The selected non-linear model is a rather simple one, with reasonable validity for compressive forces:

Mooney-Rivlin with two coefficient, $W = C_{10}(I_1 - 3) + C_{01}(I_2 - 3)$ with incompressibility condition $I_3 = 1$, or even the one-coefficient neo-Hookean behaviour with $C_{01} = 0$ (Marc, 2000, Chapter 4: Nonlinear Material Behaviour). I_1, I_2 and I_3 are the invariants $I_1 = \sum_{i=x,y,z} \lambda_i^2$, $I_2 = \lambda_x\lambda_y + \lambda_y\lambda_z + \lambda_z\lambda_x$, and $I_3 = \lambda_x\lambda_y\lambda_z$. W is the strain energy function and $\lambda_{x,y,z}$ are the principle stretch ratios. An initial trial value for the one-parameter Mooney model was $C_{10} = 100$ kPa. Which is based on a simple, unreported experiment where the skin was impressed 1 cm by a square of 1 cm². This required a load of approximately 1 N. Currently the soft tissue was assigned the same material properties as the skin, but in coming experiments the two tissues will be distinguished.

Boundary conditions and contact properties
To simulate the assumptions that the bony parts are practically undeformable and that the hip joint is fixed, the bony 'substances' were removed from the model. The surface nodes, that now form the inner closure of the body, were fixed in space. The original geometric measurements and the construction of the model is explained in more detail in (Moes, 2002a).

Two bodies are involved. One body is the deformable FEM, the other is a flat rigid surface parallel to the x-y plane, and positioned somewhere below the deformable body. The boundary conditions for the deformable body include the fixed spatial positions of the bone nodes, the reduced freedom of movement of some boundary surfaces to simulate body symmetry or tissue continuation. After positioning the rigid surface against the deformable body a load is applied on the surface, equals to the weight of the upper body. Since the material properties are non-linear the load is applied in a series of increments.

Implementation II: Optimization aspects

The scheme of figure 2 shows two optimization loops. The first is the search for the material properties of the soft tissue and the skin by matching the predicted values of the maximum pressure and the pressure gradient (Moes, 2002b). If the FE analysis is done for a few, carefully selected values of C_{10}, the estimation of the best C_{10}, and in a later stage the two parameters of the Mooney model, can be done by interpolation. The second is the search for the shape of the seat that gives optimized $\mathcal{E}$. This search executes an iterative procedure that varies the pressure distribution in the contact area according to a intermediately compiled $\mathcal{E}$-pressure gradient (Moes, 2001). The $\mathcal{E}$ is the objective functional for the modification of the shape of the deformable body. It reflects the relevant criteria for the ergonomics goodness. Examples of such criteria: the flow of blood, interstitial fluid and lymph, and the transmission of nerve pulses. Weight factors are attributed to these criteria. The value of these weight factors are determined by the product type, usage, circumstances, user group, etc. In general the $\mathcal{E}$ is defined as $\mathcal{E} = \sum w_i f(v_i, v_i^\phi)$, where v_i is a calculated value of a variable, and v_i^ϕ is the physiologically determined limit for that variable. The assessment of the value of the weights and the selection of the variables to be included in $\mathcal{E}$, which requires a deep understanding of the designer on the fundamental product properties, constitutes eventually the basic foundation of this methodology. In the first stages of the research these quantities will only refer to the calculated pressure inside the body and the physiological pressure limits as they are proposed by, among others, (Landis, 1930; Kosiak, 1985). The definition of the $\mathcal{E}$ is discussed in detail in (Moes, 2001).

Discussion

A FEM counts many variables and adjustments. A few of them were touched in this paper and will be elaborated in part II. For the one parameter Mooney model of the material properties a small series of analyses is required to obtain an estimation of C_{10} by simple interpolation, see part II; the criterion is the predicted maximum pressure. To match the maximum pressure and the pressure gradient at least two parameters are needed, for which purpose the two-parameter Mooney model is applied. Contact conditions of the current model do not include glue or friction. In future including a layer of clothing should be considered in addition.

References

Bevington PR (1969). *Data Reduction an Error Analysis for the Physical Sciences.* Mc-Graw-Hill.

Hobson DA (1988). *Contributions of posture and deformity to the body-seat interface conditions of a person with spinal cord injuries.* PhD thesis, University of Strathclyde, Glasgow, Scotland.

Horváth I (1998). Shifting Paradigms of Computer Aided Design. Delft University of Technology, Faculty of Industrial Design Engineering. Inaugural Speech.

Kosiak M (1985). Etiology of decubitus ulcers. *Archives of Physical and Medical Rehabilitation,* 42:19–29.

Landis EM (1930). Micro injection studies of capillary blood pressure in human skin. *Heart,* 15:209–228.

Manschot JFM and Brakkee AJM (1987a). Characterization of in vivo mechanical skin properties independent of measuring configuration. *Bioengineering and the Skin,* 3:1–10.

Manschot JFM and Brakkee AJM (1987b). Seasonal variations in the mechanical properties of the human skin. *Bioengineering and the Skin,* 3:25–33.

Marc (2000). MSC.Marc Advanced Course. Course material, MSC Software, München, Germany.

Moes CCM (2001). Mathematics and Algorithms for Pressure Distribution Controled Shape Design. In Culley S, Duffy A, McMahon C, and Wallace K, editors, *Design Methods for Performance and Sustainability, Proceedings of the 13th International Conference on Engineering Design, ICED01,* pages 99–106, Glasgow. Professional Engineering Publishing, Bury St Edmunds, UK.

Moes CCM (2002a). Finite Elements Model of the Human Body for Application in the Ergonomics Optimization of the Shape of Sitting Supports. In Horváth Imre, Peigen L, and Vergeest Joris S.M., editors, *Proceedings of the TMCE2002,* pages ??–??, Wuhan, P.R. of China. Fourth International Symposium on Tools and Methods of Competitive Engineering, ??

Moes CCM (2002b). Modelling the Sitting Pressure Distribution and the Location of the Points of Maximum Pressure for Body Characteristics and Rotation of the Pelvis. *Ergonomics.* Status: submitted nov. 2001.

Moes CCM, Rusák Z, and Horváth I (2001). Application of vague geometric representation for shape instance generation of the human body. In Mook DT and Balachandran B, editors, *Proceedings of DETC'01, Computers and Information in Engineering Conference,* pages (CDROM:DETC2001/CIE–21298), Pittsburgh, Pennsylvania. ASME 2001.

Rusák Z, Horváth I, Kuczogi G, Vergeest JSM, and Jansson J (2000). Discrete Domain Representation for Shape Conceptualization. In Parsaei Hamid R., Gen Mitsuo, Leep Herman R., and Wong Julius P., editors, *Proceedings of the 4TH International Conference of Engineering Design and Automation - EDA 2000,* pages 228–233, Orlando, Florida. CD-rom; Integrated Technology Systems, Inc.

Staarink HAM (1995). *Sitting Posture, Comfort and Pressure.* PhD thesis, Delft University of Technology, Delft, the Netherlands.

OPTIMIZING THE PRODUCT SHAPE FOR ERGONOMICS GOODNESS INDEX. PART II: ELABORATION FOR MATERIAL PROPERTIES

Niels C.C.M. Moes and Imre Horváth

Delft University of Technology, Dept OCP/DE
Section Integrated Concept Advancement
Jaffalaan 9, 2628 BX Delft, The Netherlands
C.C.M.Moes@IO.TUDelft.nl
I.Horvath@IO.TUDelft.nl

Abstract: In part I the developed methodology of the ergonomics shape optimization was introduced briefly. Part II presents the construction of the surface and the solid Finite Elements Models and the first results of the analysis. The presented results consider the estimation of the material properties of the part of the human that is involved in sitting. The analysis was done for a flat sitting support impressing a FE model of the human buttock area. Comparison with measured data should enable a reliable estimation of the stiffness of the FE model.

Introduction

A solid Finite Elements Model (FEM) is derived from a closed FE surface model. The following components are currently reflected in our final solid FEM: (i) the skin of the upper leg and the buttock area, (ii) the bony parts including the femur, the pelvis and the sacrum, and (iii) a matrix of isotropic, homogeneous soft tissue in between. The derivation of the shape of the skin is discussed in (Moes et al., 2001). The selected skin shape was for a male subject, body mass 77 kg, and ectomorphic index 6. These data were also used to estimate the maximum pressure below the seating bones (Moes, 2002b). Since the shape of the bony parts is not accessible in vivo without medical risk, they were obtained from the male Visible Human Data set (VHP, 1997). Fitting of the bones in the skin shape was done using measured reference points of specific landmarks of the subject, followed by translation, rotation and scaling of the bone geometries (Moes et al., 2001). First the essentials of the shape measurement procedures and the construction of the FE model are explained. A detailed overview of the model was presented in (Moes, 2002a). Then the focus is on the estimation of the material properties of the model.

Finite Elements Model

Figure 1 shows a cross section of the sacrum and the iliac bones of one of the VHP slices. The scan point density[1] is such that shape singularities of local landmarks are sufficiently

---

[1] The scanning software was Surf Driver, Rapid and Reliable 3-D Reconstruction, version 3.5.5, http://www.surfdriver.com

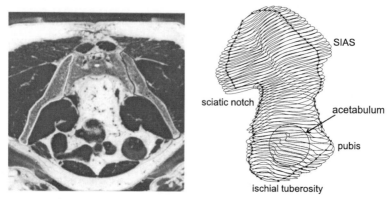

Figure 1.   Left: example of a cross sectioned view of one of the VHP slices at the level of the promontory. Right: example of the resulting scanned contour lines of the pelvis.

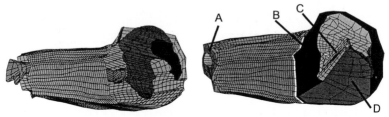

Figure 2.   The assembly of the surface FE models

Figure 3.   The solid hexmeshed model.   Left the total model is shown.   The right picture shows some details of the internal structures.

represented. The right figure 1 shows a lateral view of the scanned contour lines for the pelvis bones. The scans of the sacrum and the femur were made in the same way.

The assembly after the geometric transformations is shown by the left figure 2. To close the surface, which is required for the solid meshing engine, auxiliary surfaces were added, see the right figure 2.

Solid meshing was done with the Mentat[2] hexmeshing utility. The optimum element size was $10 \times 10 \times 10$ mm. Some coarseness was needed to get the smaller parts of the surface model filled. The result was a solid model consisting of ca 30,000 elements, see figures 3. The detailed model definition was done via sets of nodes and elements, see the shadings in figures 3. The support was a rigid body, represented as a flat surface.

The two main boundary conditions are the spatial fixation of the bone-nodes and the

---

[2] MSC Mentat2001/MARC2001

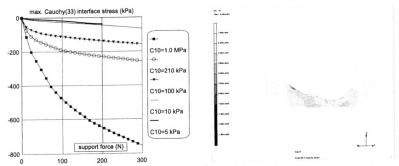

Figure 4.   Left: The applied support force and the maximum interface Cauchy stress, σ_{33}^{max}**. Right: the** σ_{33} **distribution in a vertical cutting plane along the ischium.**

applied force on the support. The support force was estimated from the regression $F = -25.5 + 7.98 \times m = 600\,\text{N}$ (Moes, 2002b). Actually 300 N was applied which is the weight on one body half.

Pilot studies showed that hardware limitations required a size reduction. In the current stage of the project this does not pose a serious problem since it still carries the character of a feasibility study. Since the expected deformations are large the elasticity is necessarily non-linear. The one parameter, C_{10}, (neo-Hookean) version of the Mooney formulation for incompressible materials was selected. Non-linearity implies that the load must be applied in a series of increments to obtain convergence.

Results

The support force and the maximum interface Cauchy (σ_{33}^{max}) stress are shown in the left figure 4 for C_{10} =5 kPa, 10 kPa, 100 kPa, 210 kPa and 1 Mpa. The stiffness was applied for both the skin and the soft tissue. The number of increments was 25 for the high values of C_{10}. For lower C_{10} four load cases were defined of 25 increments each so that in the first stage of the loading the step size is smaller, and convergence improved. The right figure shows a typical σ_{33}^{max} distribution in a vertical cutting plane along the length of the ischium. From this picture it is clear that the maximum stress occurs at the bone-soft tissue interface, (which explains the first development stages of decubitus at the boundary of the ischial tuberosity). The curve for the lowest C_{10} was not completely analyzed because the stiffness matrix became singular.

This was the result of the extremely large deformation of the elements below the ischium. Local adaptation of the element size of these elements (subdivision in $2\times2\times2$ elements) gave no improvement.

Figure 5a shows how the z-displacement is related to the exerted force. As expected the displacement decreases with increasing stiffness. Figure 5b shows the relationship of the maximum Cauchy stress in the interface and the stiffness. The increase of the stress in the right part of the diagram is clearly supported. The expected decrease in the left part is not yet investigated, but will be discussed at conference time.

Discussion and conclusions

It was proven that the measured skin contour data and the bone scans can be used to successfully create a FEM. Earlier research showed that crisp and vague instances can be generated

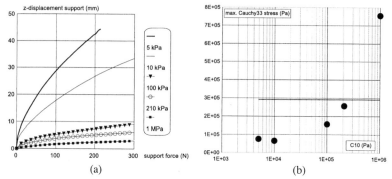

Figure 5.   **Left: the maximum interface Cauchy pressure, σ_{33}^{max}, as a function of C_{10}. Right: the displacement of the support and the force exerted by the support.**

from a generic vague geometric model. Despite the currently limited hardware conditions, and therefore the necessary reduction of the size of the FEM, it was proven that the neo-Hookean elasticity constant can be evaluated and estimated. Further research is needed to explain the mechanical properties of the skin and underlying tissues by age, gender, physical constitution, etc. This will enable a more accurate match with the predicted variables that describe the aspects of a pressure distribution.

The stiffness of the skin and the soft tissue were equal in this report. Future research will treat them differently. To enable the low-elasticity analysis an upper boundary condition, that was originally present in the large FE model, an elastic foundation should be applied on top of the model, see (Todd and Tacker, 1994). Moreover, local adaptivity of the elements of high pressure gradient will improve the results of the analysis.

The results give confidence to proceed with the model refinement using extended and more advanced hardware. The choice of the FE modeller and analyser was a serious item of consideration, but it has shown its ability to cope with all the requirements for model parametrization.

References

Moes CCM (2002a). Estimation of the Nonlinear Material Properties for a Finite Elements Model of the Human Body. In Horváth Imre, Peigen L, and Vergeest Joris S.M., editors, *Proceedings of the TMCE2000*, pages ??–??, Wuhan, P.R. of China. Fourth International Symposium on Tools and Methods of Competitive Engineering, ??

Moes CCM (2002b). Modelling the Sitting Pressure Distribution and the Location of the Points of Maximum Pressure for Body Characteristics and Rotation of the Pelvis. *Ergonomics*. Status: submitted nov. 2001.

Moes CCM, Rusák Z, and Horváth I (2001). Application of vague geometric representation for shape instance generation of the human body. In Mook DT and Balachandran B, editors, *Proceedings of DETC'01, Computers and Information in Engineering Conference*, pages (CDROM:DETC2001/CIE–21298), Pittsburgh, Pennsylvania. ASME 2001.

Todd BA and Tacker JG (1994). Three-dimensional computer model of the human buttocks, in vivo. *Journal of Rehabilitation Research and Development*, 31(2):111–119.

VHP (1997). The Visible Human Project[TM]. URL: http://www.nlm.nih.gov/research/visible/visible_human.html.

OBSERVING USER-PRODUCT INTERACTION

H. Kanis, M.J. Rooden, F.H. van Duijne

*School of Industrial Design Engineering,
Delft University of Technology,
Jaffalaan 9, 2628 BX Delft,
the Netherlands*

Usage oriented design exploits insights into the activities, experiences and judgements of future users. Such insights are typically gained by observation of actions, together with individuals' self-reports of their reactions to external phenomena (e.g. what is perceived), of internal activities (e.g. ways of reasoning, on the basis of what experience etc.) and of internal references (e.g. assessment of activities as (in)convenient). This paper explores opportunities to extend and reinforce the observation of user product interaction, and to support the analysis of recorded observations from users' perspectives rather than being restricted by theoretical preconceptions.

Introduction

Actual usage of products often comes as a surprise to designers. What is worse, in many cases the surprise is an unwelcome one as unanticipated users' activities undermine designed functionalities or even lead to accidents. The difficulty for designers is that user activities tend to be largely unpredictable on the basis of theoretical considerations. Data rather than theory seems to be needed in order to anticipate future usage. This paper deals with the various data which may, to a lesser or greater extent, support usage oriented design. Data involved will be discussed in terms of their observability/reportability and their relevance for designing usability.

Data on user-product interaction

As a guide, Figure 1 presents a graphical representation of user-product interaction. Over the years this representation has been formulated and continuously adapted on the basis of insights from observational studies. The heart of the matter is the ongoing concatenation in context of user activities (perception/cognition, use actions including the experienced effort)

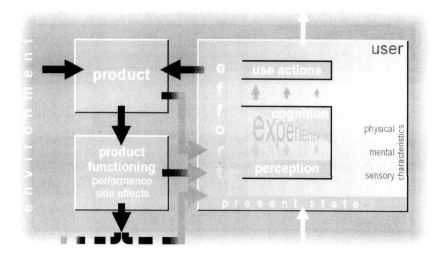

Figure 1. Product operated by user.

➡ product functioning (performance, side-effects) as the result of a technical/physical process, i.e. as the outcome of the co-occurrence of a product with use-actions in a physical environment.

➡ product functioning due to user activities, i.e. perception of featural and functional product characteristics, cognition, use actions, with the effort involved as low as possible as the default, which tendency is indicated by the increasing thickness to the left of the vertical arrows between 'perception' and 'use actions'; experience as situated knowledge of users on the basis of association and recognition, and also as a psycho-motoric condition activated in product usage (e.g. typing skill).

additional paths via the environment for the possible effects of product functioning on users and vice versa; present state as the current physical condition (e.g. wet hands), sensory (e.g. not using necessary aids such as glasses) and mentally such as a particular mood (e.g. being irritated, in a hurry, pleased).

and featural and functional product characteristics. These characteristics, i.e. a product's appearance and its performance including possible side-effects, constitute what the design of a product is about. See Kanis & Green (2000) for some further underlying considerations.

User-product interaction as presented in Figure 1 gives rise to different types of phenomena. To begin with, the left side of the representation depicts a physical process in the co-occurrence of a product activated by use actions in a physical environment. Both the constituents of this co-occurrence and its outcomes offer numerous possibilities for measurement. As to the intended functioning of a designed product, the measurables at issue in a technical sense involve the input a product requires and the output it delivers in various environmental conditions. As a rule, design requirements are set specifying levels of functionality which should be met in different circumstances, for example the force required to operate a product, the limitation of the number of steps for using an interface, various performance levels, and also side-effects (noise, heat, vibrations, waste). With respect to these functionalities, the present paper does not deal with their measurement in a technical/physical sense but rather with the identification of their role in user-product interaction, that is: as perceived, understood, experienced and reacted upon by users. Thus the focus is on the right hand side of Figure 1.

Types of data, relevance and availability in design

Different types of data. With the user as the possible source of insights to support usage oriented design, three types of human involved data can be distinguished:

(*i*) descriptions of human characteristics, e.g. anthropometrics, sensorical thresholds, general knowledge and memory capacities, that is: properties of human beings which are not tied up to particular operations such as product usage,

(*ii*) descriptions of user activities in operating a design at hand (model, product) comprising perception/cognition including the activation of experience as situated knowledge, use actions, postures adopted, movements made, produced effort, and,

(*iii*) description of what is called 'present state' in Figure 1, involving local/temporal peculiarities in a specific user-product interaction, such as deviations from usual conditions or of being in a particular mood (which may not necessarily be induced by the interaction with the product, see caption of Figure 1).

Design relevance of data. For insight into product usability, the significance of descriptive data can be conceived in terms of differences in these data being aligned with differences in usability as experienced by people in practice (for instance in terms of effectiveness and efficiency, cf. ISO, 1994). In a design context, this concept boils down to the responsiveness of a negotiated functioning of a design in progress to differences in these data (Kanis, 1998). This means that, in order to be design relevant, data should be sufficiently detailed (rather than being global, summative measures) so that links are specified between user involved descriptions (see above) and the featural and functional product characteristics to be (re)designed. This implies that user activities (see category *ii*) stand out for their potential design relevance. Time and again, empirical studies into actual usage have shown that the design relevance of human characteristics (see category *i*) is largely limited to setting boundaries for user activities. This is to say that these characteristics mainly indicate what, on average, users will not do since they are unable to. For the actual user activities (*ii*), both at and within the boundaries set by human capacities and limitations, human characteristics tend to be uninformative. As to the third category of data, descriptors of the 'present state' (see category *iii*) can be seen as providing additional insight into the origin and situatedness of actual user activities.

Availability. By definition, data on user activities (*ii*) and 'present state' (*iii*) have to be gathered by somehow observing users' operations of a particular product or of a design model in context. This is different for human characteristics (*i*). Generally, this type of data consists of means and standard deviations produced on the basis of uniform procedures and tasks (cf. Boff and Lincoln, 1988), and is available apart from any design consideration. Unlike observation involving user activities and the 'present state', research into human characteristics tends not to be carried out in order to support specific design decisions. As a rule, there is no way of linking summative descriptions of human properties to featural and functional product characteristics (Kanis *et al*, 1999). The over-estimated importance of these data for design (cf. Chapanis, 1988) seems to originate from the misconception that knowledge of human universals can drive usage oriented design.

Generating data for usage oriented design

Data which are most needed in usage oriented design (*ii* and *iii*, see above) can only be produced in some realistic environment, preferably on the spot, with a product or design model involved. Two extremes may be identified as to possible interference in the observation (rather than 'measurement') of user-product interaction: non-reactive registration versus invited self-reports of participants. The former is the ideal, in order to observe users' behaviour in as natural a context as possible. The latter, recording of self-reports (in thinking aloud and retrospective interviewing), often not only happens to be the only viable way (i.e. to establish perceived featural/functional aspects, activated experience, produced effort, 'present state') but may also reveal biasing tendencies emerging from participants being aware of a research setting as well as of the social setting involved. In retrospective interviews, participants regularly turn out to be considerably more positive about experienced functionalities than seems warranted in view of all kinds of difficulties they were observed to face shortly before. Examples from empirical studies involve the denial by participants of ('clumsy') actions that were actually undertaken, the contended perception of cues which were non-existing or imperceptible (e.g. only visible from the rear), and, more generally, the adoption of an overzealous cooperation demonstrated by the selection of those topics which participants presume as relevant in the study. This playing down of use problems may be a token of social desirability, the deranging effects of which have been extensively discussed already some decades ago (Wieck, 1968). However, other 'mechanisms' may also play a role, think of being impressed by a research setting or by the newness of a product design.

Taking for granted that participants are aware of a research setting (e.g. in the case of a simulated environment, of non-hidden recording devices, or in asking participants for self-reports) does not imply that one stands empty-handed in countering bias such as that emerging from social desirability. The distinction between intrusive and unobtrusive observation is not clear-cut, see below.

Observation of user-product interaction: obtrusive, inconspicuous, ...
There is no problem in the rare case that use actions can be recorded candidly with no instruction required while people involved do not, in hindsight, object to having been a participant. If this cannot be achieved, possible bias in demonstrated behaviour may be attenuated by putting participants on a wrong track provided, again, that they do not afterwards object to having been (somewhat) mislead. Obvious examples are the postponement of revealing the objective of a trial such as offering a cup of coffee prior to the introductory talk in the study of coffee creamer containers (Kanis, 1998), drawing the attention to the taste of water in a study of the opening of cans, confronting participants with a product in the wrong default (too high/low, upside down, in the wrong place) or with parts missing (e.g. batteries) or needing replacement (e.g. blunt knives) and, more generally, focussing on overall performance of products as a general, side-tracking task while subroutines in usage are the issue

Thinking aloud and retrospective interviewing inevitably emphasise a research setting. This emphasis may be reduced with the user being regarded as the expert speaker and the researcher in the role of an active listener. This is a departure from Ericsson and Simon (1993) whose generally acclaimed guidelines seem inappropriate for design supportive observational research (Rooden, 2001; Boren and Ramey 2000). In the case of thinking aloud, this social setting with an expert speaker and an active listener may prompt extra information which otherwise would not become available. With subtle prompts (e.g. 'Mm hmmmm' and 'Uhuh') the listener keeps the conversation going. The information revealed in this way tends to suffer less from possible biases that may contaminate retrospectively collected data (e.g. due to memory processes, rationalisation, social desirability).

Intrusive methods of recording human activities involve the measurement of physiological parameters such as heart rate, blood pressure, O_2-consumption and eye movements. Such measures may provide additional evidence, e.g. for effort involved in doing physical work. For product use the effectiveness of eye movement data is questionable: where people look (i.e. the centre of their field of vision) does not necessarily reveal what people see. This means that observation of the centre of the field of vision does not imply that information possibly available on that spot is picked up by the viewer (Coren *et al.*, 1999; Goldberg, 2000). Similarly, opportunities to pick up information outside the centre of the visual field may have been exploited. These uncertainties about the significance of measuring eye movements added to the intrusivenes of this kind of measurements raise the question whether other additional evidence may be generated which may be less problematic.

Additional utterances and expressions
The observation of use actions and the registration of talk both in thinking aloud and retrospective interviewing are not the only means to gather insight into users' activities and their 'present state'. People tend to be expressive in many small ways, often inadvertently it seems, whatever the (un)obtrusiveness of the research setting. Examplas are gestures, postures, facial expressions, silence interrupting talk, and sounds such as sighs and response cries (cf. Goffman, 1980). Such 'productions' tend to be accounted for in the study of interaction between humans, particularly in conversation analysis. In the study of user-product interaction, these utterances and expressions may help to identify users' perspectives, as can be revealed in so-called data sessions (cf. Jordan & Henderson, 1995). Data sessions are an effective means to breach seemingly obvious interpretations and to compare plausible alternatives rather than sticking to preconceived categorisations or taxonomies. As a result, outcomes of a study can be better calibrated by the actuality as perceived by participants. This is not to say that the significance of utterances and expressions indicated above is easy to pin down. In some explorative investigations we found, as expected, many differences, not only between participants but also intra-individually. For instance, the one participant tends to laugh every now and again, apparently on completely different occasions, while the other goes about straightfaced. In addition, it seems fruitful to consider a combination of facial expressions, gestures and postures as indices of nonverbal communication.

The evidence which may be raised, especially by data sessions, on the basis of various utterances and expressions of participants, also seems to address individual predispositions in responses and will therefore require individual referencing. Presumably, such evidence will always be circumstancial, providing supportive indications for plausible interpretations or (if not being vague or multi-interpretable) revealing counter-evidence.

Discussion

Usage oriented research is substantially reliant on the exploitation of 'subjectivity' in terms of individual accounts. Searching after 'objectivity' by reducing user involvement in the production of data would also reduce or even eliminate the design relevance of such data. In a previous paper (Kanis, 2001) the topic was raised of scientific credentials, particularly the paradox of grafting research into user-product interaction on positivist methods. A point of discussion was the peculiarity, in a design context, to resort to summative quantifications assessed in terms of reliability and validity, rather than to an appreciation of user-product interaction as situated and its observation itself as interactive. The present paper indicates several opportunities to reinforce and support the observation of user-product interaction. These opportunities involve research procedures, e.g. in striving for a natural context by unobtrusiveness, taking account of various (non-verbal) observables produced by participants in addition to actions and verbalisations, and analytic approaches such as applied in data-sessions with an emphasis on data-driven conceptualisation from a user's point of view rather than imposing preconceived categorisations and taxonomies.

References

Boff, K.R. and Lincoln, J.E. 1988, *Engineering Data Compendium: Human Perception and Performance*, AAMRL, (Wright-Patterson AFB, OH)

Boren, T.M. and Ramey, J. 2000, Thinking aloud: reconciling theory and practice, *IEEE Transactions on Professional Communication*, **43**, 261-277

Coren, S., Ward, L.M., and Enns, J.T. 1999, *Sensation and Perception*, (Harbourt Brace College, New York)

Chapanis, A. 1988, Some Generalisations about Generalisation, *Human Factors*, **30**, 253-267

Ericsson, K.A. and Simon, H.A. 1993, *Protocol Analysis: Verbal reports as data*, (MIT Press, Cambridge)

Goffman, E. 1980, Forms of Talk, (Blackwell, Oxford)

Goldberg, J.H. 2000, Eye movement-based interface evaluation: what can and cannot be assessed? *Proceedings of the IEA2000/HFES2000 Congress*, 6-625–6-628

ISO 1994, ISO DIS 9241-11 *Ergonomics requirements for office work with visual display terminals (VDTs) – Part II: Guidance on usability*, (International Standard Organisation, Geneva)

Jordan, B. and Henderson, A. 1995, Interaction Analysis: Foundations and Practice. *The Journal of Learning Sciences*, **4**, 39-103

Kanis, H. 1998, Usage centred research for everyday product design, *Applied Ergonomics*, **29**, 75–82

Kanis, H. 2001, Scientific credentials for qualitative and quantitative research in a design context. In M.Hanson (ed.) Contemporary Ergonomics 2001, (Taylor and Francis, London), 389-394

Kanis, H. and Green, W.S. 2000, Research for usage oriented design: Quantitative? Qualitative? In *IEA-HFES Proceedings*, 6-925–6-928

Kanis, H., Weegels, M.F. and Steenbekkers. L.P.A. 1999, The uninformativeness of quantitaive research for usability focused design of consumer products. *Proceedings HFES Annual Meeting*, 401-405

Rooden, M.J. 2001, *Design models for anticipating future usage*, Thesis, Delft University of Technology

Weick, K.E. 1968, Systematic Observational Methods. In Lindzey, G. & Aronson, E. (eds) *The Handbook of Social Psychology. Volume 2. Research methods*, Second Edition, (Addison-Wesley, Reading), 357-451

WHEN ASYNCHRONOUS COMMUNICATION IS BEST

Stephen AR Scrivener, Andrée Woodcock and Joseph Chen

VIDe Research Centre, The Design Institute, Coventry University, Coventry, UK
s.scrivener@coventry.ac.uk , a.woodcock@coventry.ac.uk, li.chen@coventry.ac.uk

A plethora of technological aids have been developed to support design teams working together at a distance, but research has focussed on synchronous remote communication. However, there are strong arguments for suggesting that domains such as design, where globalisation and the need to produce designs quickly and cheaply, will rely on asynchronous communication. This paper presents the results of an Internet survey which reveals that computer-mediated asynchronous communication is already a significant component of both local and remote design, is beginning to supercede conventional modes of remote communication and is likely to becomes increasingly important. It also reveals why, sometimes, asynchronous communication is the preferred means of communication.

Introduction

To reach global markets, manufacturing companies and designers are working across national and continental boundaries to develop products (e.g., companies in Taiwan wishing to manufacture for the European market are commissioning work from European designers). These collaborations are constrained by many factors including separation in time and place. Increasingly, Computer Supported Collaborative Work (CSCW) and Computer-Mediated Communication (CMC) technology are being explored as a means of overcoming temporal and physical separation in distributed design (Saad and Maher, 1996).

Much of this work has focused on support for synchronous interaction with asynchronous exchanges being provided as a secondary means of communication available when neither sender or recipient are available to discuss design ideas. In many respects, asynchronous communication has been treated as something that is only needed or used when synchronous communication is not available. However, Scrivener and Clark (1994) have argued that asynchronous communication is likely to be significant in industrial design as multinational companies or SME networks employ designers, experts and manufacturers in different regions to produce goods cheaply and quickly, and exploit global markets.

The time zone difference is a barrier to communicating at the same time in the global context, thereby necessitating greater reliance on asynchronous communication. Furthermore, it is reasonable to assume that asynchronous communication may sometimes be the preferred rather than the tolerated mode of communication. To explore these assumptions we undertook an Internet survey of design communication. In this paper we report the results of this research with a focus on the use and value of asynchronous communication in collaborative design.

Asynchronous Modes of Group Communication

Asynchronous communication systems include the Bulletin Board System (BBS), email, voice mail, the World Wide Web (WWW) and Short Message Service (SMS). Electronic mail is arguably the most widely studied and used asynchronous group communication technology. It is shown efficient for group communication (Palme, 1995) because of reductions in the cost and effort of traveling and gathering everyone in the same place at the same time; each participant has greater control over their communication - what, when and how to read, annotate or comment and reply; written communication is more efficient if the size of the group is larger than about five people.

The difference between email and face-to-face meetings has been previously investigated (e.g., Zack, 1993). Email has been found to be advantageous because:

- You can give and take information at your own convenience (e.g. not during meetings),
- You can participate more easily in communication when you otherwise cannot be easily reached, as when you are travelling or on holiday, etc.
- It is easier to give precise factual information.
- The recipient gets the information in a written format that can be reused or archived.
- Equality between people increases, more people are allowed to have their say and there is less risk of one single person dominating.

On the other hand, email has disadvantages. For example, it is more difficult to conduct a formal decision process by email and more difficult to persuade others and reach consensus. With email, difficult and controversial issues often lead to a 'war of positions' that can only be resolved in a face-to-face meeting. The lack of body language, voice inflections and facial expressions help explain this effect. Thus, negotiations can be difficult to conduct via electronic mail. Additionally, although asynchronous communication systems, such as email, WWW, BBS, voice mail and SMS, have been shown to be useful, they are also seen as being inferior to synchronous communication because of limited channel capacity and lack of interactivity.

Thus, although there are many tools available for asynchronous communication, with the exception of email, our understanding of their use, advantages and disadvantages is limited, particularly, from our perspective, in regard to remote design collaboration. Furthermore, the perception that asynchronous communication is inherently inferior to synchronous communication has tended to discourage investigation of the use and potential of the former.

The Design Communication Survey

Motivated by the above we undertook an Internet survey of design communication. The objectives of the survey were to establish the current character of design collaboration, to verify the findings reported in the literature, to gather more information about how remote design collaboration is undertaken, and to assess the current and future importance of asynchronous communication in remote design collaboration. The questionnaire comprised both open and closed questions concerning demographic information; design knowledge and experience about design process, design task, and remote design; and the usability of remote communication, specifically during concept design. A copy of the survey questions was placed on a web site and potential participants were chosen at random from the European Design Innovations Ltd. (EDI) web site. In total, 1766 British design companies were invited by email to participate in the survey. There were 174 responses of which 81 were found to be complete and valid.

Respondent Details

To summarise, respondents' ages ranged between 21 and over 50, with 61% being in the age range 26-40. Their occupations covered more than 21 categories, e.g., junior designer, senior designer, design manager, design, consultant (at 31%, the greatest percentage of respondents). Of these, 72% were employed in companies with less than 5 designers, working in a diverse range of design domains, e.g., graphic, product, interior, fashion, architectural design, and involved from planning through to detail design.

The majority of respondents worked in concept design and all stages of a project (i.e., 62% and 70% respectively) in teams of less than six people, on designs commissioned mainly by external clients. The length of concept design varied, but was usually less than six months.

The majority of designers worked on between 2 and 5 project simultaneously and with other designers during concept design. They identified a wide range of other professionals involved in concept design, most commonly, the client, marketing professionals and mechanical engineers. Most identified the following activities in concept design: establishing design requirements, developing possible solutions, analysing the design brief, presenting possible solutions, exchanging information, resolving design problems, and evaluating possible design solutions. The most frequently used modes of communication in concept design were, in descending order: email, telephone, face to face, fax, web, post, voice mail, video conferencing and bulletin boards. The most frequently used materials in concept design were, again in descending order: sketches/drawings, text documents, photographs, CAD files, physical models and videos.

Remote Design

This part of the questionnaire focused on aspects of remote concept design. Most designers reported that they needed to communicate with the client remotely and to a lesser extent with other designers (both inside and outside their organisation) on a daily basis or several times a week. Others communicants included manufacturers, marketing professionals and engineers. The methods used for communication were, in order of use: email, telephone, fax, the web, face to face, post, video conferencing, voice mail and

BBS. Text, images, verbal messages, and numerical data files were common accompaniments to these exchanges, used to support all concept design activities.

Usability of Remote Communication

This section sought to appraise the functionality of the different modes of communication and their likely use in the future, especially asynchronous communication. Most of the respondents agreed that asynchronous remote communication is important now and will be in the future. Most were familiar new computer-based modes of communication such as email and the web. Moreover, the majority felt that telephone, email and the web were the three most important modes of communication. Against a range of attributes, such as efficiency and familiarity, email and telephone were rated most highly, followed by fax, post and the web. Similarly, rated on negative aspects, post and telephone were identified as the slowest, and telephone, video conferencing and post the most expensive. Together, these results suggest a significant growth in the use of the Internet for remote design communication and also indicate that computer-mediated modes of communication are beginning to overtake conventional aids to remote design such as, telephone, fax and post.

The Role of Asynchronous Communication in Remote Design

The questionnaire included open-ended questions. One of these asked why specific media were preferred and therefore provided insight into the benefits of email and web-based communication. Reasons (verbatim) for selecting email and the web (for asynchronous communication) included:

- Email is quick, informal and to the point plus you can send stuff, and ease of communication, versatile fast and inexpensive.
- Many of our clients are based overseas and so synchronous communication is usually not the best way to communicate, therefore we use e-mail and fax rather more than the telephone.
- I consider personal meetings and telephone calls to be time-wasters. Rarely can a client explain what he or she wants with spoken words alone. Personal face-to-face meetings simply are not needed because [the] materials they would provide during such meetings can be more efficiently shared with me using electronic means or the post. With email, clients can send me documents outlining their ideas. They can attach logos, photos and other images.
- Easier to communicate ideas through visuals such as PDF files e-mailed to client or through face to consultations to view proof artwork and make alterations on the spot, e-mails and posted proofs are followed up by telephone to discuss client's thoughts before going further with project.
- Web/email for speed.
- I ... use the Web to show clients what I've come up with and let them suggest changes.

Earlier we noted that the respondents were invited to rate the importance of current modes of remote communication. One of the open-ended questions invited them to comment on the extent to which their response pattern was representative of remote communication during design. Over 80% affirmed this to be an accurate or very accurate representation of the present pattern of communication. Finally, they were asked to comment on the present and future importance of asynchronous communication. The responses confirm that asynchronous communication is a valued means of communicating between design teams and with clients. As a mode of communication, email has the benefit, compared to telephone, that conversations can be logged, and they can also include attachments. Another benefit of email over synchronous forms of communication is that they allow people to work at their own pace and schedule their activities, also, importantly, users can reflect on the contents of the emails and their answers before they respond. The need for reflection was the most prevalent reason cited for the highly rated importance of email. When working to tight deadlines, emails provide windows of opportunity for communication, which would be impossible in other forms (for example, a partner might not be in the office or able to meet due to other commitments).

As for the future importance of asynchronous communication in the future, 84% of respondents felt that asynchronous communication would increase in the future. The main reasons for this related to:

- Changes in working patterns - in relation to global working, flexibility in time and place of work. Asynchronous communication 'allows work schedules to be less dependent on others'.
- Improvements in communication infrastructure leading to greater speed and bandwidth, making it possible to transmit larger files and providing increased Internet access.
- Perceived modernity, e.g., 'fear of being seen as "dinosaur" business/company'.
- Functionality offered by email was liked. For example, 'store and forward principle is polite; no missed calls', also ease of use and speed.
- Email afforded features not present in other forms of communication. For example, '...when working on large projects there are many qualified staff working on the job in different technologies. This form of communication helps as it lends a hand to time and the ability of educated answers other than on the spot guessing that can happen in meetings. Also, 'it saves time by removing the elements of sociability and small talk and enables the designer to get on with the job in hand -design'
- Enhanced contactability, e.g., 'people's time availability is decreasing in proportion with technological affordability and availability-expectations are increasingly higher and faster'. Asynchronous communication provides a way of managing time and teams.

Discussion

The respondents represented a wide range of design roles (e.g., junior designer, design consultant, etc.) and domains, and they participated throughout the design process. In concept design, there is clear evidence of team working and working on multiple projects, involving communication on a wide range of tasks using a wide range of materials. This illustrates the complexity of contemporary design. Surprisingly, email emerged as the most frequently used mode of communication with the web following closely. For remote design, telephone, email and the web were rated as the most important modes of communication. The rating of positive and negative features suggests future growth in the use of email and the web for asynchronous communication and as replacements to telephonic communication. Asynchronous communication is important in concept design now and its importance is expected to increase in the future. The reasons identified for the importance of email confirmed previous findings in the literature but revealed the additional benefits of email and web-based communication. First, it provides an effective way of 'saying and showing'. Second, it provides a way of controlling social interaction - to minimise 'small talk'. Third, in an increasingly pressured and complex working environment, it offers a new way of managing time and teams. Fourth, and perhaps the most important from the point of view of design innovation, it affords the opportunity for carving out time for reflection and informed and considered decision making. Finally, at the root of all of this is the changing pattern of designing demanded by global working and the need for flexible working.

Conclusion

The results of the survey support the argument that asynchronous communication is and will become increasingly important in design communication. Furthermore, it should not be regarded as an impoverished form of a wished-for synchronous mode of remote collaboration. Instead, sometimes it is better than synchronous working. Indeed it may prove to be a vital element in enabling design companies to meet the challenges of global design. Nevertheless, there are bound to be problems with current aids to asynchronous communication. We would argue that if research is to assist design companies to meet the challenges they face, then the research agenda needs to shift to providing answers that will enable more effective and efficient asynchronous communication.

References

Palme, J. 1995, *Electronic Mail*, (Artech House, USA)

Saad, M. and Maher, M.L. 1996, Shared understanding in computer-supported collaborative design, *Computer-Aided Design*, **28**, 3, 183-192

Scrivener, S.A.R. and Clark, S.M. 1994, Experiences in computer-mediated communication, In *Proceedings of the ISAT School/ IFIP TC 8/WC 8.5 Workshop*, (Poland: Biblioteka Informatyki Szkol Wyzszych,), 122-154

Zack, M.H. 1993, Interactivity and communication mode choice in ongoing management groups, *Information Systems Research*, **12**, 3, 207-239

USABILITY TOOLS AND THE DESIGN PROCESS

K. Tara Smith[1] and Lynne Coventry[2]

[1]*HFE Solutions, Canmore House, 31 Canmore Street
Dunfermline, Fife KY12 7NU*
[2]*NCR, St Davids Drive
Dalgety Bay, Fife KY11 5NB*

There are a wide variety of tools and methods currently available to aid in the design process, however all of these have some associated problems. Although many of them help in the physical aspects of usability, none of them adequately addresses the areas of users' expectations and potential design changes, though some may go some way towards this.

The acid test is the product's resultant place in the market, but in terms of usability the questions that should be asked to judge the effectiveness of these tools are:

- How well does it support the design process?
- How well does it allow you to predict usability?

This paper presents and discusses these issues in the context of how ergonomics is applied and developed in real situations.

Introduction

This paper discusses the issues related to acceptable usability and the tools needed to support this. It approaches acceptable usability from the perspective of end users' adoption of the system i.e. "the acid test".

Human Factors practitioners already employ a number of predictive models to assist them during the design process. These models typically focus on issues such as workload, training support and even levels of usability but there is still a big logical jump between the ideals of low workload, good training support, a usable product and a success in the market place.

There are some macro level tools emerging that sit above/alongside everything else. They:

- Identify data that you need
- Co-ordinate the full design process
- Fill in gaps within the data - in a meaningful manner

And to some extent address the issue of how to:

- Understand users' expectations
- Predict potential future changes
- Understand the base design elements and how these carry over to expectations.

Before we can address this issue, we have to answer the following questions:
- Usability in Design - What is it?
- How do we judge a tool's effectiveness?

Users often have high expectations of the usability of a product. So how do we predict what those expectations are likely to be, to enable us to design for their satisfaction?

The main problems in designing for usability are how to:
- Understand users' expectations
- Predict potential future changes
- Understand the base design elements and how these carry over to expectations

The key issue for most if not all businesses is the bottom line. That is, does the product perform in the market place, is it adopted? Achieving good usability is a means to this end.

What is design?

We need to start by understanding the use of the term "design". For the purpose of this paper we are restricting this to the following areas.

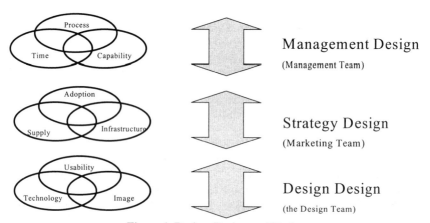

Figure 1. Design, Design and Design

There are other uses of the term "design" such as the management of the design documentation which are equally important but are not relevant to this discussion.

If for the purpose of the paper we can assume that our overall design process, capability and timing are well established and detailed design is well under control, we can consider the strategic level. We can see that adoption and uptake is still a key factor. If we examine the strategy design level from a Human Factors point of view we should be able to identify and measure these factors to build a predictive model.

Problem 1 - Who's who?

In the Human Factors world it is common during design to specify the target users. It is also becoming more common to look at more that one group, however Human Factors practitioners have a dilemma; design for a target group or design for likely users. What we need is a crystal ball to tell us who will be the real adopters. But even having a crystal ball we need to be careful: the groups below are all real adopters, but they have different characteristics.

Early adopters – (generally less than 3% for a successful product) their use of the product is not likely to be typical. Undertaking user testing with this group can be highly dangerous.

Second wave adopters – These can provide more meaningful subjects for a trial but may not have adopted critical behaviours.

Sustained user base – Typically on examination this user group is not one coherent group. A Human Factors practitioner's understanding of the variation in this population is usually a key to understanding the design aspects.

Unfortunately it is still common to define a "General Users" group, tautologically, in terms of the people who use the product; this tells you nothing that can be helpful to the design process.

Problem 2 – Expectation

Within the Human Factors world it is not common to identify expectations other than by cataloguing past behaviour. The introduction of a new technology or the combination of disparate technologies often induces user expectation. This is often stimulated by the marketing effort. A good example of this would be the advert for WAP phones which shows an individual finding a local free concert (at the Blue Lounge) within the time it takes to take a lift to get to the ground floor.

Case 1 - Old image v new image = good acceptance
A study was undertaken comparing two ATM (Automatic Teller Machine) systems, the trial took the form of a comparison of two designs; one traditional in appearance and one with a radical modern looking design. It was stressed that the functionality of the two systems was identical. There was however a very significant result in favour of the new design and people reported that from the image alone they had an expectation that it would be easier to use.

Case 2 - Introduction of new technology
This study introduced a retina scanning security feature to an ATM. The expectation results for this raised issues related to perceived security and perceived health hazards relating to lasers in the eye (though the technology was not laser based). To use this it required the user's retinal image to be scanned by the system. The expectation was that users would have to change their behaviour. This was not actually the case – the retina could be scanned without the user having to stand still, gaze into a lens, or even remove (clear lens) glasses.

Proposition 1

A key aspect of these systems was the introduction of a new technology. The introduction meant that users envisaged aspects related to the technology that affected their opinion of the design. However, some technology comes with over-positive expectation and some with over-negative.

Case 3 – Speech system

This study looked at the introduction of a speech interface to an ATM. Here the results indicated that although the experience and success rate during the trial had been negative the subjective response was positive.

Results from studies indicate that speech technology at present for public systems is good for limited command-based interfaces and bad for full vocabulary interfaces. Unfortunately users' expectations are higher than the technology can achieve (the Star Trek effect).

Proposition 2

A key theme in predicting if a technology is seen positively is whether these systems were perceived to introduce a radical change in behaviour. That is, for the first system the user had to stand still and not wear sunglasses, whereas for the speech system you just had to talk. So to evaluate the likely success and "adoption" of a design we need to understand user's expectation and perceived behavioural change.

Problem 3 – What is behavioural change?

Initially we need to define behaviour. In our example above we have three very different aspects expressed;

- In the first example we have a change by simplification. From a visual inspection of the interface it looked like the user would just miss steps and complications out of the interaction. This was deemed to be minor behavioural change by the researchers.
- In the second we have an envisaged behavioural change of standing still and an additional task of removing sunglasses. This was reported as a major behavioural change.
- In the last we have one behaviour type being replaced by another. This was reported as a behavioural improvement.

We can look at others within the banking and IT sector in terms of their uptake:

Cash back - uptake fast - minimal behavioural change in that it adds one extra step to one interaction and removes the necessity for another task.

Debit cards - uptake slow - on the face of it simplifies one interaction but potentially requires the user to undertake additional tasks of juggling their finance.

Biometrics - uptake slow – additional tasks seen as no advantage, however this looks like it could change because of the increased perception of the need for robust security measures.

Video Phone – uptake slow - requires extra effort to prepare for call in that you may wish to look respectable for the call.

Proposition 3
A model of behavioural change would use some sort of resource model akin to those utilised in cognitive workload.

Conclusions

This paper is not intended to answer all the questions regarding the Human Factors tools and where they fit within the design process or present a coherent model of a solution that is in need of applying.

It identifies three areas:
- Understanding the adopters
- Understanding user expectations
- Understanding the pros and cons for behavioural change

References

Clegg, C.W., Icasti-Johanson, B. and Bennett, S., 2001, E-business: boom or gloom? *Behaviour and Information Technology*, **20**, 293-298

Damodaran, L., 2001, Human Factors in the digital world enhancing life style – the challenge for emerging technologies, *International Journal of Human-Computer Studies*, **55**, 377-403

Parush, A., 2001, Usability Design and Testing, *Interactions*, **VIII.5**, 13-17

Jain, A., Hong, L., and Sharath, P., 2000, Biometric Identification, *Communications*, **43, 2**, 90-98

Boyce, S. J., 2000, Natural Spoken Dialogue Systems for Telephony Applications, *Communications*, **43, 9**, 29-34

DESIGNING FOR A WORLD POPULATION

K. Tara Smith[1] and Lynne Coventry[2]

[1]*HFE Solutions, Canmore House, 31 Canmore Street*
Dunfermline, Fife KY12 7NU
[2]*NCR, St Davids Drive*
Dalgety Bay, Fife KY11 5NB

This paper presents the findings of a project carried out by HFE Solutions for NCR. The aim of this project was to identify the ergonomic constraints that need to be considered by NCR when designing hardware products for use by the general public. The overall objective of the project was to develop a method to be used by NCR designers when designing future systems and by usability engineers to ensure their usability by a "global" population. The method took the form of an Excel Data Book that calculated the level of population coverage given that a particular Automatic Teller Machine (ATM) module was placed at a particular height. This Data Book ensures a consistent, early evaluation of the physical design of self-service products and provides volumes to design within for new products.

Introduction

The overall objective of the project was to develop a "design checklist plus" to be used by NCR designers when designing future systems, to ensure their usability by a "global" population.

This "checklist plus" took the form of a Data Book, which defines a process of assessment that looks across multi-national anthropometric data. It provides an easy method for assessing NCR systems across a variety of anthropometric populations and can either be used to assess an existing design or to provide input during the design process. It allows NCR to be able to specify the required level of accessibility for each of the modules used by their systems in a simple manner and once it has been specified, to provide sufficient data to the designer to allow them to assess the consequences of a design selection. The basis of the standard test is that:

- There is a standard set of anthropometric models to use
- The models are expressed as comfort oriented reach zones
- The reach zones integrate different physical operations (e.g. inserting card, taking a receipt)

The results of this test can then be related to different populations that represent the user group.

In order to produce the Data Book, a large number of assumptions had to be made. These assumptions are based on common-sense generalisations and have now been validated. This paper presents the models constructed, the validation process and the consequences for design – illustrated with case studies.

Project aims

The aim of this project was to provide anthropometric data for a global population in a form useful to NCR designers. For logistical reasons, the project only considered the data required for standard ATMs and public information terminals. This resulted in the core user tasks under consideration being limited to:

- Viewing a screen
- Using a keyboard and/or touchscreen
- Inserting/removing a card
- Removing cash and/or receipt
- Inserting cash/cheque/envelope.

Initial approach

There is a vast amount of data available on anthropometric characteristics such as reach ranges and lifting strength. Unfortunately, most of this data is concerned with the limits of what people can see, reach or lift, rather than what it is comfortable for them to do. In addition, due to the methods of data collection used, it is difficult to directly compare the different data sets; for example, some height data is grouped by age whereas height data referring to a different country may be grouped by socio-economic status. However there are four main issues here;

- How do we predict what is comfortable for the end user?
- How can you provide the designers with an indication of design tolerance?
- How do we create a standard test that designers can apply?
- How can we account for different tasks?

To address these issues the project took the following approaches.

The comfort issue
To provide the data required by designers, we had to merge this information from several sets of data. Much of the data was obtained from PeopleSize, as this tool has already amalgamated data from a large number of sources.

To determine what will be **comfortable** for people to use, we first combined pure anthropometrics (height, arm length etc) with strength and lifting characteristics and bone articulation, using three parameters:

- Is it within reach?
- Is it within a comfortable lifting zone?
- Are there any restrictions caused by the articulation of the limbs in a particular orientation?

An assumption was made that ease of operation is related to working within your strongest zone. It was also assumed that bone articulation provides a limit for some movements and will affect the appropriate angle range for that action.

The design tolerance issue
To help designers to determine what would constitute an acceptable design, three levels of data are provided for each area considered. These levels have been defined as:

- Comfortable - i.e. users will be easily able to physically interact with that aspect of the system
- Manageable - i.e. users will be able to physically interact with that aspect of the system without too much difficulty
- Possible - i.e. users will be able to physically interact with that aspect of the system, but it is likely that they will stretch, stoop or otherwise adjust their position to make it more comfortable.

Although we have defined these zones, there is actually a continuum of comfort level moving away from the optimum. Within this data we have made no attempt to estimate that gradient.

The standard test issue
Here we took a radical approach; we derived a set of eleven standard anthropometric models that covered the world population. These were then mapped back on to the different populations so that model 1 may represent 10 % of a male USA population, 2% of a German and 0% of a female Japanese. This approach means that the designer can apply a standard model and see what impact it would have for the different populations of the world. Data are also provided for a range of heights for a selection of regional populations (e.g. UK, USA, Chinese, and Dutch), "frail" and wheelchair users. The movement ability of the "frail" population, i.e. the ambulant disabled and elderly, were assumed to be a further restriction to the comfortable zone. Disabled wheelchair users were considered separately.

The different tasks issue
An assessment of the tasks that users will need to carry out when using an ATM generated five areas to be considered, these are categorised into the following:

- General vision   Cat A
- Vision and touch   Cat B
- Pinch grip insert and withdraw: palm down   Cat C
- Pinch grip insert and withdraw: palm up   Cat D
- Pinch grip insert and withdraw: palm side on   Cat E

An assumption has been made that all of the things held to operate an ATM (cards, cash, receipts) are held in a way that equates to a pinch grip. For each anthropometric model a series of points are plotted as shown in Figure 1 - example height categories for touch screen interaction and anthropometric model A1.

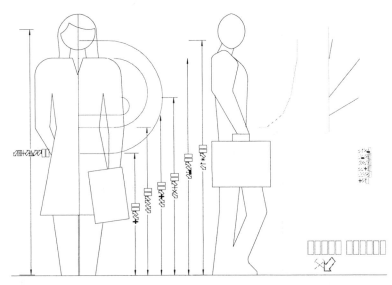

Figure 1. Example height categories for touch screen interaction and anthropmetric model A1

The following table shows which of the activities involved in interacting with an ATM will utilise the different data tables. One tick indicates that some use will be made of the tables, two ticks that considerable use will be made.

Table 1. Applicable Data for Each Different Operation

	Cat A	Cat B	Cat C	Cat D	Cat E
Viewing a screen	✓✓				
Using a keyboard and/or touchscreen	✓	✓✓			
Inserting/ removing a card	✓	✓	✓✓	✓✓	✓✓
Removing cash and/or receipt	✓	✓	✓✓	✓✓	✓✓
Inserting cash/cheque /envelope	✓	✓	✓✓	✓✓	✓✓

Initial model

The initial model helps the designers understand the compromises being made when designing to physical dimensions required by accessibility legislation.

The following figure presents a screen shot from the Data Book; it indicates that for a male and female population made up of UK, Dutch, German and US and where the A4, 5 and 6 models have been satisfied, the design satisfies approximately 60 % of this population.

Overall percentage covered by models passed for population included **59.41**						30	22	8		
Male & Female Population		A1	A2	A3	A4	A5	A6	A7	F1
	Models Passed				**1**	**1**	**1**		
Percentage of the included Population per model	Include V	2	12	25	30	22	8	1	0
Able Bodied									
UK	1	100	97	80	53	21	3	0	1
Chinese									
Dutch	1	99	89	63	31	7	1	0	0
German	1	100	97	79	49	17	2	0	0
Japanese									
US	1	100	96	79	50	19	3	0	0
Elderly and Frail									
UK									
Canada									
Japanese									
Netherlands									

Figure 2. Screen shot from Data Book

The creation of this tool has provided valuable insight into the full range of constraints on the physical design of public technology. This information will play a key role in research into the future of public technology and in placing requirements on the next generation of ATMs. It has also:

- Increased consistency of hardware evaluation
- Provided consistency of data on which to base hardware evaluation
- Defined reach zones for evaluation
- Created 11 anthropometric models to represent worldwide able-bodied anthropometrics.

This data book allows for testing against usability and accessibility product hardware requirements. The following chart illustrates the actual reach heights for each of the anthropometric models. There are 8 main categories A1 – A7 (able bodied) and F1 (frail users) and 3 wheelchair user heights.

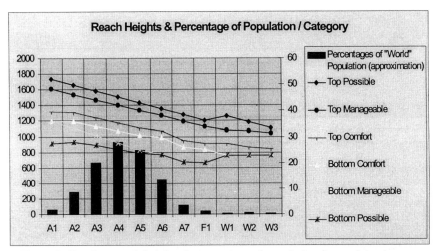

Figure 3. Comfort reach heights & percentage of population by model category.

Validation

Up till this point we have been discussing a theoretical model based on assumptions regarding reach and lifting capability. The next part of this project validated this model with a public trial.

Eighty two participants were asked to place low fidelity ATM modules made out of foamboard in their ideal comfort zone and then in the upper and lower comfort zones. Once the participant had placed the module they were asked to simulate its use. The heights of the modules, together with any awkward postures and the orientation of the hand were recorded in order to validate the model.

Conclusions from Validation

The data collected during the user trial showed that, for those who took part, the model provided a fairly accurate prediction of the "comfort zones" for using an ATM. However, strictly speaking, it fails, as it is over-optimistic in predicting that ATM users will be comfortable accessing a wider range of heights than was found to be the case.

It was also noticeable that taller participants often placed their lower limit slightly below that predicted in the model; this is possibly because they are used to stooping.

References

Tilley, A. R., 1993, *The Measure of Man and Woman*, (Henry Dreyfuss Associates, New York)
Open Ergonomics Ltd, 2000, *PeopleSize 2000*
Le Corbusier, 2000, *The Modulor and Modulor 2*, (Birkhauser Verlag AG)

TRENDS IN HUMAN FACTORS:
NEW OPPORTUNITIES THROUGH INCLUSIVE DESIGN

Alastair S. Macdonald[1] and Cherie S. Lebbon[2]

[1] *Product Design Engineering, Glasgow School of Art,
167 Renfrew Street, Glasgow G3 6RQ*

[2] *Helen Hamlyn Research Centre, Royal College of Art,
Kensington Gore, London SW7 2EU*

The next twenty years will see significant changes in population demographics, lifestyles and technological developments. These changes will provide a new context for design and human factors, and much significant work has already been initiated in relation to ageing and capability issues as more 'inclusive' products and services are – and will be - demanded and that are less 'exclusive' of significant sectors of the population. During the Royal College of Art's Helen Hamlyn Research Centre 'Include 2001' conference these changes in demographics, lifestyles and technologies were mapped onto an 'inclusive futures' timeline to help anticipate prospective opportunities. It is to share and reflect on some of the findings from this timeline that forms the focus of this paper.

Introduction

Many useful 'ageing' and 'inclusive' initiatives are already underway, defining needs, providing data and tools, identifying priorities, developing guidelines and policies, and encouraging research. Data available for older and less able people has trailed behind that for other sectors of the population, however, new material is increasingly being made available. The design needs of older and less capable people are being surveyed. Important and readily usable information tools are providing designers and HF practitioners with strategies and guidelines to consider the effects of the ageing process. The DTI Foresight Ageing Population Panel has defined priorities for action in its 'The Age Shift' publication, and the Design Council has recently published a series of recommendations in its 'Living Longer: the new context for design' publication. The GENIE (European Education Network in Europe) network has also been exploring gerontechnology - the impact of technology and ageing to ensure an optimal environment for people up to a high age. Research is increasingly and significantly being funded, e.g. through the

EPSRC's EQUAL (Extending QUAlity Life) initiative, and encouragingly, the National Collaboration on Ageing Research is a joint venture between four research councils that aims to develop a coordinated approach to ageing research. A wider-still initiative is the European Resolution on Inclusivity - ResAP (2001).

Helen Hamlyn Research Centre – Include 2001

One relatively long-standing area of age-related design research has been the Royal College of Art's (RCA) 'Designing for our Future Selves' initiative, run for the past eight years, firstly through its DesignAge programme, and more recently as a part of the research undertaken by its Helen Hamlyn Research Centre (HHRC). In April 2001 the HHRC ran the 'Include 2001' international conference, the objective of which was 'to build bridges between researchers, practitioners, and the companies that are beginning to see inclusive design as a strategy around which social and commercial objectives can converge.' In addition to the parallel sessions of papers by researchers and practitioners, and the business breakfast briefings for industry and companies, a number of workshops were held including the 'Timeline Workshops'. These comprised three interactive timelines: the purpose of each was to capture and display 'at-a-glance' during the run of the conference, information and knowledge to help identify or fill gaps in knowledge or awareness at the HHRC, and to identify opportunities for further research or new product or service development. Conference delegates were asked to contribute ideas, information or strategic knowledge from their own particular field by placing large-format post-it notes at appropriate positions on the timeline.

The first timeline, the 'History' room charted the past 50 years of inclusive design activities set against the context of world-wide events and legislation. 'Obstacles and opportunities' to the up-take of inclusive design formed the timeline theme of the second room, focussing on design, research, industry and societal factors. The final room, the 'Mapping Inclusive Futures' (MIF) timeline - was concerned particularly with mapping forward design opportunities that will be created by emerging materials and technologies as they converge with changing lifestyles and demographic trends over coming decades. For example, it would try to identify future products and services that would enhance work and play, social interaction and independence, in the context of changing abilities that come with ageing, illness or disability. The content and concerns of each timeline related to, and overlapped to an extent with the other two.

The MIF timeline itself was a large wall-based chart projecting from 2001 into the future to 2020 and beyond. This allowed for the extrapolation of information, data and trends from a number of sources and in a number of areas such as technology, materials, lifestyle scenarios, population and demographics, and the mapping of information and scenarios from a number of initiatives, some of which have already been mentioned above. The MIF timeline also acknowledged the structure of the DTI's Design for Living Taskforce's four cornerstones of design for inclusion: design for social interaction, design for independence, design for stimulation, and design for flexibility.

Delegate contributions

Information and ideas from delegates were highly varied and extremely valuable and was added to the timeline on large-scale 'post-it' notes. For example, one contribution from a Dutch delegate revealed that her government's goal was for all public buses to be accessible and usable by people with impairments by 2010, and that by 2030 all trains would be accessible. These offer great design and human factors opportunities. Another raised the issue of the need to model and articulate the scale of the cognitive challenges arising from changing demographics. The Future Foundation contributed information on changing lifestyle, income profiles, and projections of technology take-up, such as mobile phones by age group.

Areas of lifestyle opportunity

It became quite clear from material accumulating on the timeline that a number of areas will impact on 'ageing' and 'capability' thinking, such as: older workers and consumers, transport and mobility, changing work patterns, technology, housing, communication and information technologies, older workers and older consumers, leisure and lifestyle, and healthcare. The DTI taskforce had identified and developed several scenarios for desirable products and services for e.g. household technology, the SMARTer House, real and virtual communities, leisure and tourism, and transport. 'Simple robust interfaces and services between the user and the technology will be individual, personalised and adaptable ... the user is able to specify the most appropriate method of interacting with the technology to suit his/her needs and preferences'...'kits could include modular bathrooms' and kitchens would be 'designed with the needs of the ageing population in mind.' Homes will not only be more flexible, but also safer, more secure, and more convenient, and will require to appeal up and down the age range.

Technological forecasts

As we age, our sensory faculties deteriorate, and so an ageing population will require sensory assistants or enhancers: technological forecasts identify ample opportunity to develop 'assistive technologies'. The exponential growth of computer processing power, 'smart' technologies and new materials such as polymers and light-weight composites will provide the basis for a new generation of sensory-assistive and mobility-orientated products and services tailored to a broad range of particular and individual needs. Projecting over the next 20 years, one can predict the omnipresence of electronic assistive technologies through the increasing power of, and dramatic drop in the cost of computation. Kurzweil (1999) provides decade-by-decade forecasts of the projected capabilities of machines arising from this type of technological development. For example he predicts that by the year 2019 'a $1000 computer device (in 1999 dollars) will have the computational ability of the human brain. Blind people routinely use eye-glass-mounted reading-navigation systems. Deaf people read what other people are saying through their lens displays. Paraplegic, and some quadriplegic, people routinely walk and climb stairs through a combination of computer-controlled nerve stimulation and exoskeleton robotic devices.' In the area of new materials, one prediction is for adaptive materials, developed for next generation aerial vehicles, which are capable of changing their shape when applied with heat, electrical signals or magnetic fluids. These materials would be able to repair themselves and adapt in shape, to operate in the same way that muscles act in birds and insects. Able to change from the flexibility of a rubber band to the stiffness of steel in under a 100th of a second, they may be combined with stress-sensitive material to prevent sports injuries, from sprained fingers to broken necks. There are many areas for this type of development in assistive technologies, such as rehabilitation robotics, enabling computer access, designing interfaces with new technologies, understanding the social acceptability of systems including barriers to adoption, and how to overcome the social isolation of those unfamiliar or afraid of digital technologies.

Concept designs from the Helen Hamlyn Research Associates

These predictions are all very well, but what do they mean in terms of tangible products and services? The MIF timeline included work from the Helen Hamlyn Research Associates (HHRA) programme which addresses concerns with 'social developments', and is a recent initiative based on partnerships between business/industry and its research centre. Design graduates of the RCA are teamed with industry partners to undertake a range of collaborative research and development projects that respond to social and demographic change. Its objective is 'to achieve greater social inclusion through innovative design thinking and its commercial application'. According to the Research Associates Programme Director, Jeremy Myerson, 'the results point towards the emergence of a new 21st century design paradigm, in which recognition and inclusion of the special and acute needs of different users in the design process will lead towards better solutions for all society – and broader markets for industry'. For example, Shaun Hutchinson's '2030 -Urban Moving' proposal investigates the potential of public/private automotive city travel up to the year 2030. Karen Adcock and Carl Turner's 'Instinctive Wayfinding at Heathrow Airport' explores inclusive design issues of navigating Heathrow Airport's Terminal 5 by means of a 'sensory landscape' and an 'information arch'. Other concepts explore, for example, the fact that many older people are DIY enthusiasts but are excluded from this activity by the problematic design of much of the equipment on sale in DIY retail stores (Matthew White), or the issue of health such as in Ellie Ridsdale's 'walking the way to health'. Each of these HHRA concept designs was placed at its likely point of entry on the timeline. The informed design concepts of these RA's had an important function on the timeline – they provided tangible visualisations for changing needs, values and expectations within the context of changing demographics, technologies and lifestyles. This conceptualisation and visualisation allows many of the design and human factors issues to be identified, and shared with potential user groups and business who have the potential to introduce these onto the marketplace. The benefit the company gains is in proving a business case for a new design concept.

DBA Design Challenge

Other tangible examples of innovative concepts placed on the timeline were the results of the Design Business Association (DBA) Design Challenge 2000. This was a special event organised by the Small Business Programme of the Helen Hamlyn Research Centre, in December 2000, with the theme 'care for our future selves' 'to demonstrate the business potential of taking an inclusive design approach to product and service development in the care and disability sectors'. This enabled four DBA member design consultancies to develop innovative design scenarios with disabled user groups in the areas of packaging, mobility and digital communications. One solution from the DBA Design Challenge was Design House's concept for an accessible website design for a sports-based web channel that appeals to all *and* includes the needs of visually impaired users. This solution could be implemented immediately. Another, the Renfrew Group's concept, centred on a vehicle for an individual with cerebral palsy and uses his feet in place of his hands for all activities including driving. Renfrew 'developed a personalised drive-by-wire control for a car that can be programmed to the individual needs of the user via a smart card'. All four consultancies involved found this approach 'rapidly eliminated any fundamental weaknesses in a proposal and stimulated their search for designs that were radical yet inclusive'. Renfrew's design principle was seen as something that could become as commonplace as optional customised sunroofs and wheeltrim features on today's standard mass-produced car. On the MIF timeline this product would start to appear from 2005, and importantly, provided a tangible example of a radical inclusive approach assisted by new technological opportunities. This view was further strengthened by a ten-year view provided by one of the Include 2001 keynote speakers, Alessandro Coda, Director of the Fiat Autonomy Programme in Turin, for developments in vehicle design which relate to the mobility of elderly and 'disabled' people throughout Europe. One human factors issue that arises from this is how to accommodate the very wide range of, for example, dimensional variables, to meet 'inclusive' aspirations.

Opportunities for human factors and design

Extending markets
While it has been possible to discuss only a few of the findings from the MIF timeline, a number of opportunities can be identified. The Design Council recommends, e.g., that markets be extended 'for mainstream products that are designed to be compatible with accessories needed by those who cannot use even an inclusively designed product – plug-in keyboards for mobile phones for example.' The examples on display on the MIF timeline, including the work of the Renfrew Group, the Helen Hamlyn Research Associates, and Fiat all give tangibility to these intentions. Since the conference, the announcement of the potential for developing the use of 'multivariate accommodation' (Porter, 2001) should provide helpful in 'designing in' a greater range of individual dimensional variables to extend inclusive markets.

Opportunities for transdisciplinary collaboration through education
The Design Council's 'Living Longer' document recommends that we 'ensure that our designers and design decision-makers have the necessary skills base to consider effectively inclusive design issues in their day-to-day decision making'. 'Living Longer' also recommends that in tertiary education we should be 'working to shape research gathered...into useful resources for use in the training of designers, managers, engineers, architects and planners.' This should also include ergonomists. Gerontology and gerontechnology should now form a highly relevant part of any design and HF curriculum (and CPD programmes in both professional fields) to plug existing knowledge gaps. 'Living Longer' also recommends 'that the DfEE, DTI and the Design Council work with the Qualifications and Curriculum Authority and other education influencers to develop a national educational programme which integrates inclusive approaches to design, and issues surrounding population ageing and capability ranges across the whole population, at all levels of the design curricula. As one aspect of this, an appreciation of gerontology

and gerontechnology would help provide a clearer understanding of some of the issues involved, and already there are some excellent exemplars such as Arcada in Helsinki (a Finnish-Swedish polytechnic) which offers a degree in human ageing and gerontechnology. Additionally, the aim of the Ingénierie du Vieillissement, Université de Paris, which offers a postgraduate degree in Ageing Engineering, is "to change the passive medicalised and socialised approach to a positive and preventative approach." GENIE (the Gerontechnology Education Network in Europe) has produced a very useful Learning Map to 'give access to knowledge about various areas related to gerontechnology, including students, tutors and professionals'. The Royal Society of Arts (RSA) runs the annual Student Design Awards (SDA). In 2000 it launched a Gerontechnology category to actively promote the aims of GENIE in the developments of gerontechnology curricula. The SDA briefing guidelines actively encouraged the formation of transdisciplinary teams – to include fields of study from outside the design education community. Entries from the first competition included engineers, and students from Arcada. What is evident from these first entries is that because the issues are complex, relevant and innovative solutions are unlikely to be generated by individuals or single professions: product and service development teams will require the combined skills of, e.g., engineers, designers, social scientists, health care and human factors specialists. The RSA initiative encourages this approach at a tertiary education level.

Tangible visions

The most useful elements on the MIF timeline in communicating the shape of future product and service opportunities, were the informed and forward-looking design visions, some of which have been described above. Design, when coupled with ground breaking research into ageing and ability issues, and seen as part of a multidisciplinary development team giving shape and form to new inclusive values, through innovative and people-centred concepts, is a potent force for communicating tangibly what is possible and clarifying complex issues. Here lies a great opportunity for the separate fields of design and human factors to work more closely together and with these other disciplines. The HHRC, DBA, GENIE and RSA initiatives show the way ahead, using the combined expertise from separate fields, including human factors, to produce satisfying and pleasurable products and services for a changing population.

References

Coleman, R. 2001 *Living longer: the new context for design*. (Design Council, London)
Design for Living Taskforce 2000 *Ageing Population Panel* (Department of Trade and Industry Foresight Programme, London)
Kurzweil, R. 1999, *The Age of Spiritual Machines* (Orion Business, London)
Helen Hamlyn Research Associates Programme 2000 (Royal College of Art, London)
Helen Hamlyn Research Centre 2000, *Innovate 1 – how 4 design teams faced the user challenge* (Royal College of Art, London)
Minutes of the Final Meeting GENIE Network in Helsinki, 22-25 August 2001.
Porter, J. M. 2001 Beyond Jack and Jill – Designing for Individuals within Populations *The Ergonomist*, Sept 2001, 3 (The Ergonomics Society, Loughborough)

ARE THE BEST DESIGNERS FROM EARTH, MARS OR VENUS? AN ANALYSIS OF GENDER DIFFERENCES

Deana McDonagh[1], Steve Rutherford[2], Ian Solomonides[2]

[1]*Department of Design and Technology*
Loughborough University, Loughborough Leicestershire, UK. LE11 3TU

[2]*School of Art & Design,*
Nottingham Trent University, Burton Street, Nottingham, UK. NG1 4BU

There is an increase in the number of female students studying industrial and product design at UK universities and a consequent future increase in the number of products designed by women. It is possible that there are fundamental differences in the approach of male and female designers to the task of designing. One obvious difference noticed by the authors is the ease with which some female designers empathise with potential users of their products. This paper uses design students to define a general set of issues or attributes to be considered during designing and looks for differences in the sets of issues / attributes addressed due to the gender of the designer.

Introduction

There are examples of proven differences between the sexes in the way they communicate and the way they think. It is possible that there is a difference between the way a male and a female designer approach a design project. When we educate designers we instil certain value systems in them and alter the way they might have instinctively proceeded with a design process. Is there a difference between the sexes and does design education get in the way of it? This is a big question, however we are only concerned here with the first steps along the way.

The products with which we surround ourselves imply something about the person we are, or indeed, the person we would like to be. Products do not exist in isolation or solely to satisfy basic functional needs. The urge to fulfil aspirations, emotional and cultural needs often attract potential users of products. Accepting that products can satisfy needs beyond function, it is alarming that within our culturally diverse British community, designers of these products are predominantly white, middle-class and male.

As highlighted by Norman (1990) designers often perceive themselves as *typical* users. Designers need to recognise when it is appropriate to rely on their own experience and understanding, and when they need to consult the user. Is the perceived empathy of female design students a pointer to the future of design?

The design lecturing experience of the authors indicates that there are differences between the sexes in terms of the type and amount of thought they put into the design process. In one of the authors' institutions, the best student in the last 3 graduating years has been female, although they are a smaller proportion of each year group. What could the effects be of the study environment? One of the few significant differences found between male and female students in a study by Solomonides (1996), was on a research inventory sub-scale described as 'fear of failure'. Elsewhere, he (Solomonides and Crisp, 2000) discussed the impact that the Faculty, their outlooks and constructs of knowledge may have on the approaches to study of engineering undergraduates.

A study by Yeh and Chen (2000) found that 'discouraging attitudes of the Faculty' were most likely to get in the way of creativity with male and more senior design students of both sexes, while 'the fear of losing or taking risks' was most likely to prevent female and junior design students of both sexes from being creative. What we can see is that the education environment and fear itself could be playing their parts in moulding the outlook of new designers.

The first step in our exploration of what exactly is 'going on in the heads' of designers is the attempt related here to highlight the important issues within the design process for a standard product, the domestic cooker, as viewed by new, under-graduate design students. This may determine if there are differences at this stage that can be attributed in some way to gender.

Method

Students on two different under-graduate courses in two universities were used as subjects for this research. It was decided to use first year, first term students. To use final year students would have resulted in subjects with more knowledge and experience of the design process, however if their education was playing some part in moulding their thinking, this would have a greater effect on the results. Principally, the first year groups were used to elicit information on what would be the most natural way for designers to think about and order the design process. The authors reasoned that design students may be more likely to act on inherent instincts if their education had so far been very short. Male and female subjects from both institutions totalled 94 and 54 respectively.

The method used was a two-stage questionnaire. The first part gathered definitions of important factors from the subjects. The second part then fed the most important ones back to the subjects for consideration of their ranking order. In the first part, the subjects were asked to put themselves in the position of a design manager who has to brief design staff on a new project to design a domestic cooker. This team talk should highlight issues that are important in the mind of design management and that the design process needs to address seriously. The subjects were asked to write single words describing these factors then a one sentence description of the factor to reduce ambiguity.

The most important factors were fed back to the subject group in the second questionnaire. They were asked to put up to six in ranking order of importance. These rankings were scored 5-4-3-2-1-0 in order to arrive at relative scores. These scores were then weighted to account for the different numbers of respondents in the four subject groups (institution A & B, male and female) in order to allow simple, visual comparisons to be made.

Results

Questionnaire 1 resulted in the initial collection of 26 different factors deemed to be important to the design process. A simple addition of the numbers produced a list of percentage citations by subject, i.e. "31.25% of female students in institution B cited 'usability' as an important factor". The 11 strongest results overall - *aesthetics; function; manufacturing costs; retail costs; ergonomics; materials; safety; sustainability; manufacturing processes; usability; target users;* were selected to go into the second stage of the study to be ranked:

The Questionnaire 2 ranking exercise produced scores for the factors. A graphical comparison of these gives a clear picture of correlation:

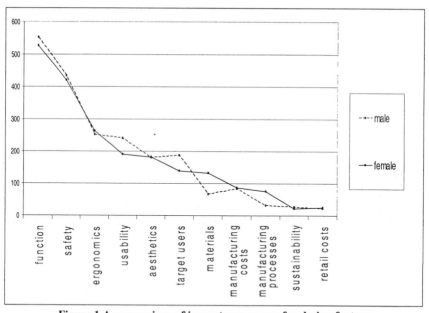

Figure 1.A comparison of importance scores for design factors

Discussion

We deduce there is little difference attributable to gender in the perception of first year design students on the relative importance of the main factors concerning industrial and product design. However, if this research was prompted by observation to the contrary, then there is still the question of what the observable differences might be. There may be differences in what the sexes meant by the words used in this study. There are some differences in the perception of students on entry to undergraduate courses (Hann & Williams, 1989) formed by their experience in the school system. As highlighted earlier, the influence of their male-dominated educational environment may be strong. Even if we delivered a perfectly balanced education the sexes could view it and use it in very different ways.

The idea that students may change the way in which they engage with a programme of study over time is one that is inherent within the so called 'approaches to study' or 'learning' paradigm based on phenomenological research into how undergraduates learn. Briefly, it can be shown that students take differing approaches depending on their perception of the task in hand and the meaning they attach to various aspects of their learning. This helps explain how students can change significantly in their qualitative approaches, levels of understanding and outcomes over a period of time. This may offer some direction for future research.

When it comes to more defined individual differences, there have been a few studies that have attempted to identify relative changes in students' approaches, focusing on factors such as age, gender or disability. Most studies have been limited and have tended to find little measured difference in gender-based approaches to study. However, this does not mean that the hypothesis that there are observable differences by the end of the programme is invalid. As Richardson (1993) points out whilst also observing that most research has ignored gender as a social issue:

"A basic research issue is whether there are differences in terms of patterns of responses... on the part of male and female students. The idea that male and female students differ in their approaches to studying is intrinsically a very plausible one."

Richardson (1993), Hayes and Richardson (1995), Trueman and Hartley (1994), Severiens (1994), Solomonides (1996) and Byrne *et al* (1997) amongst others have subjected their data sets to analysis for differences in approach by gender. All the authors found small differences in some constitutional elements of approaches to study between males and females, but these tended to account for the effect in no more than 30 to 40 percent of each of the cohorts studied. However, Jones and Hassall's (1997) studies have shown significant differences in approach to study between male and female accounting students. Some authors (Julier, 2000) have gone even further by questioning the epistemology of design education itself, suggesting that the focus of design (in particular critical or contextual studies elements) is largely based on an individualised, linear, iconic, modernist and male history as preserved in the tradition handed down from Pevsner and Banham. This is a point made generally by Richardson (1993):

"... it remains the case that female students are expected to study what are in most cases codified versions of men's experience."

All of these studies would suggest therefore, that further research is needed into the identification and meaning of different approaches to study within specific subject areas before firm conclusions can be established. Such a study, perhaps longitudinal, profiling the same cohort of students from programme entry to exit, may reveal data that would help explain the current authors' empirical observations. It is possible that, with some modification, the assessment procedures of student projects could be designed to elicit information on the success or otherwise of students' answers to the various important factors they have highlighted in this study.

How these factors of epistemology and approach combine and relate to each other is for further examination and debate, but for now it would seem that the approaches to study paradigm may be useful in future research. We would suggest however, that the study be quite refined as the research reported to date is largely based on quantitative analysis of attitudinal questionnaires. As such it may not be sensitive enough to reveal the underlying issues inherent in the findings so far. It is preferential therefore that a combination of a quantitative and qualitative methods be used, over a period of time, and that these be interpreted against the local context of study.

Conclusions

The authors' opinion is that the initial answers may lie in the different meanings that the sexes attach to the words used in the above study. If male and female designers both think 'target users' are important, does it follow that they both accord them the same weight of emphasis in the design process? If they accord them the same weight are they putting the same kind of work into the research and problem solving? Could it be the fine detail of the designer's vocabulary or feelings of ownership of the design process and product that defines the difference?

Effective design solutions should respond to users needs. If we can discover evidence of a more female way of looking at the design process and users, we may increase *all* designers' sensitivity to users. We can equip our designers of tomorrow with skills to ensure more appropriate products cater for the variety of needs in the community we live in. Proactively opening up the profession to females will begin to address an imbalance by integrating more female users throughout the design process.

Bruce and Lewis (1989) argue that design education professionals could be more active towards '[rendering] the male/female ratio more acceptable' within the confines of the education system, which would contribute to the longer-term goal of social change towards gender inequality. Certainly encouraging more girls to take relevant subjects at school and look at their career options more widely is beginning to have an effect. However, in the end, would it be right to produce an 'androgynous designer' with differences 'bred' out, or should we celebrate the differences between the sexes? It is, after all, diversity and difference that makes our world stimulating.

References

Bruce M. and Lewis J. 1989, Cherchez Les Femmes. *Design*: 58-59 May

Byrne M. Flood B. and Willis P. 1997, Approaches to Learning of Irish Students Studying Accounting. *Dublin City University Business School Research Papers.* **No 36** 1997-1998

Hann J. and Williams S. 1989, *Gender Issues in Industrial Design.* Coventry Polytechnic, Coventry

Hayes K. and Richardson J.T.E. 1995, Gender, Subject and Context as Determinants of Approaches to Studying in Higher Education. *Studies in Higher Education.* **Vol 20 No 1**

Jones C. and Hassall T. 1997, Approaches to Learning of First Year Accounting Students: Some Empirical Evidence. In Gibbs G. and Rust C. (eds.) *Improving Student Learning Through Course Design.* Oxford Centre for Staff and Learning Development

Julier G. 2000, *The Culture of Design.* (London, Sage)

Norman D. A. 1990, *The Design of Everyday Things.* (London, The MIT Press)

Richardson J.T.E. 1993, Gender Differences in Responses to the Approach to Study Inventory. *Studies in Higher Education.* **Vol 18 No 1**

Severiens S.E. and Ten Dam G.T.M. 1994, Gender Differences in Learning Styles: A Narrative Review and Quantitative Meta-analysis. *Higher Education.* **Vol 27**

Solomonides I. 1996, *Learning Intervention and the Approach to Study of Engineering Undergraduates.* Unpublished PhD thesis. Nottingham Trent University

Solomonides I. and Crisp A. 2000, Environment in Engineering Education – New Approaches to Curricula Content. *Proc. 22nd SEED Annual Design Conference & 7th National Conference on Product Design.* University of Sussex

Trueman M. and Hartley J. 1994, Measuring the Time-Management Skills of University Students. *Proc. SRHE Annual Conference.* University of York

Yeh W.D. and Chen C.H. 2000, *Obstacles to Creativity for Design Students.* The 2000 IDSA Conference on Design Education. The University of Louisiana Lafayette, USA

INTERFACE DESIGN

HMI ASSESSMENT OF THE NIMROD MRA4 FLIGHT DECK

Dave Hogg, Fiona Sturrock and Katharine Wykes

BAE SYSTEMS, Nimrod Programme, W427e, Warton Aerodrome, Preston PR4 1AX, Lancashire UK

This paper outlines the process adopted by BAE SYSTEMS HF specialists to ensure successful HMI integration on the Nimrod MRA4 aircraft. Focusing on the flight deck design, it details how the Flight Deck Assessment Rig was developed for use in a large-scale risk reduction assessment of the flight deck displays, moding and workload. In answering the question 'is the design acceptable ?', workload and performance measures have been used as key determinants. The success of this approach, supported by a custom-made Data Collection and Analysis Tool, is detailed. The paper also details how assessment data were translated into a set of design changes that balanced product improvement against cost and timescale constraints.

Introduction

BAE SYSTEMS is prime contractor for the Nimrod Maritime Reconnaissance Aircraft (MRA4). The flight deck for this aircraft poses several key challenges for HMI design. Firstly, crew composition to undertake flight deck tasks has been reduced from four on the Nimrod Mk2 to two on MRA4. It is therefore necessary to ensure that the integration of advanced flight deck systems has maintained an acceptable level of crew workload. Secondly, the flight deck is fitted with Commercial Off The Shelf equipment from different suppliers. Modification is required to ensure the different pilot interfaces form an integrated flight deck design that supports the operational role of Nimrod. Thirdly, these modifications are to be undertaken within the cost constraints of a fixed price contract. A key concern within this is the management of change to the HMI in order to ensure cost is balanced against meeting contractual requirements with a functionally effective and timely product. In order to meet these challenges, a robust process for HMI integration has been adopted. This paper outlines the process used for assessing the flight deck design, focusing on those aspects that are innovative in comparison with similar large scale projects.

The HMI Assessment Process

Prior to Nimrod, the benefits and necessity of HMI integration had already been proven within BAE SYSTEMS on projects such as Eurofighter. Nimrod MRA4 therefore started out with a project culture that viewed HMI integration as an essential requirement. In order to foster this, a structured approach to assessing the flight deck design was developed that incorporated lessons learned from other projects and had the following key drivers:

1. *Management of change.* There is acceptance within the project that requests to change the HMI design are an inevitable consequence of design assessments. Therefore, an efficient process for managing design recommendations was required that involves all the relevant stakeholders, at the correct level of the organisation, balancing cost and timescale constraints against product improvement. The process had to ensure the full impact of HMI recommendations on the system design is considered.

2. *Progressive Customer agreement.* Obtaining the Customer's progressive agreement to the flight deck design throughout the design process. It was important within this to obtain positive design statements, as well as recommendations for change, so that the design baseline could be frozen and progressed into manufacture.

3. *Performance based assessment criteria.* Traditionally the subjective opinion of pilots has tended to be the sole source of recommendations for design change. Whilst pilots are subject matter experts, this approach can easily lead to requirements creep beyond what is contractually and operationally necessary. In order to respect the Nimrod project's cost and timescale constraints, emphasis was placed on moving away from reliance on pilot opinion to an approach that determines acceptability of the design on the basis of workload and performance measures. These measures are used to substantiate or refute the necessity of design recommendations made by the pilots in order to resolve a particular safety, performance or workload issue. Following on from this, prioritisation of the design recommendations themselves was based on strict safety and operational criteria in order to embody them within appropriate timescales for development flying and production aircraft delivery.

4. *Efficient data collection and analysis.* In order to shift the emphasis away from reliance on subjective opinion, the selection and management of assessment techniques was a key area for development. The approach adopted had to be accepted by assessing pilots, the Customer and the Nimrod senior management as a robust method of assessing the design upon which significant cost decisions could be based. Allied to this, the assessment techniques had to provide relevant and meaningful data within very short timescales for collection and analysis. In the case of the assessment detailed in a later section, data had to be collected on-line during a two day assessment and be analysed in time for a debrief on the third day.

5. *Traceability.* To provide full traceability of information from data collection during the assessment through to technical implications of recommendations and the eventual decision on change embodiment.

The sections which follow outline how this approach was successfully adopted in a large scale assessment of the Nimrod flight deck displays and moding.

Flight Deck Assessment Rig

The Flight Deck Assessment Rig (FDAR), see Figure 1, was commissioned for the sole purpose of undertaking a risk reduction HMI assessment of the Nimrod flight deck workload, displays and moding design. It is a high fidelity simulator that allows pilots to fly a wide range of Nimrod's operational missions and be presented with the full range of alert conditions. Both the hardware and software are built to rapid prototyping standards using commercially available equipment to represent the production standard flight deck design. This approach minimises cost, allows the paper-based design to be assessed dynamically in advance of production equipment being frozen and delivered, allows rapid change and affords the flexibility to create scenarios and alert conditions that can not be created with aircraft standard equipment.

The FDAR was developed by a large, multi-disciplined team led by HF specialists and involving aircrew, simulation engineers and system design engineers from both BAE SYSTEMS and the Customer. Pivotal throughout the FDAR requirements, rig development, training and assessment process has been a detailed mission analysis. A forcing mission was derived which was representative of the Nimrod's operational roles of anti-submarine warfare and search & rescue. It encompassed the full mission from start-up through to take off, transit, on-task and landing and included key workload drivers such as emergency handling, changes to the mission objective whilst on-task and airfield diversions upon landing.

During assessments the FDAR was situated in an outside world dome which helped create a psychological environment for the pilots to immerse themselves in the mission and demonstrate representative levels of performance and workload. The assessment was controlled from the adjacent HF control room where miniature cameras and an intercom system were used to monitor and record the crew during the mission. From the control room the mission crew role-played tactical and ATC communications and controlled the outside world environment including enemy manoeuvres.

Figure 1: The Flight Deck Assessment Rig

Assessment of crew workload and the integrated flight deck design

Five Nimrod flight deck crew (ten pilots) participated in the assessment. Each crew received extensive training prior to the assessment:

- 2 weeks ground school
- 1 week dedicated Flight Management System training
- 1 week glass cockpit training
- 8 half day structured training missions on the FDAR
- Training on the assessment techniques

The formal assessment lasted three days per crew. On Day 1, following a briefing the crew flew the forcing mission. A range of measures were used during the mission to tap into the crew's workload and performance. These were as follows:

- Pre-defined measures of effectiveness upon which the crew's performance was assessed. There were two types of measure: 1) Maintenance of flight performance limits such as requested altitude and heading; 2) Successful completion of particular tasks and actions as judged by an expert observer. Tolerance limits for the measures of effectiveness were pre-defined and varied for each phase of the mission to reflect differing safety and operational restrictions.
- NASA-TLX subjective workload rating scale. This was completed by the crew at appropriate points during the mission (approximately once an hour) and rated workload across the phase of mission that had just been flown. To our knowledge, there is no established red-line for determining acceptable workload level scores from NASA-TLX. Red-lines are however essential if design decisions are to be based on acceptable / unacceptable workload. In light of this expert judgement derived the following for use in Nimrod assessments: a score of 0 – 60 was considered acceptable workload, a score of 61 – 80 was considered unsatisfactory workload that should be further investigated and a score of 81 – 100 was considered unacceptable workload that would probably lead to a design change.
- Bedford Rating Scale subjective workload measure was administered along with NASA-TLX to provide a more task-based indicator of workload. Scores of 1 – 3 were considered to be acceptable workload, 4 – 8 required further investigation and 9 –10 were unacceptable.
- To complement these workload measures, Instantaneous Self Assessment (ISA) was used to indicate the crew's workload 'here and now'. With this technique, the pilots selected one of five colour coded buttons at 4 minute intervals (in response to a signal) to rate current workload. Any missed responses were recorded as a maximum workload rating. Analysis rules were developed that led to investigation of ratings of very high workload and sustained periods of moderately high workload.
- The Crew Awareness Rating Scale, an in-house developed Situation Awareness (SA) technique, administered alongside NASA-TLX and Bedford Rating Scale. Responses that indicated difficulty in maintaining SA were investigated further.

Prior to the assessment there had been some concern that the techniques would be overly intrusive. However, in terms of the rating scales, the fact that they were administered at the end of a mission phase, were quick to complete and the aircrew were trained on their use found that intrusiveness was not an issue. The same was true of the ISA technique which was also readily accepted by the aircrew.

On Day 2 of the assessment, the crew completed a design acceptance questionnaire (DAQ). This led the crew through the mission they had flown and systematically investigated their acceptance or non-acceptance of the flight deck displays and moding used at each stage of the mission, rating this acceptance in accordance with standard test pilot definitions for safety and operational effectiveness.

On Day 3, a debrief session was held in which comments made on the DAQ were linked to performance, workload and SA data from the mission. Together with the crew, HF specialists investigated whether the cause of unsatisfactory or unacceptable performance / workload / SA scores was directly attributable to design deficiencies at that point during the mission or to other factors such as aircrew procedures. The product of the debrief was a set of design recommendations from the crew, some of which were substantiated by performance / workload / SA data, others of which were not.

The Data Collection and Analysis Tool

As can be gathered from the description above, extreme timescale constraints were placed on the collection and analysis of assessment data. In order to meet these timescales, the Data Collection and Analysis Tool (DCAT) was developed. Created in Microsoft Excel and Visual Basic, DCAT collected all the data during the mission, including the automatic logging of simulation parameters. The mission analysis was entered into DCAT in advance of the assessment and pro forma were prepared to record all the scores from the subjective rating scales and the expert observer's measures of effectiveness at the appropriate point. Therefore, the HF assessors had to undertake minimal data input during the mission other than to cue the administration of rating scales and to time-stamp when each phase of the mission changed. Post-mission, the logged simulator parameters and ISA data were automatically transferred into DCAT.

Once all the data had been collected, DCAT automatically analysed the data through comparison against the pre-defined tolerance limits / red-lines for each technique and highlighted unsatisfactory or unacceptable 'hot spots' to investigate further in the debrief session. This data analysis task was completed on Day 2 of the assessment in parallel to completion of the DAQ. As the DAQ had followed the mission analysis, once completed it too was entered into DCAT and the comments automatically linked to the performance / workload / SA hot spots.

DCAT's user interface has been designed to direct the assessment team to design hot spots; this can be done at an overall summary level with the capability to link down to the raw data as required. As well as being invaluable in reducing analysis time, this approach ensures that the data from the various techniques can be viewed in an integrated manner, related directly to how the crew performed and experienced the mission as it progressed in order to answer the overall question of 'is the design acceptable ?'. Other DCAT features include filters that can be applied to investigate issues such as whether an aspect of the design was problematic only during a particular mission task or across all tasks for which it was used. Post-assessment DCAT allows for storage of the assessment data to ensure essential traceability on long term design projects such as Nimrod.

Had DCAT not have been available, it is estimated that the data analysis would have taken up to two weeks per crew which would have prohibited the use of performance, workload and SA data in the debrief due to project timescales. Whilst it was developed specifically for the FDAR assessment outlined in this paper, DCAT's structure is deliberately generic and has already been adapted to suit other HMI assessments within BAE SYSTEMS.

Post-assessment processing of recommendations

During the debrief, over 1,000 recommendations were generated from across the five crew. A rationalisation process was undertaken to group duplicate or similar recommendations together both within and across crew. These were entered into a database and tabled at a technical wash-up meeting with the aircrew and HF specialists. The purpose of the wash-up was to technically agree a consolidated set of recommendations across the crew, including the operational / safety justification and a categorisation of severity. This process reduced the agreed set of recommendations to around 200.

The next stage was for the system design and HF teams to investigate the technical, contractual, cost (both in-house and supplier) and timescale implications of the recommendations. This information was collated onto a database managed by the HF specialists who also ensured that the link back to performance, workload and SA measures was maintained. The data was then analysed to determine a valid set of 71 recommendations that could be demonstrated as necessary to ensure acceptable crew performance and that fulfilled our contractual requirements. The recommendations were tabled at a senior management review that included the Nimrod Chief Engineer and Programme Director. There was total endorsement at this senior level to the robustness of the assessment and processing of recommendations, including the reliance on performance and workload measures to underpin the necessity of a recommendations. The outcome of this review was that all 71 recommendations were accepted and are now being embodied into the baseline design.

This process will ensure that the Nimrod flight deck design has a manageable crew workload. Indeed, the aircrew's overall conclusion was that, having flown the missions in the FDAR, they believe the flight deck design will be acceptable for 2 crew operation once the resultant recommendations are embodied. This will be confirmed during workload qualification testing with aircraft standard equipment. Once again, performance and workload measures will be used as part of the qualification assessment, supported by DCAT.

Conclusion

Successful integration of HF into the Nimrod MRA4 flight deck design has been facilitated by a robust process for undertaking design assessments. HF specialists led the development and use of the FDAR in a way that generated total committment from aircrew, system engineers, senior management and the Customer. This process has produced a set of recommendations to change the flight deck design that will have significant impact on ensuring the integrated flight deck design is acceptable during qualification, whilst at the same time respecting the stringent cost and timescale constraints within the Nimrod MRA4 project. Key elements within the process have been the acceptance of non-intrusive performance, workload and SA measures as a means of substantiating the necessity for design change, the DCAT which allowed a wealth of assessment data to be analysed and managed quickly and efficiently, and a multi-disciplined approach to investigating design recommendations which ensured the full impact of change was determined.

APPLYING A COGNITIVE ENGINEERING FRAMEWORK TO RESEARCH: A SUCCESSFUL CASE-STUDY ?

John Long[1] and Andrew Monk[2]

*[1]Ergonomics & HCI Unit, University College London,
26 Bedford Way, London WC1H 0AP
(now at University College London Interaction Centre)*

*[2]Department of Psychology, University of York,
York YO1 5DD*

This paper reports a case-study, which applies a Cognitive Engineering framework to telemedical consultation research. The study is generally considered a success. All the research concepts are informally classified by the framework. Additional framework concepts suggest advances to the research. Such framework applications help researchers build on each other's work, the better to support Cognitive Engineering discipline progress.

Introduction

Human-computer interaction researchers have been criticised for not building on each other's work (Newman 1994). Elsewhere, Long (1996) has claimed that poor discipline progress resides in the failure of research to validate its design knowledge. Cognitive Engineering (CE) frameworks better specify the relations between research and the design of human-computer interactions (discipline (Long and Dowell, 1989); general design problem (Dowell and Long, 1989 and 1998); and knowledge validation (Long, 1996)). Application of the frameworks indicates the support of research for the design of human-computer interactions. They thus help to meet the criticisms of Newman and Long. This paper illustrates the application of such frameworks to research on video-based medical consultation.

The paper is a case-study, whose relevant dimensions are definition, complexity and accessibility (Stork, Middlemass and Long, 1995). Audio-video links, for telemedical consultation, can be categorised as: implicitly defined, simple and accessible. A successful application would inform other similar research. The research is implicit – the researchers started with no explicit goals. It is simple, the design space is small and constrained. It is accessible, because a researcher was an informant. The case-study is unsuccessful, if there are research concepts, or relations between them, required but, unclassifiable by the CE framework.

Telemedical Consultation

A description follows; ('technical') concepts are highlighted and numbered (on first citation). The **research** (1) is that of **telemedical consultation** (2). **Field studies** (3) were carried out in 1996. An **audio-visual link** (4) connected **medical specialists** (5) at one location with a **GP** (6) or **nurse practitioner** (7) in a **treatment room** (8) at another. In general, there was a **patient** (9) in the room, to whom the **consultant** (10) talked, at the same time as the GP or nurse practitioner. There were different facilities to send video images. Two locations had **portable cameras** (11) for sending images of the problem (e.g. a wound). One site had **X-ray facilities** (12) for sharing images. The study involved **interviews** (13), but also some **observations** (14) and **video analysis** (15). The aim was to understand the **problems** (16) and **advantages** (17) of using this **equipment** (18) for this **work** (19). The full **method** (20) and **results** (21) appear in Watts and Monk (1997; 1998). These results comprise a **work description** (22) as a sequence. The text elaborates the **goals** (23) of the sequences as task characteristics. Each task characteristic has design **implications** (24).

CE Discipline Framework

The concepts of Figure 1 in bold constitute a CE discipline framework (Long and Dowell, 1989 and 1998). This framework is now used informally to structure a 'neutral' description of the telemedical consultation research.

Research Knowledge: this knowledge acquires design knowledge. The researchers used **ideas** (25) in the field studies, derived from their knowledge of the field. Ideas about the use of **shared artefacts** (26) as **communicative devices** (27) led them to expect problems when views, at the two ends of a link differed. Ideas from **task analysis** (28) led them to question the work process. Research knowledge included interviews in situ with equipment at hand for prompts to make it easier for **users** (29) to explain problems. This experiential **knowledge** (30) is difficult to make explicit.

Research practices: these practices apply research knowledge to acquire design knowledge. Live consultations were viewed and videotaped. This information was used to reason about the effects of the **communication facilities** (31). A tabular representation, a **Comms Usage Diagram** (32) (CUD), was invented. This diagram helps to identify advantages or disadvantages. The latter generally correspond to CSCW **issues** (33). These practices have a **scope of application** (34).

Practice knowledge: this knowledge is acquired by research practice and supports design practice. Watts and Monk (1997) identified six issues that telemedical consultation link should **consider** (35). Issue 1: High quality multi-party sound is a primary **requirement** (36); and Issue 4: Parties at each end of the link need to have the same image of the problem.

Practice Practices: these practices are the design practices that apply the design knowledge in **system development** (37). The research involved no such application. It might, however, be the next step, involving writing **guidelines** (38) for **purchasing** (39) and setting up equipment. The guidelines would be refined by **iterative testing** (40) with GPs etc.

General Problem: the general problem is system development as the purchasing and setting up of audio-visual links.

Particular Scope: the particular scope is audio-visual links for medical consultation.

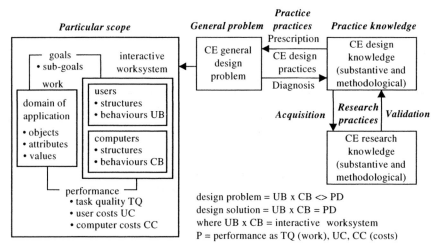

Figure 1. Cognitive Engineering Framework following Dowell and Long (1989 and 1998).

The CE discipline framework of Long and Dowell (1989 and 1998) and the earlier description of telemedical consultation together suggest: first, all telemedical research concepts and their relations can be informally classified by the framework. Thus, no concepts or relations are beyond advance by the framework. Second, although the research claims to acquire knowledge as: results (21), more research is needed to advance this knowledge, such that it supports design (as system development (37)). Third, with no explicit framework, there are only implicit relations between: design knowledge as: results (21); and design practices as: problems (16); between design knowledge and practices and between particular scope as: medical specialist (8) etc.

The CE framework specifies additional, relevant concepts and more explicit relations. Design knowledge as substantive and methodological, together specify design practices, as diagnosis and prescription. CE knowledge and practices together specify the general design problem. As an example, diagnosis is a design practice identifying poor performance. Such was the aim of the interviews (13) and what defines a problem (16) in the comms usage diagram (32) analysis. The concept of diagnosis could be used to advance the research by making this objective explicit.

CE General Design Problem Framework

The non-bold concepts of Figure 1 constitute the CE general design problem framework of Dowell and Long (1989 and 1998). This framework can be informally used to organise the telemedical consultation concepts (see earlier). Framework concepts precede telemedical concepts (although the latter are not always present), as follows: 1 Work – Work (19); 1.1 Domain of Application – telemedical consultation (2); 1.1.1 Objects; 1.1.2 Attributes; 1.1.3 Values; 2 Interactive Worksystem; 2.1 Users – medical specialists (5); GP (6); nurse practitioner (7); patient (9); consultant (10); work description (22); users (29); 2.1.1 Behaviours; 2.1.2 Structures; 2.2 Computers – audio-visual link (4); treatment room (8); portable cameras (11); X-ray facilities (12); equipment (18); shared artefacts (26); communicative devices (27); communication facilities (31); 2.2.1 Behaviours; 2.2.2

Structures; 3 Goals; 3.1 Sub-Goals – goals (of work) (23); 4 Performance – advantages (17); 4.1 Task Quality; 4.2 Worksystem Costs; 4.2.1 User Costs; 4.2.2 Computer Costs; 5 Design Problem – problems (16); issues (33); requirements (36); 5.1 UB x CB -> PA <> PD; Design Solution – implications (24); 6.1 UB x CB = PA = PD.

Figure 1 and the above suggest: first, all the concepts and their relations associated with the scope of the telemedical research are informally classified by the framework. Thus, no concepts or relations are beyond support. Second, few concepts are used to describe the work – work (19) and telemedical consultation (2) – other than as users' behaviours. Third, only one concept is used to describe performance – advantages (17). Last, although research concepts relate to the framework concepts of design problem (5), as problems (16); issues (33) etc. and of design solutions (6), as implications (24), there are no explicit concept relations and decompositions.

The CE general design problem framework (Figure 1) can be seen to support better advance by providing additional, relevant concepts and their relations. The problem (16) specified in the comms usage diagram (32) could identify the performance attributes of task quality (4.1), how well the medical consultation is performed and the user costs (4.2.1), the workload of the medical specialists (5), GPs (6) etc., incurred in performing the task that well. Such integration would provide more explicit support for the concepts of problems (16), issues (33), etc. and remains to be done by the research.

CE Knowledge Validation Framework

The CE knowledge validation framework (Long, 1996) presumes the CE discipline and the general design problem framework (Figure 1), while conceiving, in addition, CE design knowledge to be: conceptualised; operationalised; tested; and generalised. This framework can be informally used to organise the telemedical research concepts (see earlier). Framework concepts precede telemedical concepts (although the latter are not always present), as follows: 1 CE Design Problem – problems (16); issues (33); requirements (36); 1.1 Conceptualised – medical specialists (5); GP (6); nurse practitioner (7); patient (9); consultant (10); users (29); audio-visual link (4); treatment room (8); portable cameras (11); X-ray facilities (12); equipment (18); shared artefacts (26); communicative devices (27); communication facilities (31); 1.2 Operationalised; 1.3 Tested: 1.4 Generalised; 2 CE Design Knowledge; 2.1 Conceptualised – results (21); implications (24); issues (33); consider (35); system development* (37); guidelines* (38); iterative testing * (40); 2.2 Operationalised; 2.3 Tested; 2.4 Generalised. (*not used in the research.)

The CE knowledge validation framework and the above suggest: first, all the concepts and their relations, associated with the design knowledge validation of the research are informally classified. Thus, no concepts or relations are beyond support. Second, the CE design problem concepts (1) are only conceptualised as issues (33) etc. as concerns users (29) etc. The concepts are not operationalised, tested or generalised. Third, the design knowledge concepts (2) are only conceptualised as results (21), etc. The concepts are not operationalised, tested or generalised. Indeed, design knowledge as guidelines (38), etc. were not conceptualised other than as the next step. Fourth, the CE design problem (1) concepts issues (33) etc. are not related to the design knowledge concepts (2) as results (21). Last, neither the design problem nor the design knowledge concepts are explicitly specified. The problem (16) specified in the comms usage diagram (32) could be operationalised with respect to the relevant issues (30) or design implications (24), for example, high-quality, multi-party sound is a primary requirement (36) in telemedical consultation (2). Operationalisation of the design problem and knowledge would support validation of the latter with respect to the former and so advance the research.

Advancing Telemedical Consultation Research

The three CE frameworks indicate that all the telemedical consultation research concepts and their relations can be informally classified. Thus, no concepts or their relations are beyond support. The main areas are associated with: design knowledge (discipline); work, performance, design problem/solution (general design problem); and operationalisation, test and generalisation (knowledge validation). The research concepts and relations are implicit. Indeed, the general picture of the research is of a craft (or craft-like) discipline, with informal and implicit knowledge and practices (Long and Dowell, 1989). The CE framework suggests the telemedical consultation research could be advanced both in its own implicit craft-like terms and in more explicit engineering terms.

Case-Study Discussion and Conclusion

The case-study can be considered successful. First, all the concepts in the research, telemedical consultation, have been informally classified by the CE framework. Second, where concepts in the framework have no corresponding concepts in the research, the case has been made (for most, if not all – space precludes all) that adding them would be of value by identifying what remains to be done by research. As concerns the scope of the case-study, the research continues to appear: implicitly defined, relatively simple and accessible. Success suggests that the CE framework shows promise for being useful for considering similar research.

References

Dowell, J. and Long, J. (1989). Towards a conception for an engineering discipline of human factors. *Ergonomics*, 32 (11), 1513-35.

Dowell, J. and Long, J. (1998). Conception of the cognitive engineering design problem. *Ergonomics*, 41 (2), 126-139.

Long, J. (1996). Specifying relations between research and the design of human-computer interactions. *Int. Jnl. of Human-Computer Studies*, 44 (6), 875-920.

Long, J. and Dowell, J. (1989). Conceptions for the Discipline of HCI: Craft, Applied Science and Engineering. In *Proc. BCS HCI SIG Conference*, Nottingham, UK, 9-32.

Newman, W. (1994). A preliminary analysis of the products of HCI research, using pro forma abstracts. In *Proc. CHI '94*, Boston, Mass., 278-284.

Stork, A., Middlemass, J. and Long, J. (1995). Applying a structured method for usability engineering to domestic energy management user requirements: a successful case study. In *Proc. HCI '95*, Huddersfield, UK, 367-385.

Watts, L.A. and Monk, A.F. (1997). Telemedical consultation: task characteristics. In *Proc. CHI '97*, Atlanta, Georgia. ACM Press, 534-535.

Watts, L.A. and Monk, A.F. (1998). Reasoning about tasks, activity and technology to support collaboration. *Ergonomics*, 41 (11), 1583-1606.

CULTURAL ISSUES IN HUMAN COMPUTER INTERACTION AND ERGONOMICS AND THE DESIGN OF SMARTPHONES

Elizabeth Hofvenschiold

35 Keslake Road
London NW6 6DJ
hofvenschiold@aol.com

Cultural background and occupational status can influence the way people interact with and perceive technology. The paper outlines research done to gather information on the possible differences in attitude to and use of mobile phones of British and German university students and British and German young professionals with the intent to design more intuitive and usable interfaces for the mobile phones of the future (i.e. smartphones). Geert Hofstede's work on culture was used as a framework in interpreting the results and as a means of testing his work's relevance to Human Computer Interaction and Ergonomics (HCI-E). The study concluded that there were a few differences in the way that the participants used and perceived their mobile phones and that the occupational groups showed more differences than the national ones.

Introduction

Many factors have to be taken into consideration when designing interactive systems. Traditionally Human-Computer Interaction and Ergonomics (HCI-E) considered the physiological issues in product development and in the past thirty years, the perceptual and cognitive issues have been addressed (Kurosu, 1997). More recently issues concerning emotion, motivation and pleasure have become increasingly important in the HCI-E world (Simon and Benedyk, 2000). Many factors related to cognitive, emotional and motivational responses might influence the way in which people will interact with a product's interface. One of these factors is culture.

Research Objectives
The general research objective was to gather information on the possible effects of a user's cultural background and occupational status on his/her understanding and use of mobile phones. University students and young professionals from the UK and Germany were chosen to make up the participants for this case study. The national cultures of the UK and Germany were picked because they are both considered part of northern Europe and therefore part of a collective that is traditionally marketed to and designed for as a group. In terms of tailoring a product's interface to a user group's needs, how far beyond mere translation do designers have to go in terms of developing products for different national groups? Comparing two similar yet different cultures

was thought to be a way of doing this. University students and young professionals were compared so that the sample of people surveyed in each country would be as similar as possible. The two occupational groups were also chosen because they are in themselves their own sub-cultures.

It was hoped that the information gathered from the case study could contribute to addressing and better understanding cultural issues within HCI-E and to address any cultural issues that might have to be taken into consideration in the development of future mobile phones or smartphones (mobile phones with added functionality such as browsing capability). Also, when the project was begun, no published works specifically dealing with cultural issues in the design of mobile phones could be found.

Two general research hypotheses were constructed to guide the study and provide a framework for the analysis of the data but in no way was the project part of a controlled experiment. The first hypothesis states that a person's culture would affect the way he/she perceives, understands and uses a mobile phone. The second hypothesis states that there will be differences in the perception and use of mobile phones between the different occupational groups. The study was essentially a quasi-experiment set in the real world in an attempt to gather meta-knowledge to fulfil the research objectives.

Theoretical Context

There are many definitions for what a culture is. More often than not a country typifies the social environment that a culture lives within but there are many exceptions to this. For the purposes of this paper, nationality *and* occupation are the differentiating variables. Choosing nationality and occupation was thought to balance any of the inherent biases in grouping individuals together into a culture based solely on their nationality.

Geert Hofstede's (1980, 1997) work on defining countries and their cultures in terms of four different dimensions was used to provide a framework of contrasting and comparing Germany and the UK (Hofstede did add a fifth dimension to his later research but only the first four dimensions are used for this paper). Hofstede's dimensions were chosen because his research has set the standard by which many people in different fields measure and define a culture (Pugh and Hickson, 1996). Plus, his work has done a lot to influence the way in which groups of countries are targeted for advertising campaigns, marketing strategies and more recently HCI-E design practices (e.g. Jordan, 2000).

Using Hofstede's work was also a way of determining its relevance to HCI-E. HCI-E has a tradition of borrowing ideas and methodologies from other disciplines to help solve usability problems. Globalization and other cultural issues play major roles in the development of future applications and HCI-E needs to be aware of their implications if the discipline is to successfully continue providing usability solutions (Barber and Badre, 1998). Designers postulate that design guidelines and heuristics are sufficient for worldwide design but more often than not, they do not realize that these guidelines and heuristics are themselves culturally biased (Smith, 2001). The internationalization or localization of software applications in particular is a way of dealing with cultural issues in the design of interfaces. They both have their advantages and disadvantages and it is important to weigh the costs of each strategy and choose the most effective one. The research done for this paper also hoped to contribute to determining how far designers would have to take any cultural differences between the four groups of people studied into consideration for the design of future smartphones interfaces.

How do Germany and the UK differ?

According to Hofstede's classification of cultures into four dimensions, the UK and Germany are both masculine, individualistic and low-power distance cultures. However, the UK's score for individualism was a lot higher than Germany's score. The dimension in which Germany and the UK differ most was in uncertainty avoidance. Germany was perceived to have a high uncertainty avoidance culture while the UK had a very low one. This was thought to be important for this paper, as there are a variety of ways in which people deal with uncertainty and these have an effect on the *technology*, laws and religion(s) of a country (Hofstede, 1997).

What does the information Hofstede provides imply in terms of mobile phone interaction? Do cultural differences influence people enough to use their mobile phones differently? For example, based on Hofstede's work, it is plausible that Germans would use their mobile phones in ways that would reflect a society that has high uncertainty avoidance. This paper does not start with fixed assumptions as to how the groups will differ. Indeed, part of the objective was to discover how British and German students and young professionals would differ, if they differed at all.

Method

The participants
78 people participated in the study. Out of these, 8 were British university students, 17 were German university students, 21 were British young professionals and 32 were German young professionals. The participants' ages ranged from 21 to 40 years old. Most of them were recruited through an email mailing list while other participants were approached 'on the street' in the UK and Germany by the author.

Data gathering techniques
The questions posed by this study could not be answered by reviewing the relevant literature alone. English and German versions of a questionnaire were developed to gather the data. A questionnaire was chosen as it was thought to be the best method of reaching as many people as possible in the time given for the project. The questions asked were divided into five general categories: (i) personal information including nationality and occupation; (ii) mobile phone functions used and their usability ratings; (iii) suggestions for changes and additions of function and display; (iv) attitudes to mobile phones in general; and (v) frequency of, reasons for and perceived usability of personal Internet use.

It was recognized that a questionnaire is a way of collecting information on the *declarative* knowledge and not the procedural knowledge of the study's participants. However, the benefit of doing an empirical study in the time given outweighed the bias that this could have caused. In an attempt to compensate for the shortcomings of a questionnaire, four interviews were done with a member of each of the groups. The data gathered from these interviews was of a purely qualitative nature as the sample size was much too small to do any statistical analysis.

Results

Questionnaire results
A complete set of results cannot be presented here but examples of the most relevant results are given so that the reader may gather the general data trends. 2 German students and 1 German young professional claimed not to own a mobile phone but were included in the sample as they did answer the questions related to Internet use. It is important to note that the study did not intentionally target mobile phone users but a cross section of the population. When asked to list the top four easiest to use functions on their mobile phones, nearly all the participants found the same top four functions (receiving a call, making a call, receiving and reading a text message and

writing a text message) easiest to use with the exception of the German young professionals who differed in opinion about the fourth easiest to use function (looking up entries in the phone book). The choices of the different groups were more varied for the functions they found harder to use. It is interesting that writing text messages appeared here as well as in the easiest to use category. Generally all of the groups found the same functions harder to use.

When asked if they could change one aspect of their mobile phone's screen, over half of the British participants specified that they would like a colour screen while a large number of German participants said they would like a larger screen. The second most popular request for the British groups was the Germans' first choice and vice versa. All groups found the same things important but prioritised them differently.

Participants were also asked to rate their opinions about 7 attitude statements relating to mobile phone use. The Mann-Whitney U test was used to check for significant statistical differences in these responses and 2 values were found to be significant at the 5% level. Students were more likely to agree to the statement, "What a mobile can do is more important than its 'look and feel'" than young professionals were (significance value was 0.026, P≤0.05). And it was found that German young professionals were more likely to disagree to the statement, "You mobile phone is a status symbol" than the British young professionals (significance value was 0.05, P≤0.05).

The responses to the Internet use questions showed that all of the participants used the Internet and found it generally easy to use. In summary, there were many similarities found between the groups. The differences found were subtle and therefore factors such as chance, individual differences and population sample bias were taken into consideration when analysing the data.

Interview results

There was only time to interview 4 participants (one from each of the groups). Their answers were understandably richer in detail than the questionnaire data. As mentioned earlier, no statistical analyses was done on the data as the sample size was much too small but it was found that the British participants were more likely to individualize their mobile phones and were more emotionally attached to them than their German counterparts were.

Discussion and conclusion

The information gathered for this study suggests that culture and occupation do, to a certain extent, affect the way in which people interact with their mobile phones. The researcher must however always be aware that correlation does not always equate to causation. The groups did respond in a similar fashion but this does not necessarily mean that the hypotheses are disproved. Some of the results did show that the groups differed in opinion about certain things. A different statistical approach might have proved more conclusive but time and resources were very restricted for the project.

Hofstede's work did provide a good base upon which differences between the German and British cultures could be studied. The results did show that there was a difference in use of and attitude to mobile phones between the groups but it was not necessarily what was expected to be found if Hofstede's dimensions were to make up the framework of analysis. For example, if someone was part of a high uncertainty avoidance culture, he/she might be more disposed to having a mobile phone for security reasons. When asked, more British than German participants said that they got and used their mobile phones for security reasons. The nationalities did vary but not as one might have expected them to after Hofstede. Some of the other findings did seem to correspond to Hofstede's dimensions (e.g. for individuality) but it was found that the nationalities interacted with their mobile phones in much the same way. Perhaps the dimensions in which the UK and Germany are similar are the ones that are more relevant in determining attitudes to mobile phones and this is why not that many differences were found.

Another way of differentiating cultures might have been more appropriate for this study and applicable to HCI-E research in general. Or perhaps it was because the predominantly text based interfaces of current mobile phones do not contain the culturally specific elements (usually icons and other graphics) that websites or software applications do. Smartphones propose to have increased functionality and graphical capabilities, which could lead to a more complex and graphic rich interface. If so, gathering user requirements from the current mobile phones may not be as applicable as was thought to addressing the cultural issues surrounding the design of future smartphone interfaces. Asking questions on the participants' Internet use was thought to somehow compensate for their interaction with the simple graphic interfaces of their current mobile phones. Ultimately both of the extant systems analyzed proved less helpful than expected.

The study was limited in the number and choice of extant systems analyzed, its data gathering method and also in the fact that it only used Hofstede's definitions of culture. Other extant systems such as personal digital assistants (briefly done in the interviews) might have proved more forthcoming with information.   Data gathering techniques such as cognitive walkthroughs and observational studies might have proved more revealing. Again, time and resources determined the choice of methods for the study.

In conclusion, the study proved that there were more similarities than differences in the use of and attitude to mobile phones between the British and German university students and young professionals. There differences were more prominent between the different occupational groups and this shows that there is a need to explore the effect of 'sub-cultures' on HCI-E issues. This is perhaps why Hofstede's research was not as insightful as was hoped. The study also showed that the issues that HCI-E practitioners have traditionally dealt with still need to be addressed as British and German mobile phone users generally encountered the same usability problems. The more traditional issues and the increasingly prevalent motivational and emotional issues in design need to combined so that HCI-E can continue to provide effective solutions to usability problems.

References

Barber, W. and Badre, A. 1998, Culturability: The Merging of Culture and Usability, *http://www.research.att.com/conf/hfweb/proceedings/barber/*

Hofstede, G. 1980, *Culture's Consequences: International Differences in Work-Related Values*, Beverley Hills, USA: Sage Publications

Hofstede, G. 1997, *Cultures and Organizations: Software of the Mind*, USA: McGraw-Hill

Jordan, P.W. 2000, *Designing Pleasurable Products: An Introduction to the New Human Factors*, London, UK: Taylor and Francis

Kurosu, M. 1997, Wind from the East: When will it blow and how? *http://www.asahi-net.or.jp/~yz2m-krs/article/97042hci2.html*

Pugh, D.S. and Hickson, D.J. 1996, *Writers on Organizations* (fifth edition), London, England: Penguin Books Ltd

Simon, J. and Benedyk, R. 2000, Addressing Pleasure in Consumer Products Through Ergonomics. In McCabe, P.T., Hanson, M.A. and Robertson, S.A. (eds) *Contemporary Ergonomics 2000* (Taylor and Francis, London), 390-394

Smith, A. 2001, Cultural Issues in HCI, *Interfaces*, **47**, 4-5

WORKPLACE DESIGN

THE IMPLEMENTATION OF ERGONOMICS INTO A MEDICAL PRODUCT MANUFACTURING COMPANY IN RELATION TO WORK RELATED UPPER LIMB DISORDERS

Suzanne C. Fowler

*InterAction of Bath Ltd
5/6 Wood Street, Bath, BA1 2JJ*

A healthcare product manufacturing company wishes to introduce a participatory ergonomics programme to reduce the incidence of Work Related Upper Limb Disorders (WRULDS). A self-help programme needs a workplace checklist designed to be compatible with the organisation's needs and culture, a team of trained ergonomics facilitators and a process co-ordinator. To improve the programme's chances of success, the 'stage of change model' was used to establish the current attitudes of the company employees and to give an indication of an appropriate implementation style. Building an awareness of the benefits of ergonomics fostered a positive attitude towards ergonomic intervention, and in one group encouraged a move from 'precontemplation' to 'action' within a few weeks. In addition, it initiated a cultural change promoting open communication between shop floor and management demonstrating the positive effect such a programme can have on organisational culture as a whole.

Introduction

The Company is a multi-national manufacturer of healthcare products. The site has 700 employees and produces a range of wound care products and surgery packs. The manufacturing and packaging processes are manual and labour intensive. Movements are repetitive and of short duration. Many tasks demand awkward constrained postures and some require considerable force. As a result some employees have exhibited symptoms of WRULDs.

The Company had decided to review its manufacturing processes and to introduce an ergonomics intervention programme to reduce the incidence of WRULDs. It is committed to self-reliance and wanted to train in-house ergonomics assessors, to evaluate every workstation. Management had requested help with evaluating their current assessment procedures, identifying and assessing of some of their higher risk processes and training a team of employees to become ergonomic assessors. The aim of the project was to establish a successful participatory ergonomics programme.

Imada and Noro (1991) defined participative ergonomics as '*the process of involving end-users in the development and implementation of solutions, by enabling people to understand and apply ergonomics to their work*'. A successful self-help programme needs a workplace checklist designed to be compatible with the company's needs and culture, a team of trained ergonomics facilitators (assessors), and a process co-ordinator (Sinclair, 2001; Imada and Noro, 1991).

Stage of Change Model

There are many barriers to the successful implementation of long-term, internal participatory programmes, particularly lack of management support and 'resistance to change' (Westlander *et al*, 1995; Tayyari and Smith, 1997). The barriers present at any particular time are related to the attitude towards, and expectations of, ergonomics intervention. *Being able to change the attitude or other factors underlying the behaviour (e.g. knowledge and skills) can change the behaviour and therefore can stimulate the implementation of ergonomic improvements (*Urlings *et al*, 1990*)*. These attitudes can be described relative to the 'stage of change' (Haslam, 2001).
These stages of change are:
1. Precontemplative – little is known or understood about ergonomics,
2. Contemplative – some is known and there is thought of change,
3. Preparation – making definite plans to change,
4. Action – actually engaging in changing behaviour,
5. Maintenance – working to prevent relapse and consolidate changes made.
(Prochaska and DiClemente, 1982)

The approach to implementation differs according to the stage of change. There is little point in extolling the virtues of ergonomic intervention if there is no understanding of what this means. According to Kok *et al* (1987) the intervention stages are:
1. Getting attention to the information. Overcome selective perception by offering attractive and effective information.
2. Understanding the information. Information must be easy enough to understand without being childish. Evaluate information for intelligibility, clearness, and readability and adapt the information based upon the results.
3. Changing attitudes. Change negative attitudes towards ergonomics into positive ones by emphasising the advantages of ergonomic intervention.
4. Changing the intention. Make participants resistant to the negative influence of their peers, by giving them strong arguments they can use in favour of the ergonomic improvement.
5. Changing behaviour. Give training in necessary skills and encourage management to make the essential funds available.
6. Maintenance of new behaviour. Give positive feedback in response to the new behaviour.

Interviews with managers (at all levels) and maintenance, engineering and shop floor employees revealed that there were several 'stages of change' evident throughout the organisation:
1. Precontemplative (Shop floor)
2. Contemplative (Section Managers, Team Leaders, Ergonomic Committee Members)
3. Action (Manufacturing Manager General Wound Care [Ergonomic Committee Chairman], Active Assessors).

In general there was a negative attitude towards ergonomics resulting from a lack of understanding. Consequently the approach adopted needed to change attitudes in favour of ergonomic improvement and provide positive feedback to maintain long term commitment (see Urlings *et al*, 1990).

Phase one – Building the Tool

The Company already had two WRULD risk assessment tools in use on site as a result of previous initiatives. These tools had been developed by the parent company. In order to evaluate these tools it was necessary to establish the main criteria for the tool. For the Company it was important that a checklist provided them with: a good level of detail; ease of use and learning; emphasis on WRULD detection; emphasis on repetition, in particular for the wrists and hands and be quick to complete, in order to ensure that a successful long running programme be established.

Further to the particular requirements of the Company, research by Li and Buckle (1999) established the following criteria as essential for a successful exposure assessment method: the method should take environmental and psychosocial aspects into consideration; the measurements must be repeatable under normal working conditions; the method should have high validity; reliability and sensitivity; assessment data should be recorded for ease of analysis and reference and the recording equipment should not interfere with the movements being recorded or the work in progress.

Checklist Evaluation Procedure
The Company's tools were evaluated for suitability against the Company's requirements, the recommended research requirements and a commercial postural analysis technique.

Quick Evaluation Check (QEC) for musculoskeletal disorders (Li and Buckle, 1999) was chosen as the commercial/benchmark tool because: it has proven reliability and validity; is based on up to the minute research; addresses the issue of repetition and it has taken the best aspects of previous tools and amalgamated them. The tool is quick to use (approximately 10 minutes per task) and can be used when assessing change in exposure before and after an ergonomics intervention.

A new task was assessed for each new tool being evaluated to ensure that familiarity with the task did not affect the time taken to complete the assessment or the level of detail captured. During the evaluation the following aspects of the tool were considered:
1. Time taken to complete the risk assessment. The task analysis and observation.
2. Time taken to complete the analysis. The calculations or judgements to establish the risk level.
3. Level of expertise required to carry out the assessment. A high level of expertise is necessary where the assessor must make a judgement on severity of posture by referring to degrees or where there is little specification on the checklist of the risk factors to be identified (Li and Buckle, 1999).
4. Level of sensitivity. How many of the risk factors are identified and to what level of detail.
5. Ease of use. The number of references to a manual required, the clarity of the layout and the difficulty of the terms used.
6. Psychosocial and environmental sensitivity. The number of questions addressing issues such as: noise, lighting, temperature, vibration, paced work, stress, communication, boredom and job satisfaction.

None of the tools satisfactorily addressed each of the criteria set out by the Company; therefore it was necessary for the tools to be modified to fit the situation. This was an iterative design process in which the newly re-established ergonomics committee participated.

The Tailored Tool
A tailored tool was developed to fit the requirements of the Company. It has four sections:
♦ General Information - a basic task analysis method which requires the assessor to record the reason for the assessment, the task to be completed, the tool used to complete it and the steps taken to complete the task. This will familiarise the assessor with the task and provide information on which to build solutions.
♦ Stage 1 - a posture, force, repetition, and action assessment. QEC was modified to address hand postures in greater detail, as these are integral to the work and a system was developed for classifying the score from minimal to high.
♦ Stage 2 - an active surveillance method to address body part discomfort, mechanical stress, environmental issues and employee stress. There was space for worker comments.
♦ Solutions - to record the problems identified, proposed solutions, the solution actually chosen, and when it was implemented.

The four sections come together to form one coherent report, which can be passed to the relevant manager.

The tailored tool forms the basis of the self-help programme. Involving Ergonomics Committee members in all stages of tool development encouraged ownership of the tool and legitimised ideas and experiences that workers had accumulated. The resulting positive attitude towards ergonomics promoted the progress of participants onto the 'Action' stage.

Phase Two – Workstation Assessment

The second phase of the project involved the selection of workstations identified as high risk for WRULDs, the analysis of each, and suggestion of solutions. The aim was to reduce the risk of WRULDs by implementing ergonomic interventions.

Four processing lines were assessed for WRULD risk using the tailored tool and the results were shared with the relevant management. Ergonomic solutions were developed in several stages and involved the ergonomist, shop floor workers, maintenance engineers, project engineers, health and safety personnel and management.

The roles of the ergonomist in this phase of the project were that of expert, facilitator and motivator. As an expert the ergonomist assessed individual workstations and recommended solutions. As a facilitator it was necessary to raise the employees' and management's awareness and knowledge of ergonomics. This ensured their support, co-operation and participation in the assessment of the recommended solutions and the generation of alternative solutions. The job of the facilitator is much more dependent upon support of management than is that of the expert. Once solutions are agreed, the ergonomist assumes the 'motivator' role. Without a driving force solutions are unlikely to be designed, tested and implemented. If management is not committed or is busy on other projects then the ergonomist has to take control. The level of involvement depends on the input of the managers. Three levels of support were encountered on this project, in different areas of the plant, and thus three levels of action were required from the ergonomist:

1. In the first area managers were preoccupied by a high priority production problem. It was difficult to start implementing the ergonomic solutions because maintenance and project engineers were not available.
2. In the second area the manager responsible for the implementation left the company, although he had already established the participation of the shop floor, project engineering and maintenance. Without his continued input the solution implementation process began to falter. The ergonomist assumed the role of project manager to ensure continued progress and successful conclusion.
3. In the third area production was being affected by the ergonomic problems. There was full support of management, driven by the need to resume production. In this case the ergonomist did not need to assume the role of motivator as the project was driven by management.

By creating visible solutions to production problems the ergonomist highlighted the advantages of ergonomic intervention. This fostered a positive belief in shop floor workers and management in the power of ergonomic intervention and spurred others into action. A move from 'precontemplation' to 'action' occurred within a few weeks.

Phase Three – Training the assessors

The third phase of the project was to establish a competent team of assessors who could carry out ergonomic risk assessments using the tailored tool, on each process in the plant, without further expert help. The objective was to enable managers, supervisors, and employees to identify the aspects of a job that may increase a worker's risk of developing WRULDs, recognise the signs and symptoms of the disorders and participate in the development of strategies to control or prevent them.

Important principles for a successful training programme include:

♦ Design to meet the comprehension of the trainees and organisational needs.
♦ Promotion of open and frank discussion between the trainer and trainees. 'Ice-breakers' (exercises which get the group interacting whilst imparting relevant information) help to encourage such discussion.
♦ Provision of opportunities for trainees to discuss ergonomic problems affecting their own job.
♦ Engage trainees in relevant problem-solving exercises during the training.
♦ Use of various types of materials to impart information to ensure continued interest.
♦ Provision of full and immediate feedback.

(NIOSH, 1997; Budworth *et al*, 2001)
These principles were used to design a classroom training session.

Evaluating Training

Evaluating the success of the training was of the utmost importance in establishing the competence of the assessor. Evaluation of the training took place on two levels. The first was a feedback form and the second was a practical session on the shop floor (see Fowler, 2001). The results of the evaluation indicated that the training met its aims in an interesting and practical way. Each assessor completed an assessment successfully with a minimum of intervention. The time taken to complete the assessment was approximately an hour including the analysis and solution discussion. With experience the time taken will be reduced, making this a viable option for risk assessment.

Conclusions

The Company is now in the position to take ergonomics forward. They have a toolkit designed to fit their requirements for assessment of WRULD risk factors, eight employees have been trained to be competent, enthusiastic assessors, and a positive attitude towards ergonomic intervention has been adopted. Management support for the implementation of the recommendations should ensure on-going improvements and the continued development of the ergonomics participatory programme.

References

Budworth, N., Haslam, R. and Brown, S., 2001, Back in Work. Project run jointly by NSK Europe Ltd, Loughborough University, Newark Hospital and the EEF East Midlands Region. http://www.iosh.co.uk/news/display_news.cfm?NewsID=147

Fowler, S.C., 2001. *The implementation of ergonomics into a medical product manufacturing company in relation to work related musculoskeletal disorders.* Loughborough University.

Haslam, 2001 proc 10th conference of the New Zealand ergonomics society, 26 –27 July, Rotorua NZ.

Kok, G.J., Wilke, H.A.M., and Meertens, R.W., 1987, A model of changing attitudes and behaviour through information. *In:* Urlings, I.J.M., Nijboer, I.D. and Dul, J., 1990, A method for changing the attitudes and behaviour of management and employees to stimulate the implementation of ergonomic improvements. *Ergonomics,33(5) 629-637.*

Li, G. and Buckle, P., 1999, *Evaluating change in exposure to risk for musculoskeletal disorders – a practical tool.* UK: HSE Books.

National Institute of Occupational Safety and Health, 1997, Elements of Ergonomics Programs. A primer based on Workplace Evaluations of Musculoskeletal Disorders. NIOSH. http://www.cdc.gov/niosh.

Noro, K. & Imada, A., 1991, *Participatory Ergonomics.* London: Taylor & Francis.

Prochaska, J.O., and DiClemente, C.C., 1982, Transtheoretical therapy: toward a more integrative model of change. Psychotherapy Theory. *Research and practice* (19), 276-288.

Sinclair, M., 2001. Content of Loughborough University Lectures.

Tayyari, F. & Smith, J.L., 1997, *Occupational Ergonomics. Principles and applications.* London: Chapman & Hall.

Urlings, I.J.M., Nijboer, I.D. and Dul, J., 1990, A method for changing the attitudes and behaviour of management and employees to stimulate the implementation of ergonomic improvements. *Ergonomics, 33(5), 629-637.*

Westlander, G., Viitasara, E., Johansson, A., and Shahnavaz, h., 1995, Evaluation of an ergonomics intervention program in VDT workplaces. *Applied Ergonomics* **26**(2), 83-92.

DESIGNING WITH PERSONAL SAFETY IN MIND

Vicky Malyon and Nigel Heaton

Human Applications, 139 Ashby Rd, Loughborough
Leicestershire, LE11 3AD
enquiries@humanapps.demon.co.uk

Many employees fulfil roles that put them in vulnerable situations. Although, most people work without incident, for some violence and aggression is a daily occurrence, often accepted as 'part of the job.'

The design of the workplace and systems of work can allow organisations to be proactive in the management of the personal safety of their employees.

This paper looks at how ergonomics can help employers to fulfil the duty of care owed to employees by maximising the safety of their work environment and working practices at both the macro and micro level.

Introduction

The Suzy Lamplugh Trust was set up in 1986, after the disappearance of Suzy Lamplugh, an estate agent who went out to meet a male client and never returned. Predominantly driven by Suzy's family the Suzy Lamplugh Trust has succeeded in becoming the UK's lead body for issues associated with personal safety.

The circumstances surrounding Suzy's disappearance and the media attention surrounding the case led many employers to start to question the risks faced by their employees whilst going about their work. Suzy did not know anything about the man that she was due to meet apart from his name. She did not leave any details of her intended visit with anybody at her office and nobody else saw the man. She met the man in a house, which she knew was locked, the windows were barred, the electricity turned off and the telephone removed. She had walked into a trap.

In hindsight there are a number of things that Suzy could have done to prevent such errors and since its establishment the Suzy Lamplugh Trust has campaigned to raise awareness of personal safety issues. They also help organisations to develop policies and practices to maximise the personal safety of their employees. Over the last four years ergonomics issues and the need to manage risk have become incorporated into much of the Trust's work.

The principles advocated by the Trust are not difficult to follow and can be implemented in a variety of situations in both our work and social life.

Risk Management

Four years ago the Trust recognised the need to put personal safety into a risk management framework. They identified that the risks to personal safety were 'reasonably foreseeable' and that it was possible to both assess and manage them.

All Organisations within the EU have a number of legal obligations relating to Health and Safety. In the UK, these obligations are detailed predominantly under the Health & Safety at Work etc. Act 1974 (HASAWA) plus other associated Regulations. The management of risks to personal safety are explicitly part of an employer's responsibilities.

Section 2 of HASAWA requires employers to provide a safe place and safe system of work to ensure the physical and psychological well being of employees. Section 7 is also important as it requires employees themselves to take the responsibility of ensuring that they take reasonable care of their own health and safety and that of others. More recently, the Management of Health and Safety at Work Regulations 1992 (revised 1999) places a legal duty on employers to carry out an assessment of the significant risks workers face whilst at work and to take any necessary preventative measures. It also requires the provision of training, information and supervision. For many professions, particularly those requiring interaction with the Public, risk assessment will need to be cover risk to personal safety.

The risk assessment must take into account the "hazard" (i.e. that which can cause harm, with an estimate of the extent of the harm) and the "likelihood" of the harm being realised. In terms of likelihood, some groups of workers are more 'at risk'. That is, they are more likely to encounter violations of their personal safety. Groups identified include those that work alone, those that deal with the public and those that work with people who are emotionally unstable.

Identifying those most 'at risk' allows organisations to prioritise resources and concentrate efforts in the risk elimination and reduction phase.

Micro Considerations

The design and layout of the physical environment can have a major impact on personal safety. Therefore, it is imperative that consideration is given to measures to eliminate or reduce personal safety risks at the design stage. This involves analysis, not only of the user groups of an environment but also the interaction that may occur between them in light of the purpose of their meeting.

An example of where such principles may be applied is in the design of an accident and emergency reception / waiting room. Given the inherent nature of waiting rooms it is important that they do everything to make the 'wait' as tolerable as possible, especially given the possible emotional instability of the people within them. Design factors include:
- adequate, well lit parking
- comfortable seating

- up to date reading material
- refreshments
- telephone
- toilets
- children's play area / toys
- information on waiting times and priorities
- security cameras
- exits for staff
- protection screens

It is important to note that very few of the above design factors are immediately related to the reception staff, the people whose personal safety we are trying to protect. Instead the focus is predominantly on ensuring that the environment satisfies those that have to wait within it, to keep their aggression to a minimum, therefore inadvertently reducing the risk to the reception staff.

There are an endless number of factors that could be 'designed in' to the work environment. For example, where face to face contact is involved it may be suitable to replace the member of staff with a machine, e.g. automatic ticket machines. However, it is important to ensure any changes made do not increase the violence and aggression e.g. customers may require information that is not provided by the machine and could get annoyed if there is not anyone there to ask.

Alternatively, where violence and aggression is predicted, measures may be taken to protect the member of staff, for example with protective screens. Screens can be helpful if there is a risk of physical attack or theft but may create a problem and aggravate matters if somebody is trying to hold a conversation and they restrict communication.

The design of the environment itself can also effect the risk to personal safety. For example, when planning the location of roads and footpaths in new housing estates or new developments it must be ensured that they are busy and overlooked. This vital design consideration is something that has interested the Association of Chief Police Officers (ACPO). They support and manage a scheme called 'Secured by Design' which has the backing of the Home Office. Secured by Design is a family of national police projects involving the design of new homes, refurbished homes, commercial premises, car parks and other police crime prevention projects.

The aim of the scheme is to ensure that personal safety is considered at the design stage. The position of hedges, walls and fences can influence where people can be seen, and subsequently where they can hide, where people congregate and how they can escape if they have committed an offence. Therefore, by having input at the design stage, a time when Ergonomists should be heavily involved or consulted, the potential risks can be 'designed out.'

The main role ergonomics can play is to ensure that personal safety is considered as a significant design factor when looking to ensure that the environment fits the user. Standard task analysis and user requirement techniques need to be modified to include consideration of personal safety.

Macro Considerations

Although, there are a wide range of measures that can be taken at the micro level it is important that emphasis is also placed at the macro level, e.g. through the development of policies and procedures, etc.

Many of the risks to personal safety actually stem from the behaviours that people adopt as a 'routine' part of their job. For example, in Suzy's case she didn't always write down the details of the clients that she was meeting, probably because she never thought it was necessary (she was also dyslexic and was worried about her writing). Many jobs require employees to go about their working day moving between locations without informing anybody of exactly where they are going, how they are going to get there or more importantly what time they are expected back.

If we have never encountered a problem with our 'routine' behaviour there has probably not been a need to question it. However, if we assess such behaviours in relation to personal safety there are very often a number of actions that we could take to eliminate or reduce the risk.

Risk management can be achieved by examining patterns of employment and working practices. For example, with the increase in the 24-hour society an increasing number of people are working shifts where they leave work at unsociable times or in darkness. For these employees the provision of transport must be considered. Whilst an employer's duty of care typical does not extend to travel to and from work, we believe that it is becoming increasingly likely that a civil court may find an employer owed an employee a duty if that employee was required to work "unsociable" hours. For example if they had to leave or arrive at work at times when public transport was not available.

Employers already owe a duty when they provide car parking. They must ensure that for those employees that drive, the car park is "safe" e.g. it is well lit, possibly patrolled, etc.

Within an environment where employees interview clients, it may be necessary to have an alarm system to alert others if they feel a threat to their personal safety. This may not be an audible alarm but perhaps a signal that is recognised by everybody, e.g. the use of a code word.

Other systems that have been found to be effective are 'buddy' systems where employees who normally work alone are paired up. They then have to maintain contact with each other throughout the day to ensure that they are both safe. They may even accompany each other for visits where they feel they may be at risk alone.

A common approach to tackling personal safety for mobile workers is to provide mobile phones. Although, a phone can be a helpful aid to maintain contact with others it is of little help when being attacked. Similarly, the provision of a personal alarm may be seen to increase personal safety. However, employers often fail to train their employees in its use, i.e. most people think that an alarm should be used to attract attention and thus get help, whereas its main use should be to distract the assailant. It is therefore important that such provision is carefully thought out and appropriate advice sought.

Once measures have been identified to eliminate or reduce the risks it is imperative that those involved are informed and trained. This training must be supported by a policy that outlines the organisation's philosophy with regard to their employee's personal safety. It must also include the responsibilities of both the organisation and the employee and refer out to any specific procedures, systems or good practices that have been identified.

In the unfortunate event of an incident occurring it is also important that the organisation offers and provides after care and support to the victim, this may be in the form of counselling or time off work. Lessons must also be learnt from incidents to ensure that measures are implemented to prevent or reduce the risk of a reoccurrence.

Conclusion

Personal Safety is an important issue that cannot be ignored. By taking a proactive approach and conducting obligatory risk assessments it is possible to identify where the risks lie. More importantly, it is subsequently possible to identify measures that can be taken to eliminate or reduce the risks. These measures may be design features of the physical environment or the development and implementation of policies, systems and procedures to promote safe working.

Whatever the intervention, as Ergonomists it is possible to make a difference. We have standard tools for understanding users and tasks. We need to modify these tools to ensure that systems of work are safe and that environments are designed with personal safety in mind.

AN INVESTIGATION OF THE IMPACT OF CLEANROOM ENVIRONMENTS ON WORKERS IN THE MEDICAL DEVICE INDUSTRY

Liliana Fernández Carro[1]; Mónica Díez Campa[1,2]; Enda F. Fallon[1,2]

[1]*Department of Industrial Engineering, National University of Ireland, Galway, Ireland*
[2]*Environmental Change Institute, National University of Ireland, Galway, Ireland*

In this paper a study which assesses the impact of the environment on cleanroom workers is presented. Twelve skilled female workers performed a sequence of standard dexterity tests in a cleanroom wearing both Class 100,000 and Class 10,000 garments. Worker performance on these tests was measured and compared against established norms. Physiological variables including, heart rate, respiration rate, core body temperature, skin temperature and sweat rate were monitored during the course of the tests. The workers' subjective responses to cleanroom work were measured using two questionnaires containing thermal sensation, thermal comfort and perceived exertion scales. The first questionnaire was administered a number of days before the dexterity tests and the second was administered during the execution of the tests.

Introduction

Personal Protective Equipment (PPE)
PPE should not be used as a substitute for engineering, work practice, and/or administrative controls. It should be used in conjunction with these controls to provide for employee safety and health in the work place, (OSHA, 1998).

In cleanroom environments the concern is more with protecting the product from the workers, as they are the major source of contamination. Human beings shed their outermost layer of skin cells every 24 hours. Cleanroom garments act as a "personnel filter" preventing human particulate matter from entering the atmosphere of the cleanroom, (Austin, 2000).

Cleanrooms
A cleanroom is a specially constructed, enclosed area, environmentally controlled with respect to airborne particles, temperature, humidity, air flow patterns, air motion and lighting. According to Federal Standard 209E cleanrooms are classified from Class 1 to Class 100,000

depending on the size of the airborne particles and their concentration. Other classifications include ISO standard 14644-1 and B.S. 5295.

Heat Stress

Under conditions of heat stress, the body absorbs and/or produces more heat than it can dissipate. Heat stress commences with a rise in core temperature, which leads to an increase in metabolism. The first line of defence against heat stress is the cardiovascular system, which increases heart rate and dilates the blood vessels in the skin leading to a rise in skin temperature. Heat is then removed by radiation and convection. Sweating is the second line of defence.

Cleanroom work requires workers to wear protective clothing to varying degrees depending on the class of cleanroom involved. Protective clothing combined with heavy physical activity and/or very high ambient temperatures can lead to heat stress (Sanders and McCormick, 1993). Although the tasks performed in this study were not expected to be very physically demanding, the possibility of heat stress and its effects was considered.

Rationale

The majority of the literature on protective clothing and thermal environments is concerned with military and fire fighting applications. Both the garments and the environmental conditions are more extreme than those encountered in cleanrooms, (Al-Haboubi, 1995; Akbar-Khanzadeh and Bisesi, 1995; McLellen and Cheung, 2000; Pandolf, *et al* 1989; Stirling and Parsons, 1999). In this paper the effects of the environment and protective clothing on cleanroom workers are investigated in the full knowledge that they are not as severe and restrictive as those encountered in previous studies. Work reported by Tangney and Gallwey (1999) adopted a similar approach to the present study, however they did not report using a cleanroom to conduct their experiments nor did they report assessing operator subjective responses to the same extent.

Methodology

The study was carried out in a biomedical device company in Galway, Ireland. Two environments were studied: a Class 100,000 cleanroom and a Class 10,000 cleanroom. Twelve workers from the first environment and six from the second completed an initial questionnaire about thermal comfort at work.

A number of performance tests were carried out in one of the company's Class 100,000 cleanrooms. This cleanroom was also used to simulate the thermal environment of a Class 10,000 cleanroom as the only way in which they differ is in the concentration of particulate matter. Twelve skilled workers, all female and employed in the company, performed a randomised sequence of standard tasks while gowned in a Class 10,000 clothing ensemble and also while gowned in a Class 100,000 clothing ensemble. The standard tasks are described in Table 1. The protective garments included a polyester coat, hair cover and shoe covers for the Class 100,000 cleanroom ensemble, and a Tyvek[R] coverall with hood, hair cover, shoe covers, mask and latex gloves for the Class 10,000 ensemble. The workers' heart rate, respiration rate, core body temperature, skin temperature and sweat rate were monitored and recorded during the course of the tests. The ProComp+/Biograph system which can monitor, record and store 8 channels of physiological data obtained through sensors placed over the subject's body was used for this purpose. Blood volume pulse (BVP) was captured with a sensor placed on either the ring finger or the middle finger of the hand, respiration with a stretch device fastened to the subject's torso, and skin temperature with a thermistor fastened to the tip of the small finger. Sensors were always placed in the non-dominant hand. An electronic scale (± 50g) was used to detect mass loss through sweating. A tympanic thermometer (± 0.1^0 C) was used to measure ear temperature, which approximates core body temperature.

A second questionnaire about thermal comfort and thermal sensation was administered during the experiment. Thermal comfort was assessed through both the Predicted Mean Vote (PMV) and the Predicted Percentage Dissatisfied (PPD) indices (Parsons, 1990).

Air velocity and air stream temperature were both measured using a thermal anemometer and a hygrometer was used to measure relative humidity.

Table 1. Task Descriptions

Task	Description	Unit of measurement
Roeder Manipulative Test	Measures hand, arm and finger dexterity. It has two parts: sorting and assembling	No. of parts assembled
Purdue Pegboard Test	Measures gross movements of hands, fingers and arms, and fingertip dexterity	No. of parts assembled
O'Connors Tweezers Dexterity Test	Requires the use of tweezers in placing a single pin in each 1.6mm diameter hole,	Time to finish the task
Hand Tool Dexterity Test	Measures proficiency in using mechanics' handtools	Time to finish the task
Steadiness Test	The subject places and then holds a metal-tipped stylus in 9 progressively smaller holes without touching the sides	No. of errors

Experiment Description

Each subject performed the tasks in turn while wearing the two clothing ensembles on the same day. The order of testing was generated by a random list of numbers.

The environment was a turbulent flow Class 100,000 cleanroom where no product was being processed. It was maintained at a mean dry bulb temperature of 20.2^0 C, a mean air velocity of 0.07m/s, a standard deviation of air velocity of 0.03m/s and a mean relative humidity of 60%.

On entering the room, the subject was weighed and core body temperature was recorded twice using the tympanic thermometer. The sensors were then placed on the individual and the cycle of tasks began. When the tasks were finished the subject was weighed and core body temperature was recorded again. The second questionnaire was administered at this point.

Results

Subjective data
The impact of the change in the environment was reflected more by the results of the subjective assessment than by the physiological or performance data.

Perceived exertion and metabolic rate were significantly greater when performing the tests while wearing class 10,000 cleanroom clothing than when wearing class 100,000 clothing (Table 2).

Table 2. Perceived Exertion

Perceived Exertion	Class 100,000	Class 10,000
Mean Value	11.33 (fairly light)	12.5 (somewhat hard)
Standard Deviation	1.87	2.06
Maximum	13 (somewhat hard)	15 (hard)
Minimum	7 (very, very light)	7 (very, very light)

Thermal sensation was also found to be significantly different, workers felt that the head and the hands were the warmest parts of the body. More than 80% of the people (n=15) working in the cleanroom affirmed that they were satisfied with their thermal environment based on the results of the first questionnaire. However during the course of the experiment 70% (n=12) of those wearing the 10,000 cleanroom clothing expressed a desire to feel cooler, (Figure 1).

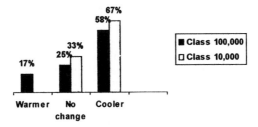

Figure 1. Desired thermal sensation after each test

The Prediction of the Percentage of thermally Dissatisfied (PPD) was greater in the Class 10,000 cleanroom, both prior to and during the experiment, (Figure 2).

Performance
The results of the performance of the standards tasks for the two cleanroom conditions were tested for significant differences using the t-test, $p=0.05$. In each case, except for the Roeder Manipulative Test, no significant differences in performance were found (Table 3).

The results obtained were also compared to the standard tables available for the Steadiness Test, the Hand Tool Dexterity Test and the O'Connor Tweezers Dexterity Test. In each case they were found to be lower than expected for very dextrous people.

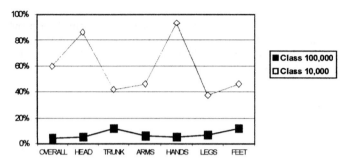

Figure 2. Predicted Percentage Dissatisfied (PPD)

Table 3. t-test results for performance

Task	Mean C. 100,000	Mean C. 10,000	Diff. Means	p
Steadiness	44.75	43.58	1.17	0.89
Hand tool	518.25	489.25	27.67	0.104
Tweezers	320.92	348.58	29.00	0.411
Purdue	68.42	65.92	2.50	0.118
Roeder	112.58	97.42	15.17	0.003

Physiological Data

The physiological data recorded during the execution of the standard tasks for the two cleanroom conditions was tested for significant differences using the t-test, $p=0.05$. Skin temperature and Pulse amplitude were found to be significantly different. A more complete discussion of results is contained in Fernandez, (2001).

Conclusions

When working in hot thermal environments, core body temperature is controlled by means of an increase in skin temperature and also by an increase in the rate of blood circulation. In this experiment, the workers bodies reacted physiologically to the change of clothing and consequently to the environmental conditions through both of the above mechanisms. The reaction did not indicate the presence of heat stress. Likewise the subjective assessment did not indicate the presence of any other heat stress symptoms. Nevertheless the subjective responses were more sensitive than the physiological responses to the change from Class 100,000 to Class 10,000 Cleanroom. This would suggest that workers subjective responses should be carefully considered when making changes in cleanroom environments.

The low level of performance achieved by the workers on the standard tasks when compared with published norms could be explained by the fact that workers were wearing restrictive clothing and for the Class 10,000 ensemble were using gloves. The performance results obtained are similar to those reported by Tangney and Gallwey (1999) who compared performance on a similar set of tasks wearing standard industrial clothing and a cleanroom suit (cleanroom class not specified). However they did not report on a comparison with published norms. Further study of the effects of cleanroom protective clothing on the fundamental motions involved in manual tasks is required.

Acknowledgement

This work has been funded by the National Development Plan for Ireland, administered through the Higher Education Authority's Programme for Research in Third Level Institutions, Cycle II.

References

Al Haboubi, M.H. 1996, Energy expenditure during moderate work at various climates. *International Journal of Industrial Ergonomics*, **17**, 379-388

Akbar-Khanzadeh, F. and Bisesi, M.S. 1995, Comfort of personal protective equipment, *Applied Ergonomics*, **26**, 3, 195-198

Austin, P. R. 2000, *Encyclopaedia of Cleanrooms, Bio-Cleanrooms and Aseptic Areas,* Third Edition, (Contamination Control Seminars, Michigan)

B.S. 1989, B.S. 5295, Environmental cleanliness in enclosed spaces

Federal Standard 209E 1992, Airborne Particulate Cleanliness Classes in Cleanrooms and Clean Zones

Fernandez, L. C. 2001, An Investigation on the Impact of the Thermal Environment on Cleanroom Wokers in the Medical Device Industry, Unpublished M.Eng. Thesis, NUI Galway

ISO 1999, ISO 14644-1, Cleanrooms and associated controlled environments. Part 1: Classification of air cleanliness

McLellan, T. M. and Cheung, S.S. 2000, Impact of fluid replacement on heat storage while wearing protective clothing, *Ergonomics*, **43**, 12, 2020-2030

OSHA 1998, 3077 (Revised), Personal Protective Equipment

Pandolf, K.B., Levine, L., Cadarette, B.S., and Sawka, M.N. 1989, Physiological evaluation of individuals exercising in the heat wearing three different toxicological protective systems. In A. Mital (ed) US Army Research Institute of Environmental Medicine, Natick, MA 01760-5007 (Taylor and Francis)

Parsons, K. C. 1990, *Human Thermal Environments*, (Taylor and Francis, London)

Sanders, M. S. and Cormick, E.J. 1993, *Human Factors in Engineering Design*, Seventh Edition, (McGraw-Hill, New York)

Stirling, M. and Parsons, K. 1999, The effect of hydration state of exposure to extreme heat by trainee firefighters, *Contemporary Ergonomics 1999*, (Taylor and Francis, London) 380-384

Tangney, K. J. and Gallwey T, J. 1999, The Effects of Clean Room Personal Protective Suits on Workers' Performance, *Contemporary Ergonomics 1999*, (Taylor and Francis, London)353-357

ERGONOMICS ANALYSIS OF DRESSMAKER WORKSTATIONS IN A TEXTILE INDUSTRY IN BRAZIL

Marcelo Márcio Soares (Ph.D.), Geni Pereira dos Santos e Germannya D'Garcia de Araújo Silva

Department of Design / Federal University of Pernambuco
Cidade Universitária / 50.670 - 420 - Recife-PE Brazil
marcelo2@nlink.com.br

This paper describes an analysis of the dressmaker workstations in a textile industry. It has used a systematic approach to the human-task-machine system which has permitted a more comprehensive approach to the work situation. In view of this it has considered several aspects such as physical, cognitive, social, organisational, environmental and other important aspects to the workers in the workstation. It has also carried out a macroergonomics analysis, informal interviews, survey, task analysis, rapid upper limb assessment and dimensional analysis using anthropometric models. The results are several ergonomic recommendations and a preliminary design of a workstation.

Introduction

The target company for this ergonomic analysis has currently 1070 dressmakers and it was accused by the workers' union of producing a large number of workers suffering from repetitive strain injuries. The workers, aged between 20 to 50 years old, are distributed over two shifts of work from 5am to 1:30pm. and 1:30 to 10pm., the rest time is 30 minutes. It is very common for an additional shift to sometimes extend the work time to twelve hours a day. Another point which attracts attention is the extreme specialisation of the tasks motivated by a system of additional salary: to obtain an increase in production, the dressmakers are encouraged to specialise in partial activities such as sewing handcuffs , sleeves or hems. This organisation, known as the Taylor system, is characterised by the parcelling of activities, and the repetition of same gestures and movements throughout the whole work shift all day long. According to Dejour (1988), by separating, radically, the intellectual work from the manual work, the Taylor system depresses the mental activities of workers.

The activity of the dressmakers can be considered as comprised of a number of static actions, with extremely repetitive movements. According to Silvestein et al. *apud* Macedo (2000), such movements are characterised by [i] a work cycle smaller than 30 seconds (superior 900 times in a workday) or [ii] when in 50% or more the work cycle is comprised of the same type of fundamental cycle (defined as a sequence of repetitive steps in a work cycle). The observed activity can be classified in the two mentioned situations. The repetitive work can present consequences such as psychophysiological problems, muscular problems and work accidents. Also, it is important to note that the activity of the dressmakers makes great demands on visual acuity, the use of detailed motor activity, and repetitive movements.

Methodology

This study used a "systematic methodology approach to the human-task-machine system" which has permitted a more comprehensive investigation of the work situation (Moraes and Mont'Alvão, 2000).. This methodology, which considered the several aspects of the work situation as part of a system, includes in its analysis the following aspects: physical, cognitive, social, organisational, and environmental.

The first step of the methodology was to identify the relationship between the "target system" (the dressmaker workstation), and its relationship with the several components of the system. It meant to identify an hierarchical position of the system, to generate a communication model of the system, to produce an action-decision flowchart and to classify the several problems in the following categories: postural, dimensional, activation, operational, architectural, cognitive, etc. The identified problems were classified considering several changes in the workstation, the tasks and an investigation of subsystems in order to characterise the human costs associated to the identified problems. Apart from this, several video register were made and an analysis of the environmental (temperature, noise, lighting).

Some identified problems

Problems of posture and dimensions (Figure 1)
- Components located out of the vision area, which contribute to the assumption of bad postures in activities of machine maintenance.
- Head movements out of comfort limit due to the actional field being out of the vision area, which contributes to the assumption of bad postures.
- Insufficient lighting to regular machine maintenance, repairs, needle changing and sewing line, in addition to bad dimensions of the workstation contribute to the assumption of bad postures.

Figure 1

Figure 2

Figure 3

Figure 4

Problems of posture and dimensions (Figure 2)
- It can be observed that the elevation of the dressmaker's left limb provokes muscle contraction in the region of the trapezium and surrounding muscles. The posture repetition contributes to the causes of muscular-skeletal diseases.
- The trunk forward bending represents a siphosis of the dorsal region of the spinal column.
- Not using the backrest characterises the inappropriateness of the seating to the task. This is an impediment to the worker obtaining adequate rest in order to relax the dorsal region.

Problems of perception and architecture (Figure 3)
- It can be observed that the sun light on the dressmaker's face can provoke dazzling with risks of accidents.

Problems of instruments and operations (Figure 4)
- The use of scissors as a tweezer is a result of the lack of an adequate instrument for this activity.
- It can be observed that the dressmaker reproduces the movement of tweezers with the ends of her fingers.

Problems of actions (Figures 5a, next page)
- Repetitive movements of arms, wrists, hands and fingers as a result of the demands of the task and the needs of action components of equipment (Figure 5a).
- Repetitive straining of inferior limbs to action machine pedals.
- Bad dimension of pedals and action angle which prevent comfort in activities.
- Too little space on the worktable to accommodate the quantity of material to be manipulated (Figure 5b).
- Some sharp points under the worktable can cause cuts and drilling.
- Some edges under the worktable can reduce the space for leg accommodation and can cause traumas.

Figura 5a

Figura 5b

Problems of operation

- Intensive work rhythm and repetition, reduced pauses, high demand of precision and reduced tolerance in the quality control program can result in high mental stress and psychopathologies of work (depression, aggressive attitudes, obsession).
- High pressure in the timetable can result and tension and anxious behaviour.

Problems of organisation

- The Taylor parcels of work (super-specialisation in some tasks) can result in lack of interest and non motivation by the dressmakers.

The extra shift of work has implications for the workers' bio-rhythms and exceeds their capability of physic and psychic recovery

Dimensional evaluation

Fourteen kinds of workstation sewing machines were identified with several variations from a number of accessories. In terms of analysis, two working stations were selected with dimensions considered extreme (the largest and the smallest one). For the dimensional analysis of each machine a model of a 2.5th and 97.5th percentile woman was used. Each one was tested at each machine. The figure below shows part of the analysis of the smaller workstation.

The smaller workstation

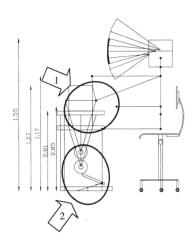

- The machine action area is out of the vision field to the woman in the 97.5th percentile (1).
- The dressmaker needs to keep a rubber band piece attached to the top of the machine and in the other extremity, an horizontal cylinder. This operation requires repetitive movements of adduction and abduction of the left arm and the trunk (2).
- It was observed that, although adjustable seats were provided, the workers are not used to adjusting them. The seats do not have upholstery which permits a better stability and weight distribution in the region of ischial tuberositis.

Data analysis regarding human costs in the workstation

Applying RULA Method

The RULA method was used (Rapid Upper Limb Assessment) (McAtamney and Corlett, 1993) in order to evaluate the repetitive strain injuries in upper limbs. This technique permits quantification from a score table (http:ergp/human.cornell.Edu/ahRULA.html), the worker exposure level to some risk factors such as number of movements/repetitions, static posture, strength, work postures demanded by equipment, furniture and workshifts and breaks.

The same workstations were used in the dimensional evaluation because they can be considered as the smallest and biggest machines in the manufacturing line. The dressmaker was observed, carrying out her task, during several work cycles and the major postures were selected from a video film. Three stages were considered from the score table: (i) identification of the work posture, (ii) application of the RULA system scores and (iii) application of a table to define the action to be carried out. The table below shows a summary of what was found. It can be observed that the smaller machine scored 6 and the biggest machine scored 4. According to Macedo (2000), such scores represent the following:

- Score 6 – The work posture is out of the movement zone, the worker carries out repetitive movements. This means that a further investigation is necessary and alterations should be done soon.
- Score 4 – Work postures are out of an acceptable zone of movement recommended by the literature, there are several repetitive activities even to those postures which are in the movement zone. Investigations are necessary.

Smaller machine	Score	Biggest machine	Score
A –Arm and wrist analysis		A - Evaluation of arms and wrists	
1.Upper arm position +15° to 45° with upper arm abduction, subject is supported or leaning and raised shoulder	4	1. Upper arm position +15° a 45°	2
2. Lowerarm elevated 90°+ with arm crossed at the medium line of the body	3	2. Lowerarm elevated 90°+ with arm out to side of the body.	3
3. Wrist positioned down 15°+	4	3. Wrist positioned down 15°+	4
4. Wrist twisted at end of twisted range	2	4. Wrist twisted at end of twisted range	2
Posture A score	6	Posture A score	5
B – Neck, trunk and leg analysis		B – Neck, trunk and leg analysis s	
1. Neck position in extension 0° to 10° with side bending	2	1. Neck position bending 0° a 10°	1
2. Trunk position 20° to 60°	3	2. Trunk position 0° a 20°	2
3. Legs and feet supported	1	3. Legs and feet supported	1
Posture B score	4	Posture B score	2
Final score	6	Final score	4
Investigate further / change soon		Investigate futher	

Survey of dressmakers

A survey was carried out in order to obtain: (i) data on physical constraints suffered by the dressmakers in their workstation; (ii) subjective opinions on equipment and work environment and (iii) suggestions to increase equipment and work environment.

Questionnaires were applied to a sample of 60 dressmakers representing the universe of 1070 workers. It was asked to point out the pains suffered in the last 12 months. The regions considered as the most painful were: eyes (17,1%, f=22), neck (14,7%, f=19), nape of the neck (13,2%, f=17), elbows (10,9, f=14) and upper back position (10,9%, f=14). These five positions of the body correspond to 66,8% of the region indicated as the most painful in the last 12 months. Curiously wrists and hands, the most painful part of the body related to repetitive strain injuries, were pointed out as the 12th. painful region of the body in these samples.

The sample were asked (n=61) to evaluate their work furniture. Most of them classified the characteristics related to the machine as good or very good, e.g. facility to operate (90%), facility to adjust (86%), machine position (86%) and facility to maintain (69%). Regarding seat/work table, only the seat width was considered as good or very good (89%). A number of other characteristics were considered as regular, poor or very poor: backrest softness (90%) and seat softness (67%), backrest (77%), work space (51%) and space for legs (43%). All these user needs must be transformed into product requirements.

In terms of work environment, most respondents classified lightening (83%), environment colours (77%) and circulation area (74%) as good or very good. On the other hand, noise and temperature were classified as regular, poor or very poor by, respectively 90 and 78 per cent of the sample.

Cardiac rhythm register

A sample of 30 dressmakers were monitored using a Polar Beat monitor. The readings were made at four times: (i) the beginning of work shift, (ii) before the meal break, (iii) returning from the meal break and (iv) the end of the work shift. The results showed that the register of the cardiac rhythm was bigger returning from the meal break than at the end of the shift, which was originally expected to be the highest level because this is the moment of production peak. The results can be explained because the workers climb a ladder to reach the second floor and had only 30 minutes for their meal, refreshment, toileting, etc.

References

BUARQUE, Lia (2000). Industrial Ergonomics. Publications to the Course of Ergonomics. Recife, Departamento de Design, Universidade Federal de Pernambuco (in Portuguese).

MORAES Anamaria de and MONT'ALVÃO, Cláudia (2000). Ergonomics: Project and applications. Rio de Janeiro, 2AB Editora (in Portuguese). .

SOARES, Marcelo Márcio (1990). Human costs in seated posture and parameters to evaluation and design of seating: 'school desck chair', a case study. M.Sc. Dissertation. Rio de Janeiro, COPPE/Universidade Federal de Pernambuco (in Portuguese).

WORK DESIGN

WORKPLACE BULLYING: AN ERGONOMICS ISSUE?

Nigel Heaton and Vicky Malyon

Human Applications, 139 Ashby Rd, Loughborough
Leicestershire, LE11 3AD
enquiries@humanapps.demon.co.uk

Failing to take into account the potential for bullying can have a major impact on the health, safety and welfare of employees and subsequently impact their performance at work. Ergonomists are often involved in the redesign of the workplace and the underlying work system and therefore have an opportunity to give advice to prevent or manage the problem.

This paper looks at the problem of bullying in the workplace, methods of identification and how to deal with complaints. More importantly it outlines a proactive approach. It looks at how Ergonomists need to take into account bullying issues when designing new systems of work. It suggests that the Ergonomist, working as part of a team, must create appropriate policies, procedures and an audit trail that demonstrates how workplace bullying is managed.

Introduction

Ergonomics is primarily concerned with fitting the task or activity to the person. Ergonomists believe in 'user centred' design, that is designing with people in mind, often via consultation and interaction. One of the problems with this approach is that the resulting system of work often focuses on the physical and even when psychological aspects are considered they tend to be at an individual level rather than an interpersonal level.

There are many aspects of an Ergonomist's job where the interpersonal could be considered. Throughout this paper the methodology behind identifying and managing workplace bullying will be explained.

There have been many attempts at defining 'bullying.' The behaviours that categorise the term are open to individual interpretation and therefore it is difficult to devise an all-encompassing definition. By examining a number of definitions it is possible to identify the elements that are shared. Bullying is usually defined in terms of the effect it has on the recipient rather than the intention of the bully. This is usually compounded by a description of the negative effects on the recipient and the fact that the behaviour is persistent. Bullying behaviour is not characterised by one-off incidents it tends to continue over a period of time, sometimes years.

One of the factors on which definitions focus is that it is a 'systematic abuse of power' (Smith and Sharp, 1994). When we think of a bullying scenario we usually imagine a subordinate being bullied by a superior. In the majority of incidents this is the case but it must not be ignored that it can happen the other way round where subordinates bully their managers!

When dealing with interpersonal or personnel issues bullying and harassment tend to be grouped together and dealt with in the same way. However, the two behaviours are inherently different in their nature. Bullying is not age, gender, race or physical ability related it is usually based on competence. It is psychological as it is rarely obvious and tends to happen behind closed doors. As a result it can take the recipient weeks to realise that they have become a victim. What the victim fails to realise is that they are actually a threat to the bully and the behaviour that they are subjected to is the bully's only way of making themselves feel better about their position, either within society, the family or the organisation.

There has also been some debate about the importance of defining the difference between bullying and strong management. Legitimate and constructive criticism and the odd raised voice can be seen to be acceptable but it is not acceptable to condone bullying under the guise of strong management.

Legislation

The difficulty with recognising bullying as a manageable problem is that there is no specific legislation under which to prosecute. However, there are a number of Acts and Regulations under which bullying may be covered.

The Health and Safety at Work Act 1974 made it a legal responsibility for employers to ensure that the health, safety and welfare of employees at work is protected. More specifically, section two requires employers to provide a safe place and safe system of work to ensure the physical and psychological well being of employees. Additionally, section seven requires employees themselves to take the responsibility of ensuring that they take reasonable care of their own health and safety and that of others.

More recently, the Management of Health and Safety at Work Regulations 1992 (revised 1999) places a legal duty on employers to carry out an assessment of the risks workers face whilst at work (this would include bullying) and to take any necessary preventative measures. There is also the requirement to provide training, information and supervision.

There is also non-discriminatory legislation. However, these Acts are aimed at harassment rather than bullying. The main difference is that harassment can be a "one-off" and is based on discrimination or abuse. The three Acts that are most important in this context are:
- The Sex Discrimination Act 1975
- The Race Relations Act 1976
- The Disability Discrimination Act 1995

It is sometimes possible to use these Acts to support a bullying claim but it is far more likely that the case will be one of harassment.

The introduction of the Human Rights Act, which states that 'no one shall be subjected to torture or to inhuman or degrading treatment or punishment', may have significance in bullying cases. Although, this would not strictly be used in an employment case its introduction highlights the increasing importance placed on such interpersonal behaviours.

Scale of the Problem

A study by UMIST (Cooper and Hoel, 2000) found that 1 in 4 people have been bullied at work and 1 in 10 have been bullied in the last six months. The TUC launched a hotline for people to call to report "bad bosses" and problems at work (including pay and conditions). The TUC were surprised to find that the number one complaint was not pay but bullying.

The costs to the organisation from incidents of bullying are huge and must not be underestimated. The main effects are an increase in absenteeism and staff turnover, demoralisation and lack of motivation and a decline in productivity and profit.

It is difficult for organisations of any size to identify the specific cause of absenteeism, (especially with the surge in musculo-skeletal problems). However, it is even harder for them to identify the cause as interpersonal problems such as bullying, violence and aggression and as a result stress related absence. It is important to consider the relationship between bullying and work-related illness. In one case, an employee successfully sued a company for "RSI" and was awarded a six-figure sum. Part of her evidence included a statement that "the company not only ruined my health but tried to intimidate me out of bringing a claim". The judge noted that the employee had been required to come in early, work during her break-times and go home late. Interestingly, after the case was lost, the manager responsible was sacked for bullying!

The negative effects of bullying on the recipient obviously play a major part in the costs to the organisation. They usually begin with a loss of confidence both in themselves and their work, which can lead to demotivation and as a result poor work quality and reduced output. Victims of bullying often resign from work with stress related ill health soon after their plight has been unearthed. Cases are often left unidentified until the recipient reaches breaking point.

In recent years there has been an increase in personal injury claims where the injury is psychological in nature. The increase in the "blame" culture and our litigious society indicates that further increases in such claims can be expected.

A personal injury claim based negligence will succeed if an individual can prove they have suffered loss, and that they were owed a duty of care, and there was a breach of the duty of care and that the injury was a result of the breach.

The awards made in recent stress claims could be seen to set a precedent for the penalty paid if interpersonal relationships are not managed within organisations. In

March 2000, the Ministry of Defence was made to pay out £745,000 to a soldier who was pushed out of a window in an extreme bullying incident 11 years before. A teacher who was humiliated, excluded, embarrassed and prevented from doing his job by the Head Teacher was awarded £101,028 in an out of court settlement after retiring in medical grounds.

Negating Ergonomics Interventions

As Ergonomists we therefore need to consider the impact of ergonomic interventions on interpersonal relationships. If we are involved in management structures and systems of work it is imperative that we consider the possibility of the abuse of power and the opportunity for bullying. Management systems are hierarchical in nature. Individuals report upwards. Much (but not all) bullying occurs between a worker and their immediate line manager or supervisor. When we design systems to make the workplace safer, more productive and more comfortable, we often assume that this relationship is a "productive" one i.e. that both parties wish to achieve a safe, productive and comfortable workplace. However, if a manager or supervisor believes that this can only be achieved by bullying the worker, either implicitly or explicitly, then our interventions fail.

We must include fail-safe mechanisms, for example bullying policies, reporting procedures and alternative "routes" for those who feel bullied to report problems and have those problems dealt with in a constructive and sympathetic manner.

Methods of Identification

There are a number of methods that can be utilised in order to gain information about the situation within an organisation. There is not the scope within this paper to go into great detail but a brief description will be included.

Employee surveys are probably one of the most common methods of attempting to reap qualitative data but sometimes their validity is dubious due to the sensitivity of the subject involved. Although, often confidential there is great stigma associated with either admitting that you have a problem or reporting that you are aware that someone else has a problem and therefore the responses may not be a true reflection of the situation.

Exit interviews and return to work interviews may also be used in order to gain qualitative data. If managed correctly they can be used to identify problems that may be causing the employee to leave.

Another popular method is to provide a hotline where employees can contact a third party in confidence and share their experiences. However, the calibre of callers must be considered as only those who feel confident will call. Nevertheless, details of incidents, however subjective can provide a valuable insight into the causes and incidence of the problem.

A lot can be said for 'gut feel' and the 'grapevine,' things that many of us rely on most of the time. If an employer has noticed that there has been a decrease in productivity or morale in certain individuals or groups they may decide to look more closely at the interpersonal relationships between those concerned.

A Risk Management Framework

Once we understand the extent of the problem, systems need to be designed that focus on prevention. Controls need to consider the potential for bullying and ensure that organisations go "as far as is reasonably practicable" to reduce the risks associated with bullying. In short, a proactive risk assessment must be undertaken.

We understand the nature of the hazard and the extent to which it can cause harm. We need to be clear about how likely it is that the harm will be realised and then to apply simple controls to manage and reduce the problem.

In practice organisations must conduct an "a priori" study of bullying, the extent and nature of the problem, design controls and then regularly conduct post hoc assessments (and amend controls as required).

In the event of a civil claim, claimants' legal experts are well aware of the need for an organisation to demonstrate that it is managing risks appropriately. Organisations will be expected to:

- Demonstrate how they have eliminated bullying throughout their organisation. For example, how have bullying cases been dealt with, what has happened to both "victims" and "offenders".
- Show how the effects of bullying have been reduced. For example by the introduction of early reporting procedures (e.g. through the use hotlines, personnel councillors, etc.). Early reporting is an important way of reducing the effects of bullying.
- Design systems of work to take account explicitly of the potential for bullying – this will include the production of a bullying policy, the procedures to manage bullying incidents and the provision of appropriate training and information.
- Provide appropriate training and information to all levels of staff. Note that this occurs at the bottom of the hierarchy and is not to be seen in isolation from the controls detailed above.
- Demonstrate a management strategy for dealing with people after they have been bullied (e.g. counselling, re-introduction strategy, etc.).
- Show how the above is audited regularly and how it changes.

If an organisation is not able to demonstrate this proactive approach and a problem goes unrecognised until an individual becomes severely harmed, then the scale of the compensation may reflect the courts view of how far the organisation has failed.

Policy and Procedure

A survey by the Andrea Adams Trust and Personnel Today in 1999 found that whilst 93% of personnel managers acknowledged that bullying occurred within their organisations, only 43% had policies to deal with it. The basis of a proactive approach is the development of a policy and procedures to ensure that all employees are aware of the organisations stance with regard to bullying and the procedures that will be followed if such behaviour is to occur.

The policy should broadly define what is regarded as bullying, although this definition should not be too prescriptive. It should outline what it is intended to achieve and the underlying philosophy that bullying behaviour will not be accepted and the commitment of the organisation to enforce that philosophy. The policy must also assign responsibilities, which should include the requirements of individuals and the organisation's responsibilities.

The policy may refer out to a procedure to be followed in the event of bullying behaviour or this could be included within. It should give recipients the confidence to report incidents and therefore should include the arrangements for the support of 'victims.'

Lastly, no policy should be without details of performance measures and arrangements for the review and monitoring of the policy in order to create an audit trail.

Inevitably, once bullying behaviour has been identified action must be taken. Firstly, it is important to identify the difference between deliberate and unconscious bullying. Due to the variation in individual interpretation and sensitivity, what the recipient may consider to be bullying the bully may see as harmless fun. In most cases, however, the bullying behaviour is rarely misinterpreted and an official route must be taken. Once reported to a third party the third party may confront the 'accused' and inform them of the complaint, this may occur without mentioning names.

It is then imperative that an investigation takes place. The alleged bully should be confronted with the facts and their response should be recorded, regardless of the disciplinary action taken some form of monitoring must also be agreed as the 'complainant' may be at greater risk once the investigation has been completed.

Conclusion

Regardless of the absence of specific legislation and the fact that it is still a 'taboo' subject bullying in the workplace is evidently a big problem that we should ensure is taken into consideration.

Systems need to be in place to ensure that

- policy and procedures are developed, measured and monitored
- all staff are trained and aware of the content and implications of the policy and procedures, and,
- if an incident does occur the correct level of support is available

As Ergonomists we need to ensure that each time we are involved in a project where there is an organisational problem we consider the interpersonal effects and the relationships within the working environment.

REFERENCES

The Andrea Adams Trust, www.andreaadamstrust.org
Cooper, C. and Hoel, H. (2000) Destructive Inter-Personal Conflict at work.
Smith and Sharp, (1994) School bullying: Insight and Perspectives. London: Routledge.

PROMOTION OF HEALTH AND WORK ABILITY
THROUGH ERGONOMICS AMONG AGING WORKERS

Clas-Håkan Nygård[1], Marita Pitkänen[1], Heikki Arola[2]

[1]Tampere School of Public Health, FIN-33014 University of Tampere, Finland
clas-hakan.nygard@uta.fi
[2]Tampere Occupational Health Center, Tampere, Finland

The objective of the present series of studies was to evaluate the effects of different programs and organizational changes ("interventions") in a food factory with special emphasis on the promotion of work ability and wellbeing of aging workers. Five programs carried out during the 1990´s were evaluated by controlled studies, before and after measurements as well as quasi-experimental follow-up studies. The age of the blue-collar workers (n= 21-512) ranged between 18 and 64 years and two third of them were women. The main outcome variables comprised perceived stress and strain, health, work ability, attitudes and learning. The interventions included follow-ups of changes in the production line, organization and training in manual material handling, training in teamwork and moving into a new factory. It was concluded that work ability and wellbeing of aging workers could be promoted with ergonomically measures at the work place. Education and training has a positive influence on work ability especially among the aging workers

Introduction

Studying the ability to work among aging workers is today of current interest because the working age population is getting older and reach its top in the next coming years in most of the industrialized countries. In the year 2000 about 40 % of the Finnish labor force was at least 45 years of age. Most of the Finns approaching retirement age, suffer from numerous diseases and disabilities (Ilmarinen, 1999). In the years 2005 to 2015 the EU countries will have the oldest work force in their histories, the mean age of their work force falling between 45 and 54 years of age (Ilmarinen, 1999).

The concept of "Promotion of work ability" has been widely used in the working life in Finland today, especially when dealing with aims to promote the work ability of older workers (Ilmarinen and Rantanen, 1999). The promotion concept is based on actions in the physical and psychosocial work environment, as well as promotion of health and lifestyle and updating of professional skills of the employee. The concept of physical, cognitive and organizational ergonomics (IEA, 2000) has been important contributors to the use of the concept.

The objective of the present series of studies was to evaluate the effects of different programs and organizational changes in a food factory on the workers, with special emphasis on the analysis of the effects on the aging worker.

Material and methods

The studied food factory employs about 830 employees today. It is situated in a small community (2200 inhabitants) in the Finnish countryside. The turnover was about 110 Million Euro in the year 2001. The main products include different sorts of convenience food (chicken, pizzas, etc).

Educational training for changes in the production (Study 1)
In 1993-1994 there was a big change in the factory from old styrox frames to new plastic transport boxes (Arola *et al* 1996). Because of this the packing and dispatch departments were rebuilt. A large apprenticeship program was introduced. The training program included both theoretical as well as practical

sessions. The target group consisted of 110 workers whose work changed. Out of these 76 workers (69%) answered both questionnaires before and after the change. A control group was formed out of those workers in the same factory whose work did not change. Two questionnaires were answered by 81 (56%) of the controls. The age range between 20 and 63 years and the mean age of the subjects was 38.8 years and of the controls 40.2 years.

Training for manual material handling (Study 2)
A study of the effect of a training program in manual material handling on work postures and perceived strain was carried out (Nygård *et al*, 1998). In all 21 female workers aged between 19 and 54 years with a mean age of 38.3 years participated in the study. Working postures and goods handled were registered, perceived rate of exertion assessed and subjective strain in the hands, neck, lower back and legs was ascertained at the end of the work shift. The subjects were trained in lifting by a physiotherapist.

Training for changes in working organization (Study 3)
In the factory a transformation process was started in order to enhance the productivity of the work by increasing personal initiative, communication skills and work motivation and also understanding of business economy (Nygård *et al*, 1999, Manka, 1999). The program was targeted specially to the broiler slaughterhouse and the broiler-packaging department because of their high sickness rates. In the present study a subsample of 114 persons were analyzed. The subjects were divided into an intervention group that participated in the training program (n=63) and a control group (n=51) that did not get any training. The mean ages in the intervention group and in the control group were 37 and 43 years, respectively (age-range; 19-62 years). Data was gathered by means of questionnaires that were delivered in the working place before the training program and after 2 years of training.

Participate planning of work place health promotion (Study 4)
In 1997 a health promotion group was set up to plan and organize the actions needed in workplace health promotion. The group included the managing director of the factory, staff managers, the occupational safety representative, representatives of occupational health services (physician, nurse and physiotherapist) and representative of the employees (chief shop steward). The members of the workplace health promotion group were interviewed in 1999. The individual theme interviews lasted about 45-60 minutes. The interviews were transcribed and the text was analyzed using qualitative content analysis (Pitkänen *et al*, 2001). With the aim to follow the changes in working atmosphere during the program, a questionnaire was carried out in the factory in the year 1999 (n=512) and 2001 (n=451) (Pitkänen *et al* , 2001).

Changes in work environment (Study 5)
In 1998 a new slaughterhouse was opened in the food factory (Nygård *et al*, 2000). The factory was designed with an emphasis on modern ergonomics. The change was evaluated by a questionnaire, which was answered by 85 workers in the old and 97 workers in the new slaughterhouse, 50 were the same in both years. The mean age of the workers was 32 (18-55) years in 1997 and 33 (19-57) years in 1999. The questionnaire included questions about perceived work stress, physical capacity, work ability, health and perceived physical exertion.

Results

Study 1
Training in connection with a change from old styrox-frames to plastic boxes improved the work community atmosphere, especially the security of the future. The perceived exertion increased significantly among older workers and perceived strain among younger workers. Perceived work ability improved both among young and old workers. No significant changes were observed in sick leaves.

Study 2

Training in manual materials handling improved the use of legs although no statistically significant changes occurred in back postures and perceived strain. The individual changes in workload after the training were independent of age.

Study 3

The biggest barriers preventing positive changes were found to be the lack of supporting systems, monotonous work, limited possibilities to influence the work and the lack of feedback. The workers had a high sense of professional competence and growth motivation among all ages. The perceived team spirit increased statistically significant compared to the controls after the training. The changes were mostly the same in all ages but in team spirit the increase was greater among the older workers. The perceived work ability did not change during the training program.

Study 4

Most members of the health promotion groups assessed that the group has good possibilities to influence on the work ability activities in the factory. The representative of all central interest groups of the factory was seen as the most important precondition for the effectiveness of the group. The know-how of the health issues and of the production processes as well as the power over the economical investments should be represented in the group. According to the interviewees the developing process of the group does not proceed without the commitment of the management of the company. The follow-up of the decisions of the group in the floor of the factory and the evaluation of the effects of the actions are also seen as necessary parts of the health promotion process. According to the results of the questionnaires, the employees' conceptions of team spirit and their possibilities to influence their work increased during the program.

Study 5

While changing from the old to new slaughterhouse the perceived mental and physical strain remained almost the same. The perception of physical capacity, work ability and health among the workers had decreased statistically significantly when moving from the old factory to the new. There were in average more changes among the workers above 40 years of age than those under.

Discussion and Conclusions

Health and work ability of aging workers can be promoted with ergonomically measures at the work place. A participate approach which includes all central groups involved in the health promotion work is essential in order to achieve good results. It may also influence central factors in the working atmosphere (e.g. team spirit, possibilities to influence ones work). Education and training has a positive influence on work ability especially among the aging workers. Changes in the production or organization should be proceeded by information and training. It is also important to follow the effects of changes in the factory on the health and wellbeing of the workers, because the aim of the changes in the organization is primarily linked to the productivity of the company.

References

Arola H, Nygård C-H, Manka M-L, Huhtala. 1996, H. Effects and feasibility of training in adaptation to production changes on health and work ability in a food factory. *Arbete och hälsa,* 16,82-88.

IEA. 2000, International Ergonomics Association, Basic documents, IEA Press

Ilmarinen J. 1999, *Ageing workers in the European Union-Status and promotion of work ability, employability and employment.* Finnish Institute of Occupational Health, Ministry of Social Affairs and Health, Ministry of Labor, Helsinki.

Ilmarinen J and Rantanen J. 1999, Promotion of work ability during ageing. *American Journal of Industrial Medicine*, 1,21-23.

Manka M-L. 1999, Topteam: towards a productive, learning and positive work community and personal well-being. *Acta Universitatis Tamperensis:668:1-316 (in Finnish, English abstract).*

Nygård C-H, Merisalo T, Arola H, Manka M-L, Huhtala H. 1998, *International Journal of Industrial Ergonomics 21; 91-98.*

Nygård C-H, Pitkänen M, Manka M-L, Arola H. 2000, Work strain, work ability and health when changing from an old to a new slaughter. *Proceedings of the IEA 2000 Congress in San Diego, USA,* 29.7-4.8.2000, 4, 13-14.

Pitkänen M, Arola H, Nygård C-H, Manka M-L. 2001, Participative approach in promoting work ability in a food factory. *Proceedings of NES 2001. Promotion of Health through Ergonomic Working and Living Conditions*; Publications 7, University of Tampere, School of Public Health, Tampere, 112-114

Pitkänen M, Nygård C-H, Arola H, Manka M-L. 2001, Work place health promotion and working climate in a food factory during two years. *Proceedings of NES 2001. Promotion of Health through Ergonomic Working and Living Conditions*; Publications 7, University of Tampere, School of Public Health, Tampere, 430-432.

EFFECTS OF MEDICATION ON THE WORKING POPULATION

S Hastings[1], S Brown[1], Cheryl Haslam[2], and R A Haslam[1]

[1]Health and Safety Ergonomics Unit, Department of Human Sciences, Loughborough University, Loughborough, Leicestershire, LE11 3TU

[2]Department of Health Studies, Brunel University, Osterley Campus, Isleworth, Middlesex, TW7 5DU

It is not clear to what extent medication taken for psychiatric illness affects a person's quality of work and productivity. Perhaps more importantly very little is known about the possible consequences for workplace safety. Twelve focus groups were conducted in this study to collect data on the personal experiences of mental health problems and the impact of psychotropic drugs on work performance from members of the working population. Information was also gathered from representatives of human resources, occupational health, personnel and health and safety from a range of work sectors, on the organisational perspectives of mental health in the workplace. The information gathered will assist in the development of guidelines for managing mental health issues in the workplace.

Introduction

The extent to which medication is used to treat minor psychiatric illnesses in the working population is unknown, although levels are thought to be considerable (Dunne, et al 1986; Potter, 1990). However, it is known that certain conditions have a substantial gender difference with, for example, women seeking treatment for depression at twice the rate of men. There is also good evidence that the incidence of psychiatric illnesses is increasing (Potter, 1990). A number of studies have highlighted the poor knowledge that exists concerning the effects of prescribed medication on work performance (Dunne et al, 1986; Potter, 1990; Tilson, 1990). Psychotropic medicines impair performance on a wide range of laboratory measures, with effects found for attention, vigilance, memory, problem solving, learning, motor coordination, gait and visual accommodation, for example (Dunne et al, 1986; Potter, 1990). However, it is not clear how these findings translate to performance in the complex setting of the workplace (Dunne et al, 1986; Nicholson, 1990). Problems with generalising from the findings of laboratory investigations to the occupational context include: laboratory investigations of performance effects are often limited to testing with young, healthy subjects; minor decrements in performance on sensitive laboratory tasks may have little relevance to real world activities (Cohen et al, 1984; Nicholson, 1990) and laboratory studies usually do not simulate the effects that workplace environments may have on pharmacokinetic pathways or physiological homeostasis (DeHart, 1990). With regard to this last point, work can affect both an illness and the action of medication used to treat it. Stress may worsen psychiatric illnesses, or shift work can precipitate mania, for example. Heavy work in the heat may lead to dehydration, affecting blood concentrations, perhaps leading to symptoms from toxicity (Potter, 1990).

It is not clear, therefore, to what extent medication taken for psychiatric illness may affect a person's quality of work and productivity. Perhaps more importantly, very little is known about possible consequences for workplace safety. A study by Mintz et al (1992) evaluated the effects of antidepressants and psychotherapy on work impairment in depressed individuals from ten published treatment studies. They found that improvement in work performance lagged behind improvement in psychological symptoms. This has important implications in terms of the advice offered by health care professionals and the expectations of patients and employers.

A further issue is that it has been suggested that lack of treatment for psychiatric illnesses may actually be a greater problem in terms of work performance than the side effects of medication (Potter, 1990). Employees suffering with depression or anxiety are likely to experience a range of symptoms that would impair performance at work, including: tiredness, lack of motivation, poor concentration and forgetfulness, poor timekeeping and attendance. Failure to seek or comply with treatment may arise from the stigma that can be associated with mental ill health or, alternatively, some patients may be reluctant to take psychotropic medication because of fears they have about possible side effects and dependency.

This research collected new and in-depth data on anxiety and depression and the use of psychotropic medication among the working population. The aim was to improve understanding of the impact of mental health problems and the treatment for these conditions on performance and safety in the workplace.

Method

The research used focus groups, for a technique well suited to elicit information about the personal experience of mental health problems and the impact of drugs on work performance. A focus group is a group interview whereby the data obtained arise from the discourse generated by a discussion (Morgan, 1997). Topics are supplied by the researcher who acts as 'moderator' for the discussion; the moderator should facilitate discussion rather than interview. Focus groups also afford participants a degree of anonymity.

A total of 12 focus groups were conducted across a broad spectrum of employment sectors. Of the twelve groups, three groups were run with participants from anxiety and depression management courses conducted by clinical psychology and counselling services. Participants had experienced anxiety and depression to the extent that they had sought professional help and additionally the groups drew participants from a wide range of occupations. A further six groups were conducted with employees from a wide range of occupational sectors in either managerial or non-managerial roles. The study strategically targeted sectors known to be high-risk for mental health problems including health care, education and manufacturing (Health and Safety Executive, 1999). These groups explored the personal experience of psychiatric morbidity; psychotropic drugs used to treat these disorders; the impact of mental health problems and psychotropic drugs on work performance and safety; relationships with colleagues and support offered by colleagues and the organisation within which they work. Organisational perspectives including: recruitment, support and rehabilitation were explored with three focus groups. Staff from human resources, personnel, occupational health and health and safety departments participated, representing a variety of organisations.

Results

Due to limited numbers of volunteers by geographical location or occupation, criteria were used to select participants who had similar levels of managerial or supervisory responsibilities in their work. A summary of the work sectors represented within the groups is shown in table 1.

Table 1. Employment sectors represented in focus groups

	Organisational groups	Clinical groups	Occupation groups
Public sector	Defence Local authority Local government National Health Service	Further education - tutor Local authority - administration NHS – Scientific Officer	Higher Education support work Local government Police support work Teachers (non-managerial)
Private sector	Engineering Heavy industry Retail Pharmaceutical	Electrician Garage foreman Office worker - SME Photographic assistant Retail – large company Telesales Vet	
		Unemployed	Voluntary sector

Symptoms and side effects
People are often unaware that they are exhibiting symptoms of anxiety and depression, it is often family, friends and colleagues who recognise the signs before the affected person.

During the build up to crisis some people began to recognise that their standard of work was deteriorating, often due to difficulties fulfilling their workload and criticism from managers. Individuals experienced confusion, dizziness and difficulties with decision-making. One of the symptoms identified by the participants was that they lacked motivation, finding everything too much for them. At work, routine tasks became difficult and would be avoided, such as making telephone calls. People felt pressured at work by the amount they had to do; often they would throw themselves into their work to meet demands and expectations becoming isolated and interacting less with colleagues.

Feelings of extreme tiredness made people feel unable to 'get a grip on things'. Sleep patterns were often erratic and disturbed, resulting in very little sleep or sleeping constantly. This created a cycle whereby people were tired at work and found the pressures intolerable. Some people were unable to concentrate, they forgot things and found it difficult to absorb information, often not being able to complete work. Individuals often blamed themselves for being inefficient rather than recognising their symptoms as anxiety and depression.

Medication

Participants were often diagnosed and prescribed medication only when they became unable to continue at work. Non-compliance with drug regimes was common, people took fewer and smaller doses than prescribed and discontinued the medication of their own volition. Addiction and dependency were major concerns of people taking medication; they would take reduced or less frequent doses because of this.

Discontinuation occurred when people felt worse taking the drugs than without them. There was also a tendency for people to stop their medication too early, when their symptoms began to improve. Some participants felt that they would be unable to return to work while they were taking medication; they felt that a prolonged period of treatment would necessitate a long absence from work and tried to manage without the medication. Side effects were often severe in the first few weeks during which time the medication was changed or discontinued by the General Practitioner (GP). Some participants felt that prescribing medication was a process of trial and error to find the right drugs for each person.

Effects on safety

Participants were concerned about safety issues regarding their responsibilities to others, e.g. teachers, GPs. Side effects and symptoms they experienced at work included confusion, dizziness and lack of concentration, which they felt made them more liable to accidents. Health professionals, including GPs and hospital doctors acknowledged that they often put themselves and their patients at risk when handling hazardous materials, i.e. blood. They also mentioned concern about administering drugs and using needles and in some cases had injured themselves.

Some people felt they were at risk when driving to and from work, because of an inability to concentrate and as a result of tiredness.

Relationships with colleagues and managers

Management methods were identified as contributory factors in work related anxiety and depression. It was felt that some managers had not been trained in management skills when work responsibilities had changed. Employees who struggled with their workload were often given time management training as the problem was seen to be with the individual rather than the organisation. Managers questioned the competence of long standing employees who started to make uncharacteristic errors without exploring the possible causes of the change in their standard of work. It was thought that training should be available to improve the general understanding of mental health issues.

Support in the workplace

When asked what kind of support they wanted at work, participants highlighted the need for practical help with the volume of work at difficult times. Many people suggested that simple but time-consuming tasks caused them unnecessary stress. It was also suggested that in-house support services such as occupational health or a counselling service should be available, or simply the opportunity for a confidential talk. Participants felt that if anxiety and depression were better understood, managers and colleagues would be more able to give appropriate support.

Organisational issues

Most organisations did not deal with mental health issues any differently to physical health issues and rehabilitation periods were often limited to one month. This could be extended when occupational health staff supported the employee and manger during the rehabilitation period. However occupational health services were not available in all organisations, and those who did have this service were not always willing to implement the recommendations made for work adaptations and rehabilitation. Support and rehabilitation of employees experiencing anxiety and depression included: reduced workload, reduced hours or a change of job on a temporary or permanent basis. Work was often target driven and some tasks were not flexible enough or too specialised to allow modification of working hours or output to enable employees to remain at work.

Feedback from organisational representatives suggested that high levels of pressure were simply a part of present day lifestyles. Managers were thought to be under pressure themselves making them less tolerant of staff with problems they don't fully understand such as anxiety and depression. Part of the role of human resources was seen to be enabling a manager to deal appropriately with those problems.

Discussion and Conclusions

Deteriorating work performance was apparent to employees and their managers however this was attributed to things such as pressure of work and poor time management. Sufferers were reluctant to seek medical advice because they were unable to identify the reason for the symptoms or due to the stigma associated with mental ill health.

Frequently medical help was only sought when sufferers reached a point of crisis and were unable to continue to work. It would seem that employees and managers were aware that an individual was having problems with their work and becoming withdrawn but had insufficient knowledge to identify these changes as symptoms of anxiety and depression. The effects of anxiety and depression sometimes made it difficult for people to act on the information they were given by their doctor, however participants felt that they were not well informed about the side effects of drug treatment or how long they should expect to continue medication, consequently non-compliance was common. Side effects from the medication and concerns about dependency were major issues, discontinuation was also the result of individuals 'feeling better' or wanting to get back to work. Many participants did not begin to take medication until they were absent from work due to anxiety and depression and they found it difficult to differentiate between the effect of the condition and the effect of the medication on their work performance, although it was thought to be due to both. Participants accepted that the effects of anxiety and depression and their medication would affect safety in the workplace both for employees and customers of the organisation. Participants were reluctant to tell their employer that they had anxiety and depression because of stigma and the negative consequences it might have on their career.

A confidential reporting system was favoured especially when management methods were seen to be contributory to the problem. Support at work was felt to be most beneficial when colleagues were able to help with the workload or offer the opportunity for a confidential talk. Human resources and counselling services felt themselves to be in a difficult position if they identified potential problems in the organisation such as bullying or harassment but were bound by confidentiality not to notify line managers.

This research has provided valuable insight into effects of medication on the working population.

Acknowledgement

We wish to acknowledge the support of the HSE who funded this research. The views expressed, however, are those of the authors and do not necessarily represent those of the HSE. We would also like to thank our focus group participants for their support and the time they gave so generously to this study.

References

Cohen A F, Posner J, Ashby L, Smith R and Peck A W, 1984. A comparison of methods for assessing the sedative effects of diphenhydramine on skills related to car driving. *European Journal of Clinical Pharmacology*, 27, 477-482.

DeHart R L, 1990. Medication and the work environment. *Journal of Occupational Medicine*, 32, 310-319.

Dunne M, Hartley L and Fahey M, 1986. Stress, anti-anxiety drugs and work performance. In: *Trends in Ergonomics of Work*, 23rd annual conference of the Ergonomics Society of Australia and New Zealand, pp 170-177.

Health and Safety Executive, 1999. *Health and Safety Statistics 1998/99*. HSE, 1999, ISBN 0 7176 1716 5.

Mintz J, Mintz L I, Arruda M J and Hwang S S, 1992. Treatments of depression and the functional-capacity to work. *Archives of General Psychiatry*, 49: 761-768.

Morgan D L, 1997. *Focus groups as qualitative research*. 2nd Edition, Sage Publications, California.

Nicholson A N, 1990. Medication and skilled work. In: *Human factors and Hazardous Situations*, Proceedings of the Royal Society Discussion Meeting, 28 & 29 June. (Edited by Broadbent, D E, Reason J and Baddeley A), (Clarendon press: Oxford), pp 65-70.

Potter W Z, 1990. Psychotropic medications and work performance. *Journal of Occupational Medicine*, 32, 355-361.

Tilson H H, 1990. Medication monitoring in the workplace: toward improving our system of epidemiologic intelligence. *Journal of Occupational Medicine*, 32, 313-319.

TEAM WORKING

DO PERSONALITY FACTORS IMPACT ON HOW MULTI-DISCIPLINARY TEAMS PERFORM?

R. Hugh Thomas

Human Factors Department, BAE SYSTEMS,
Advanced Technology Centre (Sowerby Building),
FPC 267, P. O. Box 5 Filton,
Bristol BS34 7QW.

Engineering design has become more and more of a team-based activity as technological systems have grown ever more complex. "Soft" human issues can therefore make or break a project, especially in collaborative ventures. Our study uses a popular psychometric test to investigate individual differences between departments and functions within an industrial technology facility. Differences are found between staff in the industrial technology facility and the general population, between scientists from different disciplines in the facility, and between functions within the facility (scientists versus administrators). The latter finding provides potential new insight into a previously identified problem with interactions between these two functions. Although the test used provides insight at an individual rather than organisational level, possible applications in several areas of human factors are suggested.

Introduction

With the ever-increasing complexity of modern technological systems, engineering design activities have become more and more of a team-based process. Design is therefore now often a social phenomenon, and the success or failure of a project can be brought about by the impact of so-called "soft" human issues.

The quality of communication and co-operation between co-workers on a project is clearly of crucial importance, and the impact of such factors can become heightened in the context of collaborative ventures, where project participants come from different technical, company-cultural and increasingly even national-cultural backgrounds.

Research interest in these issues has increased greatly of late. A variety of approaches has been adopted. The one taken here is to address the issues from the perspective of individual differences. This focuses on differences between people, especially styles of working, using psychometric questionnaires to measure personality variables and/or sort people into types.

One of the most widely used of such tests is that of Keirsey & Bates (1984). It offers several advantages over other candidates, including that it is non-proprietary, does not require special training to apply or interpret, and is designed to be quick and easy to run and score. The test itself and information about it are available in the open literature and on-line.

We decided to evaluate the test by applying it in an industrial technology facility, to investigate whether there is a different distribution of individual differences between the different departments and different functions, which might impinge on their potential effectiveness in the context of collaborative work.

Keirsey & Bates' Approach

Keirsey & Bates draw upon the 1950s work of Myers & Briggs, whose Myers Briggs Type Indicator distinguishes 16 different "patterns of action" (colloquially, "personality types") and allows them to be identified by a questionnaire-based tool. Keirsey & Bates' method uses just such a questionnaire, comprising 70 bipolar-choice questions. The 16 types are arrived at by the respondents revealing their "preferred attitudes" on four dimensions:

i) Extraversion (E) or Introversion (I): Extroverts are social, being energised by contact with other people, and are characterised by a preference for breadth. Introverts are territorial, requiring solitude to recover their energy, and prefer depth.

ii) Sensation (S) or Intuition (N): Sensors are practical and realistic and are grounded in reality, preferring experience, facts and "perspiration". Intuitives are innovative, preferring visions, intuitions, hunches and inspiration.

iii) Thinking (T) or Feeling (F): Thinkers prefer an impersonal basis for making choices, Feelers prefer to make choices on a personal basis. Feelers tend to be good at persuasion; Thinkers are good at argument. A gender difference is noted, with 60% of men being T and 60% of women preferring F.

iv) Judging (J) or Perceiving (P): This dimension should not be assumed to be "judgmental" versus "perceptive" in the conventional uses of those terms. Judgers are people who prefer closure, being for deadlines and conclusions. Perceivers prefer fluidity and open-endedness, with options being kept open.

The 16 types therefore are the possible combinations of these preferences: ESTJ, ISTJ, ESFJ, ISFJ, ESTP, ISTP, ESFP, ISFP, ENFJ, INFJ, ENFP, INFP, ENTJ, INTJ, ENTP and INTP. For Keirsey & Bates, the 16 types can be resolved into four temperaments:

a) Epimethean Temperament (or "Guardians"): Those of Epimethian temperament are the SJ (Sensing-Judging) types. They are typified by a longing for duty, existing primarily to be useful to the social units to which they belong. They are compelled to be bound and obligated, have a desire for hierarchy and an urge to conserve, and are typified by pessimism, preparedness, tradition, and opposition to change. They accept responsibility, are deferential to elders, and are prone to depression.

b) Dionysian Temperament ("Artisans"): People of Dionysian temperament are the SP (Sensing-Perceiving) types, and can be summed up as people who do what they want, when they want. They must be free, act impulsively on whims, enjoy crises and believe that today has to be enjoyed. Dionysians also have a hunger for action, are optimistic (tending to leap before looking) and are not goal-orientated.

c) Appollonian Temperament ("Idealists"): Appollonians are intuitive-feeling (NF) types, who pursue extraordinary goals, typically a circular and perpetual search for self and self-realisation. Integrity and meaning are all-important to them. They are interested in the social sciences and humanities, preferring to work with words, and have a yearning for intimacy.

d) Promethean Temperament ("Rationals"): The Prometheans are the NT types and are characterised by a desire for power over nature. Perpetually searching for competencies, they have a compulsion to improve. They regard abilities as ends in themselves, are very self-critical, are individualists, and sometimes seem arrogant. They love intelligence, live in their work, and are compelled to rearrange the environment. They are the "eccentric geniuses" of the world, not always sensitive to complexities of personal relationships.

Method

Sixty participants from the industrial research facility took part. They comprised 12 people from the administration (admin.) function (librarians, secretaries and clerical staff) and 12 people each from the following scientific departments: Human Factors (HF), Advanced Information Processing (AIP), Optics & Laser Technology (OLT) and Materials Sciences (MS). Sample size was approximately 50% of the total population of each group.

Participants were approached and invited to participate after a brief explanation of the nature and purpose of the study and an assurance that it was being performed on a confidential basis. Completed questionnaires were collected and analysed by the experimenter.

The study employed a simple design whereby the questionnaire was completed in an identical fashion by all of the participants. The questionnaire and marking scheme included in Keirsey & Bates (1984) were used. The results were then analysed by department and function to enable any differences between the groups of participants to be examined.

Results

Table 1 shows a summary of results obtained on the four basic personality dimensions. The second column of the table shows the distributions within the general population as given by Keirsey & Bates (1984).

Table 1. Results on the Four Underlying Dimensions (%ages to nearest whole %age)

The Four Dimensions	Percentage of general population	HF Dept.	AIP Dept.	OLT Dept.	MS Dept.	Overall Scientific Depts.	Admin. Function
Extraversion	75	63	42	50	54	52	46
Introversion	25	37	58	50	46	48	54
Sensation	75	33	33	29	58	38	50
Intuition	25	67	67	71	42	62	50
Thinking	50	58	88	50	75	68	25
Feeling	50	42	12	50	25	32	75
Judging	50	71	67	63	83	71	67
Perceiving	50	29	33	37	17	29	33

A number of trends are apparent in this table:

- All groups in the samples contain more Introverts than the general population.
- All of the scientific departments except AIP department contain more Extroverts than the admin. function, the latter having more than double the incidence of Introversion of the general population.
- The department containing most Extroverts is Human Factors, although still less than the proportion in the general population.
- All of the groups also exhibit a higher incidence of Intuitiveness than the general population. The admin. function shows double the incidence of Intuitiveness of the general population, and the scientific departments an even higher incidence overall.
- With regard to Thinking versus Feeling, this dimension shows a marked difference between functions. The admin. group contains three times as many Feelers as Thinkers, while overall the scientific departments contain more than twice as many Thinkers as Feelers.
- Finally, on the Judging versus Perceiving dimension, all groups contain a higher proportion of Judgers than the general population. All groups except OLT contain at least twice as many Judgers as Perceivers, MS dept. showing the greatest disparity.

The results as analysed into the four temperaments and 16 types are shown in Table 2.
A number of trends with regard to temperament are apparent in this table:

- People of the Dionysian (Artisan) temperament are almost absent from the samples.
- Over-represented temperaments in the sample are Epimethians (Guardians) in the admin. function, Appollonians (Idealists), who are more than double the incidence in the general population in both functions, and Prometheans (Rationals), especially in the scientific function.
- Comparing the two functions, people in the science function are about twice as likely to be Prometheans (Rationals) as are those in admin.

- There are also some indications of inter-departmental differences within science, over half of AIP being Rationals, over half of MS being Guardians, and Idealists being the most common temperament in OLT.

Table 2. Results as Temperaments and Types (%ages to nearest whole %age)

The Four Temperaments	The 16 Types	HF Dept.	AIP Dept.	OLT Dept.	MS Dept.	Overall Scientific Depts.	Admin. Function	Percentage of general population	
Epimethian (SJ Types) (Guardians)	ESTJ	4	4	17	46	18	4	13	
	ISTJ	21	21	8	13	16	4	6	38
	ESFJ	8		2		3	8	13	
	ISFJ			2		1	33	6	
Dionysian (SP Types) (Artisans)	ESTP		4			1		13	
	ISTP							6	38
	ESFP		4			1		14	
	ISFP							5	
Appollonian (NF Types) (Idealists)	ENFJ	4		13		4	8	5	
	INFJ	13	4	4	8	7		1	12
	ENFP	13	4	10	4	8	8	5	
	INFP	4		19	13	9	17	1	
Promethean (NT Types) (Rationals)	ENTJ	21	13	4	4	10	8	5	
	INTJ		25	13	13	13		1	12
	ENTP	13	13	4		7	8	5	
	INTP		8	4		3		1	

Discussion

In view of the limited time available for the study, the research that underlies the Keirsey & Bates approach was not investigated. However, the nature of some of the results obtained would tend to give support to the method's likely validity. For instance, the finding that the INTJ (referred to as the "Scientist" by Keirsey & Bates) is the statistically most over-represented type in our actual scientists, compared to the general population, is reassuring.

Several differences between the technology facility's staff and the general population are noted, especially an almost complete absence of people of Dionysian (Artisan) temperament.

Differences between the two functions are also apparent (although note that the sample size for scientists was four times that for admin.). The company of which the technology facility is part has already recognised that scientists are not always understood or appreciated by colleagues in other functions. It has instigated a project to analyse and remedy the problem.

The differences found between the two functions in the present study may provide a new insight as to a possible cause of this problem. A common source of misunderstanding between Thinkers (T) and Feelers (F), according to Keirsey & Bates, is that the F person tends to make his or her emotional reactions more visible, while the T individual may seem cold and unemotional. However, this may not be an insurmountable problem, as Keirsey & Bates add that Thinkers can react emotionally with just the same intensity as Feelers, and that T and F types tend to be able to complement each other more than, say, S and N people.

There are also indications of differences between departments within the science function. In our own field, it is interesting to note that Human Factors is the group with the highest level of Extraversion, albeit still less than the incidence in the general population.

In so far as the behaviour of a group can be accurately portrayed by the characteristics of the average or modal type or temperament of the people within it, then the Keirsey & Bates approach might provide insight into the likely effectiveness of that group in the context of collaboration. However, the potentially powerful impact of social or organisational factors, such as the way the culture of one company differs from another, not to mention well known international differences in culture, are not addressed by Keirsey & Bates.

Nevertheless, the study has offered some interesting insights into individual differences at work. If these differences are genuinely as well defined and predictable as is suggested, then there are potential implications for several areas of human factors, for example in:

- Workplace layout (perhaps Extroverts being energised by the social environment of an open plan area while Introverts prefer the solitude and quiet of their own office);

- Training (with different types and temperaments having different preferred learning styles);

- Adaptive Automation (intelligent systems allocating functions or information to the operator according to the latter's type or temperament).

References

Keirsey, D. & Bates, M. (1984) – *Please Understand Me: Character and Temperament Types*. Prometheus Nemesis, Fifth Edition.

AN INTEGRATED APPROACH TO SUPPORT TEAMWORKING

[1]Andrée Woodcock, [2]Christoph Meier, [3] Rolf Reinema, [4]Marjolein van Bodegom

[1]VIDe Research Centre, School of Art and Design, Coventry University, Gosford Street, Coventry, UK. A.Woodcock@coventry.ac.uk
[2] Fraunhofer Institute for Industrial Engineering (FhG-IAO), Nobelstrasse 12, D-70569 Stuttgart, Germany. christoph.meier@iao.fhg.de
[3] Fraunhofer Institute for Secure Telecooperation (Fhg-SIT),COR, Rheinstrasse 75 Darmstadt, D-64295, Germany. reinema@sit.fraunhofer.de
[4] Marjolein van Bodegom, Pentascope, Javastraat 1, Den Haag, Holland. vanbodegom@pentascope.nl

Developments in information technology, globalisation, organizational patterns, economic climate and movement away from manufacturing have contributed to new working practices, with a rise in task based temporary teams, distance working, and emphasis on knowledge management. Current systems do not necessarily support task based, distributed working. This paper reflects, from a user's standpoint, on the approach and the results of the first year of the UNITE project, which aims to produce an integrated system to address the challenges posed by mobile teamwork.

Introduction

Today's 'global village' is populated with international companies and employees, who work on projects in virtual teams. They may be distributed within one or many geographically dispersed organisations. Such teams consist of domain experts, who are relative strangers, coming together for a specific purpose. Traditional offices, and their associated infrastructure, do not allow smooth working on such projects. For example, systems may be incompatible, international phone calls may be difficult to co-ordinate, and any form of conferencing may require movement to different rooms, and the scheduling of shared resources, such as technical support staff. A new type of working environment needs to be created to allow geographically separated people to collaborate naturally and effectively (Reinema et al 1998 and 2000). This should reflect their shared context of work, that of a joint project. This should:

- allow geographically separated people to collaborate naturally and effectively.
- support the simultaneous working on and switching between multiple projects.
- appropriately reflect the shared work context – that of a joint project. When teams are co-located it is relatively simple to gain knowledge of one's colleagues, their working

habits and preferences, project status, current activities etc. When team members are separated such information has to be made explicit.

- provide an environment that seems familiar by exploiting people's skill in navigating and working in the real world, so enabling smooth transition between the real and virtual world
- reveal the resources available that team members are working from or have available.

Rather than develop more CSCW (computer supported co-operative working) applications (see Figure 1) to support geographically dispersed teams and their mobile members, the UNITE solution is to take existing products as a reservoir of building blocks and encapsulate them in a unified, integrated system, which meets these new user requirements. Such a system would allow people to use their preferred applications but embed them in an integrated system where applications seamlessly interweave with each other. This will remove the technological incompatibilities that blight communication especially in projects of short duration (where team members may spend longer trying to set up a workable communication infrastructure than they do in task related activities). With the mushrooming of applications and communication devices the only valid solution is dynamic integration of technology.

Figure 1 Functional Blocks/Modules of co-operative systems (Meier, 2001)

To this end, the UNITE development team (with members from Steria, IBM, the Fraunhofer Institutes SIT and IAO, ADETTI and Coventry University) are developing a web based, co-operative workplace that can enable people to switch from one project to another in a highly flexible working environment. This means that at any time, wherever they are working, be it on company premises, at home, or on the move, UNITE users will be able to focus on the context of one particular project and task, and be freed from physical locations. For companies, such a system will bring increases in flexibility, efficiency and effectiveness, e.g. the amount of time spent travelling and trying to make contact with colleagues will be reduced, the flow of information around the organisation will be increased. For individuals, problems of mobile working (Woodcock, 1999) such as isolation, lack of informal communication, removal from the locus of control, poor support by central management, inability to create affective ties will be reduced. The rest of the paper describes the user requirements for such a system.

User Requirements Capture

At the start of the project, four target customers for UNITE were identified: Start-Ups, especially in the field of new service companies (e.g. consultancies, multimedia agencies); Networks, including multiple different small and micro firms especially from the service sector; Medium sized enterprises with growing geographical distribution and increasing internationalisation requirements; and large companies, where functional units and teams are increasingly geographically distributed.

Our intention is to develop a collaboration platform (the UNITE platform) for these target customers. Given the duration of the project (2 years), we needed to arrive at a good understanding of the work processes of potential users. To do this we worked with a user company (Pentascope), considered to be typical of the target customers identified. Once the user requirements were gathered, we abstracted from these, using the collective knowledge of the consortium, to arrive at a collection of general user requirements, features and functions that would satisfy the needs of all identified target customers.

Pentascope is a knowledge based network organisation (KBNO) of highly mobile workers, who require flexibility and control over their work environments. Pentascope consultants usually work simultaneously on 3-4 internal and 2-3 external projects, of different duration. They occasionally need to share physical office resources and often need access to the expertise of others in the organisation when working with clients. It is the requirements of such organisations that form the basis of the cooperative working solution we are developing. They have the following characteristics:

- Their employees work at multiple places: at company or customer premises, at home, or on the move. This requires connection to tele-workers at different physical locations;
- The number of their employees in such organisations often larger than 250. If these employees are not co-located it is hard to establish a shared understanding of what is happening at individual, team, or organisational level;
- Their business is highly predicated by teamwork, i.e. project-driven, timely access to experts, close communications between team members, document sharing;
- The organisation is matrix-based, with tasks executed by dynamically-formed subgroups;
- They rely on responsibility and discipline at the level of their employees;
- They form multidisciplinary organisations, in which different domain experts have to interact and depend on each other. This frequently gives rise to communication breakdowns.

A three-step approach was taken to capturing the user requirements.

1. Familiarisation with users - Initial two day visit to trial users (Pentascope) during which interviews were conducted in order to familiarize the UNITE consortium with the users, their organization and ways of working.
2. Capturing user requirements - An intense two day workshop with members of Pentascope to gather in-depth information on work contexts and procedures. This included exercises such as drawing images relating to aspects of work (e.g., "What do you enjoy about working here?", "What is teamwork like?", "Yesterday what was your

work environment like?"). In asking the participants to first draw and then verbally comment on their work experience, experientially rich accounts were produced that went beyond practiced verbal descriptions. In later sessions, the focus narrowed down on project related work. Accounts of steps and activities over the life-cycle of a project were collected and subsequently worked through and detailed. Of especial importance were "doing work in the course of meetings" and "communication issues."

3. From requirements to specification - The information was reflected on by the project partners, in the light of their experience with supporting cooperation at a distance. The outcome of this was a detailed documentation of usage scenarios, user requirements and functionality iteratively co-produced by developers, users and evaluators.

User Requirements

The user requirements capture process (led by FhG-IAO) produced insights into the work processes, the particular culture of this company and led, finally, to an appreciation of particular problems members experience (e.g. with respect to internal communication). The outcome provided three sets of requirements for the UNITE platform.

General user requirements
These requirements refer to the global needs of users. They include: firstly, the integration of work contexts. Users can be in one of multiple work contexts: 1) an individual context, when they work in isolation; 2) a project context, when they work with the other team members of a given project. If they work on multiple projects, they are in one of many project contexts as well as an organisation context, when they behave as an employee of an organisation. Secondly, specific support of project style of work. Users need project and task management, document management, and communication with others in either the aforementioned work context, and thirdly, shielding from the heterogeneity of the underlying IT infrastructures and the devices used by the team members, e.g. fixed workstations and phones, notebook-computers, mobile phones.

Business processes requirements
The following are indicative of the requirements the UNITE platform system should support for work tasks to be completed. These include bulletin boards and discussion forums, personal address book and calendar, a repository for sharing documents and bookmarks, an electronic whiteboard for text, drawings, pictures, audio/video clips that can be used for socialising (e.g., as "fun corner"), a single point of entry with easy log-in for all uses of the platform (e.g., an individual portal leading to those project portals a particular member is involved in)

If we take the scenario of trying to arrange a meeting users need status displays indicating availability of other team members and physical resources (e.g., equipment and rooms) where the meeting could be held, a means of finding dates and times for joint meetings (by comparing calendar information of different team members), a unified messaging service (e.g., forwarding of e-mails to mobile and transformation into voice-message) to facilitate contact, easy publication of meeting related information on a website available to all members, a way of linking meeting-related materials to one "meeting object" and automated notification of changes for all participants.

DO PERSONALITY FACTORS IMPACT ON HOW MULTI-DISCIPLINARY TEAMS PERFORM?

R. Hugh Thomas

Human Factors Department, BAE SYSTEMS,
Advanced Technology Centre (Sowerby Building),
FPC 267, P. O. Box 5 Filton,
Bristol BS34 7QW.

Engineering design has become more and more of a team-based activity as technological systems have grown ever more complex. "Soft" human issues can therefore make or break a project, especially in collaborative ventures. Our study uses a popular psychometric test to investigate individual differences between departments and functions within an industrial technology facility. Differences are found between staff in the industrial technology facility and the general population, between scientists from different disciplines in the facility, and between functions within the facility (scientists versus administrators). The latter finding provides potential new insight into a previously identified problem with interactions between these two functions. Although the test used provides insight at an individual rather than organisational level, possible applications in several areas of human factors are suggested.

Introduction

With the ever-increasing complexity of modern technological systems, engineering design activities have become more and more of a team-based process. Design is therefore now often a social phenomenon, and the success or failure of a project can be brought about by the impact of so-called "soft" human issues.

The quality of communication and co-operation between co-workers on a project is clearly of crucial importance, and the impact of such factors can become heightened in the context of collaborative ventures, where project participants come from different technical, company-cultural and increasingly even national-cultural backgrounds.

Research interest in these issues has increased greatly of late. A variety of approaches has been adopted. The one taken here is to address the issues from the perspective of individual differences. This focuses on differences between people, especially styles of working, using psychometric questionnaires to measure personality variables and/or sort people into types.

virtual office as if they were physically present in a single office. The virtual office captures the project's context, by including its input and output, resources, tools and services.

The virtual office will present its users a context oriented view of its contents and members. This will be achieved by importing the necessary basic tools, services, and resources into the co-operation platform, relating them to each other where meaningful, embedding them in the project context, surrounding them with a context oriented user interface and thus creating value-added project services. Phone and fax calls in a virtual office for example, would only be those related to the project, conference sessions would automatically be tied to the context of project data and cost, any collaboration input and output would be matched against the project database, cost file, security rules etc.

Current and Future Work

At the time of writing, the Basic Platform is undergoing acceptance tests and usability inspection in preparation for the usability trials at Pentascope in 2002. The evaluation will take the form of a 6 week trial in which a team will use UNITE to support all synchronous and asynchronous project related activity. It will include pre and post trial reflections, and an intensive 2 day, mid project video-ed trial to gain additional insights into usability problems (e.g. learning, errors, memorability, efficiency and satisfaction). Evaluation will focus on, overall system usability (using SUMI (Software Usability Measurement Instrument) and in-house questionnaires), the manner in which the system was used during the project (using diary studies), the extent to which UNITE aided or changed working practice (as compared to current modes of working) and the extent to which UNITE offered users valuable support.

Acknowledgements

This project is funded by the EU under the Information Society Technologies (IST) Programme (Project No. IST-2000-25436). It represents the view of the authors only.

References

Klöckner, K.,Pankoke-Babatz, U. and Prinz, W. 1999, Experiences with a cooperative design process in developing a telecooperation system for collaborative document production. Behaviour & Information Technology, 18 (5), 373-383

Meier, C. 2001, Virtuelle Teamarbeitsräume im WWW. Wirtschaftspsychologie, Heft 4

Reinema, R., Bahr, K. Baukloh, M., Burkard, H.-J., Schulze, G. 1998, Cooperative Buildings–Workspaces of the Future. In: Callaos, C., Omolayole, O., Wang, L. (Eds.), Proceedings of the World Multiconference on Systemics Cybernetics and Informatics (SCI'98), Orlando, Florida, Vol. 1, 121-128

Reinema, R., Bahr, K., Burkhardt, H.-J., Hovestadt, L. 2000, Cooperative Rooms -- Symbiosis of Real and Virtual Worlds; In: Proceedings of 8th International Conference on Telecommunication Systems, Modelling and Analysis, Nashville, Texas, March

Woodcock, A. (1999), Human factors perspectives on information technology and globalisation. *Information Society '99*, Vilnius, Lithuania, 166-177

WARNINGS

DESIGNING MORE HABITUATION-RESISTANT WARNING SYSTEMS: SOKOLOV'S CORTICAL MODEL

Joseph V. Lambert

Department of Social Sciences
University of the Sciences in Philadelphia
600 S. 43rd Street
Philadelphia, PA 19104-4495

Meister (2000) is of the opinion that HF/E has no theory of its own, arguing that theories found in our literatures are, for the most part, psychological ones. These theories are oriented, he says around the human response to stimuli, rather than, more appropriately, around the relationship between human and technology. I argue that it is entirely appropriate to consider the human's response to stimuli, i.e., "psychological theories" in designing systems, e.g., warning systems. I will illustrate the importance of doing so by discussing the issue of habituation in the design of effective backup warning systems for forklifts and the suggestions that Sokolov's cortical model of attention makes regarding habituation-resistant design features.

Introduction

Each year, forklift accidents are responsible for about 100 deaths and 35,000 injuries, in the U.S, many involving pedestrians and backing-up forklifts. I believe this problem is, at least partially, the consequence of design characteristics of currently used back-up warning systems.

Duchon and Laage (1986), one of the few studies dealing with human factors in the design of such systems, noted that the conventional automatic backup warning system (with its constant, predictable, one-beep/sec sound) is subject to *habituation*. '…the ground crew frequently stop "hearing or "paying attention to" the sound of the backing-up alarm.' p. 263. This habituation to alarms is one of the most common reasons manufacturers give for not providing a backup warning device as standard equipment for their vehicles.

Stimulus Habituation

Habituation is defined as the waning of a response to constant or repeated stimulation, Flaherty, 1985. It is a form of learning, a relatively permanent change in behavior that results from an individual's experience. As Mazur

(1986) notes, habituation is most evident in the body's automatic responses to new and sudden stimuli. It serves an important adaptation function in that an organism learns the inconsequentiality of a stimulus. The neurological basis of the phenomenon is a decrease in the amount of neurotransmitter substance released from sensory neurons, that is, a reduction in the transmission of a neural impulse.

One of the more complex response patterns--and one especially germane to the present issue--occurs in the 'orienting response' (OR), the stopping of some current activity and the orienting, that is, turning toward, and paying attention to, some source of moderate intensity, novel stimulation, e.g., a sound. Accompanying this response are a variety of physiological changes mediated by the sympathetic nervous system (the 'emergency" system'). E.g., a sudden decrease in heart rate, a decrease in skin resistance, a decrease in respiration, changes in the constriction of peripheral blood vessels, changes in muscle tone, and changes in electrical activity of the cerebral cortex, i.e. an increase in arousal.

If this stimulus is repeated several times with no untoward consequences to the organism, all these physiological and brain responses will eventually habituate (Sokolov, 1963). Another component of the OR is the electroencephalogram arousal response (e.g., Sokolov, 1960); it has, also, been shown to be subject to habituation with repeated presentations of an auditory stimulus.

Principles of Habituation

In their germinal paper on the topic, Thompson and Spencer, 1966 list some general principles of habituation:

1. Habituation occurs whenever a stimulus is repeatedly presented.

2. If, after habituation, the stimulus is withheld for some period of time, the response will recover. This is 'spontaneous recovery.' The degree of recovery depends on the amount of time that has elapsed since the habituated stimulus was presented.

3. Whereas habituation may disappear over a long time interval (as just described), it should proceed more rapidly in a second series of stimulus presentations.

4. The more rapid the stimulation, i.e. the higher the rate of stimulation, the more rapid and/pronounced the habituation.

5. Habituation proceeds more rapidly with weak stimuli.

6.   The more trials there are to the habituated stimulus, the more enduring the habituation.

7.   Habituation is stimulus-specific.

8. Presentation of another stimulus prior to presentation of the habituated stimulus results in the recovery of the habituated response. This is called dishabituation.

Implications for Design of Warning Signals

Several of these principles are important in the design of habituation-resistant warning signals. In many operational environments we have the sound of backup warning alarms from forklifts and other trucks. Repeated and continued exposure to this sound could be expect to result in habituation, and a resulting loss of its attention-grabbing (and maintaining) capacities (Principle 1).

To the extent that sounds from different forklifts are similar, we would anticipate mutual transfer of habituation among them (Principle 7). If they were turned off during the course of operations, we could expect some spontaneous recovery of their attention-arousing capacities (Principle 2).

Over the course of several days working in this environment we could expect 'relearning' (Principle 3), with further loss of alerting capacity. If another stimulus had been presented during the course of operations we could expect 'dishabituation' (Principle 8).

Several of the principles outlined above suggest that the commonly used, back-up warning signal with its constant, predictable, one-beep-per-second sound is a stimulus that is particularly prone to habituation.

What design characteristics does the habituation literature suggest for warning alarms to make them less prone to habituation? An examination of Sokolov's work (1960, 1963) in the field of attention can give us an answer to this question.

Habituation and Sokolov's Cortical Model

Using electroencephalographic (EEG) recordings, E.N. Sokolov, 1960, conducted a study of the neural basis of attention. He showed that with repeated presentations of a pulsating tone, brain arousal activity, as measured by an electroencephalogram (EEG) disappeared. After subjects had been habituated to a series of tones that were spaced at regular intervals, Sokolov skipped one of the tones. Subjects exhibited an OR to the absence of the tone. That is, the brain's attention-arousal pattern reappeared when the expected tone was not forthcoming.

These findings fit nicely into Sokolov's 'cortical model' of habituation, as a diminution of attention. According to this model repeated presentation of a stimulus allows the formation of a 'neuronal model' of it. Subsequent incoming stimuli reaching the brain's cortex are compared with this model of 'expectation.' If there is a match between the cortical model and the incoming stimulus the arousal system (in the reticular formation of the brain stem) is inhibited, and no OR will occur. If there is a mismatch this activating system is not inhibited and its resultant excitation activates the brain's cortex and the attention level is heightened.

The point of Sokolov's notion is that response strength (i.e., attention arousal) is directly proportional to the degree of discordance between stimulus input and the cortical model. Sokolov demonstrated the validity of this model in a study in which subjects were presented with a 600 Hz tone every 2-sec. After a while, tones no longer produced ORs, i.e.; subjects were habituated to them. However, irregularities in pitch (i.e., frequency) of the tone would activate subjects' attention. This effect also occurred with shifts in the sound's intensity, that is "loudness." These findings were not specific to the auditory modality. In another study, analogous changes in light-stimulus presentation produced similar results, (Sokolov, 1960).

The neuronal model encodes all aspects of the stimulus; this encoding becomes more precise with repeated stimulation. Accordingly, habituation is total only when a fully accurate model has been formed. In other words, if the cortex is 'kept guessing' by varying stimulus input, attentional levels to incoming stimulation can be heightened. That is, varying stimulus inputs can prevent the formation of an accurate expectancy, thus retarding the habituation process. Sokolov (1966) also asserted that there should be a direct, positive relationship between the strength and duration of OR-elicitation and the extent of initial 'uncertainty.' That is, uncertain stimuli should, initially, not only evoke larger ORs than "certain" stimuli they should also be more resistant to habituation than certain stimuli.

Numerous other research findings also support the prediction that stimulus uncertainty--i. e., stimulus change, either intermodal or intramodal, will retard habituation, e.g. Forbes and Bolles, 1936 and Hare, 1968. All such findings converge to provide evidence in support of Sokolov's model of attention. They also provide a strong argument concerning design characteristics habituation-resistant, warning systems should possess, i.e. complexity/uncertainty. Namely, warning signals that 'keep the cortex guessing' because an accurate neuronal model of them cannot be readily formed should be habituation resistant.

The bottom-line implication is that backup warning stimuli should vary (in an unpredictable fashion) along the dimensions of light and sound intensity and frequency. Thus, "psychological theory" can be useful in systems' design

References

Berlyne, D. E., Craw, M.A., Salapatek, P.H., and Lewis, J.L., 1963, Novelty, complexity, incongruity, extrinsic motivation, and the GSR, *Journal of Experimental Psychology*, 66, 560-567.

Duchon, J.C. and Laage, L.W. 1986, The considerations of human factors in the design of a backing-up warning system, *Proceedings of the Human Factors Society 30th Annual Meeting,* 261-264.

Flaherty, C.F. 1985, *Animal Learning and Cognition (*New York: Alfred Knopf).

Forbes, T.W. and Bolles, M.M. 1936, Correlation of the response potentials of the skin with "exciting" and "non-exciting" stimuli, *Journal of Experimental Psychology, 2,* 273-285.

Groves, P.M. and Thompson, R.F. 1970, Habituation: A dual-process theory, *Psychological Review, 77,* 419-450.

Hare, R.D. 1968, Psychopathy, autonomic functioning, and the orienting response, *Journal of Abnormal Psychology Monograph Supplement, 73,* 1-24.

Mazur, J.E. 1990, *Learning and Behavior*, 2nd Ed, (New Jersey: Prentice Hall).

Meister, D. 2000, Theoretical issues in general and developmental ergonomics. *Theoretical Isuues in Ergonomics Science, 1,* January-March, 2000.

Sokolov, E.N. 1960, Neuronal models and the orienting reflexes, in M.A. Brazier (Ed.), *The Central Nervous System and Behavior*, (New York: J. Macy, Jr. Foundation).

Sokolov, E.N. 1963, Perception and the Conditioned Reflex, (New York: Macmillan).

Sokolov, E.N. 1966, Orienting reflex as information regulator. in A. Leontiev, A. Luria, & A. Smirnow (Eds.), *Psychological Research in the USSR.*

CAN FOOD LABEL DESIGN CHARACTERISTICS AFFECT PERCEPTIONS OF GENETICALLY MODIFIED FOOD?

A. J. Costello, E. J. Hellier, J. Edworthy & N. Coulson[1]

Department of Psychology, University of Plymouth
[1]*School of Health and Community Studies, University of Derby*

A substantial body of research has investigated the effect of warning label design characteristics on hazard and other product perceptions in a variety of domains. However the extent to which these findings will generalise to food labelling and to communicating the presence of genetically modified ingredients is not known. The study reported here investigates how label design features influence the perception of genetically modified foods. The effects of label colour, wording style, and information source on hazard perceptions and purchase intentions were measured. Analysis of variance revealed that purchase intentions and hazard perceptions were influenced by label design. Of particular interest is the finding that any reference to genetic modification, even if the label is stating that the product is free of genetically modified ingredients, increased hazard perception and decreased purchase intentions relative to a control.

Introduction

The genetic modification (GM) of food is a hotly contentious issue that attracts much media coverage. Furthermore, previous research indicates that many consumers have negative attitudes to GM foods (e.g. Frewer 1997). Central to the debate surrounding the acceptability of GM foods has been the adequacy of the labels that indicate GM content, and hence the extent to which consumers are able to choose whether or not to buy these food products. The National Consumer Association, for example, has expressed concern that the British Government has mandated that only certain GM products require a label (Dibb & Lobstein, 1999). Even when GM content is indicated on food labels, they may not be informative. The United States approach to labelling is quoted as being 'based on the belief that, even were the product labelled as GM, the consumer would be unable to make a rational choice based on that information' (Select Committee on the European Communities Report, 1998). In contrast, food producers lobby to resist the mandatory labelling of GM foodstuffs for fear that it will negatively impact on trade (Newspeg 1998). Clearly then, the British Government is under pressure to both promote the GM foods authorised for sale in this country for economic reasons, whilst also satisfying the demand for consumer choice through labelling. Little published research is available on the nature of consumer's reactions to labels indicating GM content. Therefore research is needed to investigate consumer reactions not just to GM foods themselves but to existing and potential GM food labels. The research presented here tackles the latter issue.

A substantial body of research charts the relationship between warning label design features (such as colour, font size and signal word), and subjective and behavioural measures. Studies have shown both that people alter their compliance levels depending upon the colour in which a warning is presented (e.g. Braun & Silver, 1995); and that subjectively estimated levels of compliance and hazard vary in predictable ways according to specific variations in design (e.g. Chapanis, 1994). More specific work on product labelling has shown that linguistic factors associated with the wording of the label and design features such as colour influence product perceptions and purchase intentions (e.g. Hyde & Hellier 1997). Generalisations to emerge from such studies suggest that colour has predictable effect on perceived hazard, with red, orange, green, blue and then white resulting in decreasing hazard connotations; and that definitive wording results in higher levels of compliance than probabilistic wording. To date we do not know the extent to which these finding generalise to the labelling of GM foods.

Previous research on consumer reactions to information about GM food has investigated the effects of the information source. The work of Frewer et al (1996), for example, suggests that trust in the information source is an important determinant of the believability of the information provided. The finding that information provided by Consumer Associations is more trusted than that provided by the Department of Health is typical. It has also been suggested that the wording of information on GM food products could affect the acceptability of the product. Runge & Jackson (2000) suggest that negative labelling (this product contains no genetically modified ingredients) is a more appropriate way to inform the public about GM content. This is compared with positive labels (this product *may* contain GM ingredients) which they argue gives little information to the consumer. However the difference between these statements is not just positive and negative. The statement 'this product may contain GM ingredients' is probabilistic, whereas 'this product contains no GM ingredients' is definite. The influence of probabilistic vs definitive wording in relation to GM food labels isn't known.

This study investigates how perceptions of genetically modified foods are affected by variations in label design. Specifically, how the important measures of hazard perception and purchase intention vary in relation to manipulations in colour, information source and wording of a label indicating GM content for a food product.

Method

Design
There were four within subject independent variables, colour (red, green, blue and white), the warning source (no source, Consumer Association, Department of Health, Manufacturer), content (GM Vs no GM) and wording, (probabilistic Vs definitive).

Participants
40 participants aged 18-40 years took part in the experiment, 29 female and 13 male, all were literate and English speaking. The participants were recruited through posters displayed at the University of Plymouth and were paid £2.50.

Stimuli and materials
Factorial combination of all levels of colour, warning source, content and wording resulted in 64 different labels. The labels were scanned onto identical pictures of a food product. A picture of the food product with a no label acted as the control.

Each label had the signal word 'Notice' as a header. The labels were coloured red, green, blue and white. The information source was indicated by the headings 'Consumer Association Notice, 'Department of Health Notice', 'Manufacturers Notice' or 'Notice' as appropriate. The informational content of the labels read 'Contains Genetically Modified ingredients'; 'May contain Genetically Modified ingredients', 'Contains no Genetically Modified ingredients' or 'Unlikely to contain Genetically Modified ingredients'. The labels were all identical in size, printed in Times New Roman font of size 10.

Procedure
Participants were presented with booklets containing 64 photographs of the food product each with a different label. When viewing the pictures, participants were told that each product was of the same price and quality and they should assume in principle that they would buy the product. The participants were asked 'How likely would you be to buy this product?' and directed to indicate the number that best represented their answer on a scale that ran from 0 to 8, where 0 represented 'Not at all Likely' and 8 represented 'Extremely Likely'. Participants were also asked 'How much hazard is indicated by the label?' and indicated their answer on a scale that ran from 0 to 8, where 0 represented 'No Hazard' and 8 represented 'Extreme Hazard'.

Results

The mean scores for hazard perception and purchase intention by label type are shown in Table 1. As the mean scores for hazard perception increased, the scores for purchase intention decreased. In addition the no label control scored highest on purchase intention, and lowest on perceived hazard. Thus the control attracted higher purchase intention scores and lower hazard perception scores even than labels stating that there were no GM ingredients in the product.

Table 1. Mean scores for label variations

LABEL TYPE	HAZARD PERCEPTION	PURCHASE INTENTION
No label (control)	1.190	5.309
Red	3.140	3.500
Blue	2.695	3.656
Green	2.631	3.760
White	2.667	3.649
Manufacturer	2.754	3.606
Dept. of Health	2.854	3.625
Consumer Group	2.679	3.716
No Source	2.845	3.619
GM+	3.842	2.767
GM-	1.725	4.516
Definite wording	2.727	3.918
Probabilistic	2.839	3.365

Hazard Perception
A 4x3x2x2 within subjects ANOVA revealed significant main effects for colour (F $(3,123) = 7.421$, $p<0.01$), source (F $(3,123) = 3.501$, $p<0.05$), and GM content (F $(1,41)$

= 95.206, p<0.01). Follow up tests (least squared difference) revealed that red labels were perceived as being more hazardous than blue and green labels (p<0.01) or white labels (p<0.05). The Department of Health and the unattributed source respectively resulted in higher levels of hazard perception than the consumer association (p<0.05, p<0.01). Labels indicating the presence of GM ingredients resulted in higher hazard perceptions than those indicating that the product did not contain GM ingredients. (p<0.01) The effect of wording was non-significant. A significant interaction between content and wording (F (1,41) = 109.726, p<0.01) showed that hazard perception was higher for definite wording when the label indicated that the product *did* contain GM ingredients. In contrast, hazard perception was lower for definite wording when the label indicated that the product *did not* contain GM ingredients.

Purchase Intention
A 4x3x2x2 within subjects ANOVA revealed significant main effects for colour (F (3,123) = 3.811, p<0.05), GM content (F (1,41) = 65.894 p<0.01) and wording (F (1,41) = 15.530, p<0.01). Follow up tests (least squared difference) revealed that green labels resulted in higher purchase intentions than either red (p<0.01) or blue labels (p<0.05) and blue labels resulted in higher purchase intention ratings than red labels (p<0.05). Labels indicating the presence of GM ingredients resulted in lower purchase intention ratings than those indicating that the product did not contain GM ingredients, (p<0.01). Definitive wording resulted in higher purchase intention scores than probabilistic wording, (p<0.01) The effect of source was non-significant. A significant interaction was found between content and wording (F (1,41) = 3.240, p<0.01). Purchase intention was higher for definite wording when the label indicated that the product *did not* contain GM ingredients. In contrast, purchase intention was lower for definite wording when the label indicated that the product *did* contain GM ingredients.

Discussion

One important finding to emerge from this data is that the control resulted in the product being rated as less hazardous and attracting higher purchase intention scores than any other label. This suggests that any mention of GM content, even a label stating the product to be GM free, increases perceived hazard and reduces purchase intentions. Of all label design features, a mention of GM is most likely to increase perceived hazard and reduce purchase intention.

Some specific findings from warning design research has generalised to the context of GM food labelling. The effect of colour on perceived hazard generalises from other warning label research and the effect of information source found here supports previous findings that sources found to be less trusted affect reactions to information about GM foods, in this case resulting in higher levels of perceived hazard.

In general, as products were perceived as being less hazardous, they were perceived as being more purchasable. This inverse relationship is unsurprising and mirrors findings elsewhere (e.g. Hyde & Hellier, 1997). The interaction displayed between GM content and probabilistic vs definite wording supports this finding and demonstrates that a definitive statement of GM content results in the highest levels of perceived hazard and the lowest levels of purchase intention. For a definitive statement of no GM content, the reverse is true. The main difference between the hazard and purchase scores is that information source had an effect on perceived hazard but not purchase intention, whereas wording had an effect on purchase intentions but not perceived hazard. Runge & Jackson (2000) suggested that definitive vs probabilistic wording might affect consumers perceptions of GM foods, and Frewer et al, (1996) suggested that information source is an important factor in informing consumers about genetically modified foods. Both authors have been partially supported by our data, but it remains unclear why hazard and purchase intention measures differ in response to these variables. If a food product is labelled as definitely containing GM ingredients, for example, rather than probably containing GM ingredients, the product does not appear more hazardous, but does appear less purchasable. The relationship between purchase intentions and hazard in relation to information source and wording requires further exploration.

In summary, these results indicate that not only do consumers have negative attitudes to GM foods, but have decreased purchase intention and increased hazard perception of products labelled as containing GM ingredients. Whilst design characteristics such as wording, colour and information source modify these reactions, content has the largest effect. As a consequence of the finding that the food product with no label was rated as least hazardous and most purchasable, the decision whether or not to label GM products is an important one. One that may have consequences for consumer perceptions and consequently the sale of GM *and* non-GM food products. In addition, it is suggested that the design and wording of such labelling will affect how well genetically modified foods are received.

References

Braun, C. C. & Silver, N. S. (1995). Interaction of signal word and colour on warning labels: differences in perceived hazard and behavioral compliance. *Ergonomics, 38,* 2207-20.

Chapanis, A. (1994). Hazards associated with three signal words and four colours on warning signs. *Ergonomics*, 37, 265-76.

Dibb, S. and Lobstein T., (1999), *GM FREE, A Shoppers Guide to Genetically Modified Food,* (Virgin Publishing Ltd., London)

Frewer, L. J., Howard C., Hedderley D., and Shepherd R., (1996), What Determines Trust in Information about Food Related Risks? Underlying Psychological Constructs, *Risk Analysis* 16, 473-486

Frewer, L. J., Howard C. And Shepherd R., (1997), Public Concerns in the UK about General and Specific Applications of Genetic Engineering; Risk and Benefit Analysis, *Science Technology and Human Values*, 22, 98-124

Hyde, C. & Hellier, E. (1997). Do Warning Labels Influence the Hazard Perception of Familiar and Unfamiliar Consumer Products? In Robertson, S. (Ed) *Contemporary Ergonomics*, 127-130, Taylor & Francis.

Newspeg, (1998), Greed or Need? Genetically Modified Crops, *Review of African Political Economy*, 25, 78, 651-653

Runge C. F. and Jackson L. A., (2000) Labelling Trade and Genetically Modified Organisms: A Proposed Solution, *Journal of World Trade*, 34, 1, 111-222

Select Committee on the European Communities report to the House of Lords, Session (1998-1999), *EC Regulations of Genetic Modification in Agriculture*, 2nd Report, (The Stationary Office, London)

ORIENTING RESPONSE REINSTATEMENT IN TEXT AND PICTORIAL WARNINGS

Paula Thorley[1], Elizabeth Hellier, Judy Edworthy and Dave Stephenson

Department of Psychology, University of Plymouth
Drake Circus, Plymouth, Devon, PL4 8AA
[1] *now at Serco Assurance, Thomson House, Warrington WA3 6AT*

An experiment is reported which demonstrates both habituation and an orienting response to visual warning signs. Using skin conductance as a response measure, subjects were repeatedly exposed to a single warning sign over twelve trials. Subjects were presented with warnings that were either pictorially- or text-based. In the experimental trials, the warning changed to a slightly different format on the tenth trial and reverted to its original form on the eleventh and twelfth. Control conditions were also used where the warning format did not change at all throughout the twelve trials. The results showed a consistent decrease in response from trial one to trial nine, followed by an increase in trial ten for the experimental subjects only. These results suggest that habitation to warnings is a measurable phenomenon.

Introduction

There is some evidence demonstrating that people can and do habituate to warnings (e.g. Thorley et al, 2001). The study reported here further investigates the effects of repeated exposure to visual warnings. Skin conductance response was measured as a determinant of the orienting response, serving as an indicator of both habituation and dishabituation.

The design of the study was similar to that of Ben-Shakhar *et al.* (2000), where the stimuli were primarily either text- or picture-based and included standard stimuli and test stimuli. It was hypothesised that there would be habituation effects following repeated exposure to a standard stimulus for both text- and picture- based visual warnings. Moreover, it was hypothesised that the orienting response would be reinstated following the introduction of a novel test stimulus subsequent to the habituation trials. Following the test stimulus, it was expected that dishabituation effects toward the standard stimulus would be demonstrated by an increased skin conductance response.

Method

Participants
An opportunistic sample of sixty (28 females, 32 males) participated in the experiment for payment. The sample was selected partly from a student and partly from a non-student population. All participants were advised of their right to decline or withdraw, although none did. No other details were recorded.

Materials

A constant voltage system and two silver-silver chloride electrodes (9mm Diameter) measured skin conductance. In the absence of availability of an electrode gel mixture to the recipe provided by Fowles *et al* (1981, as cited in Ben-Shakhar *et al.* 2000) a substitute saline gel was used; 'Johnson & Johnson KY Jelly'.

The experiment was conducted in an air-conditioned, soundproof laboratory at the University of Plymouth and was monitored via an adjacent laboratory housing the recording equipment. A PC was used to control the stimulus presentation and another computer recorded skin conductance changes. The stimuli were displayed on a Hewlett Packard HP71 colour monitor, approximately 70cms from the participant's eyes.

Stimuli

All stimuli contained 3 components based on the recommendations of Rogers *et al.* (2000). The text-based stimuli contained a signal word, instructions of how to avoid the hazard, and the potential risk. The picture-based stimuli contained a generic warning sign, instructions on how the hazard should be avoided and the potential risk. The experimental stimuli differed from the control only in either the signal word or sign. The instructions and potential risk remained the same for both groups.

Design and Procedure

The stimulus sequences used in this experiment were comprised of text-based (signal word) and pictorially based (icon) visual warnings. A test stimulus (TS), created by substituting either the signal word or icon on the standard stimulus (SS), was introduced after nine repetitions of SS, followed by two additional repetitions of SS. There were thus 12 exposures to the stimuli in all.

The between-subjects factor was the components of SS that were substituted to create the TS (0, where no change was made, or 2, where both the signal word/symbol and colour were altered in order to create a warning with higher perceived urgency). The dependent variables were the skin conductance responses (SCR) elicited by TS (OR reinstatement) and by the SS immediately following TS (dishabituation).

Participants were allocated randomly to the 4 conditions, 15 in each group (text control, text experiment, picture control, picture experiment). All participants were fully briefed and debriefed. Each participant was presented with only one of the stimulus sequences. Two electrodes were attached to the volar side of the index and middle finger on the participant's non-preferred hand using surgical tape applied with a comfortable pressure. An earthing strap was also attached to the inside lower arm of the non-preferred hand in order to maintain accuracy of measurement. Participants were requested to sit at ease and told that the computer program, to which they should pay attention, would begin in a few minutes' time, and that they would not be told when it was about to begin.

All participants were given a three-minute rest period before the program began. For all groups a flash of light was presented on the screen for 500ms, this was the range correction stimulus to eliminate individual differences. The control groups were presented with 12 visual warnings that differed in neither appearance nor intensity from one another. The warnings were displayed for 5 seconds each with a fixed inter-stimulus interval of 15 seconds. The experimental groups saw the same as the controls except in Trial 10, where the warning differed in both appearance and intensity to the previous nine. At the end of the experiment, a thank you screen was displayed, participants were debriefed as the equipment was removed from their person and they were then paid for their participation.

Results

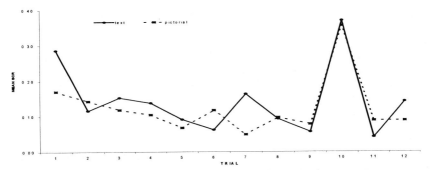

Figure1. Comparison of text and picture based visual warnings

Figure 1 compares the mean skin conductance responses between the text and picture based visual warnings. Habituation effects are evident across both modalities as is demonstrated by the downward trend in responses between Trial 1 and Trial 9. The increase in mean SCRs at Trial 10 indicates an increase in the orienting response for both modalities; however, dishabituation effects are not obvious at Trials 11 and 12.

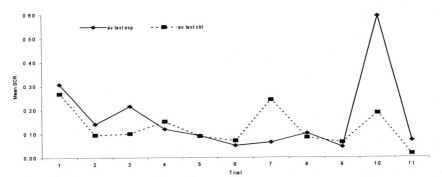

Figure 2. Comparison of text control and text experimental groups

Figure 2 compares the mean skin conductance responses between both control and experimental conditions for the text based visual warnings. Only Trials 1 to 11 are represented here, as they are the main trials of interest. Again, habituation effects between Trials 1 and 9 are evident across both conditions and disparities in responses have been recorded at Trial 10. Trial 11 does not indicate dishabituation effects.

Figure 3 compares the mean skin conductance responses between both control and experimental conditions for the picture based visual warnings. Habituation effects between Trials 1 and 9 are evident across both conditions and disparities in responses have been recorded at Trial 10. Dishabituation effects do not appear to have been demonstrated at Trial 11.

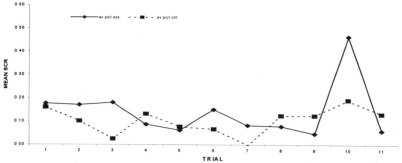

Figure 3. Comparison of pictorial control and pictorial experimental groups

In order to determine habituation effects, Trials 1 and 9 were subjected to a repeated measures mixed analysis of variance (ANOVA) (trial as the within subjects factor and modality and condition between subjects factors). Descriptive statistics are summarised in Table 1.

Table 1. Descriptive statistics for habituation

| Trial | TEXT BASED WARNINGS | | | | PICTURE BASED WARNINGS | | | |
| | Control | | Experimental | | Control | | Experimental | |
	Mean	SD	Mean	SD	Mean	SD	Mean	SD
1	.794	.455	1.13	1.31	.660	.723	.627	.477
9	.475	.955	.163	.488	.523	1.08	.441	1.08

The ANOVA revealed a significant main effect of trial, $F(1, 36) = 4.796$, $p<.05$. There was neither a trial by modality nor a trial by condition interaction.

In order to determine orienting response reinstatement Trials 9 and 10 were subjected to a repeated measures mixed analysis of variance (ANOVA) (trial as the within subjects factor and modality and condition between-subjects factors). The ANOVA revealed a significant main effect of Trial, $F(1, 36)$ 9.332, $p<.01$. There were no interaction effects across any of the variables and the test of between subjects effects revealed no main effect of either modality or condition, nor any interaction effects. Descriptive statistics for both modalities are summarised in Table 2.

Table 2. Descriptive statistics for orienting response reinstatement

| Trial | TEXT BASED WARNINGS | | | | PICTURE BASED WARNINGS | | | |
| | Control | | Experimental | | Control | | Experimental | |
	Mean	SD	Mean	SD	Mean	SD	Mean	SD
9	.475	.955	.163	.488	.523	1.07	.441	1.08
10	.451	.646	2.68	2.57	1.32	2.61	1.38	1.50

Discussion

The hypothesis stating that repeated presentation of the standard stimulus would result in habituation was supported, as the results indicated a significant decline in skin conductance response to the standard stimulus between Trial 1 and Trial 9 across all groups. It has been specified that for habituation to be identified it must exhibit certain characteristics. For example, if a stimulus elicits a response, repeated application of the stimulus results in decreased response strength (Thomson and Spencer 1966, as cited in Petrinovich, 1973) or, habituation of the orienting response to the standard stimulus is evident during the first presentations of the standard stimulus (Zimny and Schwabe, 1965). The data collected here clearly displays those features, therefore it has been concluded that habituation is a consequence of repeated exposure to a visual warning.

The hypothesis that the introduction of a novel stimulus would reinstate the orienting response was only somewhat supported. Although the orienting response reinstatement was demonstrated in the overall effects, this was clearer for the text-based warnings than for the picture-based warnings. The findings here would support Sokolov's (1960, as cited in Lipp, 1998) scheme for orienting response elicitation, where a mismatch in the comparator system results in an increased orienting response. The results also conform to two of the hypotheses that underlie Sokolov's theory as identified by Zimny and Schwabe (1965). Specifically, that presentation of a test stimulus produces a return of the orienting response and the orienting response to the test stimulus is greater than that to the standard stimulus immediately following the test stimulus. Therefore, it could be concluded that following habituation, changing components of that warning reinstates the orienting response.

Acknowledgements

This research was supported by a grant from the Economic and Social Research Council (ESRC).

References

Ben-Shakhar, G., Gati, I., Ben-Basset, N. & Sniper, G. (2000). Orienting response reinstatement and dishabituation: Effects of substituting, adding and deleting components of nonsignificant stimuli. *Psychophysiology, 37*, 102-110

Lipp, O.V. (1998, August 2000). A description of Sokolov's comparator theory of habituation and the orienting response. Available http://www2.psy.uq.edu.au/~landcp/PY269/habituation/habituation.html

Petrinovich, L. (1973). A species-meaningful analysis of habituation. In H.V.S.Peeke & M.J. Herz (Eds), *Habituation I. Behavioural Studies.* (Academic Press, London)

Thorley, P., Hellier, E. & Edworthy, J. (2001). Habituation effects in visual warnings. In M. Hanson (Ed), *Contemporary Ergonomics.* (Taylor and Francis, London)

Rogers, W.A., Lamson, N.L. & Rousseau, G.K. (2000). Warning research: An integrative perspective. *Human Factors, 42*, 102-139

Zimny, G.H. & Scwabe, L.W. (1965). Stimulus change and habituation of the orienting response. *Psychophysiology, 2*, 103-115

PERCEIVED PERSUASIVENESS OF PRODUCT MANUAL WARNINGS AS A FUNCTION OF STATEMENT TYPE

Kirk Grotjohn and Michael S. Wogalter

North Carolina State University
Dept. of Psychology, 640 Poe Hall, CB 7801
Raleigh, NC, 27695 USA

The perceived persuasiveness of warning statements derived from a power sander product manual was investigated. The content of the statements varied in several ways: the presence of a consequence statement, the form of injury statistics (percentages, frequencies, or none), the magnitude of the statistical value, and the quality of the statement. Participant ratings showed that a consequence statement together with directive and instruction statements were the most persuasive. For injury frequency statistics, high quality statements increased persuasion compared to low quality statements. Larger percentages produced greater persuasion ratings compared to smaller percentages. Implications for the design of product manual warnings are discussed.

Introduction

Most warning literature has examined differences in physical form and location (e.g. color, size, layout, presence of symbols, and placement). Although some research has examined aspects of content such as signal words, color and symbols (e.g. Laughery *et al*, 1994, Wogalter *et al*, 2001), much less research has been conducted on statement content. For example, prior research on statement content has noted that messages including hazard, consequence and instruction statements were rated more effective than without these statements (e.g. Wogalter *et al*, 1985). Statements with greater implied injury severity were rated as connoting greater hazard than statements with lower implied severity (Wogalter and Barlow, 1990). Moreover most of the prior evaluations of statement content have used ratings of perceived hazard, perceived effectiveness, or willingness to comply. One measure that has not been used in warning research is the degree to which the message is persuasive. Warnings are in some sense persuasive communications. Greater persuasion may change beliefs and attitudes which could help motivate compliance behavior (Wogalter *et al* 1999).

Most research on warnings has concerned signs and labels. Very few studies have systematically manipulated components of product manuals. Research has found that highlighting and including symbols benefits memory and comprehension (Young and Wogalter, 1990) and priority ordering of statements facilitates subsequent recall (Vigilante and Wogalter 1999). Product manual warnings are different than other kinds of warnings because they are embedded in the context of a large amount of non-warning information. For example, many power tool manuals include general work-related statements (e.g. Keep work area clean) or technical information. This non-warning information is likely to be less persuasive than warning information. However, the extent to which a warning is persuasive may depend on several factors. One is whether it contains all necessary information such as consequences (Wogalter *et al*, 1985). Another potential factor affecting warning persuasiveness is whether it contains statistical (quantitative) information about accidents and injuries. Conzola and Wogalter (1998) found that warnings with quantitative information were perceived as more important, vivid and explicit. However, that study did not measure the persuasiveness of the messages. In the present study, the form of statistical presentation was manipulated. The statistics either used numerical frequencies (e.g. approximately 2,500 persons suffer eye injuries each year) or numerical percentages (e.g. approximately 35% of all power sander injuries involve injuries to the eyes).

Another potential factor affecting warning persuasiveness is whether the statement contains high quality, relevant information versus low quality, irrelevant information. In the present research, a high quality warning is operationally defined as one that implies a large number of power sander accidents having occurred in the past. A low quality warning statement is one that either implies that there have not been very many power sander accidents, or gives irrelevant information (e.g. the number of accidents with power tools in general). If the quality of the warning statement is a factor, then the number of power sander accidents implied by the warning statement (low versus high) would affect perceived persuasiveness. Additionally, the persuasiveness of a statistic may depend on its magnitude, independent of the quality of the statement. Large statistical values do not necessarily imply a large number of accidents (e.g. if the statistic is irrelevant to the number of power tool accidents).

The goal of this research was to investigate the persuasiveness of several different types of warning statements derived from a power sander product manual that were manipulated as a function of the above-named factors.

Method

Participants
Eighty-seven North Carolina State University undergraduate students (35 males and 52 females) enrolled in introductory psychology classes participated for research credit towards a laboratory participation requirement. There was considerable variation in the degree programs with which the students were affiliated.

Procedure and Materials
The present research was part of a larger study. Participants were shown a list of different warning statements derived from existing power sander product manuals, and asked to rate the statements on how convincing each was in supporting the claim that the power sander is potentially hazardous and caution should be taken when using it (using similar methodology as Petty and Cacioppo, 1986). All statements were rated on a 7-point Likert-type scale (1 = not at all convincing, 7 = extremely convincing).

Table 1. Examples of warning statement types

Statement type	Example
Directive, consequence, and instruction (what to do, why and how)	Secure work. Unsecured work could be thrown towards the operator causing injury. Use clamps or vice to secure work.
Technical information only (no information about hazard)	To avoid damage, do not exceed a +/- 10% voltage variation or a +/- 3% frequency variation
General work-related statements (no information about hazard)	Know your power tool. Read operator's manual carefully.
High quality statements using percentages (larger and smaller)	Approximately (35%/4%) of (eye injuries from power tools/all eye injuries) are suffered while using power sanders.
Low quality statements using percentages (larger and smaller).	Approximately (35%/4%) of all power sander injuries involve injuries to the eyes.
High quality statements using numerical frequency values (larger and smaller)	Approximately (2,500/100) persons suffer eye injuries each (year/month) while using power sanders.
Low quality statements using numerical frequency values (larger and smaller)	Approximately (2,500/100) persons have suffered eye injuries (since 1975/each year) while using power tools.

There were a total of 99 different statements representing 13 different statement types (grouped according to statement content). However, each participant only rated a subset of the total statements to avoid fatigue. The statements were divided into four different groups, and each participant rated only the statements from one group. The order of the statements within each group was randomized.

The power sander product was selected based on the results of an earlier, related study showing it to be a product that this population of participants was not familiar with and perceived to be moderately hazardous.

The warning statements mainly concerned eight common hazards associated with this power tool (hair/clothing getting caught in moving parts; unsecured work being thrown; dust or foreign objects injuring eyes; dust and debris injuring lungs; electrical shock; hearing damage; lacerations from accidental starting; and fires from sparks). For statements using statistics, the larger and smaller values were designed to be as far apart as possible while still remaining credible. Table 1 shows example statements. Table 2 gives the entire set of 13 statement types.

Results

The mean rating for each statement type was computed across all participants. An analysis of variance conducted on the mean convincingness score showed that statement type was significant ($p < .0001$). Table 2 shows the means and standard deviations for each of the 13 statement types. The subscripts in the table show significant differences between the statement types. Only a subset of the comparisons are described in this section.

Table 2. Mean convincingness ratings for statement types

Statement type	M	SD
Directive, consequence, and instruction	4.64$_a$	1.66
Statistical information using numerical values (larger) that suggest a high number of overall power sander accidents	4.22$_b$	1.65
Statistical information using percentages (larger) that do not give any information about the overall number of power sander accidents	4.10$_b$	1.48
Statistical information using numerical values (smaller) that suggest a high number of overall power sander accidents	4.02$_b$	1.58
Statistical information using percentages (larger) that suggest a high number of overall power sander accidents	3.95$_b$	1.45
Only technical information	3.56$_c$	2.05
Only the consequence	3.51$_c$	1.61
Directive and instruction (no consequence)	3.50$_c$	1.83
Statistical information using numerical values (smaller) that do not give any information about the overall number of power sander accidents	3.41$_c$	3.41
Statistical information using numerical values (larger) that do not give any information about the overall number of power sander accidents	3.22$_{cd}$	1.62
Statistical information using percentages (smaller) that suggest a high number of overall power sander accidents	2.98$_d$	1.37
Statistical information using percentages (smaller) that do not give any information about the overall number of power sander accidents	2.94$_d$	1.44
General work-related statements	2.52$_e$	1.76

Notes: 1 = Not at all convincing, 7 = Extremely convincing
        Means with similar subscripts are not significantly different

Comparisons showed that statements with the three components (directive, consequence, and instruction) had the highest ratings ($M = 4.64$, $SD = 1.66$), and were rated as significantly more convincing than statements with (a) only a directive and instruction ($M = 3.50$, $SD = 1.83$), (b) only the consequences of the hazard ($M = 3.51$, $SD = 1.61$), or (c) only technical information ($M = 3.55$, $SD = 2.05$). The latter three statement types did not differ from each other ($p > .05$), but all were rated more convincing than general work-related statements ($M = 2.52$, $SD = 1.76$).

With frequency statistics, the high quality statements were rated more convincing than low quality statements, but there were no differences as a function of numerical magnitude (for high quality: small frequency, $M = 4.02$, $SD = 1.58$, and large, $M = 4.22$, $SD = 1.65$, and for low quality: small frequency, $M = 3.41$, $SD = 3.41$, and large, $M = 3.22$, $SD = 1.62$).

The pattern was somewhat different when statistics were presented as percentages. Statements using larger percentage values were significantly more convincing than statements using smaller percentage values. There were no significant differences between high quality (for large percentages, $M = 3.95$, $SD = 1.45$, and small, $M = 2.98$, $SD = 1.37$) and low quality (for large percentages, $M = 4.10$, $SD = 1.48$, and small, $M = 2.94$, $SD = 1.44$) statements.

Discussion

The results showed that the most effective warning statements were those that included a directive, consequence and instructions, and that removing the consequence, or using only the consequence had a negative impact on the perceived persuasiveness of the statement. This finding is similar to Wogalter et al (1987) who found that that when the signal word, hazard statement, consequent statement, or instruction statement was removed, the warnings were perceived to be less effective.

When using injury statistics, results were slightly different depending on the type of statistic. When frequency statistics were used, it appeared that the quality of the information being conveyed (e.g. how many power sander accidents it suggested) was more important than the magnitude of the statistic. However, when the statistics were percentages, it was the magnitude of the percentage that seemed to be more important than the quality of the information conveyed by the statistics. This suggests that in some situations, including statistics with small values may reduce the persuasiveness of the warning statement.

General work-related statements were significantly less persuasive than all other statement types. Technical statements were also not very persuasive. Both of these types of statements are commonly included in product manual warnings.

Note that the statement with the highest ratings had no statistics at all. However, the next group of statements all had some sort of statistic. Future research could examine the effectiveness of combining the directive, consequence and instructions with statistical information to determine whether the combination produces a higher level of persuasiveness than those used in the present study.

References

Conzola, V. C. and Wogalter, M. S. 1998, Consumer product warnings: Effects of injury statistics on recall and subjective evaluations, *Proceedings of the Human Factors and Ergonomics Society*, **42**, 559-563

Laughery, K. R., Wogalter, M. S. and Young, S. L., Eds. 1994, *Human Factors Perspectives on Warnings: Selections from Human Factors and Ergonomics Society Annual Meetings 1980 - 1993*, Human Factors and Ergonomics Society, Santa Monica

Petty, R. E. and Cacioppo, J. T. 1986, The elaboration likelihood model of persuasion. In L. Berkowitz (ed.), *Advances in experimental social psychology* (Academic, New York), 123-205

Silver, N. C. and Wogalter, M. S. 1989, Broadening the range of signal words, *Proceedings of the Human Factors Society,* **33**, 555-559

Vigilante, W. J., Jr. and Wogalter, M. S. 1998, Product manual safety warnings: The effects of ordering, *Proceedings of the Human Factors and Ergonomics Society*, **42**, 593-597

Wogalter, M. S. and Barlow, T. 1990, Injury severity and likelihood in warnings, *Proceedings of the Human Factors Society*, **34**, 580-583

Wogalter, M. S., DeJoy, D. M. and Laughery, K. R. (Eds.) 1999, *Warnings and Risk Communication*, (Taylor and Francis, London)

Wogalter, M. S., Desaulniers, D. R., and Godfrey, S. S. 1985, Perceived effectiveness of environmental warnings, *Proceedings of the Human Factors Society*, **29**, 664-668

Wogalter, M. S., Young, S. L., and Laughery, K. R., Eds. 2001, *Human Factors Perspectives on Warnings, Volume 2: Selections from Human Factors and Ergonomics Society Annual Meetings 1993 - 2000* (Human Factors and Ergonomics Society, Santa Monica)

Young, S. L., and Wogalter, M. S. 1990, Comprehension and memory of instruction manual warnings: Conspicuous print and pictorial icons, *Human Factors,* **32**, 637-649

LIST VS. PARAGRAPH FORMATS ON TIME TO COMPARE NUTRITION LABELS

Michael S. Wogalter, Eric F. Shaver, and Linda S. Chan

Ergonomics Program, Department of Psychology
North Carolina State University
Raleigh, NC 27695-7801 USA

How textual information is presented may affect how easily information can be extracted. Formatting might benefit the search for particular types of information. The present study examined times to compare differently formatted food nutrition labels. Labels were either in one of two list-type formats with or without horizontal lines ("rules") or in a paragraph-type format. The results indicate that both types of list format produced significantly faster comparison times than paragraph format. The two list formats did not significantly differ. Also, participants who were non-native English users and who were not students were significantly slower than native English users particularly for the paragraph format label. Implications for list format over paragraph, and formatting, in general, are discussed.

Introduction

How information is presented may affect how well information can be extracted. Most of the textual information is in continuous prose with full sentences comprising parts of paragraphs. A much smaller percentage of text is presented in a list-type format with numbers or bullet points to denote different component information. Most of the relatively limited amount of research comparing the usability of these two types of text formatting has shown that list-type format is beneficial in reducing search or comparison times (e.g. Galitz, 1989, Goldberg *et al*, 1999, Wogalter and Post, 1989, Wogalter and Shaver, 2001). One reason for the benefit of list format over paragraph format in information search tasks is that the former has a lower print density than the latter. Greater print density has more characters in a given area and requires more search time than lower print density (e.g. Galitz, 1989, Tullis, 1983).

While there is research suggesting that list format can be beneficial to users viewing product manuals (e.g. Wogalter and Shaver, 2001), the issue whether it is better than paragraph format in food labels has not been addressed. Users' ability to search for important information on nutrition components (e.g. amount of fat and sodium) might be aided when information is presented in a list-type format as opposed to a paragraph-type format.

It is important to consider the disadvantaged populations when designing a food label. For example, when people get older, their eyesight and other senses decline (e.g. Coren *et al*, 1999). These deficiencies can make it difficult for older adults to read printed material. Also, older adults tend to have increasing comprehension and memory difficulties (e.g. Morrow and Leirer, 1999). Food product nutrition labels need to be designed so that they facilitate readability and understandability of the information on them. Research shows that list format facilitates older and younger adults reading of labels (e.g. Hartley, 1999, Morrow and Leirer, 1999). Instead of decreasing the print size to allow list format use, other label designs like tags increase the available surface space (e.g. Wogalter *et al*, 1993, Wogalter and Young, 1994).

The present research examined times to compare differently formatted nutrition labels for food products. Labels were either in one of two list-type formats or in a paragraph type format. In the U.S., the Food and Drug Administration (FDA) requires that the "Nutrition Facts" label be present on all packaged food products produced after 1994 (e.g. FDA, 2001). The law specifies a list-type format, but allows a paragraph version (linear "string" format) on food products with 101.6 square centimeters or less total surface area available for labeling. The list- type format presents the dietary component names on the left side column and the corresponding numerical information on the right side column. In the present study, there were two versions of the list-type label. One simulated the actual FDA label that has thick and thin rules ("horizontal lines") separating sections of the label. The other list format was otherwise identical but lacked the rules.

This experiment sought to determine whether the paragraph format hindered performance compared to list format. The participants' task simulated how these labels might be used in the real world, that is, in comparing two labels with regards to health benefits to the individual. Also, basic demographic data (e.g. age, gender, occupation, etc.) was collected to determine if they affected comparison times and if they interacted with label format.

Methods

Participants.

Thirty-six participants (M = 30.5, SD = 14.5) from the Raleigh, North Carolina area participated. The sample was composed of 22 males and 14 females. Half of the participants were college students and the other half were non-students from the surrounding area. The former received experimental credit for an introductory psychology course for participating, while the latter received remuneration of five dollars.

Apparatus

Each participant received three booklets containing 15 pairs of food labels positioned side-by-side. Figure 1 provides examples of each of the three types of food labels (list with rules as per style required by the FDA, list with no rules, and paragraph). Each of the list format labels used an abbreviated label format (e.g. Goldberg *et al*, 1999).

Procedure

All participants were given three booklets; each with one of three format conditions (list with horizontal lines, list with no lines, and paragraph). Each booklet contained 15 pairs of food labels positioned side-by-side. All three booklets had the same information content in the labels such that there was a label pair with identical information content in each format condition. Within each pair of labels, only one number of one of the labels was different for one of the nutrient conditions. Order of the booklets was counterbalanced across participants using a Latin-square of six orders. The font and size of print were identical in all conditions.

All participants were initially asked to assume that their physician had told them that they should eat foods with greater amounts of Calcium, Iron, and Vitamin C and to reduce intake of foods with Fat, Sodium, and Cholesterol. These components were manipulated in the labels, such that one of the two labels of each pair was more in compliance with these instructions. For each pair, one number was slightly higher or lower for one of the relevant nutrients. The participants' task was to determine which of the two labels on a page indicates a healthier food based on the above-mentioned dietary/health recommendations of their physician. Participants were told to evaluate the pair of labels on each page and to mark their answers on a response sheet. A timer was started when the participants began the first page of their booklet and was stopped when the participant gave a response for the last pair in the booklet.

Nutrition Facts
Serving Size: 1 cup

Amount Per Serving

Calories 190	Calories From Fat 90
	% Daily Value*
Total Fat 7g	15%
Saturated Fat 4g	10%
Cholesterol 18mg	6%
Sodium 200mg	8%
Total Carbohydrates 27g	9%
Dietary Fiber 3g	4%
Sugars 16g	
Protein 2g	

Vitamin A 0%	Vitamin C 7%
Vitamin E 0%	Potassium 2%
Calcium 4%	Iron 2%

*Percent Daily Values are based on a 2,000 calorie diet

(a)

Nutrition Facts
Serving Size: 1 cup

Amount Per Serving

Calories 190	Calories From Fat 90
	% Daily Value*
Total Fat 7g	15%
Saturated Fat 4g	10%
Cholesterol 18mg	6%
Sodium 200mg	8%
Total Carbohydrates 27g	9%
Dietary Fiber 3g	4%
Sugars 16g	
Protein 2g	

Vitamin A 0%	Vitamin C 7%
Vitamin E 0%	Potassium 2%
Calcium 4%	Iron 2%

*Percent Daily Values are based on a 2,000 calorie diet

(b)

Nutrition Facts Serving Size: 1 cup (16oz), Amount per serving: **Calories** 190, Calories From Fat 90, **Total Fat** 7g (25% DV), Saturated Fat 4g (10% DV), **Cholesterol** 18mg (6% DV), **Sodium** 200mg (8% DV), **Total Carbohydrates** 27g (9% DV), Dietary Fiber 3g (4% DV), Sugar 16g, **Protein** 2g, Vitamin A (0% DV), Vitamin C (7% DV), Vitamin E (0% DV), Potassium (2% DV), Calcium (4% DV), Iron (2% DV) Percent Daily Values (DV) are based on a 2,000 calorie diet

(c)

Figure 1. (a) List with horizontal lines ("rules"), (b) list with no rules, and (c) paragraph

Results

Using a repeated measure analysis of variance (ANOVA), the results indicated an overall effect of label format, $F(2,70) = 30.7$, $p<.0001$. Comparisons among the means, using Tukey's Honestly Significant Difference (HSD) test, showed that the list-type labels with the rules (M = 182.2 seconds, SD = 63.4) and without (M = 195.4 seconds, SD = 62.9) produced significantly faster comparison times than the paragraph-type label (M = 291.4 seconds, SD = 110.0). The list label with the rules produced somewhat faster comparisons than the list label without the rules, but the difference was not significant.

In addition, a set of analyses was conducted including several demographic variables to determine if they produced a main effect or interaction with label format. The demographic variables included age, gender, participant's first language (English vs. non-English), and occupation (student vs. non-student). Thus, a set of 2 (demographic variable) x 3 (label format) mixed-model analyses of variance (ANOVAs) with the former a between-subjects factor and the latter within-subjects factor were conducted. The only significant effects in the analyses were produced by the variables of first language and occupation.

There was a significant main effect of participants first language, $F(1,34) = 4.5$, $p<.05$, as well as a significant interaction with label format, $F(2,68) = 7.1$, $p<.01$. Non-native English users ($M = 275.9$) were significantly slower than native English users ($M = 214.7$). Table 1 shows the interaction means. The non-native English users were much more impaired by the paragraph format relative to the two list formats than the native English users.

Table 1. Mean search times as a function of first language and label format

	Nutrition label format (seconds)		
Language	List with rules	List without rules	Paragraph
English	162.9	170.2	233.4
Non-English	202.6	220.6	349.4

There was a significant main effect of occupation, $F(1,34) = 15.1$, $p<.001$, as well as a significant interaction with label format, $F(2,68) = 4.0$, $p<.05$. Non-students ($M = 257.5$) were significantly slower than students ($M = 188.9$). Table 1 shows the interaction means. The non-students were much more impaired by the paragraph format relative to the two list formats than the students.

Table 2. Mean search times as a function of occupation and label format

	Nutrition label format (seconds)		
Occupation	List with rules	List without rules	Paragraph
Student	179.3	194.0	270.7
Non-student	204.4	203.8	419.6

Discussion

The results confirm that list-type format facilitates the ease with which comparison times are made between nutrition labels. The data shows that paragraph format is much more difficult than the list format to make the same comparisons probably because it take more time to locate comparable parts in former than in the latter format. In the paragraph format the component parts are not linearly aligned in column as in the list format. The lack of a difference between the two list formats is probably due to the relatively small physical difference in the appearance of the two conditions.

The results also demonstrate that the paragraph format is particularly disadvantageous for certain populations of users. Although search time performance was generally lower for non-students and non-native English language users, these groups performed much more poorly with the paragraph format relative to the list formats. In fact, non-native English users took more than twice as long to make comparisons with the paragraph labels compared to both sets of list labels. A similar, but not as extreme, trend was found for non-students. One possible reason the student/non-student search time differences relates to age. The non-students ($M = 41.3$, $SD = 13.6$) were older than the students ($M = 19.7$, $SD = 1.5$). Possibly some of the non-students were experiencing age-related declines (e.g. difficulty in reading printed material).

The FDA currently allows a paragraph-format in nutrition labels when surface area for labeling is limited. However, the performance reduction with the paragraph-format labels in the present study was so dramatic that it strongly suggests that the FDA should reconsider allowing its use in favor of more usual list-type label. While the list label takes more space than a paragraph label (holding type size constant), sometimes labels can be re-arranged to provide more space. Additionally, the lack of space can sometimes be remedied by the use of alternative label designs (e.g., the addition of tag or wrap-around labels) to avoid a reduction of print size (e.g. Young and Wogalter, 1993). A question that remains is whether list format is better than paragraph when print size is reduced and/or surface area is held constant (e.g. reducing print size and line spacing in the list format version to match the footprint of the paragraph format version).

The current research has implications for other kinds of applications involving text format. Given the present results and earlier studies (e.g. Galitz, 1989, Goldberg *et al*, 1999, Wogalter and Post, 1989, Wogalter and Shaver, 2001), list formatted text appears to aid information acquisition performance across numerous kinds of information display domains compared to traditional paragraph presentation.

References

Coren, S., Ward, L.M. and Enns, J.T. 1999, Perceptual change in adults. In *Sensation and Perception*, Fifth Edition, (Harcourt College Publishers, Fort Worth)

Food and Drug Administration (FDA). April 2001, *Nutrition labeling of food,* 21 CFR 101.9, (U.S. Government Printing Office, Washington, DC)

Galitz, W.O. 1989, *Handbook of Screen Format Design,* Third Edition, (QED Information Sciences, Inc., MA)

Goldberg, J.H., Probart, C.K. and Zak, R.E. 1999, Visual search of food nutrition labels: *Human Factors, 41,* 425-437

Hartley, J. 1999, *What does it say? Text design, Medical Information, and Older Readers*, (Lawrence Erlbaum Associates, Inc., NJ)

Morrow, D. and Leirer, V.O. 1999, *Designing medication instructions for older adults,* (Lawrence Erlbaum Associates, Inc., NJ)

Tullis, T.S. 1983. The formatting of alphanumeric displays: A review and analysis: *Human Factors, 25,* 657-682

Wogalter, M.S., Forbes, R.M. and Barlow, T. 1993, Alternative product label designs: Increasing the surface area and print size: In *Proceedings of Interface '93,* (The Human Factors and Ergonomics Society, CA)

Wogalter, M.S. and Post, M.P. 1989, Printed tutorial instructions: Effects of text format and screen pictographs on human-computer task performance: In *Proceedings of Interface '89,* (The Human Factors Society, CA)

Wogalter, M.S. and Shaver, E.F. 2001, Evaluation of list vs. paragraph text format on search time for warning symptoms in a product manual: In A.C. Bittner Jr., P.C. Champney, and S.J. Morrissey (ed.) *Advances in Occupational Ergonomics and Safety IV* (IOS Press, Amsterdam), 434-438

Wogalter, M.S. and Young, S.L. 1994, The effect of alternative product-label design on warning compliance: *Applied Ergonomics, 25,* 53-57

SLIPS TRIPS AND FALLS

WHY DO PASSENGERS GET HURT WHEN BUSES DON'T CRASH?

Rachel Grant[1], Alan Kirk[1], Richard Bird[2]

[1]Vehicle Safety Research Centre, [2]formerly of ICE Ergonomics
Research School in Ergonomics and Human Factors
Loughborough University
Loughborough, Leicestershire, LE11 3UZ. UK

Two major bus safety reports have been completed at the Research School in Ergonomics and Human Factors at Loughborough University. It has become evident during these projects that non-collision incidents are an important part in the injury experience of bus casualties, especially for elderly occupants. By consideration of both national statistics and in-depth cases a picture has been formed of the bus and coach casualty population and the types of incidents in which these people are injured. A brief summary is given of these statistics, along with possible reasons for such a high proportion of casualties occurring in non-collision incidents and recommendations have been made that would lessen the risk of these injuries occurring, through better design and operational changes.
Keywords: Non-collision, DDA, DIPTAC, PSV, Bus, Coach

Introduction

The first study was undertaken on behalf of the Department of Transport, Local Government and the Regions (DTLR) and was entitled the 'Assessment of Passenger Safety in Local Service PSVs'. This study assesses the impact of the Disability Discrimination Act (DDA) and the Disabled Persons Transport Advisory Committee (DIPTAC) regulations on bus travel. The second was a study undertaken for Task 1.1 of the Enhanced Bus and Coach Occupant Safety (ECBOS) project, funded by the European Commission 5th Framework Programme (project no. 1999-RD.11130). 'Real World Bus and Coach Accident Data from Eight European Countries' is a collation of European data that identifies the important issues in bus and coach occupant safety.

Both studies used British national road accident data to investigate bus and coach accidents. Unfortunately there is no way to distinguish between a 'city' bus or coach and a 'touring' bus or coach in the coding. The analysis therefore covers all buses and coaches that have 17 or more seats (regardless of whether or not they are being used in stage operation). (RAGB, 1994 to 1998 and STATS 20).

As part of the study undertaken for the DTLR, physical designs of the current bus fleet were examined during a market review. This provided information on the types of designs currently in use within the UK and the hazards associated with these designs. A task analysis was undertaken of the actual bus journey from the passenger's point of view. This identified the extent of which passengers would be exposed to any hazards during the journey including such activities as boarding and alighting. As well as investigating the bus design, passenger issues were considered. These included the effects of sensory disabilities; slips, trips and falls; and the characteristics of the bus user population. This work has been used to identify how and why injuries occur.

The casualty population

Passenger casualties on buses and coaches represent 1.4% (644 out of 47,652) of all killed or seriously injured (KSI) casualties in Great Britain. Whilst this percentage is low, and an analysis of exposure indicates that bus travel is one of the safest modes of transport, this study identifies issues that should make local bus transport even safer. Also, as new low floor buses make travel more viable for less physically mobile passengers it is vitally important to make sure that these people are not suffering injuries inside the vehicle which will make the overall proportion of bus casualties higher.

Of these KSI casualties 62.6% occur when the vehicle suffers no impact and 57.1% occur when the passenger is not seated. Overall 48.8% of KSI passenger casualties are not seated and the vehicle does not have a collision. These are large proportions of the bus casualty population. When they do receive an injury, passengers who are not seated at the time are more at risk of receiving a serious or fatal injury (5.5 compared to 10.0%).

Looking at just the non-collision population it is found that 93.9% of all casualties occur on roads with a 30 mph speed limit and 3.6% on 40 mph roads. These roads are defined as built up areas by the UK government. In the data it is not possible to separate local buses and coaches but this high figure in built up / urban areas indicates that the great majority of non-collision incidents occur on local service buses.

The non-collision casualty population: gender and age
The gender distribution for KSI casualties, injured when no collision takes place, shows that there are over twice as many females (71.6%) as males (28.4%). This is likely to be both a function of greater bus use by females and a lower tolerance to injury. In figure 1 a peak is seen for school age males and there is an obvious increase in numbers amongst elderly females. The mean age for female casualties is 15 years higher than for males.

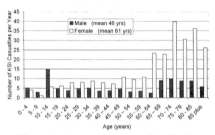

Figure 1. Age and gender distribution

Overall there is a marked increase in the likelihood of a serious or fatal injury to a female occupant as their age increases. Governmental surveys show that generally women travel more on local buses than men for most types of area and age. This goes some way to explaining why women have a much greater representation as bus or coach casualties than men on the database. Overall it is estimated that in the 16 to 59 years old age group women travel 47% further on local buses than men. Women of all ages also make more local bus journeys than men, whilst travelling further, giving higher exposure to injuries that occur whilst standing, boarding or alighting the vehicle, as they get on and off more often.

How and why do these injuries occur?

This work has shown that 63% of all KSI bus passenger casualties are in non-collision incidents with a shift towards elderly female passengers. This section will discuss problems on buses that cause these injuries. Generally it is felt that most of these types of non-collision injury are taking place on local service buses, borne out by 93.9% of these injuries occurring on 30 mph roads. The rest of this paper will therefore concentrate on these vehicles.

Slips, trips and falls on the vehicle

Caused by:

Slippery floors,
Weather conditions,
Uneven floors (fig 2)
Unexpected or high steps (fig 3),
Steep slopes,
Lack of visual cues,
Physiology in the elderly.

Figure 2   **Figure 3**

Slips, trips and falls whilst boarding or alighting

Caused by:

Step to the kerb can be too high,
Riser steps of different heights,
Passengers can be encumbered.

Figure 4

Operational issues or heavy braking

Falls can occur from the mechanisms mentioned above but the operation of the vehicle can also initiate a fall on a bus.

Caused by:

Acceleration, vehicle pulls away before passenger reaches seat,

Deceleration, passenger stands to get off bus before bus has come to a halt,

Vehicles sometimes need to turn sharply into and out of recessed bus stops,

Emergency manoeuvres.

Driver issues

In the work carried out for DTLR one operator said 90% of complaints from injured passengers put the blame on the driver, but it is important to recognise workload is high.

Caused by:

High levels of traffic congestion,

Pressure to keep to timetables,

Single operator buses.

Interior design

What are the dangers when a passenger does slip, trip or fall? Why are injuries caused? These pictures give examples of interior design that can lead to injuries when passengers make contact with internal parts of the bus. These are typical of the bus fleet.

Figure 5

Figure 6

There are unprotected metal grab rails in the areas where seated passengers' heads will naturally fall forward and passengers' upper extremities may hit if they fall over (figures 5 and 6).

Figure 7

Figure 8

Figure 9

Figures 7 and8 show ticket machines with very hard metal edges that a standing passenger could easily fall forwards and hit, for example, during hard braking . Likewise a boarding passenger could trip and strike the machine. Generally ticket machines, card readers, and bins are not integrated into the design of the bus, they appear to be bolted on afterwards depending upon the requirements of the operator. This inevitably causes them to encroach on the standing area. Also shown is an example of the hard metal joints used for the interior grab bars (figure 9).

New legislation

Public Service Vehicles Accessibility Regulations legislate on guidelines from the Disability Discrimination Act (DDA), Disabled Persons Transport Advisory Committee (DIPTAC) bus regulations. Under the Public Service Vehicle Accessibility Regulations, which have been in force in the UK since January 2001, the previous guidelines of the Disability Discrimination Act (DDA) have been adopted. These are in line with the European directive on bus safety. Generally these guidelines make access on and off vehicles easier and vehicle interiors safer. They have significant advantages on the ease of access for all passengers but especially the less mobile. New buses will have low floor access, priority seats and crucially space for wheelchairs and push chairs (figure 10). The improved overall design also includes straight stairs on double-deckers (figure 11), better lighting and better visual marking (figure 12).

Examples of new design

Figure 10   **Figure 11**   **Figure 12**

The continuing relevance of non-collision injuries

Even though new legislation has been introduced recently, Great Britain will still have older buses for some time to come and all buses in service will not have to comply until 2015 (coaches, 2020). In fact in 1999 the average age of the public service vehicle fleet in Great Britain was 10 years old, with 10,000 being up to 18 years old. (DETR 1999). Therefore the authors believe it is very important to still consider the access issues raised in this paper as they will affect bus users for at least another 10 years. Also whilst these access regulations generally improve the interior design of the bus, interior contacts must be kept in mind during the vehicle design. In fact an unfortunate by-product of some of these regulations is that the number of seats are reduced, which means that more people may be forced to stand or move upstairs, it is therefore just as relevant to consider falls, especially from bus operation, on these new buses as on older buses.

Conclusions and recommendations

- The majority of killed and seriously injured bus passenger casualties in Great Britain (63%) occur when the vehicle is not involved in a collision.
- It has been found that there is a high proportion of elderly female passengers in this casualty population who, when injured, have an increased risk of serious injury.
- Legislation is changing the design of buses and the authors obviously support those changes which make public transport more widely available. However legislation is improving access for all, enabling more extremes of the population and therefore the less mobile, to travel on buses. These people will be both more susceptible to falls, and to injuries if they fall, whilst on the vehicle. New regulations are in force but they do not place requirements on good operating practice. Also the vehicle fleet includes a large proportion of older vehicles and these new bus designs will not be commonplace for many years to come.

Recommendations

- Regulations have improved access but better interior design is needed, especially around the ticket/driver area and near to the doors to minimise contact injuries. Maintenance procedures should also ensure there is no compromise on safety.
- There should be less pressure on operators and therefore drivers to achieve stricter timetables in mounting congestion at the expense of safety.
- Systems need to be in place to ensure that drivers are aware that a seated passenger wishes to alight at the next stop, and passengers need reassurance that the driver is aware they wish to alight. Bell pushes to achieve this should be in easy reach of all.
- During busy times on busy routes it would be beneficial to have a conductor accompanying the driver, collecting fares, helping passengers (especially with wheelchairs) and dealing with unruly passengers, leaving the driver to concentrate on driving.

References

Transport Statistics, Department of the Environment, Transport and the Regions, *Road Accidents Great Britain - The Casualty Report*, London, editions 1994 to 1998.
Technical University Graz, Austria, ECBOS report Task 1.1. www.dsd.at/ecbos.htm
Transport Statistics, Department of the Environment, Transport and the Regions, *Focus on Public Transport, Great Britain*, London, 1999 edition.
Transport Statistics, Department of the Environment, Transport and the Regions, *A Bulletin of Public Transport Statistics, Great Britain*, London, 1999 edition.
Coach and Bus Week Magazine, published by Emap Automotive.
Reports for Task 1.1 of the Enhanced Bus and Coach Occupant Safety project (European Commission 5th Framework Project no. 1999-RD.11130) www.dsd.at/ecbos.htm
Vehicle Standards and Engineering, Department of the Environment, Transport and the Regions, *Assessment of Passenger Safety in Local Service PSVs*, London.
Transport Statistics, Department of the Environment, Transport and the Regions, *STATS 20, Instructions for the Completion of Road Accident Report Form STATS 19*, London.

National (STATS 19) road accident data is collated by the Department of Transport, Local Government and the Regions (DTLR) and supplied to the Vehicle Safety Research Centre in electronic form by the UK Data Archive at Essex University.

REDUCING FALLS IN THE HOME AMONG OLDER PEOPLE – BEHAVIOURAL AND DESIGN FACTORS

C L Brace, R A Haslam, K Brooke-Wavell, P A Howarth

Health and Safety Ergonomics Unit,
Department of Human Sciences,
Loughborough University,
Leicestershire,
LE11 3TU
UK

Falls in the home are a major problem for older people. Although personal and environmental risk factors for falling among this group are well understood, less is known about how these risks are influenced by behaviour. The research reported in this paper addresses this problem, while also giving consideration to the practicalities of everyday living for older people. Interviews were conducted with 177 community dwelling people from 150 households, representative of the UK population (aged 65+). The study has highlighted many behavioural and design factors involved in fall risk. These include footwear, walking aids, storage, use and design of domestic products in combination with behaviour such as hurrying, carrying objects, and keeping pets. Reducing falls in the home among this population requires a holistic approach, with attention both to design and behaviour.

Introduction

At least two decades ago it was recognised that a third of individuals over 65, and nearly half of those over 80, fall each year (Prudham and Evans, 1981), with little impact on the scale of the problem during the intervening years. Approximately half of all recorded fall episodes that occur among independent community dwelling older people happen in their homes and immediate home environments (Lord *et al*, 1993). The most recent HASS data reveal that in 2000, 330,000 older people in the UK received injuries from a fall in the home severe enough to require attendance at a hospital A&E department (DTI, 2001). This does not include patients seeing their GP or those not seeking treatment. The consequences of falling for older people can be traumatic and seriously disabling.

Falls pose a threat to older persons due to the combination of high incidence with high susceptibility to injury. The tendency for injury because of a high prevalence of clinical diseases (e.g. osteoporosis) and age-related physiological changes (e.g. slowed protective reflexes) makes even a relatively mild fall particularly dangerous (Josephson *et al*, 1991).

Falls can lead to three types of impairment: injury, restriction of activity, psychological distress (Cwikel *et al*, 1990), including anxiety of falling again, restrictions in activity/mobility, and increased need of assistance. The cost to individuals and society is great and likely to increase in line with general ageing of the population.

There are numerous risk factors in falling. Intrinsic factors are age and disease related changes within the individual that increase the propensity for falls. Extrinsic factors are environmental hazards that present an opportunity for a fall to occur. Individual fall incidents are generally multifactorial

Intrinsic factors involved in falls among older people include decreased balance ability, disturbed gait, cognitive impairment, reduced strength, impaired vision, illness, and side effects from use of medication (Askham et al, 1990). With regard to vision, depth perception and judgement of distance may both be involved in falls (Davis, 1983; Cohn and Lasley, 1985). General psychological state and experience can also have an effect on the individual, affecting confidence and fear of falling. Issues here include fall history, previous falls, length of lie on floor/ground surface, range of activities of daily living, and degree of social interaction and support (Nelson and Amin, 1990; Tideiksaar and Kay, 1986).

Extrinsic causes are extensive, and include floor surfaces (textures and levels), loose rugs, objects on the floor (e.g. toys, pets), poor lighting, problems with walking aids and equipment, lack of hand rails on stairs, badly repaired stairs, ill-fitting footwear, unlaced shoes, high heels, slippers without soles, sensory surround and feedback (audio and visual), placement of furniture, and required activities in the physical environment (Nelson and Amin, 1990; Burleson, 1993).

Previous research (Hill et al, 2000; Haslam et al, 2001) has identified important behavioural factors which affect the risk of older people falling on stairs, e.g. rushing, carrying objects. The research reported here is extending this work into other areas of the home.

Method

Semi structured interviews were conducted with 177 older people (150 households), aged 65 – 99, in their own homes. Participants were recruited through existing subject lists held by the researchers and through contacts within the local community. Participants were sampled according to age and gender using estimated population figures from the UK, and according to their accommodation. Properties were selected by both age and type of housing, using national estimates of housing stock.

Issues explored by the interviews included factors affecting risk of falling in the home, embracing age-related aspects and self-perceived safety. The immediate and longer term consequences of having a fall and the value and acceptability of preventative measures were also discussed. The interviews involved detailed discussion of different areas of the home, with regard to specific risk factors, and the interviewee's fall history. In addition, standard anthropometric dimensions of interviewees were recorded, along with other measurements including grip strength, ability to get off a stool without using hands, spectacle wear and measures of visual acuity and depth perception.

Interviewees were briefed both verbally and in writing about the study prior to participation. They were informed that the discussions would consider falls in the home (including the garden), examples of falls, and risk factors and safety issues that might be involved. However, they were not given any further information prior to the discussion, to avoid leading responses in any particular direction. Each interview lasted approximately two hours, with all interviews conducted by the same researcher.

Results

Table 1. Participant characteristics (n=177)	
age	mean 76 years (sd 7.3) (range 65-99)
male	27%
female	73%
living alone	47%
eye sight test in last 2 years	90%
use bifocal spectacles	38%
have a condition that affects vision (e.g. cataracts, glaucoma, macular degeneration etc)	35%
take at least 1 prescribed medication daily	79%
take 4 or more prescribed medications daily	23%
fallen in home since age 65	48%
experienced 2 or more falls in home	21%

Table 2. Household characteristics	
property age	median 40 years (range 3-110)
detached	23%
semi-detached	26%
terraced	5%
flat	16%
bungalow	26%
other types of properties (e.g. bed sit)	4%

Behaviour Affecting Fall Risk

Behaviour involving direct use of the home environment, perceived by the participants as affecting fall risk, included: hurrying, aspects of house maintenance (e.g. changing light bulbs, using stepladders, 'clutter') and gardening. One of the most hazardous activities was thought to be getting in and out of the bath.

Participants accepted that actions affecting the home environment also affect safety. These included fall risks introduced by occupants, e.g. leaving things on the floor, maintenance and type of floor, use and condition of low lying furniture, and the provision of lighting. Figure 1 illustrates some of these issues. Cohabitation, visitors and pets were also believed by study participants to be factors. Other inhabitants and visitors (including grandchildren) were reported to result in an increase in 'clutter', especially unexpected objects, which were not usually present. Pets underfoot and pet items including bowls and toys were also reported as an issue for fall safety, due to tripping. Individual capability affected by behaviour was also discussed as amplifying risk of falling. Inappropriate footwear and spectacles, use of medication, use or non-use of lighting, and a lack of regular exercise were examples of this.

Qualitative evidence on age-related factors thought to lead to increased risk of falling highlighted the negative effects of decreased mobility, reduced balance and strength, and deteriorating vision.

Risk Perception

With regard to self-perceived fall safety, the most perilous areas of the home were considered to be the garden, kitchen, bathroom and stairs, due to the nature of the tasks performed in these places (bending, reaching, etc.) and the environmental hazards present, encompassing changes in level and surface type and texture, slopes, objects left on the floor and rugs.

Figure 1: Household 'clutter' (Household 22)

Fifty eight percent of the home environments visited had been altered in the last 5 years to improve safety with regard to falling. Changes included the use of compact fluorescent (long life) light bulbs, anti slip mats, grab rails, and low-maintenance garden designs. Participants from 40% of the households had ideas for changes that they wanted to make to their homes in the future to make them safer. These included downstairs toilets, walk in showers, and extra grab rails around the home and garden.

Product Design and Falling

The design of some domestic products may contribute to falls, including oven and dishwasher doors that open downwards forming a trip hazard, or cleaning equipment that is heavy and difficult to hold. The design of stepladders was clearly an important issue for many of the interviewees and improved designs would be welcome. Features suggested were an additional high handrail and a tool holding compartment, to reduce the need for continual movement up and down the ladder to fetch items. The design of light shades for ease of removal during bulb changing was another feature that participants felt would be beneficial, and again would ease a task performed off the ground.

Footwear design, especially slippers, was felt to be a factor in fall safety, particularly the quality, thickness, grip and durability of the sole.

Walking aids were often reported to be unsuitable for use in the home environment, due to the changes in floor surface and texture, and limited room for manoeuvre, predominantly for 'zimmer frame' type appliances. The difficulty in storage of these aids was another problem raised in relation to falling and tripping.

Fall alarms were generally regarded by the sample as a good idea, particularly for people living on their own, although this often did not translate into actual usage. Problems were reported with these products including comments that, 'they get in the way', 'they're uncomfortable to wear', and 'only old people wear them'.

Building design may also introduce risks. For example, additional steps within the house or garden (particularly in older properties) or difficult to access storage, such as kitchen or other cupboards that are too high to reach. Lack of storage space was another problem highlighted, resulting in objects being left on the floor, particularly if an occupant had moved into a smaller property than lived in previously.

Discussion

This survey has found the risk factors for falling to be widespread, especially with regard to fundamental design features and interviewees' behaviour. This research supports previous suggestions (Hill *et al*, 2000; Haslam *et al*, 2001) that despite many risk factors for falling in the home being apparent to older people, this awareness does not necessarily influence how they behave in practice.

In some cases a risk may be recognised, but without behaviour making an allowance for it. For example, some of the interviewees did not, or did not want to, acknowledge that their physical abilities had deteriorated with a decline in health status, and therefore continued with activities regardless of this factor. Unfortunately, it is difficult to say what is the best advice to give to this population. It is important in terms of health, autonomy and independence to remain active for as long as possible in old age; conversely, it is not prudent to advise older people to continue with activities that could be a danger to health and well being. On the other hand, some people are very sensitive to the issue of fall safety and have made modifications themselves to their homes. In some instances, interviewees appear to be 'ageing gracefully', by putting in place new mechanisms and methods for completing activities of daily living, and being sensibly cautious in their behaviour.

In other cases, a risk may not be recognized or understood. A lack of awareness of risk by others may also lead to a dangerous situation. For example, objects left on the floor might cause difficulties for another member of the household, who is unaware the hazards are there.

The accuracy of risk perception is also an interesting aspect. For example, interviewees report taking extra care in the bathroom because they think that they are at increased risk of falling there, although, according to HASS data (DTI, 2001), almost twice as many falls occur in the bedroom compared to the bathroom. Few interviewees mentioned the bedrooms as areas of concern for falling.

The serious immediate and long-term implications of having a fall were generally well understood by the sample, due to personal experiences or those of friends or relatives. However, less than 5% of the sample could recall ever seeing or receiving any advice or information about fall safety.

Confronting the problem of older people falling in the home requires a holistic, ergonomics approach that addresses design as well as behavioural issues. The effects of improved building regulations and standards will direct improvements in the design of future housing and the home environment. Improvements in the design of household appliances and personal products (walking aids, footwear etc.) could make a useful contribution to safety and general ease of use. Many of these design improvements would benefit every user, not just older people.

Conclusions

Opportunities exist to reduce the risk of older people falling in and around the home, both with respect to behaviour and the design of products and buildings. Improvements to products and the built environment will need to be longer-term initiatives. Meanwhile, there are more immediate measures that can be taken by older people and their carers to improve fall safety. Most importantly, there is a need to raise awareness of the problem and provide practical fall prevention advice.

Acknowledgements

The authors wish to acknowledge the support of the Department of Trade and Industry (DTI) who sponsored this research. The views expressed, however, are those of the authors and do not necessarily represent those of the DTI.

References

Askham, J., Glucksman, E., Owens, P., Swift, C., Tinker, A. and Yu, G., 1990, *A Review of Research on Falls Among Elderly People* (Department of Trade and Industry: London).

Burleson, L.K., 1993, Parkinson's disease: Relationship between environmental design and falls risk. Unpublished doctoral dissertation, Texas Tech Univ, Lubbock, TX.

Cohn, T.E. and Lasley, D.J., 1985, Visual depth illusion and falls in the elderly, *Clinics in Geriatric Medicine*, **1**, 601-620.

Cwikel, J., Fried, A.V., and Galinsky, D., 1990. Falls and psychosocial factors among community-dwelling elderly persons: a review and integration of findings from Israel. *Public Health Review 1989/90*; **17**: 39 – 50.

Davis, P.R., 1983, Human factors contributing to slips, trips and falls, *Ergonomics*, **26**, 51-59.

Department of Trade and Industry (DTI), 2001, *Home accident surveillance system including leisure activities: 23rd annual report 1999 data* (DTI: London).

Haslam, R.A., Hill, L.D., Sloane, J.E., Brooke-Wavell, K., and Howarth, P.A., 2001, Safety of Older People On Stairs. In: Proceedings, RoSPA, Home Safety Congress, 12-13 November 2001, Stratford-Upon-Avon.

Hill, L.D., Haslam, R.A., Howarth, P.A., Brooke-Wavell, K., and Sloane, J.E., 2000, *Safety of Older People on Stairs: Behavioural Factors*. (Department of Trade and Industry: London). DTI ref: 00/788.

Josephson, K., Fabacher, D., Rubenstein, L., 1991, Home Safety and Fall Prevention. *Clinics in Geriatric Medicine* 1991; **7** (4): 707 – 731.

Lord, S.R., Ward, J.A., Williams, P., and Anstey, K.J., 1993, Physiological factors associated with falls in older community-dwelling women. *Australian Journal of Public Health*; **17** (3): 240-5.

Nelson, R.C., and Amin, M.A., 1990, Falls in the Elderly. *Emergency Medicine Clinics of North America* 1990; 8, 309 – 324.

Prudham, D., and Grimley-Evans, J., 1981, Factors associated with falls in the elderly. *Age and Ageing* **10**:101.

Tideiksaar, R., and Kay, A.D., 1986, What causes falls? A logical diagnostic procedure. *Geriatrics* 1986; **41** (12), 32 – 50.

FORENSIC ASPECTS OF PEDESTRIAN FALLS

Daniel Johnson PhD CPE[1] and H. Harvey Cohen PhD CPE[2]

[1]*President, Daniel A. Johnson Inc.*
6221 Swayne Drive NE
Olympia, Washington, USA
[2]*President, Error Analysis, Inc.*
5811 Amaya Drive, Suite 205
La Mesa, California, USA

When adults fall serious, even fatal, injuries can occur. The forensic ergonomist may be called upon to investigate the fall and to render an expert opinion as to whether the fall resulted from inadequate design, improper maintenance or incautious or unpredictable behavior on the part of the injured person. The forensic ergonomist must be ready to gather the appropriate information, take necessary measurements and photographs, analyze the collected information and compare it to building codes, safety standards and scientific data about what causes falls and how those falls can be avoided. This information must then be presented to the trier of fact. Data gathering, taking measurements, interview techniques and preservation of evidence are discussed as related to trips, slips, missteps and falls.

Introduction

Each year in the United States falls cause thousands of deaths, millions of injuries, and cost billions of dollars (Pauls, 1991). And, with the aging of the population, deaths, serious injuries and costs may be expected to increase.

But recent ergonomic research has resulted in improved designs of walkway surfaces, such as ramps and stairways, and greater awareness of the need for good maintenance and lighting. The results of this research, properly implemented, should counteract the trend toward more falls.

Ergonomic research has led to improved safety standards that are often codified into law. These Codes are used as benchmarks in court to determine if a property owner has liability for a fall that occurred on the property. Often at issue is whether the behavior of the injured party contributed to the fall or the severity of the resulting injuries.

Forensic ergonomists investigate, write reports and provide testimony as to whether a given fall resulted from inadequate design, improper maintenance or incautious or unpredictable user behavior. To be effective a forensic ergonomist must gather the appropriate information, take the necessary measurements and photographs, analyze this data, compare it to what is known about what causes people to fall and how those fall can be avoided. This information must then be presented to the trier of fact (Cohen and

LaRue, 2001). An effective forensic ergonomist can help the court determine if a plaintiff's injuries were the result of inadequate design or maintenance, or were due to inappropriate or unexpected user behavior. In the long term, an effective ergonomic effort can result in improved walkway design and fewer falls.

Gathering appropriate information

Typically, a forensic ergonomist is contacted by an attorney for either the property owner (the defendant) where a fall has taken place, or by the attorney for the injured party (the plaintiff). Occasionally, plaintiff's attorney will contact the forensic ergonomist soon after an injury to find out if a lawsuit should be filed. But more often the forensic ergonomist is contacted months, sometimes years, after a suit has been filed. The forensic ergonomist must then gather and review all the relevant information.

Photographs
Since photographs taken of the fall site at about the time of the fall can be very useful, large color prints from original film (e.g., the negatives) should be requested.

Statements
Accident reports and eyewitness statements filled out shortly after a fall can give insight since memory of the event generally degrades over time (Robins, 2001).

Medical reports
Medical reports should be reviewed with some caution since medical personnel seldom witness the fall. They are less concerned, and rightly so, with how or why a fall occurred than with how to resolve the injury. They record only what they have been told and are not necessarily accurate as to how, or even where, a fall took place. (In one case a surgeon recorded that the plaintiff fell from a ladder. When the plaintiff brought to his attention that he said he actually fell on a ramp, the surgeon checked his files, found that a ladder was never mentioned, and wrote a note about this error that was included in the patient's medical records.)

Medical reports are, however, very important in that they can shed light on how a fall occurred. Knowing that a fractured tibia was displaced medially can help determine how a fall occurred. Medical reports often mention minor injuries a plaintiff may forget but which can be important clues to the mechanism of a fall.

Sworn testimony
Deposition transcripts are a major source of information. Attorneys for either side elicit specific and detailed information not readily expressed by plaintiff or eyewitnesses in accident statements. Unfortunately, such information is gathered only after a suit has been filed and this may be months or years after the event when memory of the specifics may have faded or been influenced.

Interviews of plaintiffs and witnesses
Often the most useful information is gathered when interviewing a plaintiff or witness.

It is helpful to fill out a pre-printed interview form to ensure that all relevant topics are queried. And it is good practice to let the person describe in his or her own manner what happened. The person should be asked not to guess about what happened, since speculation can act to modify a person's memory of an event (Robins, 2001).

When the person starts describing the fall the forensic ergonomist should refrain from interrupting but, instead, simply take notes. It is important to understand what the person remembers. Leading questions, which can lead to modification of the person's memory, should be avoided (Robins, 2001). When the event is fully described the forensic ergonomist should then review the notes with the person and try to resolve ambiguities.

Once the forensic ergonomist is satisfied that the specifics of the event are understood, then measurements and photographs can be taken. Occasionally, taking measurements and plaintiff or witness interviews occur simultaneously so the order is not overly important. It is important, though, that the forensic ergonomist knows how and where the fall occurred so that appropriate measurements and photographs are recorded.

Similarly, interviews may be held with the most knowledgeable people associated with the defense regarding issues of safety, standards of care being met, policies and procedures, as well as customs and practices being followed.

Taking measurements

What measurements to take and how they are taken determine the effectiveness of a forensic ergonomist.

Surface friction and how it is measured

The traction between a shoe and the surface can be measured by various methods. Since high-speed photography shows that a shoe often appears to stop for an instant before starting to slide (Perkins, 1978), the static coefficient of friction (sCOF) is considered, in the U.S., appropriate. The accepted minimum standard is for the surface to provide a sCOF = 0.5 (Sacher, 1993). A sCOF = 0.6 is recommended in facilities expected to be peopled by those who are frail or who have difficulty walking.

On a dry, clean surface the sCOF can be measured using drag sled devices such as the Horizontal Pull Slipmeter or the American Slip Meter (ASM). If the interface between the shoe and surface is wet, however, drag sled devices tend to give spuriously high readings. On wet surfaces, the Variable Incident Tribometer (VIT) and the Brungraber, Mark II are devices that reportedly give readings consistent with subjective ratings of the surface slipperiness (e.g., English, 1996).

On the other hand, both drag sled devices and people are influenced by gravity. In pulling a drag sled down a grade it registers a lower sCOF than when pulling it up grade. A device such as the VIT, which uses compressed gas as an energy source rather than gravity, is not affected by gravity. So, when using the VIT on sloped surfaces, the angle of the ramped surface must be measured and combined with the VIT readings to determine its relative slip resistance (Templer, 1992).

At least four measures should be taken at different angles in the vicinity of the fall. If excessive variability is noted more measures should be taken. Statistical analysis should reveal whether the surface produced results that were reliably lower or higher than the minimum standard for a surface with acceptable slip resistance characteristics.

Ramp steepness and how it is measured

Steep ramps increase the chance of a slip, and may also increase the chance a frail person will lose balance and fall. Pedestrian walkways sloped in excess of 7 degrees are not allowed by U.S. Building Codes. A convenient measuring device is an inclinometer that allows readings with smallest gradation units of 1 degree. The device can take readings from a surface that may be only 8cm long. Another device, an electronic level that measures linear areas of 61cm or more, provides readouts to the 0.1 degree.

Most people who fall on ramps report that their forward foot slipped on heel strike. (The walk cycle is made up of toe off, swing phase, heel strike and stance phase.) Few, however, remember exactly where the slip commenced. If visual examination reveals some areas steeper than others multiple measurements and photographs should be taken.

Trip hazards and how they are measured

A trip occurs when the toe (usually) of the shoe in swing phase contacts an obstruction such as a sharp rise in a walkway. Vertical rises of as little as 1.3cm can stop a foot in flight since the toe may come this close to the surface during swing phase (Winter, 1979; Cohen, 2000). Such a small rise, if not readily visible, may be considered a design defect.

One method for accurately measuring small changes in level makes use of a carpenter's profiling tool – a device containing about 180 small (0.8mm diameter) stiff wires, aligned in a row at a density of about 12 per cm. The device is about 15cm long. When the wires are pushed down onto a surface, such as a sharp rise in a sidewalk, they hold the shape. The profile of the surface can then be traced onto paper.

Larger trip hazards, such as a berm in a pedestrian path, can be measured with a straight edge or measuring tape.

Stairway hazards and how they are measured

Rise heights and run lengths have traditionally been taken by the "quick and dirty" method in which a measuring stick or tape is placed at the back of the tread, and the vertical and horizontal dimensions taken. Codes specify the tread depth as the horizontal distance between successive step nosings (Pauls, 1998). But this method gives inaccurate results when the treads are not consistently level, which they often are not.

Pauls (1998) has described a more elegant method made possible by the availability of accurate electronic levels. First, lay a 600mm long electronic level on successive pairs of step nosings and record the angle from the horizontal. Then measure the distance between the two nosings. Rise height is the sine of the angle times the nosing-to-nosing distance. Run length is the cosine of the angle times that distance. This method provides more reliable measures of rise heights and run lengths, and more accurately defines the stairway as the user experiences it.

Results should be compared to Code requirements, which in the U.S., state that the total variation within a flight – the largest minus the smallest -- for either risers or runs, be no more than 9.5mm. Carpeted stairs are more difficult to measure than hard surfaced stairs, since the forward projections of the compressed surfaces are sometimes difficult to determine.

Handrails and how they are measured

Research has shown that both the height of a handrail and its graspability affect whether a person is able to arrest a fall, or at least to exert enough force to the handrail to reduce injury. Maki reports that the greatest forces can be exerted if a handrail is from 91cm to 102cm above the surface (Maki, 1984) and has a circumference of about 120mm (Maki, 1985).

Handrails that are too large may allow only a "finger pinch grip" that prevents a person from exerting the force necessary to stop a fall. Both the size and profile of handrails can be measured by a tape measure or a carpenter's profile tool.

Lighting and how it can be measured

Light falling on a walkway can be readily measured with a photometer. In U.S. Codes the minimum amount of light falling on a walkway considered as an exitway (i.e., one to be used in emergency evacuation) is 1 foot candle (10 lux). Safety recommendations (IES, 1993) are that up to 10 times that illumination be provided especially where, such as on a stairway, a fall can result in serious injury.

Unfortunately, Codes call out the amount of light falling on a surface rather than the reflected light that we see. Many stairs have dark carpets (they do not readily show soil) but which reflect only 5 or 10 percent of the incident light. So, most of the light called for by Codes is not be seen.

Photography

The camera is perhaps the forensic expert's single most important tool. It can be used to show the general area where a fall occurred, what luminaires were present, different views of the potential hazard, how and where the measurements were made, and even to record the measurements themselves.

One safety issue that U.S. Codes do not address but which the forensic ergonomist should, is when confusing carpet patterns or grout lines mask stair nosings. This masking effect can best be demonstrated photographically.

In order to sufficiently document a scene it is necessary to have a still camera with, at a minimum, both normal and wide-angle lenses. The normal lens (e.g., a 50mm lens for a 35mm camera) gives a representation similar to what the human eye sees. The wide-angle lens (e.g., a 28mm lens for a 35mm camera) captures a larger view of the scene but one that appears somewhat distorted. A zoom lens is a convenient device for capturing both of these views as well as intermediate views.

A video camera is a useful adjunct. When some dynamic event needs to be recorded, such as how other people may be using a walkway, the video camera is useful (see e.g., Cohen, 2000).

By using a digital camera, or by scanning and digitizing a photographic print, one can input the image into a computer where it can examined in detail. The image can even be modified to make it appear as it did when a fall occurred. For example, if the forensic ergonomist wishes to show how a single step riser may have appeared at the time of a fall, before it was highlighted, say, with a contrasting paint applied after the fall, photographic manipulation can be helpful. A full explanation of what photo-manipulation was conducted, and how it was performed, should be provided in subsequent reports or testimony.

In addition to the cameras and lenses described, the forensic ergonomist should arrive at the scene with adequate artificial lighting.

Photographs often give inaccurate representations of what a surface looked like due to variations in such factors as exposure time, film type, processing and age of the photograph. Color swatches can be used to match colors at the scene for later referral. One such set of swatches (Munsell Charts) gives, for many different colors, the percent luminous reflectance factor under different lighting conditions.

Presentation of Opinions

The forensic ergonomist's educational qualifications and experiential background should be given when rendering written reports and sworn testimony. In addition, the following should also be described: the materials examined, the data collected, and the standards considered. Then the conclusions about how the fall occurred should be set forth. Finally, the opinion as to whether the fall resulted from violations of safety standards or Codes, from inappropriate or unpredictable behavior on the part of the plaintiff, or a combination thereof should be rendered.

References

Cohen, H.H. 2000, A field study of stair descent, *Ergonomics in Design*, 8(**2**) (Human Factors and Ergonomics Society, Santa Monica, California) 11-17

Cohen, H.H. and LaRue, C.A. 2001, Forensic human factors/ergonomics. In *International Encyclopedia of Ergonomics and Human Factors*. (Taylor & Francis, London)

English, W. 1996, *Pedestrian slip resistance: how to measure it and how to improve it* (William English, Alva, Florida)

IES, 1993, *Lighting Handbook*, 8th Edition, (Illuminating Engineering Society of North America, New York)

Munsell Charts for Judging Reflectance Factor For Illuminating Engineers, Television Engineers, Architects, Designers and Decorators. (Macbeth Division, Kollmorgen Corp., Baltimore, Maryland)

Maki, B. E., Bartlett, S. A., and Fernie, G. R. 1984, Influence of stairway handrail height on the ability to generate stabilizing forces and moments, *Human Factors*, 26(**6**), (Human Factors and Ergonomics Society, Santa Monica, California) 705-714

Maki, B.E. 1985, Influence of handrail shape, size and surface texture on the ability of young and elderly users to generate stabilizing forces and moments (West Park Research, Toronto, Ontario, Canada)

Pauls, J. 1991, Cost of injuries in the Unites States & the role of building safety, *Safety News*, (Human Factors and Ergonomics Society, Santa Monica, California)

Pauls, J. 1998, Techniques for evaluating three key environmental factors in stairway-related falls, Idea Fair Poster, Human Factors and Ergonomics Society Annual Meeting 1998, (Human Factors and Ergonomics Society, Santa Monica, California)

Perkins, P. J. 1977, Measurement of Slip Between the Shoe and Ground During Walking, in Walkway Surfaces: Measurement of Slip Resistance, ASTM Special Technical Publication 649 (ASTM, West Conshohocken, Pennsylvania)

Robins, P. 2001, *Eyewitness reliability in motor vehicle accident reconstruction and litigation*, (Lawyers & Judges Publishing Company, Inc. Tucson, Arizona)

Sacher, A. 1993, Slip Resistance and the James Machine 0.5 Static Coefficient of Friction -- Sine Qua Non, ASTM Standardization News 1993, (American Society for Testing and Materials, West Conshohocken, Pennsylvania)

Winter, D. 1979, *A Review of Biomechanics of Human Movement*, (Wiley & Sons, New York)

Templer, J. 1992 *The Staircase: Studies of Hazards, Falls and Safer Design*, (MIT Press, Cambridge, Massachusetts)

METHODOLOGY

HUMAN SYSTEMS ENGINEERING:
NEW CONCEPTS FROM OLD PRACTICES

Iain S. MacLeod

Aerosystems International
West Hendford
Yeovil, BA 20 2AL
UK

The consideration on systems architecture is changing from one that is hierarchical and 'pipeline' to one that is flat and that of a System of Systems. The primary consideration on pipeline systems is that engineering solutions and technology are seen as the driver to the achievement of the required performance of the system. In a System of Systems, technology is seen as an enabler to promote the overall performance of many collaborating systems, this through the use and handling of system wide knowledge resident both in the system machines and in the 'society' associated with the system. Thus, System of Systems is a socio-technical system where its architecture takes into account pertinent issues and their implications related to participating organisations, cultures, and pipeline systems with the aim of promoting their joint operation as a coherent whole. To assist this process, Ergonomics must develop to allow its consideration to go beyond its traditional emphasis on the work of the individual.

Introduction

Gradually, the emphasis on the consideration of systems is changing from a technological viewpoint to a viewpoint that is based on socio-technical system considerations. This is partly driven by an increased awareness of the new effectiveness of work production that can be supported by the latest Information and Communications technologies. However, there is a change of consideration in that technology is no longer seen as a driver or dictator within design but as an enabler.

By a socio-technical system is meant a system that is composed up of many systems both social and technical. The design and development of such a system considers the overall architecture, processes, diversity, and dispersion its many constituent sub systems. Part of this consideration is the technical build, performance, and effectiveness of the sub systems. However, a major part of the consideration is related to the organisation, culture, and teamwork involved with the system [MacLeod, 2001a]. A socio-technical system involves human-machine systems at a larger scale than hitherto and parallels the concept of System of Systems.

The concept of System of Systems is that the various constituent sub systems are no longer considered purely on their own merits as 'pipeline' systems. A pipeline system is build on its own principles in that it is designed primarily to serve its own purposes without any consideration on the use of its products by other systems. It may be designed under standards promoting its interoperability with other systems, but it will not be designed specifically to operate interactively with other systems to make a larger all inclusive System of Systems where the product of the whole is greater than the sum of the parts. Thus, Pipeline systems normally rely on externally imposed procedures and communication protocols to allow any joint operation with other sub systems.

In contrast, for a System of Systems the overall effectiveness of the combined performances of the contributing systems is considered from the onset of design and development under terms of interoperability of multiple agents. An agent defined as an entity capable of autonomous operation that is nevertheless reliant on timely quality data and information from other agents and can cater for the reciprocal needs of these other agents. The combined performance of agents, whether they be engineered or social entities or combinations of both, relies on a common shared knowledge and the use of the knowledge. Thus the structure of a Systems of Systems must be more flat than hierarchical in that it requires a common form of Cognitive Functionality (i.e. system functions related to the use and handling of knowledge – [MacLeod, 2000]) as the functional 'glue' of its operation and task performance. Some of the constituents of a System of Systems are illustrated in Figure One.

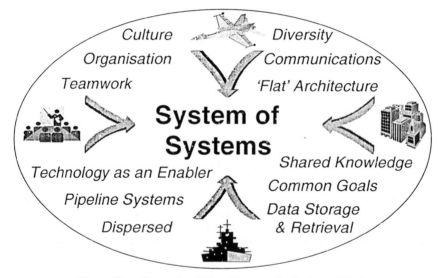

Figure One: Example of Constituents of a System of Systems

The socio-technical and System of Systems perspectives present new problems to the effective application of Ergonomics within the design and development of such systems. Firstly, Ergonomics has a tendency to focus on a 'one to one' consideration of human and machine. Only over the last decade has Ergonomics attempted to address the needs and problems associated with human teamwork. In addition, the System of Systems concept will also require a consideration of man-machine teaming, and understanding of machine-machine teaming, these both considering diverse teams and teams often distributed geographically

Problems of Concept

To often the narrow and traditional Ergonomic perspective on the individual human and system results in the slavish maintenance of traditional design and development practices regardless of the form of the system to which they are applied. Improvements in system performance are largely sought by improvements through use of new technology rather than improvements within design and development practices. Improvements in technology generally result in more increased automation of systems; presently it can be argued that automation is technology. Therefore, overall system design and performance, considering the human as a component, is too often poorly addressed.

As an outcome of the above, each purchased system tends to have its own specialized support systems, and has problems of interfacing and interoperability with other associated systems. This narrow approach results in large cost and time overheads in the development of procedures to allow interoperability between systems, in their individual maintenance, in the training of their associated personnel, and with relation to the overall through life costs of their operation. One avenue to the amelioration of the above has been the investment in large efforts to achieve the standardization of interfaces and to achieve agreed routes to interoperability through such as represented by the production of North Atlantic Treaty Organisation (NATO) STANdardisation AGreementS (STANAGS).

The operation of a System of Systems involves an assisted appraisal of communicated information and knowledge, throughout the overall system, or the retrieval of information and knowledge from forms of easily accessible storage. A major problem in such retrieval and communication is to find the information and knowledge for an intended purpose, enrich it to ensure its pertinence, and communicate it in a timely fashion to the correct recipient. Such activity can be performed automatically by the system provided the information or known location and type can identify knowledge. If the information has to be searched for, advice based on knowledge of the states of the System of System could be used, this considering the System of Systems sensed influences of the current environment, context, and the locations of similar previous retrievals.

As a socio-technical system, the form of command and team structures within a System of Systems must have an important influence on the effectiveness of control and communications. Thus, it is important that all command, control, and team structures share a clear set of strategic goals and can co-operate to co-ordinate the communication of the expected outcomes with the planned recipients. The sharing of knowledge in any dispersed system must rely on System of System nodes of teamed Agents being responsible for particular tasks, the control of work, and the node's possession of relevant forms of information and knowledge. Within an effective System of Systems, the formation of these nodes is dynamic depending on the stability of the goals of the whole, the health of the participating agents or sub systems within nodes, and the availability of agents to participate. The promulgation of the outcomes of the activities of any node is likely to be in clear forms of information, knowledge, and advice that depend on organisation, culture, situation and context.

Thus the team structure, relationships, direction of communication flows between the nodes, and the leadership of teamed nodes, all will all need established, flexible but effective processes to cater for changes in demand dictated by external influences. This suggests that the overall management and communication structure of a Systems of Systems will be flat rather than hierarchical as compromises will have to be continually strived for to maintain a balance within the workings of the system regardless of the influences on that system. Thus the relationship between control and performance will be close but may have to be constrained to prevent facets of the System of Systems becoming unmanageable and out of control. For these latter reasons, a careful but dynamic choice and amalgam of all forms of Agents must be possible depending on the influences of situation, context, and goal. The actioning of such choice must depend on the effectiveness of the System of Systems architecture, processes, and the reliability / redundancy of communications (i.e. as secured by the completion of transmission, reception, and acknowledgement of the understanding of the contents of a communicated message).

However, in addition to any appreciation of the influences of environment and context on system direction and control, there are many other elements that effect the concept of operations of a system and, therefore, its ultimate designed performance. For example, there are interactions between elements existing outside logical and physical design practices, elements that influence the efficacy of any approach to the design of a system or System of Systems. Figure Four illustrates some high level issues resulting from the interactions between the elements of Traditional Practice, the influences of the elements of existing knowledge on Technology (whether that be on its form, capabilities, or maturity of its associated engineering practices), and the understanding and promulgation of the concept of the system including its expected capabilities.

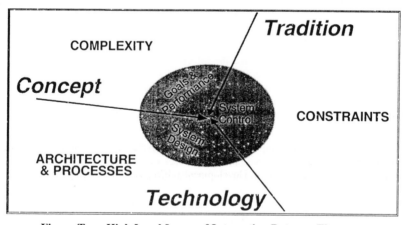

Figure Two: High Level Issues of Interaction Between Elements

The issues considered in Figure Two are design Complexity, Constraints imposed by the elements, and their effects on considerations of the adopted Architecture and Processes imbued on the system or System of Systems.

To briefly discuss the issues, Traditional Practices will place Constraints on the effectiveness of the adoption and application of any new technology. These practices include those of Ergonomics. Technology is invented through research but the methods and practical constraints applicable to its application are usually left to the system build engineers and are seldom developed until after many applications of that technology. As a partial resultant of the above, the Concept and functional capabilities required of the system are seldom matched to the adopted Technology through effective consideration on the designed optimisation of architecture and processes. In turn, there is a tendency to design over complex systems that include unneeded functions for the satisfaction of the required capabilities, the development of such systems exacerbated by the inappropriateness of traditional practices. [MacLeod, 2001b].

Conclusion

Ergonomics has traditionally addressed the problem of human–machine interaction with 'Simple' digital avionics systems, namely systems where one person operates one sub system. Increasingly, digital avionics systems are becoming 'Complex' in that one operator may be interacting with many sub systems through a single workstation or many operators may be geographically dispersed but still working as a team. With a System of Systems there can be many organisations and systems involved, plus there can be many associated 'Complex' workstations.

The attention of Ergonomics tends towards the traditional application of the discipline and thus to lag behind in its consideration of new technologies and system concepts. Ergonomics is not alone in this aspect. There are many causes of this lag varying from biases introduced by education, to the limitation placed on the use of HF practitioners through the traditions of bureaucratic organisations, or to the slow revision / update of engineering methods to cope with the changes introduced by new technology to the workplace. However, the rapid changes in system concepts must be accompanied by parallel changes in Ergonomics if the discipline is to remain as a viable influence on promoting quality design.

References

MacLeod, I.S. (2000), *A Case for the Consideration of System Related Cognitive Function Throughout Design and Development*, Systems Engineering, Vol.3, No.3, Wiley, pps. 113-127.

MacLeod, I.S. (2001a), *'Reliable' Military Aerospace System Development and Performance: Influencing Factors,* in Proceedings of 4th International Workshop on Human Error, Safety, and System Development (HESSD01), Linkoping, Sweden, June 11th-12[th.]

MacLeod, I.S. (2001b), *The Inexorable Link between System Control and Designed Performance,* in Proceedings of Conference on Cognitive Science Approaches to Process Control (CSAPC01), Munich, 23rd –25th September.

The views expressed are those of the author and do not necessarily represent those of their employers or any other agency.

FATIGUE: CAN WE NOW ASSESS IT?

Trevor F. Shaw & Colin J. Mackay

Human Factors Unit, Health & Safety Executive,
Magdalen House, Stanley Precinct, Bootle, Merseyside, L20 3QZ, UK

Fatigue has been recognised as a significant factor in worker performance for at least a century. With the present trend towards increased 24/7/365 working its importance is further enhanced, particularly with respect to health and safety. But if we are to be able to intervene successfully we must first be able to assess levels of fatigue with some degree of accuracy. The first step is to decide upon a suitable definition of fatigue so we know what it is we want to assess. The next step is to decide the most appropriate approach. This paper discusses the problems associated with this objective and recent work aimed towards meeting it, with a particular emphasis on the needs of employers with responsibility for risk assessment.

Introduction

Fatigue is an issue whose time has come with calls being made for it to be taken seriously (e.g. Feyer, 2001). With recent trends such as downsizing, many organisations employ fewer workers but working harder and for longer hours than previously and often with additional pressures such as fear of further job losses. This, combined with the increase in 24 hour, 7 day-a-week, 365 days a year workplace operations, means more shiftworking and more opportunities for working overtime or in additional jobs. Indeed about one in five workers in Europe now work some form of shiftwork, more than one out of two work at least one Saturday each month and one in four work one Sunday per month (European Foundation, 2001). There are perhaps fewer opportunities for employees to relax and recuperate from work than in the recent past. With mounting evidence that shiftwork and long working hours may have a detrimental impact on health (e.g. Spurgeon et al, 1997), and recognising the role fatigue can play in the occurrence of accidents such as road traffic (Feyer, 2001) and major hazards (Rogers *et al*, 1999), it seems imperative that we should make more effort to take the topic forward. In particular we need practical tools for assessing levels of fatigue if employers are to control it in the workplace.

A brief history

Fatigue was first considered as a factor affecting worker performance towards the end of the nineteenth century but was not taken seriously in the UK until the early years of the twentieth century (McIvor, 1987). The establishment of the Health and Munitions Workers Committee during the 1st World War, which became the Industrial Fatigue Research Board in 1918, produced the first concerted efforts to understand and tackle the issue which had been recognised as important as a result of the shift systems associated with the war effort.

By the late 1940s when the IFRB was wound up and the Ergonomics (Research) Society was founded, most considered that a test of fatigue was impossible (Broadbent, 1979) and much work on fatigue remained to be done. However, Welford (1953) at a 1952 Symposium on Fatigue summed up the state of progress in understanding the psychological aspects as having reached "the end of the beginning." The physiological mechanisms underpinning 'psychological' fatigue had not been established, but some possible psychological processes taking place were proposed. These were largely related to the theories of attention and selection in the emerging information processing paradigm. However, motivational factors were also acknowledged.

Work continued, and Donald Broadbent's conclusions in his Ergonomics Society lecture at Oxford, on the likelihood of developing a test of fatigue, suggest he was fairly optimistic that such a test might soon be possible (Broadbent, 1979). A number of other significant articles were published, in particular Grandjean (1979) which developed a coherent theoretical model of the neurophysiological processes involved in fatigue, centred around the activating and inhibitory systems of the brain. Since then however, fatigue research appears to have become less prominent. To some extent this can be explained by the research effort having become redirected into various facets of the topic such as workload, cognitive effort, sleep, shiftwork, and specific types of fatigue such as visual.

Defining and measuring fatigue

Early on it was concluded that fatigue was an unhelpful term and that any attempts to measure it without it being well-defined should be abandoned (Muscio. 1921). The attraction of the term 'fatigue', rather like that of 'stress', is perhaps that it has some meaning for everyone, based on personal experience.

So to successfully assess fatigue we need to be clear about how we define it. Disregarding chronic fatigue syndrome for the moment, there are four principal descriptors in mainstream use: general fatigue, physical or muscular fatigue, visual fatigue (which has arguably much in common with muscular fatigue) and mental fatigue Localised physical or muscular fatigue is probably the best understood since it has a clear basis in the functional physiology of the body (Åstrand & Rodahl, 1986). To some extent the same is true of general whole body fatigue especially where it is associated with prolonged hard physical effort. However, this type of fatigue was more prevalent in classic manual work systems and is not addressed further in the present paper. The focus here is on general fatigue associated with hours of work as well as mental fatigue which tends to be characteristic of modern technologically-based work systems (Kumashiro, 1995).

General fatigue can be described variously as a state of malaise or tiredness resulting from sustained exertion (mental or physical) or lack of rest, or from a lack of stimulation or motivation. It is typically characterised by reduced arousal or alertness. Mental fatigue has many features in common with general fatigue and for most practical purposes the distinction may be of little use since both include a similar subjective element. However, mental fatigue is often used to describe the fatigue associated with the prolonged intensive mental demands typified by many safety-critical tasks.

Information processing and attention-related errors which may be attributable to mental fatigue have tended to be tackled by focusing on mental workload (although more work is clearly still needed in this area). This is perhaps analogous to the approach taken in dealing with physical fatigue, except that subjective assessment of mental workload (e.g.

using tools such as the NASA TLX or SWAT) tends to be favoured over more objective measures, probably because they are easier to administer outside the laboratory. However, measurement of both mental workload and fatigue present similar problems since the two topics are closely related. Mental fatigue is only considered further within the context of general fatigue in this paper.

General fatigue associated with hours of work has mainly been understood through research into circadian rhythms, sleep loss/quality, and shiftwork. Where measures of fatigue have been developed they are normally concerned with the individual and are to a greater or lesser extent intrusive upon the tasks being performed. As Broadbent (1979) made clear, such techniques have limited validity if they end up inadvertently stimulating the individual. The range of objective techniques currently advocated as useful include critical flicker fusion (CFF), concentration maintenance function (CMF) and heart rate variability (HRV) (e.g. Kumashiro, 1995). They are concerned with assessing an individual's state of fatigue or alertness/arousal at a specific time, or for monitoring it over a period, and generally require some specialist knowledge to administer, analyse and/or interpret. Therefore they are of limited value to most employers who wish to carry out a risk assessment for particular jobs or work activities.

Clinical aspects

Alongside the typologies of fatigue in the general working population there has been speculation about the extent to which acute fatigue may develop into chronicity and so become persistent, incapacitating idiopathic fatigue - a syndrome or condition which will be familiar to many occupational physicians (Mounstephen & Sharpe, 1997). Medically unexplained fatigue of at least six months duration and of new onset is a principal feature of chronic fatigue syndrome. Somatic fatigue is also a common characteristic of many psychological disorders such as depression. Although the epidemiology of fatigue is complex (Lewis and Wessely, 1992) large scale population surveys show that the number and severity of fatigue-related symptoms are continuously and monotonically distributed in the population. Perhaps, therefore, to control both acute and chronic types of fatigue believed to be caused by exogenous factors, a population rather than individual-based strategy seems appropriate (Rose,1992).

A pragmatic approach

Given the difficulties in measuring fatigue in a particular individual at work at a given time, a more fruitful approach would be to predict the level of fatigue expected based upon known contributory factors such as circadian rhythms, time of day and time on task. This would allow the relative risk associated with the predicted level to be assessed and where necessary appropriate steps could be taken to reduce the level of fatigue. Some recent work has taken this approach. Three examples are briefly discussed.

Some current fatigue risk assessment tools

Fatigue Index (FI)

The mental fatigue index (Lucas *et al*, 1997; Rogers *et al*, 1999) was originally created to provide guidance for UK railway companies on undertaking a fatigue risk assessment under the Railway (Safety Critical Work) Regulations 1994. In this approach individual factors have been deliberately excluded to simplify the assessment and because it was intended it should be used to assess groups of safety critical workers. This makes the approach less useful in considering individual cases, for example to later determine likely levels of fatigue in an individual at the time an accident occurred.

The first version received generally positive feedback from a group of experts so it was decided to further develop it. This led to some changes in the factors included, but retained a simple additive method of obtaining an overall chronic fatigue score or index (Rogers *et al*, 1999). The five factors now included are: time of day the shift starts, shift duration, timing and duration of pre-shift rest periods, length of periods of continuous critical concentration (high mental workload - if relevant) before a break occurs, cumulative fatigue (for consecutive shifts).

The model may sound relatively straightforward but is quite complicated to use if data is analysed manually. Development of a software implementation should enhance the model's usability. The index is also limited by not being able to cope with work periods which exceed 16 hours and assumes an individual has had an adequate rest (sleep) period between work periods/shifts. Unfortunately this is not always a safe assumption (e.g. in emergency scenarios where shifts may be extended and since sleep quality can depend upon the timing of the sleep period). It may be considered fair from the employer's perspective provided they have designed shift schedules to allow for rest. Whether an employee uses the period to take their full requirement for sleep is often outside the employer's control.

The model has the advantage of being in the public domain and can be applied to non-safety critical workers by leaving out the fourth (concentration/break) factor. In its present stage of development it appears best suited to comparing different shift schedules. Further work is underway to build a database of cases which may allow the index to indicate critical risk thresholds.

FRATE

Fatigue Risk Assessment (FRATE) software, has been developed as a commercial tool to help assess or predict fatigue risk in the rail transport industry (Circadian Technologies Ltd., 2001). The development has entailed using a large database of 24-hour fatigue-related data (e.g. self-reported sleepiness, sleep/wake logs, wrist actigraphy and EEG) collected from various train crews in North America and the UK and applying it to a generic circadian alertness simulation model (CAS) in an attempt to optimise each stage of data conversion. The result is a tool which converts actual train driver work-rest schedules into probable sleep-wake patterns, which are then used to calculate probable alertness level fluctuation across the day. Finally, these two sets of intermediate data are used as the basis from which to derive seven parameters which are weighted and used to calculate an overall chronic fatigue score covering a period of either a week or month. Individual work schedule data is fed into the model but it is intended to be used for comparing group data, or changes to group schedules, by calculating the mean fatigue score and other statistics.

FRATE can be used to model an individual's level of fatigue at the time of an incident, provided detailed information is available about their usual sleep characteristics (e.g. short or long sleeper, and 'lark' or 'owl' type) and their sleep/wake pattern over the previous week. It can also take into account three different duty types which reflect different levels of demands (Goodwin, 2001).

The FRATE tool perhaps has both the advantage and disadvantage of being optimised for a particular working population. This may be an advantage since train drivers as a working population are likely to share certain characteristics (e.g. personality) which differ from other worker populations. But it could mean that FRATE is not a valid tool for use with other populations. However, a sister tool has been developed for the road transport industry (FRARE) and both are based on CAS which has been subject to verification, validation and testing in the USA (Goodwin, 2001).

Dawson & Fletcher's model

Another recent predictive model has been published (Dawson & Fletcher, 2001; Fletcher and Dawson, 2001). This apparently simple model quantifies work-related fatigue by taking into account the timing and length of shifts, recent work history and circadian influences on sleep. A 'token economy' analogy is used to describe the model. This entails calculating an individual's total number of fatigue and recovery tokens with the balance representing the current fatigue score. The tokens are credited to either the fatigue or recovery accounts, calculated according to the function assigned to the factors listed. Hence the timing and duration of a period of work or recovery will determine the token value for that period with tokens from previous periods (up to 7 days) being taken into account. An upper limit on the recovery account simulates the fact that there are limits on the benefits of extra sleep. The model looks promising and further work is proposed.

Comparison

All three of these methods use basically the same factors to calculate daily levels of fatigue: shift timing and length, and circadian influences. What differs is the way in which the factors are used in calculating levels of fatigue (i.e. the transfer functions). All use basic work schedule information as input data and so potentially have practical application in all 24/7 workplaces, for example to compare different shift schedules by comparing resulting estimates of fatigue (e.g. as illustrated in Dawson & Fletcher, 2001).

For safety critical work the most appropriate are the FI and FRATE. The FI tries to take account of periods of sustained vigilance while FRATE can allow for three levels of task load (i.e. different duty types). These factors should make them better able to assess safety critical work than the Dawson & Fletcher model (in its current form).

The most complex model to use is perhaps the FI because it requires more data for input (e.g. on timing of sleep periods) whereas the other models infer this (in the case of FRATE, based on the train crew database). However, FRATE can also take account of individual sleep characteristics and has the benefit of being developed from CAS. What would be useful is a comparison of these three models (or at least the two non-commercial ones and any others not identified) using the same data to produce fatigue estimates, particularly where these can be related to known fatigue-related accidents or incidents. A tool which could produce absolute ratings of fatigue, rather than relative estimates, would be extremely useful, but may prove less easy to develop.

Conclusions

If employers are to take responsibility for carrying out fatigue risk assessments in 24/7/365 working then they need appropriate tools for the job. Clearly the specialised tools of the human factors practitioner may be either too complex, too intrusive or too limited in scope for use by most employers. However, there are signs that a useful pragmatic tool for assessing levels of general fatigue amongst groups of employees is now possible, based on a reasonable understanding of the major influences upon general fatigue. Any such tool needs to be free or available at low cost, it should be accessible to the non-specialist and should provide valid and reliable results. Perhaps none of the tools discussed here meet all these requirements yet, but hopefully further development will result in such a tool shortly.

References

Åstrand, P-O. & Rodahl, K. 1986, *Textbook of Work Physiology: Physiological bases of exercise 3rd Edition*, (McGraw-Hill)

Broadbent, D.E. 1979, Is a fatigue test now possible? The Ergonomics Society Lecture 1979, *Ergonomics*, **22**, 1277-1290

Circadian Technologies Ltd. 2001, Fatigue Risk Assessment in Railway Operations: Development and Validation of FRATE as a Fatigue Benchmarking Tool. (Circadian Technologies, Ltd., London)

Dawson, D. & Fletcher, A. 2001, A quantitative model of work-related fatigue: background and definition, Ergonomics, 44, 144-163

European Foundation for the Improvement of Living and Working Conditions, 2001, Ten Years of Working Conditions in the European Union, (Office for Official Publications of the European Communities, Luxembourg)

Feyer, A-M. 2001, Fatigue: time to recognise and deal with an old problem, British Medical Journal, 322, 808-809

Fletcher, A. & Dawson, D. 2001, A quantitative model of work-related fatigue: empirical evaluation, Ergonomics, 44, 475-488

Goodwin, S. 2001, Personal communication, Circadian Technologies, Ltd., London

Grandjean, E. 1979, Fatigue in industry, British Journal of Industrial Medicine, 36, 175-186

Kumashiro, M. 1995, Practical measurement of psychophysiological functions for determining workloads, IN J.R.Wilson & E.N.Corlett (eds), Evaluation of Human Work: A Practical Ergonomics Methodology 2nd ed., (Taylor & Francis, London), 864-837

Lewis, G. & Wessely, S. 1992, The epidemiology of fatigue: more questions than answers, Journal of Epidemiology and Community Health, 46, 92-97

Lucas, D.A., Mackay, C., Cowell, N. & Livingstone, A. 1997, Fatigue risk assessment for safety critical staff, IN, D.Harris (ed), Engineering Psychology and Cognitive Ergonomics Vol. 2: Job Design and Product Design, (Ashgate, Aldershot), 315-320

McIvor, A.J. 1987, Employers, the government, and industrial fatigue in Britain, 1890-1918, British Journal of Industrial Medicine, 44, 724-732

Mounstephen, A. & Sharpe, M. 1997, Chronic fatigue syndrome and occupational health, Occupational Medicine, 47, 217-227

Muscio, B. 1921, Is a fatigue test possible? British Journal of Psychology, 12, 31-46

Rogers, A.S., Spencer, M.B. & Stone, B.M. 1999, Validation and development of a method for assessing the risks arising from mental fatigue, CRR 254, (HSE Books, Sudbury)

Rose, G. 1992, The strategy of preventive medicine, (Oxford University Press, Oxford)

Spurgeon, A., Harrington, J.M. & Cooper, C.L. 1997, Health and safety problems associated with long working hours: a review of the current position, Occupational and Environmental Medicine, 54, 367-375

Welford, A.T. 1953, The Psychologist's Problem in Measuring Fatigue, IN W.F.Floyd & A.T.Welford (eds), Fatigue, (H.K.Lewis, London), 183-191

The views expressed in this paper are those of the authors and do not necessarily represent HSE policy.

HUMAN FACTORS INVESTMENT STRATEGIES WITHIN EQUIPMENT PROCUREMENT

Roger S. Harvey[1], Ray Wicksman[1]*, Duncan Barradale[2], Keith Milk[3] and Norman Megson[3]

[1]*QinetiQ Ltd, Centre for Human Sciences*
Farnborough, Hants GU14 0LX
[2]*Dstl, Centre for Defence Analysis*
Farnborough, Hants GU14 0LX
[3]*HVR-CSL, Selborne House*
Mill Lane, Alton, Hants GU34 2QJ

*Now at CCD Design and Ergonomics Ltd

The introduction of the Human Factors Integration (HFI) methodology within the procurement framework of the UK Ministry of Defence (MoD) has been a contributor to improving military operational effectiveness and reducing whole-life and performance risks. However, there have been limited quantitative data to permit comparative judgements to be made about the success of Balance of Investment (BOI) strategies within procurement. This study was initiated to develop an "Influence Model" of HFI activities to describe the interactive influences of HFI and other aspects of a system. The study developed a simple methodology for conducting HFI BOI strategies, and tested the potential for the use of Systems Dynamics modelling, in conjunction with the software tool POWERSIM, as the means to support HFI BOI studies in procurement.

Background to the Study

The application of the HFI methodology to the MoD procurement framework has arisen because of an increasing recognition of the need to match available technology with users and maintainers and their operating environment. By encouraging efforts to enhance this match, HFI has made an important contribution to improving military operational effectiveness and reducing whole-life costs and performance risks. A recent scoping study (JJB Ltd, 1998) found that whilst HFI *did* have a measurable effect upon the design and acquisition of military systems there were only a few techniques available to identify, quantify and analyse the benefits of the process and its effects. However the study concluded that there was merit in the use of *Influence Diagrams* as a means to capture and describe linkages between HFI cost and benefit factors. Following this preliminary study a more detailed, two-year, HFI Balance of Investment (BOI) study programme was initiated and reported on in this paper.

This work was carried out as part of the Human Systems Technology Research Domain of the MOD Corporate Research Programme.

The HFI BOI Study: Aims and objectives

Arising from the preliminary study outlined above, the main HFI BOI study was initiated to develop an Influence Model of HFI activities to describe the interactive influences of HFI on the system and the procurement process; to assess the potential costs and benefits of a manned system; and to develop a methodology for conducting HFI BOI studies including recommendations for an appropriate software tool specification to support HFI BOI studies in equipment acquisition.

The Year One programme of work

Year One of this study was conducted in two phases. The first phase established the means of understanding and identifying relevant HFI data and sources together with the selection of candidate military equipment systems for initial data gathering. The second phase concentrated on detailed data gathering at Defence Procurement Agency Project Office level, together with database structure development and analysis. Military and civilian Subject Matter Experts (SMEs) were consulted throughout this period. Special attention was given to the need to obtain both cost *and* non-cost data. Thus it was hoped that quantitative cost data would be supported by descriptive non-cost data, with the latter relating to HFI activities and their perceived benefits.

As Year One of the study progressed it became evident that the quantity, format and availability of historic data would be insufficient to support the traditional approach to the construction of a Balance of Investment Model. The absence of a *generic* Equipment Breakdown Structure (EBS) for use in project cost recording severely hindered the data compilation process. US Military Handbook 881 (1998) could provide this generic EBS, which would be essential to making the database meaningful in achieving the goals of the BOI model because data compiled in a generic format lend themselves easily to comparative estimating and, ultimately, identification of Cost Estimating Relationships. (The prototype database developed during Year One used Military Handbook 881 as the basis of collating project cost and non-cost data for subsequent analysis.) However, it was considered that building a BOI model suite in this way would be time-consuming and expensive, and therefore beyond the scope of this short study. Nevertheless the use of Influence Diagrams was successful in focusing attention on the sorts of data required. Most importantly they captured the essence of the flow between the *investment* and *benefits* (a value chain) of a system, in this instance military equipment acquisition. This was achieved by identifying influences, and the value added, by points or nodes in the chain.

A Human Factors SME Workshop was convened to gain agreement upon the approach to Influence Diagrams and their content, and to construct *generic* Influence Diagrams for each HFI domain (Manpower, Personnel, Training, Human Factors Engineering, Health Hazard Assessment and System Safety). These Diagrams were generic in that they could ultimately be applicable to any equipment programme with suitable customisation (as subsequently occurred in Year Two of this study). Elucidation of the generic Influence Diagrams can be found in Canipel, Newman and Shepley (2000). An example is shown in Figure 1.

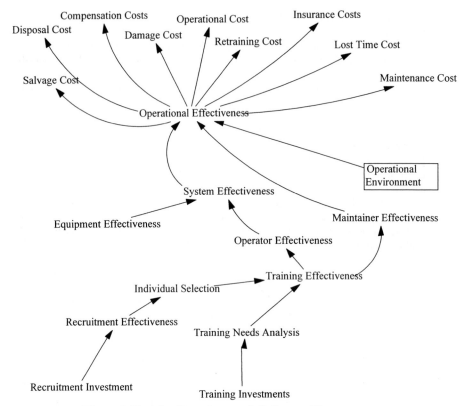

Figure 1. Training Domain generic Influence Diagram

The Year Two programme of work

The principal difficulty during Year One lay in gaining sufficient quantitative data related to HFI issues. For Year Two it was therefore decided that a separate "Conceptual" Model of the HFI environment should be constructed using System Dynamics (SD) techniques and SME consensus opinion. This Conceptual Model could then be validated with available historic data at a later date. It could be anticipated that the Model would move from a Concept, with a large number of qualitative factors, to a more representative BOI model with increasing levels of objectivity with quantitative factors. The Year Two programme of work was a "Proof-of-Principle" approach to analysing the linkages between HFI and costs and benefits associated with military systems. It also addressed ways of modelling the links between differential investment in the HFI domains and the outcomes of these changes in investment strategies. It was therefore decided to generate Influence Diagrams for an "exemplar" weapon system. After an extensive review of various candidate weapon systems the study team selected a generic Light Anti-Armour Weapon (LAW) as the "exemplar" system.

Thus the Year Two programme of work further expanded and refined the Influence Diagrams from Year One by quantifying the nature of the linkages between entities within the Influence Diagrams (as applied to the agreed generic exemplar system LAW), and developing a Systems Dynamics model that would subsequently permit meaningful investment trade-offs to be proposed, tested and validated. Full details of the Year Two programme of work and its findings are in HVR Consulting Services Ltd (2001).

The generic single-domain Influence Diagrams were reviewed in a further "Workshop-style forum" comprising HFI and weapon system SMEs so as to be specifically applicable to the agreed exemplar system LAW. Most importantly this forum considered *interactive influences* and by this means a complex series of enhanced Influence Diagrams was generated, representing the continuous and interactive application of Human Factors (HF) specialist activity *for all six HFI domains*. Limited space within the written paper does not permit reproduction of the complex diagram showing the substantial number of interactions between the many HFI issues, but the focal point of the diagram was a measure of the total number of HFI concerns remaining within the exemplar system design. The impact upon military capability was expressed through a shortfall in the number of systems operating. This occurs as a consequence of changes in task completion times, error rates, personnel requirements or personnel numbers when HFI concerns are not fully resolved. The interrelationships between HFI domains indicated that determining an optimum balance would not be simple because non-linear behaviour could be expected.

The Proof-of-Principle SD model was developed using the software-modelling tool POWERSIM. This produced a 'stock-flow' type of diagram, which is a more detailed derivation of the Influence Diagram. In the stock-flow diagram the measurable quantities (eg hazard concerns, safety issues, financial budgets, and charge-out rates) were delineated as separate, but linked, flows. The additional factors needed to allow the computation of rates of change were also included. The program input parameter values for the Health Hazard Assessment (HHA) domain are shown in Table 1. Sample results from the model illustrated the considerable potential for a dynamic interaction between investment strategies and options and the resulting military capability.

Hazard Domain		
Effort to identify hazard mitigation	1	person-month
Monthly spend on hazard identification	13.4	£K per month
Effort per hazard identified	1	person-month
HF specialist cost	6.7	£K per person-month
Initial hazard mitigation budget	2500	£K
Average cost of hazard mitigation measures	50	£K
Monthly spend on hazard investigation	13.4	£K per month
Hazard HF budget increment	161	£K per annum
Effort per hazard ETO	1	person-month
Average time to mitigate hazards	1	month
Proportion of hazards to be mitigated	0.9	fraction of hazards for which mitigation measures may be identified
Number of initial hazards	50	

Table 1. Sample Program Input Parameter Inputs for HHA domain

No specific meaning was attached to the results for the "exemplar" system, beyond showing that the model produced an appropriate response when input data are changed. These data were indicative only. A "baseline" condition was chosen so as to create the situation whereby the funding of HFI effort was sufficient to resolve all the known domain issues for the proposed system, with a 66-month development programme culminating in 200 fielded systems as shown in Figure 2 below.

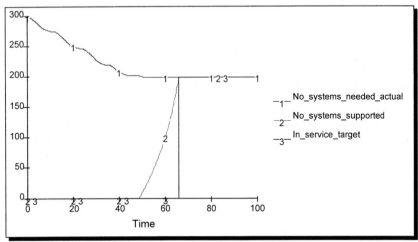

Figure 2. Baseline Results Example

A variation from this baseline is shown in Figure 3 where the funding of the Hazard and Safety domains has a 90% reduction such that unresolved hazard and safety issues are translated into an increased number of systems (228) needed to deliver the military capability. More personnel are required with a delay while these are trained.

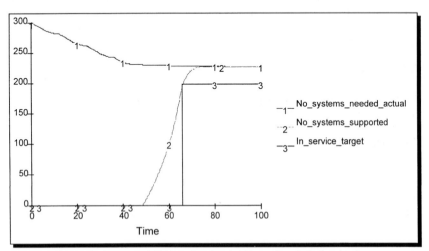

Figure 3. Impact of 90% reduction in Hazard and Safety budgets

Conclusions

The SD modelling approach was highly suited to the additional prototype development needed to gain better understanding of the requirements and limitations of a BOI methodology. It was also encouraging to note that the early prototype model was able to portray appropriate behaviour and responses. The first objective for additional development should be to establish greater confidence in the model and data for use in support of BOI analysis. This will involve continued scrutiny of both the Influence Diagrams and the Proof-of-Principle model so as to confirm the modelling approach. It will also be necessary to seek out missing interactions, consider the need for additional modelling of training and recruitment processes, add time (ie project phase) dependence to parameters and make simplifications where possible so as to resist excessive detail or complexity that could make model behaviour difficult to understand.

As in the Year One programme of work, data collection was again highlighted as the major obstacle to model development and use. The high-level, aggregated approach adopted by the SD modelling reduces the need for much detailed data. However, it should be recognised that SD modelling introduces a new approach for assessing investment in HFI and leads to a need to capture new data concerning the specialist tasks undertaken by the HF community so as to provide model input parameters. Definition of the model has sought to use terms for which data should, in principle, be collectable even if records are not kept at present. As the impact of the HFI methodology becomes more apparent in equipment acquisition there will be an increased requirement to capture data pertinent to HFI in the acquisition cycle. The consequence may be a need to compromise, modifying the model as well as seeking the new data required.

Further work is required for testing, calibration and validation using historic data, where available, and additional SME opinion. Once this has been achieved, it will be appropriate to consider adding functionality to allow, for example, optimisation of BOI strategies as an automated feature. Adoption of a standard interface and improved 'user friendly' features could complete the software development needed to deliver a packaged HFI BOI methodology.

Reference list

Canipel, P., Newman P. and Shepley N. 2000. Human Factors Integration: Balance of Investment Investigation. Contract Report carried out for Centre for Human Sciences, DERA Farnborough.
HVR Consulting Services Ltd, 2001. Cost analysis data and collection and research – Final Report. Contract Report carried out for Centre for Human Sciences, DERA Farnborough.
JJB Ltd, 1998. A study on "HFI Cost-Benefit Analysis". Contract Report carried out for Centre for Human Sciences, DERA Farnborough.
US Mil-Hdbk-881 1998. Department of Defense Handbook, Work Breakdown Structure.

WHAT CAN FOCUS GROUPS OFFER US?

Joe Langford[1] & Deana McDonagh[2]

[1] *Human Factors Solutions, Englefield, Broadway,
Harwell, DIDCOT, OX11 0HF, UK*
[2] *Department of Design and Technology, Loughborough University,
LOUGHBOROUGH, LE11 3TU, UK*

This paper describes the key advantages and limitations of focus groups, and the main purposes for which they can be used within the ergonomics and product design professions. It concludes that the method offers great benefits and is a valuable addition to the standard 'toolkit', although for some applications, adaptations to the basic format are helpful.

Introduction

The fields of ergonomics (human factors) and industrial/product design share a common interest in gaining a sound understanding of users and the ways they use products or systems. There are many ways of gaining this information, including trials, surveys and observation. More recently though, focus group methodologies have emerged in support of these methods.

A focus group is a carefully planned group discussion, designed to obtain the perceptions of the group members on a defined area of interest. Typically there are between five and twelve participants, the discussion being guided by a moderator. The group-based nature of the discussions enables the participants to build on the responses and ideas of others, thus increasing the richness of the information gained.

The advantages of focus groups

A key benefit of focus groups is their ability to provide in-depth understanding and insight into the topic being explored. Researchers can interact directly with participants. They can explore the responses given and thus discover more about what people are thinking, including perceptions and aspirations. If necessary, questions can be added or modified in 'real time' to make the most of unexpected information or occurrences. Whilst this is true for all interview techniques, focus groups have an additional advantage in that people can build upon the comments of others. This synergistic effect often leads to the emergence of information or ideas that would otherwise have remained hidden. A great deal of data and information can be gained relatively efficiently and quickly with focus groups.

Focus groups offer flexibility. They are applicable to a wide range of topic areas and types of participant. Where necessary, the focus group can be taken to the participants, which is very helpful in cases where participants find it hard to travel.

The disadvantages

Group settings are more difficult to manage than one-to-one interactions. Discussions can veer off into irrelevant areas and dominant participants can bias the discussions. A skilled moderator can reduce the effect of such problems, but they are unlikely to be completely eliminated. It is usually necessary to run a number of separate sessions to provide confidence in the results.

Focus groups cannot reflect real-life scenarios, so participants use memory, and this can limit the quality of the outputs. Also, there may be sub-conscious differences between what people say they do, and what they actually do in practice.

The data resulting from focus groups is largely open-ended, with little structure to it. Analysis can be time-consuming and results biased by selective use of data. Care must also be taken to ensure that focus group findings are not given more weight than is actually warranted. Transcribed comments or video recordings of discussions may be more influential than a table of statistics, but are not necessarily more valid.

Focus groups in ergonomics and design

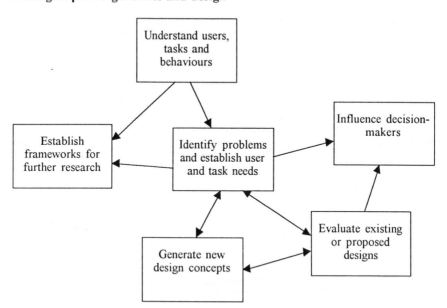

Figure 1. Uses of focus groups in ergonomics and design

Designers and ergonomists have not been slow to see the potential for using focus groups to support their work. Examples in published literature start to appear in the mid 1980's, but there is evidence that similar methods were in use even earlier. For example, O'Brien (1981) describes the use of 'shared experience events' which involved elements of the focus group method in the design of control rooms.

Discussions with those working in design and (particularly) ergonomics and have revealed that the use of focus groups and similar group discussion methods is fairly common, but interestingly, often not reported in the literature. This has led the authors to conclude that focus groups have simply become part of the 'toolkit', being seen as a no-nonsense, practical way of gathering data.

The flexibility of the focus group method has been widely exploited by ergonomists and designers. The most common uses are shown in Figure 1. As this diagram shows, many of the uses are closely related, or overlap to some extent, and focus group sessions are therefore often set up to fulfil two or more different purposes at the same time. For example, a focus group aimed at evaluating or generating new concepts for a product may also be designed to include discussions about general problems or needs that the users have concerning that type of product.

Understanding users, tasks and behaviours

Those working in design and/or ergonomics appreciate that a good understanding of users and their activities is vital. Focus groups can provide an excellent way of gaining this. Immersion in sessions means that researchers have an opportunity to empathise more with the people they are designing for.

This use of focus groups is relatively common. For example, when investigating job hazards for youths working on farms, Bartels *et al* (2000) used focus groups to identify activities that might lead to problems and determine the participants' perceptions of the risks. Ede (2001) used focus group studies to learn about system administrators' jobs. Contrary to initial impressions that their jobs were very technical, she found that they actually spent most of their time dealing with people.

Identifying problems and establishing user and task needs

Focus group studies such as those described above will often identify problems experienced by users, and hence help determine their needs. For example, from their study on fire safety for older people, Lerner and Huey (1991) were able to identify key problems with existing products and technologies. They were also able to determine a set of fire-safety product needs and desirable features.

Similarly, Parsons and Wray (2000) ran a series of user group workshops to investigate postal workers' perceptions of their current safety footwear. They identified areas for improvement and features considered most important, thus helping to define specifications for future equipment.

In their work on online financial advice systems, Longmate *et al* (2000) ran discussion groups with potential users to understand their experience and attitudes towards advisers, the Internet and on-line advice. The findings helped form the basis of a set of requirements used to develop a concept interface for an online system.

Establishing frameworks for further research

Ergonomists and designers have used focus groups when starting research in new areas. Group discussions provide a rich source of background information and help prioritise areas for further investigation. They also help with the design of research, for example, the methods to be used or details such as the correct terminology to use in questionnaires or interviews.

The focus groups carried out by Barrett and Kirk (2000) on information needs of elderly and elderly disabled people were set up specifically to guide more detailed research. The focus groups were used to identify major themes and aid in the creation of a questionnaire for a national survey. Similarly, Hide *et al* (2001) used focus groups to gain background understanding about accidents in the construction industry, enabling detailed follow-up studies.

Evaluating existing or proposed designs

Opinions on the validity of focus groups for evaluation are divided. There is little doubt that for gaining users' 'impressions' of a new product or system, focus groups are extremely valuable. In many cases, this level of detail is perfectly adequate, such as when exploring emotive attributes or where the usability implications of the product are not too complex. For example at the start of a design project, Dolan *et al* (1995) used focus groups to gain feedback on a range of telephone handset designs.

Some have questioned the suitability of focus groups for evaluation purposes, particularly with respect to more complex usability issues. Nielsen (1997) sums up the main concerns:

> *Because focus groups are groups, individuals rarely get the chance to explore the system on their own; instead, the moderator usually provides a product demo as the basis for discussion. Watching a demo is fundamentally different from actually using the product. There is never a question as to what to do next and you don't have to ponder the meaning of numerous screen options.*

There are however, attractions in being able to carry out evaluations of designs in, or in conjunction with focus groups. Caplan (1990) provides examples of this in his work on photocopier design, for example, 'twinning', where focus groups are used to discuss outcomes of user trials carried out prior to the discussion sessions.

Generating new design concepts

The generation of new ideas is a natural step on from the evaluation of existing or proposed designs. A problem or design deficiency identified during a focus group provides a springboard that can lead directly to a range of solutions. Focus groups are particularly useful in the early stages of concept generation.

With participatory design, focus groups and similar methods are often employed to help users identify new designs for equipment, workplaces and systems for themselves. This involves adapting basic discussion group methods by incorporating 'creativity' tools such as drawing and 3d modelling, as in 'design decision groups' (Haines and Wilson, 1998).

Influencing decision-makers

Reports, models and drawings have their part to play, but do not necessarily enable decision-makers to appreciate the full importance of key issues. Those using focus groups have found that the evidence produced has a strong impact, and helps to 'sell' their suggestions. For example, Caplan (*op cit*) describes how he used videotapes from discussions to put over key points, and in some cases, involved the decision-makers in the focus group process to help them understand and accept the results.

While not adding to the quality or content of the work, the ability to present key findings in these ways can help to achieve the objectives of the project. This fact alone may well be a compelling reason to include focus groups as part of the project approach.

Conclusions

Focus groups provide a means to discover information that might otherwise have remained hidden with other research methods. They are flexible, adaptable and relatively easy to set up and run. There are of course some disadvantages, including the potential for bias, particularly due to dominant group members. Also, what people say they do can be markedly different to what they actually do.

Ergonomists and designers have much to gain from using focus groups. Indeed, there is plenty of evidence that they are actively used in these fields across a very wide range of areas including product and equipment design, human computer interaction, health and safety, ageing and job/task design.

Limitations of the basic group discussion format have led ergonomists and designers into making adaptations to extend their usefulness. These modifications include 'twinning' focus groups with evaluation trials and integrating creativity tools to aid generation of new ideas.

There is no such thing as a 'perfect' research method. All have strengths and weaknesses, and focus groups are no exception. For designers and ergonomists however, they are an extremely versatile technique to add to the toolkit.

References

Barrett, J. and Kirk, S. 2000, Running focus groups with elderly and disabled elderly participants. *Applied Ergonomics*, **31**, 621-629

Bartels, S., Niederman, B. and Waters, T.R. 2000, Job hazards for musculoskeletal disorders for youth working on farms. *Journal of Agricultural Safety and Health*, **6**, 191-201

Caplan, S. 1990, Using focus group methodology for ergonomic design. *Ergonomics*, **33**, 527-533

Dolan, W.R., Wiklund, M.E., Logan, R.J. and Augaitis, S. 1995, Participatory design shapes future of telephone handsets. In Proceedings of the Human Factors and Ergonomics Society 39th Annual Meeting, 331-335

Ede M. 2001, Focus groups to study work practice. www.useit.com/papers/meghan_ focus groups.html, 31 October 2001

Haines, H. and Wilson, J.R. 1998, *Development of a Framework for Participatory Ergonomics*. HSE Books, Sudbury. ISBN 0 7176 15731

Hide, S., Hastings, S., Gyi D., Haslam, R. and Gibb, A. 2001, Using focus group data to inform development of an accident study method. In M.A. Hanson (ed.) *Contemporary Ergonomics 2001*, (Taylor and Francis, London), 153-158

Lerner, N.D. and Huey, R.W. 1991, Residential fire safety needs of older adults. In Proceedings of the Human Factors and Ergonomics Society 35th Annual Meeting, 172-176

Longmate, E. Lynch, P. and Baber, C. 2000, Informing the design of an online financial advice system. In S. McDonald (ed.) Proceedings of HCI 2000, 103-117

Nielsen, J. 1997, The use and misuse of focus groups. www.useit.com/papers/ focusgroups.html, 12 September 1997

O'Brien, D.D. 1981, Designing systems for new users. *Design Studies*, **2**, 139-150

Parsons, C.A. and Wray, A. 2000, Specification of footwear for postal workers. In McCabe, P.T (ed.) *Contemporary Ergonomics 2000*, (Taylor and Francis, London), 344-348

DEVELOPMENT OF ERGONOMICS

WORK AND HEALTH:
AN HISTORICAL REVIEW OF THE LITERATURE

Sheila Lee

7, Antrim Grove
London
NW3 4XP

The problems of work-related illness and injury have occupied the minds of medical men and thinkers over the ages. To some, these problems have represented inconvenience, to others, the subject has presented medical problems to be solved, yet others have used this for political ends, whilst a compassionate few have been struck by the unfairness. The fact remains that the problems were observed and recognized. The upper classes, the employers, have traditionally been slow to act upon this knowledge, whilst the lower orders, the employed, have for the most part accepted injury as part of their lot.

Introduction

From the study of writings on occupational health it has become clear that this subject which has captured the imagination of a few members of past generations over many thousands of years, has direct relevance to the problems of today. Yet it has taken the 19th century upsurge in industrial output to overturn some of the entrenched ideas and start to bring about legislation for change. The 20th century notably produced The Factories Act 1961, The Offices, Shops and Railway Premises Act 1963 and the Health and Safety at Work Act 1974. These have all contributed to the establishment of the Management of Health and Safety at Work Regulations 1992, which has given rise to the familiar Six Pack, its standards and guidance.

The Vedas BC 1000

Three thousand years ago Agnavisa wrote in the Carake Veda that disease may be caused by misuse of the body. He went on to explain that the problem is usually three-fold, involving body, mind and spirit being out of harmony with the environment. This statement has echoes of modern thinking on sick-building syndrome. (Pheasant, 1991). Agnavisa laid emphasis on spirit, which is something not usually seen as part of the equation, but is nevertheless of vital importance. The physical aspects of this state of being out of harmony with the environment, are well documented. The psychological aspects are now receiving a lot of attention i.e. stress at work. But where does spirit fit in to this model? We know and speak of the importance of the spirit (the morale) of a group, we understand that a demoralised group cannot function properly, we are aware of a broken spirit of an individual. But what does all this mean? It is easy to dismiss the ancient teachers as esoteric, out of touch with today, not in line with modern thinking. Yet without the right spirit, both collective spirit and individual spirit, a country, a company, a small business, a person, can sink without trace; can perish.

'It is more dangerous to change from idleness to work than from work to idleness'

Hippocrates

Hippocrates BC 460

What Hippocrates meant by work is open to interpretation. Yet when he wrote this all those years ago, Hippocrates identified a problem which he considered a danger and which has only recently started to become generally recognised in industry. Nonetheless this has been known amongst athletes and dancers for many years. Hippocrates even described in great detail the need for gradual increase of activity amongst athletes (Hippocrates 435BC). Generally when people return to work nowadays, after a break due to injury or illness, even from holiday, they are expected to resume their tasks at the same level as before the break. This practice contains the strong possibility of injury or re-injury. It has been shown that inactivity produces changes which begin to take place at a cellular level (Troup and Videman 1989). In the case of manual workers or those involved in repetitive actions this should be clearly understood. A graduated return to optimum level of activity has been advocated over recent years by more enlightened scientists and ergonomists. (Carter et al 2000). Physiological changes may be slight and easily reversed in the case of a few days off after a holiday. However it becomes a much more serious matter after a longer break following injury or illness.

The Dialogues of Plato BC 390

The significance of rigorous preparation by athletes both physically and psychologically was emphasized by Socrates in The Dialogues of Plato (Jowett 1920). He advised his followers in this way, to help them optimise performance and avoid the risk of injury. In his explanation of the athlete's responsibilities, Socrates states "those taking part in contests are required to prepare themselves, by exercise and practice". Athletics was an important part of the Athenian life of the Citizen upper classes. However the rules were very different for the ordinary people. These were the artisans who performed the necessary tasks to keep Athens functioning so that the philosophers could lay back and discuss ideas and the athletes bring glory to the state. These artisans were people who needed to earn their living. Socrates is credited with stating: "the mechanical arts carry a social stigma and are rightly dishonoured in our cities, for these arts damage the bodies of those who work at them." Far from addressing the problem in those Athenian days, the accepted practice was to blame the workers for working and therefore neglecting their duties of citizenship. For all his erudition, Socrates had clearly not through this through. He would have been lost without those who baked his daily bread. And at the very lowest end of society, those individuals who were slaves were not even considered. They must have suffered many injuries but they were simply there to perform the lowliest tasks and the laws for them were very different.

De Triplica Vitae 1466

Father of the Italian Renaissance Marsilio Ficino, in his work De Triplica Vitae (Ficino 1466) he became preoccupied by the problems of those who led a sedentary and studious life.   He was concerned that constant sitting, guttering candles, a smoky and chilly atmosphere brought about respiratory conditions and failure of the body,   inducing digestive and musculoskeletal injuries .   He advised learned men to live near a healthy environment, to pay attention to their diet, take exercise and try to avoid melancholy, which he believed was often brought about by a solitary and studious life.   This particularly applied to the friars and those concerned with producing illuminated texts. In other words he advised them to bring some variability into their lives.

De Morbis Artificum 1700 & 1713

About two hundred years later Bernardino Ramazzini became intensely interested in the diseases of workers.   At a time when university life was dominated by the nobility and the church, Ramazzini, who initially held a chair at the University of Modena and later became Professor of Practical Medicine at the University of Padua, spent many hours in the squalid workshops of these towns, sitting down with artisans, talking to them, gathering and collating information which culminated in his well known work De Morbis Artificum, which was first published in Modena in 1700.   The later edition published in 1713 in Padua contains more detailed information connected with with fifty three different occupations (Ramazzini 1713). Initially he was concerned by the blindness that was suffered by men who emptied cesspits, then he moved on to the problems of inhalations from poisonous fumes of workers in metals.   Later the musculoskeletal disorders of manual and sedentary workers occupied his attention. His well known description of work related upper limb disorders [WRULDs] suffered by scribes and notaries, has been widely acclaimed in recent years.

The Conditions of the Working Class 1844

Many of the pioneers in occupational health have found themselves swimming against popular opinion and Friedrich Engels was no exception. At the time of the Industrial Revolution when steam power had just arrived and some people were getting very rich, Friedrich, the son of a German part-owner in a Lancashire cotton mill was sent by his father to learn at first hand about English business methods.   He found conditions amongst the Lancashire cotton mills to be based on profit to such an extent that the health of workers, including children, was not even considered and was seriously compromised. This gave rise to his famous book The Conditions of the Working Class (Engels 1845). He became aware of the workers' many chest complaints from inhaling fibrous cotton-dust and the long hours they worked in these unhealthy conditions.   In 1831 the twelve hour day had been introduced for those under eighteen, which included the the very young children who were employed.   Older workers were compelled to work longer hours.   Machines were unguarded giving rise to many accidents, loss of limbs and fatalities.   Engels was shocked by the deformities and malformations of children's limbs which came about because of working postures.   He wrote of young women bent and diseased with constant headaches and watery eyes and who were expected to return to work three days after giving birth.   He noted the permanent damp sogginess on the clothes of young children who were occupied with 'wet-spinning' and the overall indifference of the bourgeoisie factory owners to the misery they were creating. Conditions which induced in the workers, overcrowded living, poverty, immorality and in many cases early death.   The ideas of Friedrich Engels were very unpopular, yet they were part of the zeitgeist of the times.   Reforms were starting.

Lights in Darkest England 1891

It would not be generally obvious to associate the Salvation Army with manufacture. Nonetheless the Safety Match was introduced by the efforts of General Booth   (the founder of the Salvation Army) and instituted a change in the way matches (Lucifers) were made; moreover it brought about a change in the price the public were prepared to pay for them.   He discovered that many workers, mostly girls, were suffering from necrosis a condition they called 'Phossy Jaw'.   This common condition amongst the young workers caused the jaw bone to gradually disintegrate, inducing great pain and disfigurement.   General Booth realised that the yellow phosphorous, the igniting element, was poisoning the girls.   He noted that traces of it remained on the hands of workers, as there was no water to wash, and it got into their food as they ate their (often frugal) lunch. He made a decision to open a factory with much better light, airy, hygienic conditions. Here only non-poisonous red phosphorous was used, although it was more expensive. This was the origin of the Safety Match which became known as 'Lights in Darkest England'. In 1906 at an international meeting in Switzerland a treaty was signed by many nations which outlawed yellow phosphorous, and Lucifers ceased to exist.

Letter published in The Manchester Guardian 1892

"We employers cannot disregard the almost universal determination of working men to preserve their bodily powers and skills for as long a period as possible throughout life, since they form their only capital. In the course of nature, this capital diminishes as the years increase.   It may be exhausted in twenty years of excessive strain, or still be available after forty years of earnest daily work.   The combinations of working men are in duty bound to preserve this capital; and if customs prevail which exhaust it wastefully leaving men prematurely broken and decrepit in body and mind, to be cared for by their fellows or by society, we are living under unnatural conditions."
WILLIAM MATHER MP

William Mather was the co-founder of Mather and Platt - Manchester firm of ironfounders.   He was also the author of a study concerning fatigue in workers and established the forty-eight hour week in his factory whilst at the same time making no reduction in wages. He did this because he believed that no useful work was achieved between six and eight a.m. Following this popular action there was less spoiled work in his factory and the company went on to become more profitable. This action set the tone for other employers to follow.

Discussion

Three thousand years ago a harmonious environment was deemed to be important by Agnavisa to avoid mis-use of the body. Three thousand years later, call centres are slowly beginning to take this on board. Some, although not all, are beginning to recognise the benefits of taking a short break after a difficult call; a psychological re-balancing time for the operator. Common sense tells us this must be helpful and that we should live and work in the most harmonious circumstances possible, yet we have allowed unsafe working practices and mis-use of the human body to prevail in some industries for centuries. Ramazzini in the eighteenth century started the process of change. Nineteenth century reformers began to think about changing working practices when some laws were passed. Twentieth century minds, have begun to concentrate more on these matters, producing Draft Documents in which ergonomics has played an important role and culminating in HSE Regulations and Guidelines. Perhaps we have arrived at a time when we can no longer afford to ignore the wisdom of the past.

References

Carter, J.T. and Birrell. L.N. (Editors) 2000 *Occupational health guidelines for the management of low back pain at work - principal recommendations.* (Faculty of Occupational Medicine. London)

Engels, F. 1987 *The Conditions of the Working Class in England* (Penguin Books)

Ficino, M. 1989 *Three Books on Life* (The Renaissance Society of America, Binghampton New York)

Hippocrates. BC460 *Airs, Waters, Places - Treatise.*

Jowett, B. 1920 *The Dialogues of Plato, Laws VIII* 830 (Random House, New York)

Pheasant, S.T. 1991 *Ergonomics, Work and Health* (Macmillan, London)

Ramazzini, B. 1940 *De Morbis Artificum* (University of Chicago Press. Chicago Illinois)

Troup, J.D.G. and Videman, T. 1989 *Inactivity and the aetieopathogenesis of musculoskeletal disorders* Clinical Biomechanics **4**: 173-178

Waddell, G. and Burton, A.K. 2000 *Occupational health guidelines for the management of low back pain at work - leaflet for practitioners.* (Faculty of Occupational Medicine. London)

WHY UK ERGONOMICS LOST ITS CHANCE OF EXPANSION IN BRAZIL: A CONTRIBUTION TO THE ENGLISH ERGONOMICS HISTORY

Anamaria de Moraes, D. Sc.

LEUI Laboratory of Ergonomics and Usability in Human Technology Systems
Pontifical Catholic University of Rio de Janeiro
Rua Marquês de São Vcente, 225
Gávea, Rio de Janeiro, Brasil 224453-900
moraergo@rdc.puc-rio.br

The aim of this paper is to make a reflection on the role of French and British ergonomics to the development of Brazilian Ergonomics and to discuss why British ergonomics has not put a strongest role in Brazilian ergonomics as French ergonomics has done. If we take a look in the first meeting hold in Brazil, in 1974, we will see that Professors from UK were the majority. But during the eighties things changed and in all the Brazilian congress researchers from France were invited by the students that had done their PhD in Ergonomics there. This had consequences for Brazilian ergonomics and while others areas of knowledge developed a strong link with UK, the same did not happened with ergonomics.

Introduction

Since the first Ergonomics Congress, in Brazil, in 1974, a French influence was built. We can attribute to Alain Wisner from CNAM the strategy of develop a relation with Brazilian students that has as consequence more than 40 ergonomists that work and teach in Brazil that made his Ph.D. in France. Brian Shackle came to the same meeting but form that did not result a strong link with research centre or universities in England. We intend to show what happened during the Brazilian meetings and try to demonstrate how it continues till today and try to find a way to change the present conditions. It has some political consequences to Brazilian Ergonomics because as powerful group of Ph.D. who made their studies in France to their best to maintain this situation sending more students to France and trying to control the Brazilian Ergonomics Association. So the tendency is to continue these disequilibria between Anglophonic and Franco phonic ergonomics. When we plan a Congress the Doctors want to invite their advisors. When a demand arrive to ad hoc consultant the majority of them had made their PH.D in France, so it is normal to send their students to continue their PH.D in France.

From the beginning

In 1974 was held the First Brazilian Congress of Ergonomics, in Rio de Janeiro. Some foreigner ergonomists were invited such as Brian Shackel and Alain Wisner. Brian Shackle presented a paper about Ergonomics in England and in Loughborough. He presented a brief history of Ergonomics and after talked about was happening in Aston, Birmingham. After he presented some examples of works developed in Loughborough: the problem of temperature for workers who had as activity the storage of products in a cold environment. He showed some experiments on heart beat and other about tracking, intensity of illumination, energy consumption, manual material handling, safety and consumer products, and public transportation. Some practical examples about a section of diamond for a company reinforced the importance of ergonomics. In a second moment he presented the curriculum and the teaching techniques using experimental methods and fitting trials. Another work was about the redesign of the Control Centre of Esso, in the Heathrow airport, done in 1967/68. But what was used as a bible with copies all over Brazil was the Brian Shackle's paper presenting in details tests – explaining the methods, techniques and instruments and measurements   - about comfort of chairs, included in the proceedings. The paper was based in Shackle, B; Chidsey, K. D. ; Shipley P. The assessment of chair comfort. It was considered as a source of reference for many years mainly to the industrial design students, which course was the first in Brazil to have ergonomics as a mandatory subject in its curricula.

From 1971 till 1972, Colin Palmer, was a visiting professor, at Federal University of Rio de Janeiro, with the support of British Council and The Brazilian University. A synthesis of his conference was published in Portuguese, in 1976. Each chapter had an extended bibliography to help student to continue their researches in a specific topic of ergonomics of his interest.

In 1995, Wisner has been invited to The IEA World Conference that happened in Brazil joint with the 7th Brazilian Congress, in this occasion he said some words about our first congress, September 9 –13, 1974.

"Between the organizer we could see the names of some ergonomists that work in ergonomics before it was established.
(...) We must emphasize that the principal contribution came from the Brazilian researchers. Before this meeting and continuing after a great number of Brazilian went abroad to make M.Sc. and Ph.D. in Ergonomics. (...) In the CNAM Lab (in Paris), we had the pleasure of receiving 30 young researchers – 20 to M.Sc. and 10 to Ph.D.

Chapanis "Human Engineering", Laville "Ergonomics", and Palmer "Ergonomia" were the only source of reference published in Portuguese language, for many years.

Professor Alain Wisner, from CNAM - Conservatoire National des Arts e Métiers, Paris – was one of the pioneers of Brazilian Ergonomics. He had been in Brazil for the first time in 1968 and had established an academic cooperation network between his ergonomics research institute - "Conservatoire National des Arts and Metiers CNAM" - and the Fundação Getúlio Vargas, in Rio de Janeiro, the first Brazilian institution to have a postgraduate diploma in Ergonomics. Their first research was the analysis of the sugar cane worker conditions in Campos, a city nearby Rio. From 1970 a growing academic link has began and many Brazilian students – engineers, designers, and physicians, psychologists – were stimulated to carry out a postgraduate study at CNAM - Paris. The Brazilian government had sponsored many of these students with scholarship and fees support. They made researches about technology transfer, studying cases as the underground o Rio de Janeiro, a French technology – the Metro of Rio, from "Metropolitan", till now has a TCO, as said by the controllers without knowing the origin of the word ("tableau de contrôle optique"). After returning to Brazil, after 1985, many of those students settled up at universities and research institutes in some Brazilians cities as Florianópolis (Federal University of Santa Catarina), Rio de Janeiro (Federal University of Rio de Janeiro), Brasília (Federal University of Brasília) and São Paulo (Fundacentro) and has started ergonomics postgraduate courses in their area of concentration. With the support of Professor Wisner, who has visited Brazil many times, they became a very strong and solid research group in this country. Even now that Prof. Wisner is retired every year we have Brazilian designer, engineers, psychologists or physicians asking for resources to go to CNAM.

A history that has no end

Only after ten years, 1984, was held the Second Brazilian Ergonomics Congress, also in Rio de Janeiro. This was held one year after the foundation of ABERGO – the Brazilian Ergonomics Association, with the support of Professor Alain Wisner. It was a small meeting and with no invited researchers.

In the third congress, 1987, in São Paulo, nevertheless, a number of French researchers were invited to the conference – Wisner, Denoit, Daniellou, Foret, Clodorré (with the support of some Brazilian ergonomists who had returned from their postgraduate studies at CNAM – Paris) - the only English speakers were Corlett and Salvendy. The CNAM has continued with his politics of persuading more and more Brazilian students to attend its courses, but England Universities has made no similar actions. It can be asked why it has been done if English language is much more popular in Brazil and disseminated at Brazilian universities than French. We might say that Corlett has been successful and motivated one student to go to Nottingham do PH.D.

The forth Brazilian Congresses, in 1989, in Rio de Janeiro, had again very few representatives of Anglophonic ergonomics in comparison with the Francophonic – Laville, Pavard, Leplat, Sperandio – and only John Wilson from UK. It marks an important fact from this moment many Brazilians considered the cognitive ergonomics as French subject. Ken Eason in 1999, in Salvador during our 9th Brazilian Congress, had to repeat that he had been working with cognitive ergonomics for several years.

The fifth Congress, in 1991, was held in São Paulo. Professor John Wilson was present again. But, again we had as majority researchers from France - Damien Cru, Dominique Dessors, and Norbert See. It happened few mouths after the 11th IEA Congress, in Paris, when ABERGO, had became a member of IEA, with a little help from our friend Alain Wisner.

The sixth Brazilian Congress, in 1993, occurred, for the first time out of the circuit Rio/ São Paulo. It happened in Florianópolis, in the South of Brazil, organized by Ph.D. that has been in CNAM. Hall Hendrick (from USA, as President of IEA) came to Brazil for the first time. But again and again we had Pierre Falzon, Yves Clot, Dominique L. Scapin, Marie-France Barthet, from France. Gabriele Bammer, from Australia, gave a very interesting conference about work-related neck and upper limb disorders, with examples from her country.

The IEA World Conference in Brazil – the only one in Latin Americas

It was only in 1995, during the IEA world Conference in Brazil, and the 7th Brazilian congress that the scenario really changed. Some of the greater researchers in Ergonomics came to the meeting:
* Argentine - José Eduardo Grosso e Pablo Leandro Frnandez; * Australia - Margaret Bullock; * German - Dieter Lorenz, Holger Luczak, Klaus Zink, W. Bauer e P. Kern; * Belgium - Jacques B. Malchaire; * Canada- Birman Das, Claude Chapdelaine, Ian Noy, Jean-Marc Robert, Shrawan Kumar; * Chile - John Chalners, Paulina Hernandez. Horacio Rivera, Pilar Sandoval, Monica Le-Fort; South Africa - P. A. Scott; * Finland - Juhani Ilmarinen; * France - Alain Wisner, Christophe Kolski, Florent Coffin, Jacques Escouteloup, François Daniellou, Marie-France Barthet, Marie –Françoise Castaing, Marie P. Lemoine, Nathalie Edo, Pierre-Henri Dejean, Pascal Beguin, Tahar Hakim Benchekroun; *Netherlands: Pyter N.

Hoekstra, Pyter Rookmaarker; * Italy - Adriana Baglioni, Alfredo Bianchi, Carlo Amati, Erminia Attaianese, Gabriella Caterina, Maria Rita Pinto, Luigi Bandini Buti, Lina Bonapace. * Japan - Iiji Ogawa, Kentaro Arai; * Poland - Ewa Nowak, Janusz Bielski, Leszek M. Pacholski, * Portugal - Francisco Rebello, Luiza Barreiros; * UK - Bekere Oladipo, Christine M. Haslegrave, J. Mark Porter, J. May, T. Horberry, John Dowell, John Long, John Wilson, Neville A. Stanton, Robert Feeney; * USA - Colin G. Drury, Hiroshi Udo, Malgorzata J. Rys, Stephan Konz, James R. Buck, Steven M. Zellers, Kambiz Farahmand, Karl Kroemer, James D. McGlothlin, Lawrence Shulze, Margareta Nordin, Ogden Brown, Stuart O. Parson, Subramaniam Deivanayagam, Thomas Waters, Waldemar Karwowski; * Sweden - Martin G. Helander, Torsten Dahlin, Walter Ruth.

Even so, with so many people from France and England we had 12 X 10 to the French. But one thing begin to be clear the relations with University College of London, Nottingham, Loughborough, and Brunnel. From this moment increase the number of Brazilian students that went for those Universities make their Ph.D.

It is important to mention that John Long had been in Brazil in 1992, and gave very important and successful conferences about usability and HCI.

From till now

In the 1997 Congress, the 8th one, in Florianópolis again, again the majority of invited researchers came form France - Maurice de Montmollin, Suzanne Sebillote, Michele Rocher – and Anders Ingelgard e Per Odenrick, from Sweden, and Geoffrey Broadbent, from England.

In 9th Congress, held in Salvador, repeated the same configuration. We had from France: François Guerin (ANACT, Montrooouge/Paris), Hakim Benchekroun (CNAM, Paris), Bernard Pavard (France), Pascal Begin (France). From UK; Ken Eason (Loughborough), Yvonne Rogers

The 10th. Congress, that happened in Rio de Janeiro had as invited speakers Hal Hendrick (Usa), Davidf Campbell (BRTC, USA), Andrew Imada (USA), Francisco Rebello (PT), Najm Mmeskhati (USA) e Hakim Benchekroun (CNAM, Paris),.

In the 11th Conngress, in Gramado, in the south of Brazil, Hal Hendrick came once more with Dave Cochran from OSHA(USA), Jean Marc Robert (Montreal, Canada) and Franciscco Rebello (PT, Canadá).

The next meeting, the 12th Congress, will be in Recife, in the northeast of Brazil.

For the meetings of LEUI Laboratory of Ergonomics and Usability in Human Technology Systems we invited in 2001, Francisco Rebello (PT) and for 2002 we intend to have as speaker Andrew Monk (UK). York, UK) and Rachel Benedik (from UCL, UK). London).   And for the meeting of Design Research and Development, last year, we invited Neville Stanton, and for the next will have Patrick Jordan. We know the contribution of English researcher for cognitive ergonomics, HCI, usability and ergo design.

An other interesting observation is that we have two books of Alain Wisner translated to Portuguese and a brand new of Guérin, Laville. Daniellou, Duraaffourg and Kerguelen

Nowadays there are only two Brazilian doctors who had obtained their doctorate studies in Ergonomics carried out at British institutions and another nearly to finish their studies. All of them had attended a postgraduate programme from Loughborough University. It can be asked why there is a lack of consistent ergonomics academic net link amongst British and Brazilian universities in opposite of what happened with French and Brazilian institutions. Almost forty Brazilian students have concluded their studies at CNAM – Paris until the present. For 2002 one more had applicated to CNAM and must obtain Brazilian resources to his Ph. D. It can be said that a significant part of Professor Wisner research of anthropotechnology was carried out with the support of Brazilian students - with studies regarding the METRO system and oil companies, for instance - and consequently Brazilian funds from the Government agencies, which sponsored these Ph.D. students. Nowadays Brazilian Ergonomics Society has almost 500 members, ergonomics is a matter of interesting in a large number of universities in Brazil. In our point of view this is an important moment to recover the lost time and try to establish links between English and Brazilian universities in terms of ergonomic researchers. In addition to help to promote English ergonomics research in this country, it can contribute to establish an academic network with postgraduate students and researchers labs in the same way CNAM have been doing successfully since the seventies – every year we have new students .in their post-graduation courses in France and more and more French researchers are invited to come to Brazil. The contribution of Alain Wisner to the foundation of Abergo Brazilian Ergonomics Association was fundamental, the same in Paris in 1991 when ABERGO became a member of IEA. In 1974 we had France and UK in the Brazilian first Congress. Now Brazilian ergonomists give papers in all IEA meetings we have a history. But the influence of UK Ergonomics is still very timid. I think that the moment arrived to change this.

GENERAL ERGONOMICS

A VIEW OF THE FUTURE?
A DESCRIPTIVE TAXONOMY OF VEs

**Alex W. Stedmon, Harshada Patel, Sarah C. Nichols, Andrea Barone,
& John R. Wilson**

*Virtual Reality Applications Research Team (VIRART)
School of MMMEM
University of Nottingham
Nottingham NG7 2RD*

Without a formal framework to underpin the description and development of Virtual Environments (VEs), confusion may arise when comparing systems and applications to meet user needs. A taxonomy which describes VEs from an 'agent-neutral' standpoint has been developed that allows VEs to be classified according to generic properties or characteristics they exhibit, rather than focusing on any particular technology or application rubrics. At the highest level the taxonomy provides a descriptive overview of VEs, however it also provides a basis for considering potential technical and usability issues in the development and application of VEs for specific purposes. The taxonomy has been successfully applied to a number of VEs and offers a consistent means of describing different VEs which is independent of the particular purpose of the VE (training, visualisation, etc). What is the possible, however, is the ability to quickly visualise differences between specific VEs.

A VIEW of the Future

Virtual and Interactive Environments for Workplaces of the Future (VIEW) is a project funded by the European Union within the Information Society Technology (IST) program. The overall goal of VIEW is to develop best practice for the appropriate implementation and use of Virtual Environments (VEs) in industrial settings for the purposes of product development, testing and training. Fundamental to this is a need to be able to compare different VEs across different applications or technical features through their clear and unambiguous description. In order to achieve this a taxonomy is currently being developed. This paper sets out the taxonomy and shows how it can be applied to two seemingly different VEs that share many common characteristics.

Towards a Descriptive Taxonomy of VEs

VEs are 'computer-generated models for participants to experience and interact with intuitively in real-time', and can be generated via a host of technologies (see Wilson, 1997) of which Virtual Reality (VR) is one of the most common (Nichols, Haldane & Wilson, 2000). Without a formal framework to underpin the description and development of VEs, confusion may arise when comparing systems and applications to meet user needs. The taxonomy defines VEs from an 'agent-neutral' standpoint, allowing VEs to be classified according to generic properties or characteristics they exhibit, rather than focusing on any particular technology or application rubrics. From this standpoint, the taxonomy allows factors to be generalised at a later stage, for analysis or development, across users, tasks, systems or applications.

Developing a framework that can be demonstrated and applied for VE description will assist developers and users alike by integrating VR technology more efficiently for particular applications. Two VEs may seem quite different (such as the Kitchen and Netcard VEs described in this paper) but share many common features; or they may appear quite similar but have many defining differences. Tackling this problem from a technical perspective would provide, at best, only a partial solution. Likewise, concentrating exclusively upon the Human Factors issues would provide a solution ignorant of technical aspects. The taxonomy provides a basis for a more synthesised and sympathetic solution to designing VEs in the future by allowing a description of VEs which takes into account technical and Human Factors issues.

A Descriptive Taxonomy of VEs

At the highest level the descriptive taxonomy provides an overview of particular VE attributes. However, it also allows for the potential usability and technical factors to be considered at a later stage, which, in turn, equips the taxonomy with a power to account for these in the development and application of VEs. This provides a much richer descriptive power within the framework and, by adopting this approach, it is easier to maintain consistent criteria when comparing VR systems and applications.

During the initial stages of developing the taxonomy a number of perspectives were taken into account which focused on different approaches to the way VEs might be described. The type of information presented in VR and the way in which such information is used and experienced provided, a basis for describing VEs.

From this premise, in developing a tool to describe VEs, it is necessary to focus on the typical properties that they exhibit. As Rheingold (1991) states, VR consists of two basic elements: navigation and immersion. The combination of these two aspects: the experience of being in another environment and having the ability to govern one's own actions in that environment, is essential for a true VR experience (Rheingold, 1991). However, describing VEs purely on factors of immersion and navigation would be quite simplistic and at the expense of other factors which provide more subtle defining characteristics of VEs.

Only recently have researchers (Eastgate, 2001; Wilson et al, 2002) begun to develop specific guidelines for designing VEs and these provide another dimension to describing VEs in more detail. As such, the level of detail; number of objects; instructions; cues and feedback; proceduralised/random actions; user movement; object manipulation; screen resolution; texturing; object intelligence; and programming interactivity, can be considered as defining characteristics of a VE. However, definitions of VR and guidelines for VE design provide only part of the basis for describing VEs, and some means of incorporating the way the VE is experienced by the user is required. As such, a notion of 'presence' within the VE provides a useful start when describing VEs. Presence, as defined by Witmer & Singer (1998), is 'the subjective experience of being in one place or environment, even when one is physically situated in another', and is based on 4 factors control, intuitiveness, interface and realism. By combining many of these aspects the taxonomy has undergone an iterative development which has been purposefully designed to provide a top down and neutral approach to describing VEs. As illustrated in Table 1, the framework consists of a top level (Level 1) that can be divided into further levels (Level 2 & 3) of detail.

Descriptive Levels for VEs

Level 1	Level 2	Level 3	Descriptions
Control	Navigation	Active	User controls movement in the VE
		Pre-Set Path	User is an observer, movement is controlled by a preset path
		Passive	User is an observer of the VE; another individual is controlling movement
	Virtuality Continuum	VR	Simulating a complete virtual world to the user
		Augmented Virtuality	Augmenting virtual feedback to the user with physical cues
		Augmented Reality	Augmenting natural feedback to the user with simulated cues
	Activity Set	Procedural	Task must be completed following a proceduralised path
		Random	Task does not require the user to follow a specific procedure
		Both	The user has to follow procedures to complete the task, however features within each subtask can be performed randomly
Intuitive-ness	User Movement	Physical	Actions or movement performed by the user in the real world map to the actions performed in the VE
		Symbolic or Representative	Actions or movement performed by the users in the VE are not mapped to actions performed by the users in the real world
	Interaction Metaphors	Real	The user can see the action that is being performed with an object
		Abstract	The action being performed with an object is assumed by the user
	Semantics	Concrete	Representations of real world objects
		Abstract	Abstract or symbolic representations
	Ecological Validity	Real	Response in VE represents response in real world
		Abstract	Response in VE does not represent response in real world
Interface	Instructions	Embedded	The instructions presented to the user are embedded within the environment
		Separate	The instructions are presented to the user as part of the outer interface of the environment, or when presented obscure the environment
		Mix	The users receive instructions that are both embedded and separate from the VE
		No instruction	The user does not receive instructions when interacting with the VE
	Viewpoint	Egocentric	Users interact with the VE through their own individualistic perspective (worm's eye view)
		Exocentric	Users interact with the VE through a broader perspective, not only in relation to themselves, but also through inter-relationships between actions and objects within the environment (bird's eye view)
		Both	The user can view and interact with the environment through both an egocentric and exocentric perspective
Realism	Look of the VE	Textures	The type of textures found within the VE which are used to detail objects. This does not include texture presented through haptic feedback
		Objects	The number of static/active objects within the VE. An object is anything that affords an action within the context of the environment
		Colours	The number of colours used in the VE
		Photo-realism	The realism of the objects in comparison to the object found in the real world

Table 1. Descriptive taxonomy of VEs

Applying the Taxonomy

To illustrate the application of the taxonomy the defining characteristics of two VEs are presented in Table 2. 'Netcard' is a training environment to assess the effectiveness of VR as a training tool for computer network card replacement, and 'VR-Kitchen' is an environment to familiarise users with learning disabilities with a number of cookery procedures.

VE Descriptors			VEs	
Level 1	**Level 2**	**Level 3**	**Kitchen**	**Netcard**
Control	Navigation	Active	Restricted	Open Space
		Pre-Set Path		
		Passive		
	Reality Mix	VR	Desktop VR	Desktop VR
		Augmented virtuality		
		Augmented reality		
	Activity Set	Procedural	Clear procedure via instructions	
		Random		
		Both		Tasks performed in a specified order. However subtasks can be performed randomly
Intuitive-ness	User Movement	Physical		
		Symbolic/ Representative	Less precision required & large objects	Extremely precise actions & small objects
	Interaction Metaphors	Real	Some actions viewed by user	Some actions viewed by user
		Abstract	Some actions not viewed by user	Some actions not viewed by the user
	Semantics	Concrete	Users more familiar	Users less familiar
		Abstract		
	Ecological Validity	Real	Actions mimic real world	Actions mimic real world
		Abstract		
Interface	Instructions	Embedded		
		Separate	Written & verbal instructions	Only written instructions
		Mix		
		No instruction	View from user's perspective	View from user's perspective
	Viewpoint	Egocentric		
		Exocentric		
		Both		
Realism	Look of the VE	Textures	Textures do not provide additional detail to the VE	Textures difficult to distinguish
		Number of Objects		
		Number of Colours		
		Photo-realism	Less photo-realism	Higher photo-realism

Table 2: Application of the general taxonomy to the Kitchen and Netcard VEs

Discussion

To date, the taxonomy has been successfully applied to a number of VEs and whilst the framework is still being developed it does offer a consistent means of describing different VEs in a neutral manner. From the table above, it does not necessarily matter if the VEs are for particular purposes (training, visualisation, etc) they can still be described according to their attributes. What is possible, however, is the ability to quickly visualise differences between specific VEs (through the shaded portions of the table) and ultimately, through further research, to evaluate defining characteristics of different VEs and how they may be exploited for different purposes.

The taxonomy has highlighted how two seemingly different VEs share many common characteristics. The main difference between the two is in the activity set and the way that tasks are se out and completed. This would suggest that any other differences in the VEs are more qualitative by their nature.

Since technical (Kalawsky, 1996) and usability (Gabbard & Hix, 1997) taxonomies exist, it was initially thought that these could be incorporated into this taxonomy structure in some way. However, after early iterations, it was suggested that such aspects do not contribute specifically to the description of VEs as they arise through actual presentation and use of VEs. It would be more beneficial, therefore, to evaluate VEs in terms of such aspects separately, after they have been described according to more generic factors, and then use the findings to develop particular VEs in terms of best practise for the purpose.

Through the application of the taxonomy to date a number of factors have been identified which may require further development or demonstration:

- Some descriptors are subjective and responses will vary accordingly. Descriptors such as the level of photo-realism, number of objects, and number of colours within a VE, are all open to varying degrees of interpretation. To overcome potential discrepancies, detailed definitions need to be developed regarding the subjective categories within the framework so that they can be applied consistently.
- Since the taxonomy is generalised to the entire VE, it may consist of a number of aspects classified under a single factor. Whilst this can be described within the taxonomy at present, at the highest level this may provide fewer specific differences between VEs, or make it difficult to abstract the pertinent factors lower down the taxonomy.
- The taxonomy is not yet fully developed and as a consequence, specific descriptors may need further refinement or others may need to be added. One example is that 'interaction metaphors' may need better descriptors for when an action is not viewed (i.e., carrying an object from one place to another) but the interface indicates that the participant is holding an object. Other terms that may be useful in describing VEs could include cues, feedback, navigational aids (i.e. map of space), and the concept of 'near and far' VEs such as a close up endoscopy VE or a navigation VE with many features in the distance.
- To date, the taxonomy has not been applied to collaborative virtual environments (CVEs) and this will be done during the VIEW project.

A main point from the application of the taxonomy to date is that VE descriptions can be very similar even though their purpose/design is different (anything from VEs for training, design and information visualisation, to education and entertainment applications). To a certain extent this is to be expected as at the highest level of the taxonomy the descriptors are more general, but it also illustrates the power of the taxonomy to detail common factors across different VEs regardless of users, tasks, systems or applications.

At the highest level (Level 1) VEs can be understood in terms of detail in the other levels. After many iterations and the use of different examples it may be that particular VE formats (that could be application/technology/user specific) may have defining characteristics). The taxonomy may also be used in conjunction with other frameworks such as the Usability Taxonomy (Gabbard & Hix, 1997) or the Technical Taxonomy (Kalawsky, 1996) to provide a more comprehensive description of VEs. This taxonomy does not define all the technical, usability, or application issues, but instead emphasises where these issues belong in relation to each other.

At lower levels the detail in the taxonomy is more specific and this allows VEs to be compared in more detail. At such levels, however, the taxonomy may appear difficult to apply generally since many aspects within the same environment can differ. For example, in one room of a VE, actions may be real but in another room the actions may be abstract. Thus, the problem may not necessarily be attributed to the taxonomy *per se*, but instead may be due to VE design inconsistencies. This highlights a problem in developing the taxonomy and applying it to existing VEs, and so a further development of the taxonomy within the VIEW project will be to set guidelines for developing new VEs upon descriptive criteria that can be generalised and ultimately exploited for different purposes through further research.

Acknowledgements

The work presented in this project is supported by IST grant 2000-26089: VIEW of the Future.

References

Eastgate, R.M., 2001. *The structured development of virtual environments: enhancing functionality and interactivity.* Unpublished PhD thesis, University of Nottingham.

Gabbard, J.L. & Hix, D., 1997. *A taxonomy of usability characteristics in virtual environments. Final report to the office of Naval research.* Grannt No. N00014-96-1-0385. Department of Computer Science, Virginia Polytechnic Institute and State University Blacksburg, VA 24061.

Kalawsky, R., 1996. Exploiting virtual reality techniques in education and training: technological issues. Prepared for AGOCG. *Advanced VR Research Center, Loughborough University of Technology.*

Nichols, S., Haldane, C., & Wilson, J.R., 2000. Measurement of presence and its consequences in virtual environments. *International Journal of Human-Computer Studies.* 52, 471-491.

Rheingold, H., 1991. *Virtual reality: the revolutionary technology of computer-generated artificial worlds – and how it promises to transform society.* Touchstone, New York.

Wilson, J.R., 1997. Virtual environments and ergonomics: needs and opportunities. *Ergonomics.* 40, 1057-1077.

Wilson, J.R., Eastgate, R.M.., & D'Cruz, M. 2002. Structured development of virtual environments. In K. Stanney (Ed.), *Handbook of virtual environments.*

Witmer, B.G., & Singer, M.J., 1998. Measuring presence in virtual environments: a presence questionnaire. *Presence* 7(3), 225-240.

ARE INTERFACE PRESSURE MEASUREMENTS A TRUE REFLECTION OF SKIN CONTACT PRESSURE WHEN MADE OVER DIFFERENT LAYERS OF CLOTHING?

Robin H. Hooper and Gary R. Jones

Department of Human Sciences, Loughborough University
Leicestershire LE11 3TU, UK

To examine whether interface pressure measurements are a true reflection of skin contact pressure when made over different layers of clothing, pressure data were recorded from 4 participants whilst wearing 12 different clothing combinations. Participants wore both single and multiple clothing layers whilst walking on a treadmill carrying a loaded rucksack. Pressure measurements were taken from 2 sensors simultaneously, one placed on the skin, the other placed above the clothing layer/s. Results showed no reduction in skin surface pressure by wearing either single or multiple layers. Thus, if confirmed with further data it can be concluded that the soldier (or leisure user) will gain no or very little relief from applied pressure by wearing garments, even in layers, when carrying a rucksack. Also, if confirmed by further data, interface pressure may adequately be assessed using a sensor placed above the clothing layer(s) rather than at the skin surface.

Introduction

Today's infantry soldier has to carry ever heavier loads, much in a rucksack supported by the shoulders. Poorly designed load carriage equipment can cause injury (Knapik et al. 1996, Bessen, Belcher et al. 1987, Wilson 1987) including contemporary military equipment. Some injuries result from prolonged high pressures at rucksack-skin interfaces. Improvement is needed.

Alleviation of high pressure zones define a focus for design improvements. To assess interfaces pressure measurements can be used and have successfully shown interface materials influence the level and distribution of pressure (Martin & Hooper, 2000). In this paper a cotton T shirt lay between the skin and shoulder straps.

Soldiers often have layered clothing, including body armour, at the interface. Does this influence the skin surface pressure? If so, designs will need to work with the full clothing ensemble. If not, studies of interface pressures can be simplified by using 'above clothing' cf. skin surface pressure measurement. This paper presents some initial data addressing these issues.

Method

Four healthy males participated under conditions approved by the Loughborough University Ethical Advisory Committee, age 23.5±3.1 yr, weight 75.2±3.1 kg, height 178.4±7.5 cm, sitting height 92.5±2.4 cm, and right shoulder width 16.2±0.7 cm.

Participants carried an evenly loaded rucksack (23.5 kg) with a stable centre of gravity. A single military rucksack (90 pattern 'long back' Bergan) was used. They walked at 5 km.hr^{-1} on a treadmill, 0% grade, for 3 min. During this time pressure was measured (FSCAN from Tekscan^{TM}) at the right shoulder interface. After each 3 min spell, participants stopped and changed the clothing combination, necessitating doffing / donning the Bergan. The buckle positions were marked so the shoulder straps were repositioned giving a similar position and tension each time.

The pressure sensors are formed from a double layer of thin, pliable plastic sandwiching a thin layer of gel. Pressure sensitive ink is laid in multiple discreet zones, or sensels, 960 per sensor. These were equilibrated and calibrated at 10 psi before each use in a proprietary calibration box supplied by Tekscan^{TM}. A new pair of sensors was used for each participant. Two sensors were used separated by a layer of one or more garments and a no garment condition leaving the sensors in contact as a control. The skin sensor was taped onto the skin surface, positioned in relation to anatomical landmarks (the clavicle, mid-shoulder and upper ridge of the scapula). The strap sensor was taped above the clothing directly under the strap. Alignment with the skin sensor was checked. Pressure was recorded once each minute, giving 3 measures, over a one second period covering at least one full gait cycle

Table 1	Individual garments	Garment layers				
		1	2	3	4	5
The garments, tested singly and in layers, are shown on the right:	Shirt	X	X	X	X	X
	Thick fleece (F)		X	X	X	X
	Combat jacket (CJ)			X	X	X
	Raincoat (RC)				X	X
	Combat body armour (BA)					X

After each garment and layer was assessed, the sensors were crossed over and all measurements repeated, reversing the sequence of layers and garments. All pressure data used for analysis were a mean of the paired, crossed-over measures to obviate any effect if sensor offset was present.

Results

The studies described above are in hand. The data from each sensor pad were used in two ways. They were averaged, using only those sensels that registered pressure. Sensels registering zero were assumed to lie beyond the edges of the interface. If a degree of imperfect alignment existed these average pressure may have not been perfectly paired. However, the alignment did ensure the same zones of highest contact pressure were measured by both skin and strap sensors. Consequently the 90th percentile pressure recorded should match closely between them. The results presented below also show the 90th percentile data (90 %ile).

An n of 4 participants leaves inadequate power for comprehensive analysis, although a one sample t-test of difference between the skin and strap sensors was applied. More data and their analysis will be presented at the annual meeting of the Ergonomics Society.

Figure 1 gives the range of pressure measured under the shoulder harness during load carriage by sensors placed (i) directly on the skin surface and (ii) above the individual garments directly under the strap. In the control condition the strap sensor lay in direct contact with the skin surface sensor, labelled "strap (skin)".

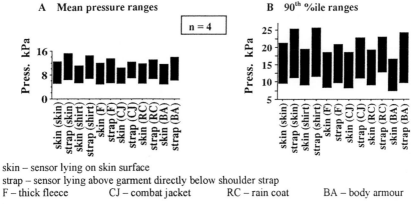

skin – sensor lying on skin surface
strap – sensor lying above garment directly below shoulder strap
F – thick fleece   CJ – combat jacket   RC – rain coat   BA – body armour

Figure 1 Pressures measured by skin and strap sensors paired beneath the shoulder strap

The difference between the mean data was calculated for each participant between the skin and strap sensors with the sensors lying in contact (labelled 'skin') and when separated by the thickness of a single garment. The average differences (± 95th C.I.) are given in figure 2A. Figure 2B shows the analogous data for the 90th percentile of measured pressure.

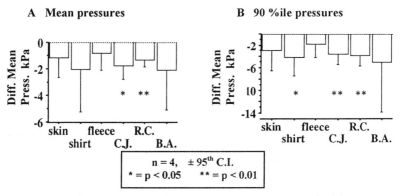

Figure 2 Pressure difference between skin and strap sensor separated by individual garments

The test conditions involving each garment separately were not different to the control condition (labelled skin in the figure) in which there was no clothing between the skin and strap sensors.

In use, the garments will be layered, depending on environmental and task requirements. So as to examine the influence of increasing thickness when layering garments, the difference between the mean data was calculated for each participant between the skin and strap sensors with the sensors lying in contact (labelled 'skin') and when separated by layers of increasing thickness (increasing numbers of garments). The average differences (± 95th C.I.) are given in figure 3A. Figure 3B shows the analogous data for the 90th percentile of measured pressure.

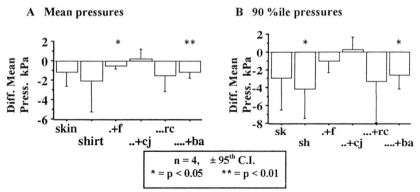

A   Mean pressures   **B   90 %ile pressures**

Figure 3   Pressure difference between skin and outer sensor when separated by increasing layers of garments

The test conditions involving each layer were not different to the control condition (labelled skin in the figure) in which there was no clothing between the skin and strap sensors.

Discussion

The data in figure 1 demonstrate the expected detection of pressure, inevitably present under shoulder straps when a rucksack / backpack is supported wholly or in part by a shoulder harness. The mean pressures shown are the data returned from the activated sensels in a sensor placed under the right strap, running from close to the axillary fold over the top of the shoulder. They are the mean from several gait cycles and so include pressure measures which range from a peak following heal strike through to a trough during the single support phase as well as all zones of contact, whether of higher or of lower pressure. The 90th percentile data are taken from this same data set. These are likely to represent the data measured during the period of highest loading following heal strike in the most loaded zone(s) within the interface.

Continuous pressure of 14 kPa has been suggested as an upper limit (Doan, 1998) to avoid tissue damage during sustained load carriage. In the conditions used here, the sensor lying on the skin detected mean pressures below this threshold (Fig. 1A) whichever garment lay at the interface. Although not shown, this was also the case when layered garments were present at the interface. The 90th percentile pressures of course show a higher range. Whether these are sustained higher pressures or whether they tend to be present intermittently during the gait cycle cannot be deduced from these data.

From figure 2 it is tempting to speculate that the pressure at the skin surface is influenced by the presence of a garment. Given that there are only 4 tests shown, the power is low but all mean differences are negative (sensor on skin less sensor above the garment, next to the strap) and two reach significance from zero. However, in the control condition when the strap sensor lies in direct contact with the skin surface sensor, there is a mean negative difference ($p > 0.05$).

If confirmed with further data, this suggests that the outer (strap) sensor returns higher data. If true, this could be due to the nature of the surfaces contacted by the sensors. One is in contact with the skin. The second is in contact with the lower sensor, a more slippery, plastic-on-plastic surface. So the latter may perhaps conform more to the strap shape than the shoulder shape perhaps affecting shear forces or curvature of the sensors. Each has the potential to alter the data returned by the sensors.

When compared to the control condition, the garment layers do not make a difference, none being significant with the small number of trials reported here (Fig. 2). This holds for the mean data and the 90th percentile data. The nature of the various materials which as a result contact the outer, strap sensor do not seem to create sufficient difference to change the measure of pressure found. Small variation is expected as the pack had to be removed and replaced when garments were changed even though the buckles were set at the same point.

The effect of layering garments is more variable (Fig. 3). There is little clear evidence of lower skin surface pressures even with multiple garment layers and these disappear when allowance is made for the control condition.

If confirmed, this refutes the starting supposition that multiple layers will affect the 'footprint' of transmitted pressure from the strap to the harness to a greater extent than individual layers – or more to the point to a greater extent than the control condition. It would suggest the soldier (or leisure user) will gain no or very little relief from applied pressure by wearing garments, even in layers, when carrying a rucksack / backpack. It also suggests, if confirmed by further data, that interface pressure may adequately be assessed using a sensor placed above the clothing layer(s) rather than at the skin surface. This would provide a more practical approach for objective assessment of the interface, especially in trials outside the laboratory.

References

Bessen, R. J., Belcher, V. W., and Franklin, R. J. 1987. Rucksack Paralysis With and Without Rucksack Frames. *Military Medicine* **152(7)** : 372-375.

Doan, J.B., Stevenson, J.M., Bryant, J.T., Pelot, R.P., and Reid, S.A. 1998 Developing a performance scale for load carriage designs *Proceedings of the 30th annual conference of the Human Factors Association of Canada*

Knapik, J. J., Harman. E., and Reynolds, K. 1996. Load carriage using backpacks: A review of physiological, biomechanical and medical aspects. *Applied Ergonomics* **27(3)** : 207-216.

Martin, J.L., Hooper, R. H. 2000. Military load carriage: A Novel Method of Interface Pressure Analysis. *Proceedings of the RTO Human Factors and Medicine Panel (HFM) Specialist Meeting, Soldier Mobility: Innovations in load carriage system design.* 22 1-8

Wilson, W. J. 1987. Brachial plexus palsy in basic trainees. *Military Medicine* **152** : 519 - 522.

REQUIREMENTS ANALYSIS FOR DECISION SUPPORT SYSTEMS – PRACTICAL EXPERIENCE OF A PRACTICAL APPROACH

Caroline Parker

Centre for Research in Systems and People (CRSP), Computing Department, Glasgow Caledonian University, Cowcaddens Road, Glasgow, G4 0BA
c.g.parker@gcal.ac.uk

Decision Support Systems (DSS) are computer-based systems designed to help users to make more effective decisions by providing information in a way that actively supports the decision process. DSS developers in UK agriculture have not found it easy to adopt a truly user centred approach to design and development because of lack of experience and the absence of an appropriate set of methods tailored to the technology. The author has argued that to produce usable DSS small scale developers, like those in agriculture, need prescriptive, user-centred and DSS applicable methods that are easy and relatively inexpensive to adopt. This paper describes the use of one such method for user requirements analysis and its use within three on-going agricultural DSS projects.

Introduction

Decision Support Systems (DSS) are computer-based systems designed to help users to make more effective decisions by providing information in a way that actively supports the decision process. Unlike expert systems, which are usually designed to supplant some aspect of an expert's role, DSS exist to complement and 'support' decision-makers rather than to replace them. DSS have been developed on many platforms in many industries and for a wide variety of uses, for example medicine (e.g. Reisman, 1996), utilities (e.g. Lindquist *et al.,* 1996) and financial services (e.g Wong & Monaco, 1995), and agriculture. They are usually based around a simulation model (i.e. a mathematical model describing the way an entity or group of entities will react under given circumstances), or a rule-base, or both. In UK agriculture, as in other countries, DSS have been promoted as a means of revitalising the knowledge transfer process in the wake of the removal of state funded advisory services.

As a human factors practitioner the starting point for the delivery of useful and usable DSS technology is the user, and the application of a user-centred design methodology to the development process. User-centred design being taken to be the involvement of users at all stages of system development from initial planning, through requirements analysis, into development and evaluation. DSS developers in agriculture have not however found it easy to adopt a user centred approach to design and development for three reasons.

First, many developers are scientists and not commercial programmers. Even now, when research centres are under increasing pressure to make money from their activities,

DSS are often developed as an afterthought or as follow on research projects and do not start out as commercial products. The market for agricultural DSS has been too small to attract commercial attention and developments usually take place outside of a commercial development structure. The DSS developer has often been someone without the benefit of commercial software development experience, who cannot afford to outsource the work because of the limited nature of research funding, and who lacks knowledge of structured software development methods. Second, even those developers who have access to software production skills will find that there are few methods that offer guidance on the involvement of users in the development process. Many user-centred design methods are considered to be impractical by mainstream developers (e.g Rossen, 1987) and methods adopted by the computing industry focus almost all of their attention on the technical aspects. Finally, a determined developer might adopt the most user-centred design approach they can find to suit a small scale project like a DSS (e.g. DSDM) only to find that it doesn't really support the capture of user requirements for the decision making task.

The author has argued (Parker, 2002) that to produce usable DSS small scale developers, like those in agriculture, need prescriptive, user-centred and DSS applicable methods that are easy and relatively inexpensive to adopt. This paper describes one such method for requirements capture, initially defined and developed within a DSS project called DESSAC project now complete (Parker, 2001), and its use within three current agricultural DSS projects: PASSWORD, a DSS for pest and disease management in Oilseed Rape, WMSS, weed management DSS and Slugs, a DSS for slug management in brassica and salad crops.

User requirements analysis for agricultural DSS

The method described in the paper is part of a suite of related approaches to the user-centred design and development of agricultural DSS discussed in Parker (1999). It is a method for the initial identification of user requirements for DSS and is based on two foundations, the Decision Enquiry approach to requirements capture developed by Arinze (1992) and the use of workshops as cost-effective means of involving users in the design process. These will be discussed in brief before the method itself is outlined.

Decision enquiry or question-based approach

Arinze reasons that the key information flow between the DSS and the user is the stream of requests from the user, i.e. the questions that the user asks of the system when using it to support decision making, and that these should therefore be the key determinant of the shape and form of the DSS. The data from the DESSAC project suggested that much of the crop protection decision task was indeed concerned with getting answers to questions about the weather, disease levels, product effectiveness etc. The focus on user questions, both from the observations of system failure and the observations of the decision making process, seemed therefore to support the Arinze argument.

Where Arinze's work is particularly useful to task analysis and requirements specification is the division of these questions or 'decision enquiries' (Arinze's term) into a functional taxonomy (op cit.). He argues that when decision-makers interact with a DSS they will invariably make an enquiry of one of three main types labelled: state, action and projection enquiries. State enquiries are made when the user is seeking information about the state of the world (or a model of it):

- entities (e.g. products, diseases)
- processes (e.g. pest and disease lifecycles, market behaviour)
- attitudes (e.g. buyer attitudes, consumer attitudes)
- policies (e.g. legislation, buyer policies)
- people (e.g. staff, customers, suppliers)

Action enquiries are requests for a plan of action to achieve a specified end state. This is a reverse 'what if' question, i.e. instead of what will happen if I do this, an action enquiry asks how do I get to this pre-specified end-state. In this type of query, it is the function of the DSS to generate actions in response to the user's goal setting. Projection enquiries are more commonly known as 'what if' enquiries. They are requests for an indication of outcome given a set of defined conditions e.g. 'How much will I lose if I delay the application of this spray for three days?'. The importance of this taxonomy is that it provides a direct link to specification. The identification of State enquiries tells the developer what data the user needs to have at hand, in DSS databases, linked programs or encyclopaedia, identification of action and projection enquiries provide a definition of the models that will be needed to support the user.

Workshops

What is the best and most cost effective method of involving users in design? Workshops, or focus groups, were adopted as an ideal way to meet this requirement for a number of reasons: their relative cheapness compared to interviews and co-opting methods, their use in human factors research and usability evaluations (e.g. Jordan, 1998) and a history of successful use within agriculture (Norton & Mumford, 1993). Another reason for the use of workshop or focus groups is that the workshop participants "stimulate and encourage one another", (Bruseberg and McDonagh-Philip, 2002) and a short workshop can generate a wealth of information and consensus on issues of importance.

User-centred user requirements method for DSS

This section of the paper outlines the method used in the three agricultural DSS project, WMSS, PASSWORD and Slugs. WMSS and Password are DSS for arable farmers and Slugs is a DSS for horticultural brassica and salad growers. The main distinction between the user groups for these systems is that arable farmers tend to focus on yield and gross margin while horticultural growers are forced by the nature of their product and their markets to focus on quality.

Each of these projects is LINK funded, i.e. partly government and partly industry funded with a consortium made up of research and commercial partners and has a heavy emphasis on basic biological research in addition to DSS development. All three projects were in the first year of their 3-4 year life span at the time of the requirement's analysis and the funds available for the identification of user requirements were very limited in all cases. While each of the three projects felt that the product they were developing was solidly based in a real need, expressed by the industry and supported by the industrial partners on the team, none of them had previously carried out any form of detailed requirements analysis. In all cases therefore there was an urgent need to identify a clear set of requirements to inform the biological and technological development, in as cost effective way as possible.

The workshops took place at different times but all within the 'slack' period for crop producers i.e. November to March. WMSS workshops took place in December 2000, PASSWORD in February and Slugs in March. In each case the lead partner in the project arranged for invitations to take part to be sent to a large mailing list of appropriate producers, both farmers and those who provide advice to them on agronomic matters i.e. independent and distributor based consultants. Workshops for each project were planned to take place over two days, with two workshops per day, one in the morning and one in the afternoon. Separate sessions were held for farmers and consultants as previous experience suggested that these groups talked more freely in the company of their peers, without the complication of a commercial relationship.

The aim of these sessions were: to identify the sub-tasks or stages within the decision process, the questions asked within them and the sources of information currently used to inform the questions. Additional aims were to prioritise requirements and to provide answers to specific questions raised by the technical partners in the projects.

All of the workshops followed the same basic structure and took between 2 and 3 hours to run, flip charts and tape recordings were used to record the data. The structure and approximate timing of each is represented in table 1 below. Eleven of the twelve planned sessions went ahead with 70 people in total taking part, roughly one third farmers and two thirds consultants. The Slug project workshops were less well attended because they had been delayed until March when spring activities are beginning to demand attention.

Table 1. Format of workshops

Topic	WMSS	PASSWORD	SLUGS	Mins (approx)
Introduction	✓	✓	✓	5
Aspects of decision making.				
Whether to act	✓	✓	✓	20
What type of action to take	✓			20
When to act	✓	✓	✓	20
What to apply	✓	✓	✓	20
Coffee break	✓	✓	✓	5
Additional support	✓	✓	✓	30
Most important problems	✓	✓		10
Availability of data	✓	✓	✓	20
Delivery mechanisms	✓	✓	✓	20
Questionnaire	✓	✓	✓	10
			Total	*180*

After the participants had settled and the aims of the project and of the workshop were explained to them they were taken through the main stages or sub-tasks in the decision process. These sub-tasks were identified by the author prior to the workshops, on the basis of past experience. The subtasks are almost identical in the three groups and problem management can be said in all cases to include the decisions: whether to act, when to take action and what type of chemical to apply if action is needed and chemicals have been chosen. Only in the case of weed decision making (where cultivation is also an option) is there any real choice between chemicals and other options in intensive crop production. The participants were asked to list the questions or issues that were most important at each stage i.e. what questions did they ask before they felt able to take the decision. Their answers were recorded on flip-charts and were visible during the discussions. After the questions were exhausted the participants were asked where they obtained the information to answer their queries. The aim of this section was to identify user enquiries which could be translated via the Arinze taxonomy into concrete requirements for model and database components.

While the issues were fresh in their minds the participants were asked to split up into groups of two or three with a cup of coffee and identify areas in which they felt more support would be useful i.e. where information was either not readily available to answer their questions or was of poor quality. After about twenty minutes they were asked to re-convene and to report back to the group. Once again their suggestions were recorded on a flip-chart in plain view. The final part of this exercise was the ranking of the items in order of importance i.e. very important, important, useful or nice to have. The aim of this section was to identify the areas that the user group considered to be most important and therefore provide a concrete means of prioritising the work of the project. In a related exercise the WMSS and PASSWORD groups were asked to identify the weeds or pests or diseases that they felt were most important to them, this was not seen as relevant to the Slug groups as few people actually distinguish between slug species.

DSS are highly data driven and in order to identify the limitations under which the software would have to operate each group was asked the degree of access they had to observation and weather data. At the time of the workshops there was serious debate in all projects about the potential use of the Internet as a means of delivering DSS. The participants were asked what type of delivery mechanisms they might prefer for different elements of the support package they'd defined. Finally participants were asked to complete a short questionnaire which contained specific questions raised by the technical partners and which obtained a more personal view of users willingness to invest in additional data gathering equipment or conduct more field level observations.

Data analysis

The questions and issues generated in the decision-making session were divided into the three Arinze (1992) taxonomies, State, Action and Projection on the basis of best fit. The questions were placed in a table alongside the information used to answer them and the groups that suggested it as important. An example of this layout is provided in Table 2.

Table 2. Example of table for decision element data

Decision element:	Do I need to act against weeds?				
Enquiry category :	State				
Question	Information sources	1	2	3	4
What are the levels/population in the crop? E.g. good, bad, horrific; low, moderate, severe	Crop walking/observation Weed map, black-grass map Distribution Field history	✓		✓	✓

The areas suggested by the groups as requiring more support were collated by the author under headings selected on the basis of a perceived natural grouping. The raw data was made available to project partners to allow alternative groupings to be selected if necessary. These headings were also tabulated and the ranking provided by each group listed. An averaging of these ranks provided a means to sort the suggestions in order of priority. Other data was summarised and reported in tabular and textual form.

Conclusions

The approach described above has been used in three different projects and performed equally well in all of them. It was simple to organise and run, the participants enjoyed taking part and despite disappointing numbers in some cases, generated a wealth of information. The approach is a formal and practical means of gathering information about the task of decision making, of organising it and using it as the basis for design decisions; and it can be used to ensure that tasks and functions are appropriately allocated within a decision support system. By identifying the 'enquiries' or questions inherent in a decision making process, it becomes possible to state the users' requirements for data and for mathematical models, the two main components of the DSS. Because the designer knows that the system has to support the posing and answering of specific questions, this knowledge also guides the development of the interface. This makes the approach consistent with the Ecological Interface Design method of Vicente & Rasmussen (1992), in that the model of mechanisms which people have for dealing with the complexity of the environment is provided by the questions they ask of it. The next stage in these projects is to obtain feedback on the emerging user interface design and the functionality it represents. Rapid prototype based activities in the context of workshops are likely to form the basis of this phase.

References

Arinze, B. (1992) A user enquiry model for DSS requirements analysis: a framework and case study. *International Journal of Man-Machine Studies,* **37**, 241-264.

Bruseberg, A. and McDonagh-Philip,D. (2002) Focus groups to support the industrial/product designer: a review based on current literature and designer's feedback.

Jordan, P.W. (1998) An Introduction to Usability.

Lindquist, K., McGee, M. & Cole, L. (1996). Tva-epri river resource aid (terra) - reservoir and power operations decision-support system. *Water Air And Soil Pollution,* **90** (1-2), 143-150.

Norton, G.A. & Mumford, J.D. (1993) *Decision Tools for Pest Management.* CAB International, Wallingford, UK.

Parker, C.G. (1999). A user-centred design method for agricultural DSS. In Second European Conference for Information Technology in Agriculture. Bonn, Germany. 27-30th 1999, pp. 395-404. Bonn: Universität Bonn-ILB. Vol A.

Parker, C.G. (2001) An approach to requirements analysis for decision support systems. International Journal of Human-Computer Studies, **55**, 423-434.

Parker, C.G. & Sinclair, M. (2002) "Why user-centred design works: the case of decision support systems in crop production". Accepted by Behaviour and Information Technology May 2001.

Reisman, Y. (1996) Computer-based clinical decision aids. A review of methods and assessment of systems. *Medical Informatics,* **21** (3)**,** 179-197.

Rossen, M.B., Maass, S. & Kellogg, W.A. (1987) Designing for designers: an analysis of design practice in the real world. *CHI '97 conference on Human Factors in Computing Systems* ACM, New York, pp. 137-142

Vicente, K.J. & Rasmussen, J. (1992). Ecological Interface Design: Theoretical Foundations. *IEEE Transactions of Systems, Man and Cybernetics*, **22**(4), 589-606.

Wong, B. K. & Monaco, J.A. (1995). Expert system applications in business: a review and analysis of the literature (1977-1993). *Information & Management,* **29** (2), 141-152.

A FIELD STUDY INVESTIGATING THE RELATIONSHIP BETWEEN VISUAL DISCOMFORT AND THE MENSTRUAL CYCLE IN VDU WORKERS: PRELIMINARY FINDINGS

Stacy A Clemes and Peter A Howarth

Visual Ergonomics Research Group
Department of Human Sciences
Loughborough University
Loughborough
Leicestershire
LE11 3TU

This paper reports the initial findings from a field study investigating whether visual discomfort in VDU workers is influenced by the phase of the menstrual cycle. The study took place at a BT Directory Assistance Centre and 52 participants have been tested to date. Participants were categorised into one of four groups, an experimental group (naturally cycling pre-menopausal women) and three control groups (oral contraceptive users, postmenopausal women and men). Visual discomfort was measured using a visual symptoms questionnaire, completed at the beginning and end of the shift, twice a week for a duration of five weeks. Variations in visual discomfort were seen over the menstrual cycle in the experimental group, with reports of discomfort increasing during the second week of the cycle. No consistent variation was evident in any of the three control groups.

Introduction

The fluctuations in female reproductive hormones that occur over the human menstrual cycle appear to have a widespread influence on many sensory functions. For example, changes in visual thresholds (Diamond *et al*, 1972; Braier and Asso, 1980; Parlee, 1983), olfactory thresholds (Pause *et al*, 1996), auditory thresholds (Elkind-Hirsch *et al*, 1992), gustatory thresholds (Than *et al*, 1994) and responsiveness to pain (Hapidou and Rollman, 1998; Riley *et al*, 1999) have all been reported to vary during the menstrual cycle. There appears to be a general trend for these changes to occur during the second week of the cycle, just prior to ovulation, with thresholds decreasing around this time.

Following the suggestion that the cyclical levels of reproductive hormones in females may influence susceptibility to motion sickness (Grunfeld and Gresty, 1998), we previously examined susceptibility to "visually-induced" motion sickness over a complete menstrual cycle in normally cycling volunteers (Clemes and Howarth, 2001). This malaise, referred to as virtual simulator sickness (VSS), occurred during the use of a virtual reality head-mounted display (HMD). Susceptibility to VSS did vary over the menstrual cycle, with an increase in susceptibility occurring during the second week of the cycle. In addition, reports of visual discomfort also varied over the menstrual cycle, following the same pattern as observed for VSS. As no previous study has linked visual discomfort to the menstrual cycle, the aim of the current study was to determine whether such a relationship exists in other situations.

This study was a preliminary investigation conducted at a British Telecom Directory Assistance centre. Directory enquiry operators were chosen because their job involves intensive visual work using a VDU, and previous field studies have shown that individuals whose employment involves concentrated visual work exhibit changes in reports of visual discomfort over the working day (e.g. Howarth and Istance, 1985).

Methods

Participants
Participation in the study was voluntary. At the beginning of the study volunteers met with an experimenter on a one-to-one basis at which time they completed a health screen questionnaire, provided written informed consent, and were assigned a number. The use of numbers ensured that volunteers remained anonymous throughout the duration of the study. To date, 52 volunteers have completed the study.

Following their responses on the health screen questionnaire participants were categorised into one of four groups: i) the experimental group, consisting of pre-menopausal women not taking any form of hormonal contraception (n = 20), ii) a control group, consisting of pre-menopausal women taking an oral contraceptive (n = 13), iii) a second control group consisting of postmenopausal women (n = 10), and iv) a third control group consisting of men (n = 9).

Participants were unaware of the true aim of the experiment, instead they were informed that the aim of the study was simply to investigate the occurrence of visual discomfort over a five week period. They were also unaware that the participants were categorised into the groups described above.

Participants in the experimental and oral contraceptive groups provided information about their menstrual cycle on the health screen questionnaire by writing down the start date of their last menstrual period. At the end of the study participants in the experimental group each received a letter explaining the real aim of the study and requesting that they inform us of the start date of their last menstrual period. This was because it was anticipated that all participants would have started a new cycle at some point during the five weeks of the study. With this information the menstrual cycle phase could be determined over the duration of the study for members of the experimental group.

The study received ethical approval from the Loughborough University Ethical Advisory Committee.

Measurement of visual discomfort
Subjective ratings of visual discomfort were collected using a standard visual symptoms questionnaire, completed at the beginning and end of the shift on designated study days (Mondays and Fridays). The questionnaire is structured so that participants first rate symptoms of visual discomfort (tired eyes, sore eyes, irritated eyes, runny eyes, dry eyes, burning eyes, blurred vision and double vision) preparing them to make an overall rating of general visual discomfort on a 7-point scale with '1' being no discomfort and '7' being severe discomfort. The *change* in this rating over the shift was taken as the primary measure of visual discomfort (Howarth and Istance, 1985).

The study was run over a five week period to ensure that those in the experimental group were tested during each week of a complete menstrual cycle. Participants each received a pack of twenty questionnaires that they kept with them in the exchange. Following the completion of each questionnaire participants detached it from the pack and placed it in a collection box provided. This procedure ensured that participants were not influenced by their responses on previous questionnaires.

Data analysis

From the information provided by the participants about their cycle, it was possible to assign a 'cycle week' to each questionnaire completed.

In the experimental group the change in ratings of general visual discomfort were compared for each week of the menstrual cycle, and with the changes reported over the "pill" cycle for the oral contraceptive group, and over the four week "pseudo cycle" for the postmenopausal group and male group. The mean changes in visual discomfort were calculated for each week of the corresponding cycle for each group of volunteers. In addition the pre and post ratings of general visual discomfort reported over each week of the cycle were compared for each group.

The data was analysed using SPSS (version 10.0 for Windows) and Meddis' system of ANOVA by ranks, on the MS DOS statistics package "Omnibus" (Meddis, 1980). From previous research (Clemes and Howarth, 2001), it was predicted that reports of visual discomfort would vary over the menstrual cycle in the experimental group, with discomfort increasing during the second week of the cycle. Using Page's Trend test in "Omnibus" the changes in ratings of visual discomfort were analysed to test for the rank order: 1, 2, 1, 1.

Results

Experimental group

Of the 20 participants in the experimental group, 95% reported having regular 28 day menstrual cycles on the health screen questionnaire completed at the beginning of the study. The average menstrual cycle length of this group was 27.8 days (SD = 2.5). Due to the nature of this research the number of participants starting the study during each week of their menstrual cycle was unbalanced. 40% started the study during week one of the cycle, 20% during week two, 15% during week three and 25% during week four. With this in mind, data collected from the experimental group over each week of the study (prior to arranging the data by menstrual cycle week) was analysed for the presence of order effects. No statistically significant order effect was observed ($p = 0.65$, Friedman test) indicating that the calendar week of study and the imbalance in the number of volunteers starting the study during each week of their cycle had little effect on the data set as a whole.

Pre and post ratings

The ratings of visual discomfort reported at the beginning of the shift for each group of participants were compared and no statistically significant differences between groups were present ($p = 0.6$, Kruskal-Wallis test). Similarly, for the experimental group, the levels of visual discomfort reported at the start of the shift did not vary significantly over the menstrual cycle ($p = 0.3$, Friedman test).

Overall, post-shift ratings tended to be higher in the experimental group than in the three control groups, however this difference did not reach statistical significance (p = 0.25, Kruskal-Wallis test). Within the experimental group, variations in post shift ratings were apparent with an increase in discomfort being reported during the second week of the cycle, however this did not reach statistical significance (p = 0.28, Friedman test). No consistent variations in post shift ratings were observed in the three control groups.

Change in ratings

Figure 1 displays the mean *change* in ratings reported during each week of the menstrual cycle for the experimental group. It also shows the change in ratings over each week of the pill cycle for the oral contraceptive group, and the change in ratings over a four week "pseudo cycle" for postmenopausal women and for males. When comparing the mean change in ratings for each group, a statistically significant difference between groups was observed (p = 0.02, Kruskal-Wallis test).

Figure 2 (overleaf) shows the mean change in ratings of visual discomfort reported for each week of the menstrual cycle in the experimental group alone. The analysis of variance by ranks applied to test for an overall effect of the cycle did not reach statistical significance (p = 0.2). However, as predicted a greater change in discomfort occurred during the second week of the cycle and the trend test applied to the data was statistically significant (predicted trend 1, 2, 1, 1, p = 0.03, Page's Trend test). In contrast, no consistent variation was observed in any of the three control groups.

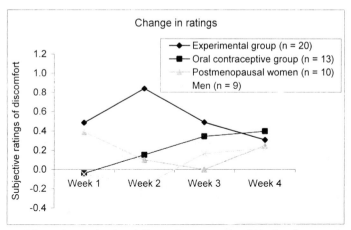

Figure 1. Mean change in visual discomfort, over the shift, for the experimental group and the three control groups.

Discussion

The aim of this investigation was to determine whether visual discomfort in VDU workers was influenced by the stage of the menstrual cycle in naturally cycling women, and to compare the responses in this group with those from three control groups.

At the beginning of the day, no differences were apparent in the ratings of visual discomfort of the four study groups. Similarly, the levels of visual discomfort reported

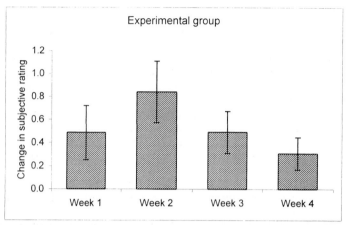

Figure 2. The mean change in ratings, with standard error bars, of visual discomfort during each week of the menstrual cycle in the experimental group.

at the beginning of the shift for the experimental group did not vary over the menstrual cycle.

However, when the post-shift ratings were examined, along with the change in discomfort over the working day, a different picture emerged. There was a tendency for the experimental group to report slightly higher ratings of visual discomfort at the end of the shift than those reported by the three control groups. When comparing the mean change in ratings over the shift for the four study groups a statistically significant difference between groups emerged, with the experimental group exhibiting a larger change than the control groups.

Variations in reports of visual discomfort were apparent over the menstrual cycle for the experimental group. From a previous study (Clemes and Howarth, 2001) it was anticipated that, if visual discomfort were influenced by the phase of the cycle, participants in this study would report an increase in discomfort during the second week of the cycle, and this was observed. Although the test for an overall effect of the menstrual cycle did not reach the level of statistical significance required, the predicted trend was statistically significant ($p = 0.03$). It is thought that statistical significance for an overall effect would have been reached had the sample size been larger.

To examine the changes in discomfort further, the mean changes for weeks 1, 3 and 4 were taken for the experimental group and compared with those of the three control groups. No statistically significant difference between the groups was present on these occasions ($p = 0.1$). Therefore it can be concluded that the experimental group of pre-menopausal women reported similar levels of visual discomfort, and similar changes in discomfort, during three weeks of their cycle to those reported by the control groups. However, they showed a larger increase in discomfort over the working day during the second week of their cycle when compared with the control groups, and also in comparison with the other weeks of their cycle.

There is a vast amount of evidence indicating that thresholds to many types of sensory stimuli decrease, i.e. there is an increase in sensitivity, in the second week of the cycle. The findings of the current study demonstrate that sensitivity to discomfort also increases during this stage of the cycle.

References

Braier, J.R. and Asso, D. 1980, Two-flash fusion as a measure of changes in cortical activation with the menstrual cycle. *Biol.Psychol.* **11**, 153-6

Clemes, S.A. and Howarth, P. A. 2001, Changes in virtual simulator sickness susceptibility over the menstrual cycle, *Presented at the 36th United Kingdom Group Meeting on Human Responses to Vibration*, held at Centre for Human Sciences, QinetiQ, Farnborough, UK

Diamond, M., Diamond, A.L. and Mast, M. 1972, Visual sensitivity and sexual arousal levels during the menstrual cycle. *J.Nerv.Ment.Dis.* **155**, 170-6

Elkind-Hirsch, K.E., Stoner, W.R., Stach, B.A. and Jerger, J.F. 1992, Estrogen influences auditory brainstem responses during the normal menstrual cycle, *Hear.Res.* **60**, 143-8.

Grunfeld, E. and Gresty, M.A. 1998, Relationship between motion sickness, migraine and menstruation in crew members of a "round the world" yacht race, *Brain Res. Bull.* **47**, 433-6.

Hapidou, E.G. and Rollman, G.B. 1998, Menstrual cycle modulation of tender points, *Pain*, **77**, 151-61

Howarth, P.A. and Istance, H.O. 1985, The association between visual discomfort and the use of visual display units, *Behaviour and Information Technology*, **4**, 131-49

Meddis, R. 1980, Unified analysis of variance by ranks, *B J Maths and Statistical Psychol.*, **33**, 84-98

Parlee, M.B. 1983, Menstrual rhythms in sensory processes: a review of fluctuations in vision, olfaction, audition, taste, and touch, *Psychol.Bull.* **93**, 539-48

Pause, B.M., Sojka, B., Krauel, K., Fehm-Wolfsdorf, G. and Ferstl, R.1996, Olfactory information processing during the course of the menstrual cycle, *Biol.Psychol.* **44**, 31-54

Riley, J.L. III, Robinson, M.E., Wise, E.A. and Price, D.D. 1999, A meta-analytic review of pain perception across the menstrual cycle, *Pain*, **81**, 225-35

Than, T.T., Delay, E.R. and Maier, M.E. 1994, Sucrose threshold variation during the menstrual cycle, *Physiol Behav.* **56**, 237-9

Acknowledgments

We would like to thank Mr Steve Longden for giving us permission to undertake the experiment at the BT Directory Assistance Centre in Truro. We would also like to thank Mrs Margo May for all of her help in organizing the smooth running of the experiment, as well as the employees at the Centre who took part in this study. This research was supported by The Vision Research Trust and the EPSRC.

INTEGRATING HEALTH ERGONOMICS WITH TEACHING RESEARCH SKILLS TO OCCUPATIONAL THERAPY STUDENTS

Joanne Pratt[1], Angus K McFadyen[2], Catriona Khamisha[1], William M. Maclaren[2].

[1]*Division of Occupational Therapy*
School of Social Sciences
[2]*Department of Mathematics*
Glasgow Caledonian University

Lecturing students on research methodology has its place in undergraduate education but participation in small group research studies is often the first time a student relates to data in a true sense. This is partly due to their feeling of "ownership" of the data and of the methodology/protocol being applied. This poster presentation will illustrate a few of the simple experiments that have been employed at Glasgow Caledonian University during the development of research skills and subsequent discussion group sessions. These experiments were undertaken by students, under the supervision of academics from the Occupational Therapy Division. The design and analyses were discussed by the students, the occupational therapists and the statisticians, who supervised the students during the data preparation and analysis stages.

Introduction

The academic preparation of occupational therapists and ergonomists has many theory bases in common which underpin the subsequent practice of these professionals. These include anatomy, physiology, biomechanics, as well as applied knowledge in tool use, work performance and basic principles of design. Occupational therapists have traditionally combined this with knowledge and understanding of disease processes and assessments of clients' functional status. Specialised roles have been developed for a range of purposes including disability management at home, sustaining functional performance through the ageing process, general rehabilitation, pre-vocational and return-to-work programmes. The Disability Discrimination Act (1995) has also created opportunities for OT's to be involved in the design of reasonable accommodation for workers with disabilities.

Increasingly therapists are required to demonstrate the efficacy and effectiveness of their interventions which requires a number of research related skills. These include critiquing the published literature for evidence, as well as methods for systematic data collection and analysis/evaluation of outcomes. In order to introduce these topics meaningfully to second year OT students, a number of experiments have been utilized in the teaching of the Research Methods and Introduction to Statistics modules over the past eight years. These include:

- Structured observation and video analysis to assess the developmental milestones of babies and toddlers
- Analysing the calorific expenditure during wheelchair propulsion in a controlled laboratory environment. Different types of wheelchairs (lightweight and standard NHS issue models) are compared. In an alternative version of this experiment different tyre pressures are assessed. The results are used to inform discussion of the practice implications for physically compromised or elderly wheelchair users who either do not or cannot maintain their wheelchairs
- Investigating the interface pressure between the ischial tuberosity and three different seating surfaces. Commercially available wheelchair cushions act as two of these surfaces. The results are used to discuss the implications of the findings for seating prescription given the range in cost associated with these and other types of cushions.
- A multi-level sorting task – by colour, letter and number - is used in an experimental design to investigate any differences between massed practice – one 3 minute trial, and distributed practice – three 1 minute trials.

This paper presents the latter two of these investigations as illustrations.

Investigating Massed versus Distributed Practice using Valpar component Sample Seven (VCWS 7)

Methodology: At the beginning of this session, students were asked to consider their role in education of rehabilitation/habilitation or retraining clients and how optimal performance can be achieved. The apparatus consisted of the Standardised Valpar Component 7, multi-level sorting. A "posting" box which has slots arranged in eight columns and seven rows, each slot being labelled with either a colour, colour and letter, colour and number, or colour, letter and number. The box is placed on a flat table, the row closest to the subject being a practice row and in front of this the mixed storage area for the one inch square chips. The chips are coloured and labelled to match the different slots on the box. A swivel chair is placed in front of the box for the participant with another chair for the rater. All participants were informed that they could use either their left or right hand to post the chips but that they could not switch a chip from one hand to the other. Subjects following the "massed" practice were asked to post the chips into the appropriate slots and had to complete as many as possible within three minutes. The distributed practice subjects were also asked to post as many chips as possible within three minutes but after each minute, they were given a thirty second rest period during which they turned away from the board. The participants were "tested" in an area cordoned off from the general working area and were accompanied by the rater who recorded, for each student, the total number of correctly posted chips and incorrectly posted chips. The other recorded variables were student identification code, gender, age and hand dominance

Results: forty-six student participated in the experiment (42 females and 4 males) with age ranging from 18 to 42 years. Each student was randomly allocated to either the "massed" or "distributed" practice group. This resulted in sample sizes of 24 and 22 for the "massed" and "distributed" groups respectively. One of the dependent variables, the numbers of incorrectly posted chips, was in general very low for both groups and the students decided that the most appropriate dependent variable for investigating whether a difference existed between the two groups would be the number of correctly posted chips for each subject.

As an initial step, a boxplot of the number of correctly posted chips was constructed (see Figure 1).

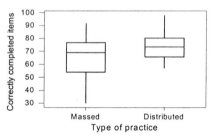

Figure 1. Number of correctly completed items, by type of practice

A table of the appropriate means and standard deviations for the dependent variable was also constructed for each group (see Table 1).

Table 1. Summary statistics of number of completed items, by type of practice

	Massed	Distributed
Mean	66.25	74.14
Standard deviation	15.26	10.95
Sample size	24	22

Given the concept of a two independent sample design, the inferential analysis was conducted using a two-sample t-test after confirmation that the dependent variable could be considered approximately normally distributed for both groups. This two sample test resulted in p = 0.052, which was not significant at the 5% level of significance. The resultant discussion centred round the slightly higher average performance of the "distributed" practice group, the lower variation of the same group and the lack of statistical significance between the two means.

Which Cushion? Comparing Interface Pressure

Methodology: two different investigations were undertaken using the Talley Skin Pressure Evaluator Model SD500. The first concerned repeatability of measurement where each of the group of 55 students measured the interface pressure three times when a 6 pound weight was placed on a laboratory bench. Three different gauges were used with each student using only one of them. This illustrated that variability is present whether the measurement is repeated by the same person or by different people. The second investigation concerned the relative effectiveness of two different types of cushions in reducing pressure on ischial tuberosity experienced by wheelchair users. Twenty-seven of the 55 students volunteered to act as wheelchair users.

Results: inter-observer variability was studied by comparing the first reading taken by each of the 55 students from the 6 pound weight. The large variation was illustrated using diagrams such as Figure 2.

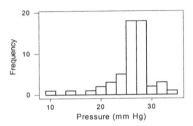

Figure 2. Pressure exerted by 6 pound weight
(first measurement made by each of 55 students)

The intra-observer variation was compared by calculating the standard deviation of the three readings taken with the 6 pound weight for each student. The resultant frequency distribution of these standard deviations, as illustrated in Figure 3, highlighted the lack of repeatability of the measurement made by some of the students.

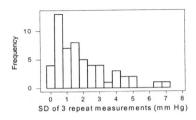

Figure 3. Standard deviations of three repeated
measurements of pressure using 6 pound weight.

A comparison of the three pressure gauges was discussed as a means of illustrating within-group variation and between-group variation which whilst beyond the scope of their current module sets the scene for ANOVA at a later stage.

For the comparison of the cushions, the 27 participants each sat on a wheelchair fitted successively with each of the cushions and without a cushion (control) and the average of three pressure readings was taken for each condition. Two descriptive comparisons were used to illustrate the difference between the two cushion data – firstly ignoring the fact that the data were paired (see Figure 4) and secondly by taking the pairing into account and thus analysing the paired differences in pressure (see Figure 5).

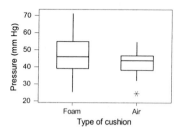

**Figure 4. Pressure exerted by 27 wheelchair users
(foam versus air cushion).**

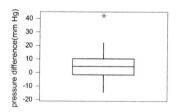

**Figure 5. Pressure, paired differences
(27 wheelchair users foam versus air cushion).**

Discussion

In each of the above detailed inclusion/exclusion criteria were set and discussions took place involving the concepts of informed consent, randomisation, design, manipulation, control and the importance of a detailed protocol. From such simple examples the students can gain insight into some conceptually difficult concepts and crucially they have undertaken the experiments and hence gain this "ownership" feeling for the data. Because they are so close to the data, the students themselves often bring to the discussion reasons why a certain experiment did not work well or even suggest improvements. Giving students "perfect" design scenarios can often lead them to believe that research in the real world is "perfect" which we all know is not the case. The development of research skills comes with experience. From the two simple examples illustrated above the students learn about:

- the use of simple tasks in assessing clients
- the importance of standardisation in a protocol
- the concept of repeatability
- simple descriptive methods of analysis
- inter and intra observer variability
- the formulation of Experimental and Null hypotheses
- the level of significance of a test
- the importance of the Normality of your data when deciding on a testing procedure
- the importance of the differences between two group designs and paired designs

and crucially

- the contextual interpretation of the results obtained both from a descriptive and inferential view point.

The development of these evaluative skills will serve students well in occupational therapy, and for those that choose to do so, in the specialisation of health ergonomics.

References

The Disability Discrimination Act (1995) DfEE, HMSO, London

Valpar(1993). Multi-Level Sorting Reference Manual., Valpar Assessment Systems. Valpar International Corporation, Tucson

THE REMOTE CONTROL OF VEHICLES AND TOOLS

J. P. Viveash, M. G. Kaye, J. L. White, J. Boughton and S. K. King

QinetiQ, Centre for Human Sciences
A50, Cody Technology Park,
Ively Road, Farnborough,
Hants GU14 0LX

Control performance with remotely controlled vehicles and tools depends on the information received by the human operator, and performance advantages can be gained by the correct presentation of visual information to the operator. In these experiments the use of monocular, stereoscopic and enhanced stereoscopic display information was investigated. Two tasks were used: one was a driving task, the other a manipulation task. The results indicated that the degree of stereoscopic advantage is task-dependent. In the manipulation task the best performance was achieved using stereoscopic presentation techniques, the enhanced technique producing a 38% improvement in performance over the monocular presentation technique.

Introduction

There is considerable military and civilian interest in the remote guidance of robotic tools and vehicles, which are required for use in hazardous and inaccessible environments. The efficiency of control and the performance achieved with these systems are highly dependent on the information received by the human operator, and performance advantages can be gained by the correct presentation of information gathered by cameras mounted on the vehicle or tool.

One aspect that has to be considered when determining the best camera layout is whether or not there will be any improvement in control performance if stereoscopic rather than monocular images are provided to the operator. This question will be of particular importance when the operator's task requires distance and depth perception.

Although two-eyed vision does provide a primary cue to depth perception there are many people with one-eyed vision who have very good depth perception. This is because there are at least seven monocular cues (overlapping contours, motion and linear perspective, texture, light and dark shading, accommodation of the eye and aerial perspective) that are also used in distance and depth perception (Boff and Lincoln, 1988).

There are two potential binocular cues to depth and distance perception: stereopsis and convergence. Stereopsis amounts to a detailed comparison of the two retinal images on the basis of parallax geometry; the two retinal images are fused by the brain and yield a vivid and highly detailed perception of three-dimensional space. Typically, the stereoscopic threshold varies from 1.6 to 24 seconds of arc. Targets with larger disparities may be seen as double images with no accompanying sensation of depth, or only one image may be seen and the other suppressed (Boff and Lincoln, 1988). For objects at different distances, the eyes converge by rotating to align their images on the fovea, and depth discrimination does vary with the convergence angle. However, although the convergence angle provides some information about absolute distance, it is a relatively weak depth cue (Gregory, 1990).

When two cameras are used to provide stereo information, varying the distance between the cameras will have a substantial effect. Placing them further apart will result in larger disparity, giving improved stereoscopic discrimination. Thus, another question that needs to be considered is whether such enhanced disparity produces a better performance in all tasks.

This paper reports the results of two trials designed to answer some of these questions for two different tasks. One trial investigated the remote control driving of a vehicle around a track, the other the remote manipulation of a tool.

Apparatus and Methods

Robotic vehicle
The remotely controlled vehicle used in these trials was based on a four-wheeled model car chassis (Tamiya model 58089, 'Bullhead') with both velocity and direction control. The vehicle was 426 mm long by 380 mm wide, and a digital radio control system (Sanwa RD 6000) was used with a single joystick to control its speed and direction. A wood and aluminium plate was fixed to the chassis and used as a mounting platform for the cameras and associated electronic equipment.

Stereo viewing system
The stereo viewing system consisted of two miniature (half inch) colour video cameras and a two-channel video multiplexer, which produced a composite video signal that was transmitted to the remote viewing apparatus. The inter-camera distance (ICD) was set by a series of parallel fixing holes in the platform and three ICDs were used: 0, 22 and 66 mm. The 0 mm condition was the 'monocular' condition in which a single camera was centrally mounted on the platform. In the 'stereopsis' condition, the 22 mm ICD was the minimum possible separation obtained with the two cameras mounted side by side, and it gave an apparently natural stereoscopic view within this model environment. The 66 mm ICD gave an impression of an 'enhanced stereopsis' condition. It should be noted that the so-called 'monocular' condition should, more accurately, be called the 'bi-ocular' condition, because the participant views the monitor using both eyes. However, to draw parallels between the camera configurations and one- and two-eyed performance, it will be referred to as the 'monocular' condition throughout this paper.

The remote viewing apparatus consisted of video receiver, de-multiplexer, high speed (160 Hz) video monitor with LCD polarising shutter, and polarised lens spectacles. The composite video signal was received and de-modulated and the outputs from the two cameras were fed to alternating rasters of the monitor. The polarising shutter was synchronised to the raster and the participant viewed the screen through spectacles with left and right eye filters of orthogonal polarisation. Thus, information was displayed on the monitor as left eye and right eye images at different instants in time, the only difference between the images being that caused by the lateral separation of the cameras.

Participants

Nine participants, seven males and two females aged between 20 and 37, took part. All had normal or fully corrected vision and they were tested with the Randot stereo test to ensure that they had normal stereopsis. Prior to the experimental trials, each participant practised controlling the speed and direction of the vehicle with the joystick until judged by the experimenter to be sufficiently competent.

Driving task design

The driving task consisted of guiding the vehicle around a circuit through seven pairs of upright wooden markers. These were placed 480 mm apart, a distance 100 mm wider than the car chassis. The car needed to be square on to the markers to go cleanly through the gap. To reduce monocular cues to depth, the markers were varied in size (diameter and height) and their bases, where they stood on the floor, were obscured from view.

Manipulation task design

The manipulation task required the participants to capture five metal rings on a probe attached to the front of the vehicle. The rings were of five different diameters and were hung in a line from a supporting bar. The vehicle returned to a starting position each time a ring was captured. The probe was asymmetrically mounted on the vehicle and rose at an oblique angle, so that it was impossible simply to establish visual alignment between the probe and the ring, and then advance the car. Instead, participants found it necessary to exert continuous control, rather than simply executing a pre-programmed movement.

Experimental design and procedure

For the two tasks, a one-way ANOVA design was used, counterbalanced for order effects. For each camera configuration, two runs were made. The first was a practice run and only the data collected in the second run were used in the analysis. In the driving task, performance assessment was based on the number of markers knocked down and the total time taken for the run. In the manipulation task, assessment was based on the number of attempts to remove the ring and the total time.

After completing both trials the participants were presented with a questionnaire that asked them to rank order their preferences for the three ICDs used in the experiment, rank order 1 being the preferred choice.

Results

The main aim of the experiment was to assess the effects of three different camera configurations on vehicle control performance. The mean data for the driving task are shown in Figure 1 and those for the manipulation task in Figure 2.

Driving Task

There was no significant main effect due to the different visual conditions. Mean times for the driving task at the three ICD separations are shown in Figure 1. The number of markers knocked down under the three display conditions varied very little (19±2).

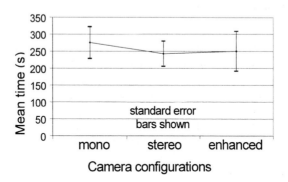

Figure 1. Mean time to complete in the driving task experiment

Manipulation task

There was a significant main effect of camera configuration on the time taken ($F_{(2, 14)} = 10.06$, $p<0.01$), and post-hoc comparison showed that monocular and stereo presentations resulted in significantly longer task durations than the enhanced disparity presentation. Mean times for the three ICD configurations on the manipulation task are shown in Figure 2. It was also noted that the participants had the greatest number of unsuccessful attempts to remove the rings under monocular presentation conditions and the least number under the enhanced disparity condition.

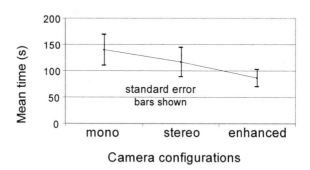

Figure 2. Mean time to complete in the manipulation task experiment

Subjective

Rankings of the preferences for the methods of visual presentation for the two tasks are shown in Figure 3, in which lower scores indicate stronger preference. Although better performance was obtained in the enhanced condition than in the stereo condition, several subjects commented that it gave them feelings of eyestrain.

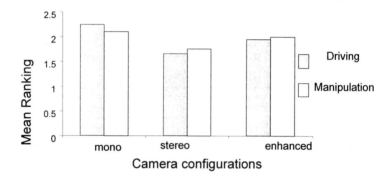

Figure 3. Subjective rankings

Discussion

The results confirm that the best method of task presentation (monocular or stereo) for a given task is task-dependent. In the driving task experiment there was no significant difference between the time taken to complete a circuit for any form of presentation. In the manipulation task experiment, however, stereoscopic presentation gave significantly better performance than monocular.

The negative outcome for the driving task experiment was not unexpected, because others have had a similar result (Gold *et al*, 1968). Such a result is thought to arise when tasks are performed in environments that are so rich in monocular cues to depth that stereopsis provides little additional information. This was disappointing in this particular trial because particular care had been taken minimise the monocular cues.

The results from the manipulation task clearly demonstrated an improvement in control performance arising from stereoscopic presentation, which reduced mean task time by approximately 16%. Moreover, there was a further improvement in performance from enhanced stereopsis, which decreased task time by 38%, in comparison with the monocular condition. This was mirrored by the numbers of attempts required for the task. The participants made most attempts to remove the rings when using the monocular presentation display and fewest with the enhanced stereoscopic presentation. Other experimenters have also shown improved performance with stereopsis in manipulation tasks (Smith *et al*, 1979).

In the subjective assessments the majority of participants preferred stereoscopic over monocular presentation. The enhanced stereoscopic presentation, although more popular than the monocular presentation, was found to cause feelings of eyestrain, which would make it difficult to use for long periods of time. There were no comments about the distortion of the scene, even though such comments have previously been reported for cameras converged to a point near to a vehicle (Nagata,

1996) and not to infinity as in this trial. Nevertheless, as expected, double images were seen in the foreground when viewing the enhanced presentation.

The fact that some subjects complained of eyestrain when using the enhanced stereo presentation indicates that a further investigation of enhanced stereo techniques and eyestrain is required. This investigation should determine the optimum camera convergence angles for a variety of tasks and the degree of disparity easily tolerated by the majority of the population.

Another point to note is that performance was never worse for stereo rather than monocular presentation and there is, therefore, no indication that there is ever a disadvantage to using stereo presentation techniques for remote control tasks. For manipulative tasks the performance gain may be as high as 38%, and such an advantage is clearly of importance for tasks such as bomb disposal and in-flight refuelling.

Conclusions

These experiments have demonstrated that: (1) The relative advantage of stereoscopic over monocular presentation is task-dependent; (2) There was no significant difference in control performance between stereo and monocular display presentations in the driving task; (3) There was a significant difference in control performance between stereo and monocular display presentations for the manipulation task; (4) In both trials, a stereoscopic display always resulted in performance that was at least as good as with a monocular display.

References

Boff, K.R. and Lincoln, J.E. 1988, Depth Perception. In *Engineering Data Compendium. Volume II: Human Perception and Performance.* Harry G Armstrong Aerospace Medical Research Laboratory, Wright Patterson Air Force Base, Ohio

Gold, T. 1972, The limits of stereopsis for depth perception in dynamic visual situations. In *International Symposium & Seminar (72E06), Volume II* (Society for Information Display, New York)

Gregory, R.L. 1990, *Eye and Brain: The psychology of seeing,* Fourth edition (Weidenfield and Nicolson, London)

Nagata, S. 1996, The binocular fusion of human vision on stereoscopic displays. Field of view and environmental effects. *Ergonomics,* **39** (11), 1273-1284

Smith, D.C., Cole, R.E., Merrit. J.O. and Pepper, R.L. 1979, Remote operator performance comparing mono and stereo TV displays. The effect of visibility, learning and task factors. *Naval Ocean Systems Center, Hawaii Laboratory*

Part of this work was carried out on behalf of Eurofighter IPT

Part of this work was carried out as part of the Technology Group 5 (Human Sciences and Synthetic Environments) of the MOD Corporate Research Programme

ERGONOMICS IN THE SECONDARY SCHOOL CURRICULUM

Andrée Woodcock

VIDe Research Centre, School of Art and Design, Coventry University, Gosford Street, Coventry, UK. email: A.Woodcock@coventry.ac.uk

This paper addresses the extent to which ergonomics and ergonomics related subjects are present in the AQA (Assessment and Qualifications Alliance) Specifications for GCSE, Advanced, and the forthcoming Vocational GCSE Levels. Previous research has indicated that some aspects of ergonomics are taught in secondary schools, but this has been largely based on reports by first year design and engineering undergraduates. With the growth in IT and multidisciplinary subjects, it is believed that ergonomics is making a significant, but largely unrecognized contribution to a number of disciplines in secondary schools. A better understanding of this could make an important contribution to reappraising the strengths of our discipline and indicate the way forward to a more proactive and creative approach to the development of curricular support at this level.

Introduction

Previous investigations have shown firstly, that children, even in primary schools can have an appreciation of ergonomics (Woodcock and Galer Flyte, 1998), secondly that ergonomics is taught at 'GCSE' and 'A' Level (Woodcock, Galer Flyte and Denton, 1999) and thirdly that when it is taught, it may not inspire children to continue its further practice in their later careers, e.g. in the design profession (Woodcock and Denton, 2001). These results have been based on observations of children and questionnaires given to first year undergraduates. The results are not especially encouraging and in an earlier paper it was suggested that an appreciation of ergonomics was 'taught out' of children during their secondary school education.

Although the scope of ergonomics can be summed up broadly as "designing for human use" (Sanders & McCormick, 1992), it is still frequently equated with anthropometry and 'knobs and dials'. Whilst these are core areas of the discipline, they may seem unattractive to today's adolescents who are far removed from the world of heavy manufacturing and industry.

Other parts of the discipline may prove more successful as an introduction to the subject. For example, Applied Ergonomics places equal importance on improved system efficiency and individual health, considering all aspects of interaction (e.g. human, tool, task and environment). More recently there has been a renewed commitment to looking at how the science may benefit society. For example, Community Ergonomics has the 'goal to improve the fit between people and their environment to achieve higher levels of self regulatory control and individual effectiveness' (Cohen and Haims, 1999). Participatory Ergonomics has similar goals reflected in methodological considerations and organisational change. Whilst Cultural Ergonomics is exploring human factors issues in relation to the cultural settings in which they occur with special emphasis on 'how cultural factors influence and interact with human performance and human interfacing in work environments worldwide' (Kaplan, 1999).

The survey outlined in this paper both complements and extends the author's previous investigations by considering the extent to which ergonomics is present in all of the AQA Specifications. The overall aim is to inform the type of ergonomics material that should be produced for school children. Each specification represents the syllabus for a subject, although they do not stipulate how topics should be taught. Analyzing the GCSE, 'A' and Vocational GCSE Specifications will enable the current status of ergonomics to be established, with respect to the most relevant subjects. This is turn will help in the tailoring of curricular support material.

Method

AQA Specifications were downloaded from the AQA site (www.aqa.org.uk). The new Vocational GCSE Specifications were included but not the GNVQ's (see footnote). Using results of previous research, attention was given initially to those subjects where respondents had remembered being taught ergonomics.

The Specifications at each Award Level have a standard format with sections covering background information, scheme of assessment (rationale, aims and objectives), subject content, key skills, spiritual, moral and cultural issues, administration (e.g. awarding and reporting, center based components) etc. Priority was given to subject content and how different methods of assessment would encourage ergonomics thinking, in order to determine the extent of ergonomics content and the potential for more.

Results

This section summarises the results for each Award Level.

GCSE Level

These are usually taken by 16 year olds. Over 30 Specifications were examined, relating to single, double and modular courses. Ergonomics did not feature as a major component in any of the Specifications, or in any of their units/modules, and was barely mentioned by name. However, ergonomics content did feature in the Specifications as summarized in Table 1. This shows the extent of, and potential for the teaching of ergonomics for the most promising subjects. The final rating was mediated by an understanding of the wider aims of the subject, its organization at the Award Level, and its relationship to ergonomics.

Table 1. Ergonomics at GCSE Level

Ergonomics*	Subjects	Potential
None	Physics, Psychology, Science, Biology (Human), Geography, General Studies, Human Physiology and Health	Discrete areas could be targeted e.g. biomechanics, cognitive psychology, metabolic rate
A little	Biology, Business Studies (A and B), Physical Education, Sociology, General Studies, Travel and Tourism	Minor enhancements possible, but ergonomics will never form a major contribution
Some	Art and Design, Business Studies (Specification B) Short Course, Design and Technology – Food Technology, Graphic Products, Systems and Control Technology	Support material could be provided for specific areas e.g. Health and Safety, documentation usability, inclusive design
A lot	Information and Communication Technology Specification B (short course), Business and Communication Systems, Design and Technology – Electronic Products, Product Design	Material could be produced to support current teaching in the form of textbooks and case studies.

* Space does not permit a full content description here. Contact author for the detailed analysis.

Vocational GCSE

Vocational GCSE's are two year courses commencing in 2002 and are equivalent to two GCSE's. Of the six draft Vocational GCSE Specifications available for inspection, attention was focused on Art and Design, Business Studies, ICT and Science.

Table 2. Ergonomics at Vocational GCSE Level

VGCSE	Units	Ergonomics
Art and Design	2D and 3D visual language	None
	Materials, techniques and technology	None
	Working to project briefs	A little
Business	Investigating businesses	Some
	People and businesses	Some
	Business finance	None
ICT	ICT tools and applications	None
	ICT in organisations	Some
	ICT and society	Some
Science	Developing scientific skills	A little
	Science for the needs of society	None
	Science at work	A little

Table 2 shows a similar pattern to that found in other Awards, however the overall depth of analysis of ergonomics related topics is less than at GCSE and A Level but with a slight increase in ergonomics teaching in Science.

Advanced Subsidiary (AS) and Advanced (A) Level
These are post 16 year old qualifications, with the 'AS' Level forming the first half of the study towards an 'A' Level, different pathways are also open to candidates. Over 20 specifications were examined. The results were derived in a similar way to that indicated above and may be summarized as follows:

- Subjects with a substantial amount of ergonomics-related study were Business Studies, Design and Technology (Systems and Control Technology), Computing, Design and Technology (Product Design), and Information and Communication Technology (ICT), with the latter three having the greatest content.
- Art and Design had some content (depending on the pathway selected), as did Design and Technology (Food Technology), Sport and PE.

For all Award Levels different types of information are needed to support the different disciplines.

- Design and Technology areas require support in terms of child-friendly, interactive anthropometrics packages, instruction in the use of capturing user requirements and evaluation techniques; additional support is required in the form of case studies.
- Business Studies focus more on organizational issues, change management, factors affecting the efficiency of the workforce (e.g. remuneration, training, motivation, management style, introduction of new technology, changing working patterns, communication and information flow).
- The computing domains cover the system development lifecycle with explicit mention of human computer integration and interaction, maintainability, usability, user support, documentation and evaluation.
- In addition to this subject specific material, all subjects, whether they 'teach' ergonomics or not require information and advice on health and safety issues.

Some of this information especially for over 16 year olds may best be communicated in the form of standard textbooks. However, more attractive, usable and task related teaching resources should also be considered. The exact requirements of such material will form the next stage of the research, but may include videos, web sites, work sheets, posters, design guidelines and interactive 3D packages.

In terms of teaching and assessment, most of the work students undertake with regard to ergonomics is of a practical nature (e.g. designing a product/software to suit specific requirements), or in the form of a case study (e.g. considering problems in an organization and proposing solutions). There is a real opportunity in the Specifications to direct teachers to resources produced by the Society. Such resources could include guidelines/best practice in conducting case studies, specimen case studies, videos or the availability of members to talk of their practice.

Discussion and Conclusions

A cursory inspection of the AQA specifications reveals little mention of ergonomics per se. However, a more in depth analysis revealed aspects of applied ergonomics (user centred design, job design, evaluation, case studies etc) in an array of disciplines at both GCSE and 'A' Level, so confirming previous work. Design and Technology related disciplines again naturally emerged as the champions of ergonomics. Previous investigations have tended to ignore other subjects taught at secondary schools. This research extends the work by providing a more thorough investigation of other subjects and has found that business related studies, ICT and computing provide considerable ergonomics content. There is also potential for ergonomics to feature more widely in some subjects, such as human biology and physics. However, if we are going to provide information targeted at secondary schools it would perhaps be most appropriate to concentrate first on those courses where recognition of the discipline already exists.

What this research has not shown is how the subject is actually taught or assessed. This is the next stage for the research and one that will help to more clearly define the requirements of the teaching material. The Specifications indicate that an action learning approach is taken – with case studies, activities and reflection on proposed solutions.

For example, in the first year of Design and Technology candidates have to understand health and safety issues of their studio and working environment, design posters to highlight these issues and how accidents can be avoided. Although the Design and Technology curriculum emphasizes the need to 'make' an artefact (electronic product, teaching aid, dish for a vegetarian), candidates are expected to conduct research and understand the ways in which their product will be evaluated. In some cases it is expected that they will attempt a user evaluation. Other case study approaches, for example in Business Studies and ICT consider analysis of the present situation, looking at ways in which it can be approved, and measuring or evaluating the impact of the solutions. This is not just theoretical. For example, at AS Level Design and technology students undertake projects concerning disability. This includes looking at the design of artificial limbs and what it means to be disabled. One project from this involves improving the school environment for disabled children, which draws on interviews, looking at the needs of children, visiting schools which are known to have better facilities, putting forward proposals for changes, and assessing their costs and benefits. These results are extremely encouraging, with the assessments indicating that candidates are expected to analyse workplace problems or design briefs, implement and evaluate their solutions and reflect on their relative merits.

A preliminary interview conducted as part of this research suggested that 'ergonomics' had an image problem and that there would be little support for the development of an 'A' Level at least at present. It was suggested that a starting point for the wider teaching of ergonomics would be to lobby for its inclusion in the National Curriculum Orders for Key Stages 2 and 3 thereby ensuring that it was taught and providing a market of students who might be interested in studying it at an Award Level. The analysis of the Specifications has shown that there may be some truth in this assertion, for although the term ergonomics hardly features in any Specification at any Award level, there is a great deal of ergonomics present. Also, given that elements of ergonomics are found in different subjects, it may well prove attractive to a number of students as a discipline in its own right.

In summary, this research has provided evidence that ergonomics has cross disciplinary potential within schools reaching across Computing, Business Studies and Design and Technology. It also has the potential to be linked to new curriculum developments in the areas of citizenship and social responsibility, and in the newer vocational courses. Unfortunately no evidence has been found to suggest that ergonomics is taught in those disciplines that provide its core foundations such as physics, psychology and physiology. Lastly, the analysis of the Specifications has indicated ways in which material might be developed to support teaching.

Acknowledgements

I would like to thank Mike Tainsh for his support and enthusiasm for this research.

Footnote

The analysis of GNVQ qualifications has been omitted from this paper for reasons of clarity. Candidates may follow different routeways through these qualifications, which makes it very difficult to provide a succinct overview. The overall trend in the coverage of ergonomics is similar to that found at A Level, but is at a greater depth. More details on this and the other analyses are available on request from the author.

References

Cohen, W.J. and Haims, M.C. 1999, Community ergonomics and urban poverty: Planning and design activities and interface solutions, in (Ed.) Scott, P.A., Bridger, R.S. and Charteris, J. *Global Ergonomics*, Elsevier, Amsterdam, 463-465

Kaplan, 1999. The cultural milieux of ergonomics, in (Ed.) Scott, P.A., Bridger, R.S. and Charteris, J. *Global Ergonomics*, Elsevier, Amsterdam, 607-612

Sanders M. S. and McCormick E. J. 1992, *Human Factors in Engineering and Design*, 7th Edition, McGraw-Hill Inc., New York.

Woodcock, A. and Denton, H.G. 2001, The teaching of ergonomics in schools: What is happening? *Annual Ergonomics Conference*, April, 2001 in Contemporary Ergonomics, 2001

Woodcock, A. Galer Flyte, M.D. and Denton H.G. 1999, Globalisation and technological change-opportunities for pro-action and integration, *Proceedings of the Symposium " Strengths and Weaknesses, Threats and Opportunities of Ergonomics in front of 2000" International Ergonomics Association (IEA) and the Hellenic Ergonomics Society (HES)*.Santorini, 1-2 September 1999 ed., Marmaras, N.,59 –74

Woodcock, A. and Galer Flyte, M.D. 1998, Ergonomics: It's never too soon to start, *Product Design Education Conference*, University of Glamorgan, 6th-7th July

COMPARATIVE STUDY OF THE PLANNERS' PERFORMANCE IN DRAWING BOARD AND IN THE COMPUTER (CAD)

**Lilian Morais Carvalho[1], Olavo Fontes Magalhães Bessa[1],
Ana Francisca de Oliveira[1], Anamaria de Moraes[2] and
José Mário B. Alves[1]**

*[1]Unifenas/FAPEMIG
Rod. MG Km 0 C.P. 23. Alfenas MG
[2]PUC-Rio / LEUI*

Summary

This study is about a comparison among the designers' performance in the use of the computer and in the use of the drawing board, comparison instruments basically used were the stages of the project process and the result of the graphic presentations.

Introduction

What is more appropriate to motivate the use of the board or to motivate the use of the computer to develop architectural projects? This discussion was frequently observed in the academic and professional milieu. Starting from these questions, it was had as objective to do a comparative analysis among the designers' performance that use the board and the designers' performance that use the computer to accomplish architectural projects. It was taken as hypothesis that the elaboration of the architectural project is influenced by the use of the CAD tool.

Methodology

Definition of the observed factors
The factors observed in the experiment were those accepted by the literature as pertinent and relative to the architect's activities, attributions and knowledge (Bessa et alli, 1998; Estephanio, 1995; Ballay, 1987; Mcginty, 1984; Silva, 1986; Broadbent, 1976;). The most subjective questions of the creativity were immediately excluded by the difficulty in evaluating them. The comparison was made considering the *design process* and the result of the *graphic presentation*.

Definition of the factors related with the design process
Ballay (1987), presents some factors that served us as base for analysis of this research, which are: (1) *planning*, when the designer acts as a person that solves the problems, registering the solutions; (2) *routine primary processes*, "while the routine processes may be interrupted for such purposes as evaluating and editing the solutions; (3) *workplace management*, this is the organization and reorganization of de immediate environment in

which the designer works. He/she also organizes the cognitive processing to restart the work. (Fig. 1).

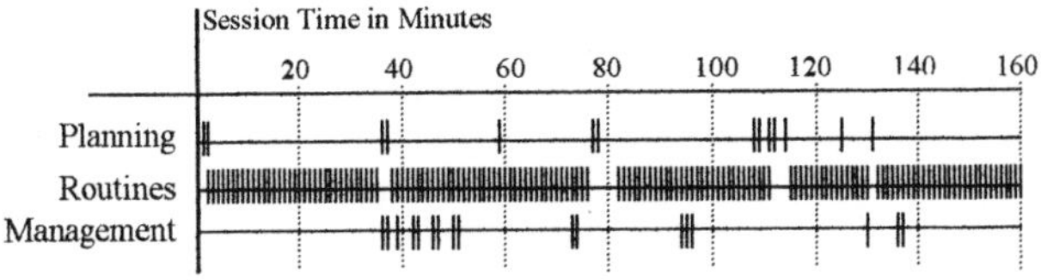

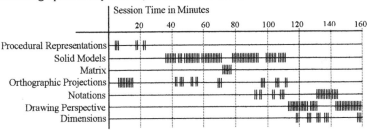

Figure 1. Annotation of the information collected during the design process

In the Ballay's experiment 7 types of external representations were noticed (Fig. 2): (1) *Procedural representations*, encoded action scenarios. The designer imagines how the operator will use the project; (2) *solid models construction*, that are three-dimensional materials that help the designer to notice the volumetric relationships; (3) matrices, that the designers builds to help him to choose which of the alternative parts he wants to work with (the rows of the matrix represent part types and the columns represent alternative models of each type); (4) *orthographic projections*, are views perpendicular to the Cartesian planes; (5) *notations*, are words, numbers or related symbols that may vary from a simple character to little words; (6) *perspective drawing*, are graphic conventions that imply a "spatial" view of the objects; (7) *dimensions*, representations that combine symbolic and graphics components (arrows and numbers).

Figure 2. Alternation of the represented models (detailed protocol analysis)

By virtue of the subjects of the experiment have not answered fully to the items indicated by Ballay (1987) some factors were increased and retired others. Of the original model of Ballay (1987) were used only the items *orthographic projections*, *notations* and *perspective*. The way how the subjects freely presented the experiment results didn't allow be observed the items: *procedural representations*, *solid models*, *matrices* and *dimensions*, so these items were excluded. It was possible, however, to include the following items: (1) *researcher's intervention*, when the researcher had to intervene in the process to explain to the subject how he/she should execute the test-project; (2) *analysis*, it was the stops that the subject did to analyze the reference plan views to elaborate the test-project; (3) *research*, it was the times that the subject sought some information in the available material to consult (newspapers, books etc.); (4) *routine 1 and 2*, they were the movements done during the development of test-project to organize the work (for the movements accomplished in the board was called *routine 1* and for the movements accomplished in the computer it was called *routine 2*); (5) *details*, they were the refinements of drawing done by the subjects.

Thus, the factors defined to observe the *design process* were as shown in the next: orthographic projections, notations, perspective, research's intervention, analysis, research, routine 1, routine 2, details.

Definition of the factors related with the results of the graphic presentation

Considering that the Technical Drawing is applied during all the project process, it was used to define the factors that should be observed for the comparison between the board and the computer. The elements that was used as reference were obtained from Hoelscher et alii (1978) and Follis & Hammer, (1979) and are as following described:

- **Uniformity**: (1) of the letters – regular appearance of types and sizes; (2) of the drawing – uniform language in the drawing, (3) of the hatching and fill uniformity – regular fill of a drawing area.
- **Precision**: (1) of the angle – relationship of distance among two competitive straight lines expressed in degrees; (2) of the scale – use of a reduced proportion of measures to accomplish the drawing.
- **Legibility**: (1) due to majuscule and minuscule combination – selection of the upper and/or lowercase characters for writing the texts; (2) due to the size of the characters – if it is out-and-out incompatible with the scale chosen for the project; (3) due to alphabet type – quality of the writing, if letter handwritten, if thecnical letter.
- Number of **Details**: (1) vertical plan view (elevation) – drawing showing the vertical arrangement (internal or external view); (2) vertical clipped plan view – horizontal section of an object that show its heights and the exact form of its internal parts; (3) Increased view – enlarged drawing to show some structure type that needs more detail; (4) special situations – as concrete armors, hydraulic facilities etc (5) 3-D view (conical or isometric perspective) – projection which shows more than one of the faces of an object; (6) layout – drawings of the furniture that are good to have dimension and space notion.

Techniques

Interviews

Some interviews were made with professionals to know about the two ways of projecting (board/computer). When declaring the preference for one or other system, they valued certain aspects that were taken as factors for observation. The professionals aimed the following factors (Table 1):

Table 1. Synthesis of the factors mentioned in the interviews

FACTOR TO BE ANALYZED	ATTRIBUTED ADVANTAGE
Easiness in modifying the project	to the computer
Easiness in detailing the project	to the computer
Capacity to observe the whole	to the drawing board
Line Tracing Regularity	to the computer
Final presentation of the technical drawing	to the computer
Easiness in doing outlines	to the drawing board
Speed in the elaboration of the project	to the computer

The experiment

It was requested that the subjects projected a chimney and a pool, using as reference the plan of a pre-definite house (equal for all the individuals). The test-project should allow to the researcher to observe the designer's performance during the project process so much in the computer (CAD system), as in the board, besides allowing that the individuals of the experiment, for free will, used the factors described in the previous items (factors of the project process, factors of the result of the graphic presentation and of the aspects pointed in the interviews). Ten individuals were chosen from the Faculty of Architecture – five to accomplish the test-project in the computer and five to accomplish it in the board. The subjects that made the project in the board, as the ones that did it in the computer were in a same position and space situation (Fig.3).

Figure 3. Disposition of the subjects.

The materials commonly used by the planners to drawing was placed in available manner. It was also disposed some material for consultation (books and magazines). It was used a video camera (V8 - Handycam), with tripod and a 14 inches television.

Application of the Test-project

Immediately, before beginning the experiment, it was requested to the subject to make the projects as they were to be given to the planning teacher (it is usually requested in the discipline the plan view, the vertical clipped plan view, the façades and the situational plan in the ground. They should, still, draw the pool with two curves and a flight of steps. It was requested, also, that the student made the project of the chimney with straight, symmetrical lines. It was illuminated to the subjects that some research material would be available for consultation. It was authorized that the subject, when beginning the test-project, organized the drawing material in the board (denominated situation 1) or organized the computer environment (denominated situation 2) and both analyzed the project of the house. The spent time in the accomplishment of these tasks was computed and appraised. In this study the individuals was submitted to the verbalization that served to aid in the annotations of the results.

Annotation of the results

The observation model was based on Ballay (1987). For capture of the information of the project process, the author observes that, any that is the **idiossincrasic** difference among the individuals, there is basic patterns that can be observed and registered (Fig. 4):

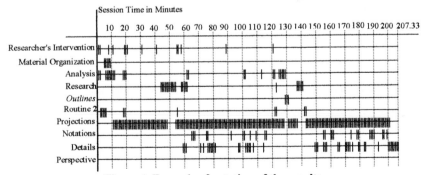

Figure 4. Example of notation of the results

Results

When comparing the results with base in the technically pre selected criterions, it can be said that in spite of similar, there was prevalence for the drawing done in the computer, in the following items: 1) uniformity, 2) legibility; and there was prevalence for the drawing done in the board, in the following item: 3) Details. When observing the concluded projects, it can be said that those that were developed in the board come, in a general way, better documented. When we compare the situation 1 (board) and the situation 2 (computer), it can come the following comparative picture (Table 2):

Table 2. Comparative table between the board and the computer

Item	RESULT OF THE PROJECT ACCOMPLISHED IN THE COMPUTER	RESULT OF THE PROJECT ACCOMPLISHED IN THE BOARD
Uniformity	Worse – it is more difficult to get uniformity (of letter format, spacing, size and drawing hatch)in the board, for the act of drawing to be manual	Better – the letters spacing, format and size were regular, including the hatching, that is more uniform in the computer
Outlines	Better – even the subjects of the computer opted for the board	Worse – the subjects that made it, drew by hand, in the board alleging that there was not any instrument adapted for that activity type in the computer
Legibility	Worse – nor all the projects presented the appropriate alphabet. As well as, nor everybody was readable.	Better – all the subjects presented an appropriate and readable alphabet, although the relationship of the letter chosen in the project, not always it was compatible.
Designer's trust	Better – the subjects demonstrated more trust, requesting the researcher's less attention (perhaps, in the board, they get to dominate the area in that are drawing).	Worse – they demonstrated little trust, requesting with the researcher's attention more frequency, (perhaps for they get not to dominate the work area very well (for being virtual).
Organiz. of the material	Worse – you lost more time in organizing the material in the board.	Better – the "work area" is practically ready, it is enough to adjust some preset data to use the computer.
Notations	Better – perhaps for they dominate the area that they are projecting, the subjects get to place a larger amount of information in their drawings.	Worse – perhaps for they get not to dominate the "work area", the subjects don't place many information in the project.
Routines	Worse – the subjects waste more time making the routines. During the project process, they needed to do more movements in search of the drawing materials.	Better – the drawing materials in the computer are placed in the "virtual drawing area", reducing the designer's movements, even so it should be considered that can have sailing problems that were not appraised for this study.
Details	Better – the subjects, present drawings with a larger number of details, leaving the most presentable and readable project.	Worse – the subjects gave less attention for this stage, presenting little detail.
Amount of drawings	Better – they presented a larger amount of plan views and increased view drawings.	Worse – they presented the possible minimum of drawings.

The psychometric tests only aimed that the subjects which presented a percentile of low abstract reasoning appealed more to the research in the consultation material. This can indicate that such individuals have the need of external references, once they have difficulties to represent them.

Discussion and Conclusions

It should be considered that the subjects that made the project in the computer, worked in an atmosphere that was not "costumerized" in agreement with their habits and that the tasks demanded by the test-project didn't favor the repetition of procedures. Of this, the most general conclusion was that, the students from Faculty of Architecture developed the project presenting a better result in the board and that, they don't still get to dominate the area of work of the computer when it is to create and not to reproduce drawings. Perhaps this is a subject of Usability to be researched.

A correct way of projecting doesn't exist, even so, important and crucial stages of project were not executed fully by the students (like: notations, development of ideas in outlines and perspective drawings). The final result, in both situations, presented little complexity. This was due to the fact of students to do little research and keep not the solutions that created systematically and the information that collected (composition of matrixes). When they wanted a solution that had already been thought, they didn't have a way to return to a previous drawing, no longer they knew more what they had drawn. When they needed an already researched information, they had to leaf the researched material again.

Because the results presented in the board, in a general way, were superior, in spite of the project accomplished in the computer present some better performances, this study checked the hypothesis that the designers that use the computer seek easier roads and they limit the expressive capacity in function of the difficulty of using the CAD tool. It was also confirmed that the designers that make use of the board feel more comfortable at the act of projecting with the conventional instruments, because they get to dominate the work area better, while in the computer, for being virtual, there is difficulty in the work area domain.

It can be ended, finally, that the software used in this experiment, assists better to the solicitations of the Technical Drawing than to the solicitations of the Project Process.

Bibliography

ALBERNAZ, Maria Paula & LIMA, Cecília. Dicionário Ilustrado de Arquitetura. São Paulo: ProEditores, 2000.

BALLAY, Joseph M. *An Experimental View of the Design Process.* Org. VENTURINO, Michael. *Selected Reading in Human Factors.* Londres: Human Factors Society, 1987.

BENNETT & WESMAN. *Manual de psicologia: teste de aptidões específicas DAT.* Rio de Janeiro: Centro Editorial de Psicologia Aplicada, 1990.

BESSA, O. F. M. et alii. *Deslocamento do pedestre no ambiente Urbano: metodologia para elaboração de uma cartilha de ergonomia aplicada à arquitetura.* Cap. 2, seção 14. Alfenas: Unifenas- FAPEMIG, 1998.

BOUDON, Raymond. *Os Métodos em Sociologia.* São Paulo: Ed. Ática, 1989.

BOGOLYUBOV & VOINOV. *Engineering Drawing.* Moscow: Editora: Mir Publishers, 1973.

BROADBENT, Geoffrey. *Diseño Arquitectónico: Arquitectura y ciencias humanas.* Barcelona: Ed. Gustavo Gili S.A., 1976.

HOELSCHER, Randolph P. et alii. *Expressão Gráfica:* desenho técnico. Rio de Janeiro:Livros Técnnicos e Cientificos, 1978.

ESTEPHANIO, Carlos. *Desenho Técnico: uma linguagem Básica.*Rio de Janeiro:C. Estephanio, 1995.

FOLLIS & HAMMER. *Architectural Signing and Graphics .* Londres: The Architectural Press Ltda, 1979.

McGINTY, Tim. *Projeto e Processo de Projeto* in SNYDER, James C. e CATANESE, Anthony. *Introdução à Arquitetura.* Rio de Janeiro: Editora Campus Ltda, 1984.

SILVA, Elvan in Comas. *Projeto Arquitetônico: Disciplina em crise, disciplina em renovação.* Organizador CARLOS EDUARDO. São Paulo: Projeto, 1986.

THE LONG AND SHORT OF IT: THE DEVELOPMENT OF THE NAVAL SERVICE REACH TEST

Emma Bilzon, Peter Chilcott & Bob Bridger

Institute of Naval Medicine,
Alverstoke,
Hampshire,
PO12 2DL

Stature is considered to be an unsubstantiated selection measure for the RN and, as such, could be considered to be against equal opportunity. In 2000, INM conducted a ship-based study, investigating the anthropometric requirements of critical and generic tasks. In 2001, as a result of the findings of this study, a selection test was designed to replace the height requirements. Entry to the Naval Service is to be based on the ability to reach and grasp the top clip of doors on RN ships. This paper describes the ergonomics progression from the tasks through to the test. The test design is based upon engineering drawings and actual measurements of ship doors. Further to this work a directive has been given to test only those candidates who are less than 151.5cm stature.

Introduction

Traditionally stature has been used within many professions (such as the Fire Brigade, Armed Forces and the Police Force) to select personnel. Selection criteria based on stature are, however, a crude method that is frequently unrelated to the job in question, and therefore, unnecessarily eliminates suitable candidates. All recruitment criteria/tests should be job-related, refer only to critical/essential job components and be consistent with business necessity and safe performance of the job (Hogan and Bernaki, 1981).

Prior to 1998 there were a number of height requirements for entry to the Naval Service. These included two requirements for entry to the Royal Navy (RN), set at 1.57m for standard entry and (from October 1993) 1.54m for "outstanding officers and ratings", 1.63m for Royal Marines Officers and Royal Marines and 1.55m for Royal Marines Buglers. In May 1998 it was decided to remove the RN 1.54m waiver whilst studies were carried out.

Stature in itself is rarely a limiting factor in design (doors heights being one of the few direct applications) so, by reviewing the correlations between anthropometric measurements and stature, it is possible to establish those measurements that stature is surrogate for in selection. In both males and females strong correlations are found with stature and functional reach, vertical functional reach, waist height, span, and inside leg length. All of these measures are related to the vertical length of the body and as such are expected to correlate strongly with each other and with stature. In contrast, measurements such as foot breadth, hand width and bideltoid breadth are weakly correlated with stature and are therefore unlikely to have had an impact on selection previously.

The Equal Opportunities Commission (EOC) advises that the setting of different entry requirements for men and women constitutes unlawful direct discrimination, contrary to the Sex Discrimination Act (1975). However, it also suggests that to set the same (fitness) criteria for both men and women *could* also result in unlawful *indirect* discrimination. Therefore, the employer must be able to show that

'*...the means chosen for achieving the objective correspond to a real need, are appropriate for achieving the objective and are necessary to that end*'.

The EOC suggests that the employer also needs to demonstrate that: the requirement is operationally essential for the job; that it could not be done by other means such as the purchase of equipment or redesign of the task; that individuals who do not pass the test could not perform the job satisfactorily.

Critical and Generic Tasks

RN critical and generic tasks were identified and agreed through the conduct of structured questionnaires and interviews with RN, representatives. It was found that *hypothesised* reasons for RN height requirement were: stature is an accurate predictor of reach; stature is an accurate predictor of strength; stature is a general measure of an individual's ability to perform RN tasks.

It was agreed in the interviews that in the RN the majority of job requirements are specific to an item of equipment or vary between vessels. The Police and Fire services have reported similar findings and have concluded that it is not justifiable to set entry criteria upon specific, non-critical situations that may never be encountered, or upon specific items of equipment that will be modified and ultimately changed within a matter of years.

At the present time the only non-physically demanding, generic and critical anthropometric requirements that all RN personnel must satisfy are: escape through an escape hatch; ability to operate kidney hatches, the ability to operate bulkhead doors and; the satisfactory fit of Personal Protective Equipment.

Vertical climb through an escape hatch (EH)
In times of emergency, personnel may be expected to use escape hatches to reach the upper decks. In these circumstances Emergency Life Saving Apparatus (ELSAs) may also be worn. An ELSA currently has eight minutes breathing time. The greatest expected number of decks personnel might be expected to climb using only one ELSA is four.

Open and secure kidney hatch (KH)

In times of action or emergency, all personnel may be required to open and ascend or descend through a kidney hatch. This is essential to ensure that the safety of the ship is not compromised.

Open and secure bulkhead door (BD)

All personnel are required to open and secure bulkhead doors as part of day-to-day activity, as well as during action. A fully-bolted bulkhead door has a multiple clips, the top clip height being 204cm at the point of rotation.

Anthropometric Requirements

A study was designed to investigate the anthropometric requirements of the three critical and generic tasks in the RN. Within each of the age bands (17-24, 25-32 and 33-40), subjects with the following attributes were selected: stature less than 163cm (any hip/chest circumference), males with chest circumferences or hip circumference greater than 100cm (any stature), females with hip circumference greater than 100cm (any stature). The anthropometric values were derived from RN anthropometry surveys of 1333 males and 136 females (Jenkins et al. 1990a; 1990b, unpublished). The anthropometric range in the sample was representative of the anthropometric range in the RN population. Prior to participation, subjects gave their informed, written consent to take part in the study.

Study Design

Subjects' stature, body mass (Sauter SD 100) and skin fold thickness at four sites: biceps, triceps, subscapular, superiliac (Durnin and Wormersley 1974) were recorded for the computation of Body Mass Index and percentage body fat. Strength at standing reach (using a grip-strength dynamometer – vertically held in the dominant hand with the bars adjusted to individual preference and repeated three times), chest circumference, waist circumference, shoulder width and standing reach were measured (in accordance with Jenkins et al. 1990) to compare body width with ability to complete escape through an escape hatch.

Each subject was required to complete the three tasks wearing each of three clothing ensembles: Fire-fighting (Fearnought) including Extended Duration Breathing Apparatus (FF), NBCD (N), and action working dress with anti-flash (AWD) in turn. The clothing order was randomly assigned to groups of subjects according to a Latin Square design. The groups were of equal number and males and females were assigned to separate Latin squares. These tasks were not dissimilar to those that would be completed by RN personnel. Whilst dressed in a particular clothing ensemble, each subject completed three tasks in the same order for their group. In all tasks the tightness of bolts and clips was standardised. Subjective assessment of the adequacy of clothing fit was established by questionnaire for each clothing ensemble. The time taken to complete the tasks was measured using a stopwatch.

Performance Criteria

19 male subjects and, 11 female subjects participated in the study. All 30 subjects completed all the tasks satisfactorily. The maximum time taken to complete an escape

hatch task in fire fighting ensemble was 5 minutes 32 seconds (average time 2 minutes 35 seconds), well within the 8-minute time limit. Wearing fire-fighting clothing increased the time taken to complete the escape hatch task and the kidney hatch task by an average of 55 seconds (maximum 175 seconds) and 10 seconds (maximum 22 seconds) respectively as compared to Action Working Dress.   Clothing did not significantly affect time taken to open the bulkhead door, which took an average of 19 seconds to complete.

The shortest stature (160cm without shoes) and the shortest standing functional reach (196cm without shoes) were recorded for different subjects. The subject with the shortest standing functional reach was 162cm in stature. The top clip of the bulkhead door measured 204cm to the centre of rotation of the clip, including shoes the subject with the shortest standing functional reach measured 199.5cm in shoes (196cm + 3.5cm shoes). This subject experienced more difficulty than the others did, going onto tiptoes to push the clip to its securing point.

No subjects reported fitting problems with action working dress or NBCD clothing. Fire-fighting ensemble was, however, described as too large, cumbersome, heavy or hot by 50% of the subjects. No single or combination of anthropometric characteristics was predictive of performance for the escape hatch and kidney hatch tasks. Strength, body size, technique and training were all contributory to the success of the tasks.

Standing overhead vertical reach (on tiptoe) does predict performance of the bulkhead door task. Personnel who cannot reach the top clip on tiptoe cannot open bulkhead doors on RN ships.

Whilst the findings of the report have been conveyed to naval architects in order that they can be considered during future ship design, it is financially and operationally not feasible to replace all doors on all ships in the short term. It is also not possible to redesign the clips to improve the task due to space constraints above the majority of the doors. Steps or boxes would create trip hazards for the personnel who currently have no difficulty, as these interventions would have to be permanently fixed in place. Also, the use of the rim around the edge of the door as a step would be suitable only in calm conditions.   To balance on this rim as a matter of course to open doors would be dangerous to the individuals concerned.

The Reach Test

As a result of these findings, the Directorates of Naval Manning and Recruiting (DNM & DNR) requested that a test be designed for use in selection based on the ability to reach and grasp the top clip of bulkhead doors

The INM reach test is a scale replica of a bulkhead door on Type 22, 23, 42 and CVS RN ships. The decision to replicate a bulkhead door was taken to ensure that the reach test is as self-explanatory as possible for both the candidate and the Careers Advisor. The base to the centre-point of rotation of the clip measures 204cm, the clip is representative of the 'Ordinary Type A' clips – the most common clip used on bulkhead doors and is inclined at 45° to the horizontal.

A mat is provided for candidates to stand on (stocking-foot) when performing the test. This is to ensure that all candidates have equal and representative conditions. The mat is designed to be the thickness of in-service footwear (35mm sole) including a thin, coloured, top layer (2mm) that will show through when it is worn and ready for replacement.

Discussion

Is standing on ledges dangerous?
It is assumed that it is not appropriate for the shorter candidate to stand on the bottom clip or ledge to reach the top. In reality this does happen. Many personnel (including those who can reach) report standing on ledges to open doors as part of day to day life. However, they are not taught to operate in this way and it is not recommended that they work this way. This is a debate similar to the debate about the construction workers hard-hat. It is the employer's duty to ensure that personnel are provided with the appropriate equipment and training to conduct their job in a safe manner. If an employee chooses to stand on the ledge (or not to wear their hat) they do so at their own risk.

It is also possible that standing on the bottom ledge is an obvious *misuse*. That is to say, it is expected that those with shorter stature *will* stand on the bottom ledge. In this instance it is necessary to ensure that it is safe for personnel to do so. These appear to be contradictory arguments, however, the resultant is a convergence of opinion. In the first instance it is recognised that there are some people who have difficulty reaching the top clip. In the second, it is argued that people misuse the bottom ledge to reach the top clip. The solution is clear – the bottom ledge should be designed such that it *is* safe to stand on to reach the top clip. This design recommendation will, however, take time to implement.

Should all people do the test, even if they are tall?
In view of equal opportunity legislation it is appropriate for all candidates to be required to show their capability of achieving set criteria, regardless of their apparent ability. Reach is a requirement, not a substitute for stature, it therefore does not matter why a person does or does not have the required capability.

Are there other ways to establish reach?
There are a number of other ways to establish whether a person could reach the top clip, however, none of these are as reliable as completion of the task itself. Predication of performance always involves an element of error. By breaking a task down into discrete elements it should always be possible to replicate the critical aspect of that task and as such be certain of an individual's ability to achieve that task.

Conclusions

The most effective form of assessment of critical and generic tasks is reproduction of those tasks. Through the conduct of trials it is possible to establish SMART (Specific, Measurable, Achievable, Realistic, Time-bound), justifiable criteria. In the case of the reach test a non-ergonomics approach may have been to use statistical analysis to establish a cut off point for stature – effectively unnecessarily eliminating a proportion of the population who could have achieved the test.

In the longer term efforts should be made to redesign tasks to be more inclusive of the population.

Note

Since the conclusion of this work a Naval directive has been given to offer the reach test to only those candidates who are shorter than 151.5cm (5th%ile female). All successful candidates and those taller than 151.5cm must complete a sea survival course that will determine their ability to achieve the required standards for sea survival duties (including escape through hatches and operation of bulkhead doors).

Acknowledgements

The authors would like to acknowledge the contributions of the following people:

Cdr David Dickens RN and Cdr Julian Malec OBE RN for their invaluable contributions to the work, Nigel Cox of the DPA for his expert advice, George Torrens and Gavin Williams of the Department of Design and Technology, Loughborough University for their swift manufacture of the test, the subjects who participated in the trials, the numerous Naval and Civilian professionals who have offered advice and guidance throughout the development of the test.

References

Bilzon, E. 1999, An evaluation of the use of stature as a deciding factor for recruitment into the Royal Navy. *INM Report* Unpublished

Bilzon, E., Chilcott, P. and Bridger, R.S. 2000, An investigation of anthropometric criteria for escape through a kidney hatch, an escape hatch and for passage through a bulkhead door. *INM Report* Unpublished

David, G.C. and Hoffman, J.S. 1996, Minimum and maximum height requirements for fire service recruits. *Home Office Report* Unpublished

Hogan, J.C. and Bernaki, E.J. 1981, Developing job-related pre-placement medical examinations, *Journal of Occupational Medicine*, **19**, 205-7

Jenkins, P.B., Pethybridge, R.J. and Hooper, R.H. 1990, An anthropometric survey of 1333 Royal Naval Personnel 1986-1990. *INM Report* Unpublished

Jenkins, P.B. and Pethybridge, R.J. 1990, An anthropometric survey of 136 Personnel of the WRNS 1986-1990. *INM Report* Unpublished

THE ALL NEW NAVY BUNK
COFFIN OR CABIN?

Mark Lowten, Emma Bilzon and Bob Bridger

Institute of Naval Medicine,
Alverstoke,
Gosport,
Hampshire,
PO12 2DL

Since the 1970's when hammocks were phased out in favour of space saving 3 tier bunks and the 1990's when women were first allowed to work at sea, accommodation has become less personal and more difficult to organise on Royal Navy ships. The need for change requires the innovative use of space, with the ultimate objective being for each person to have their own cabin. The POD concept (somewhere between a miniature cabin and enclosed bunk) aims to address this problem. Despite suggestions that it may be claustrophobic, the design is such that each individual is essentially provided with their own cabin. This paper discusses the issues surrounding innovative space-optimisation solutions for future RN ships and presents the ergonomic implications of these innovations.

Introduction

Over the last hundred years there has been a huge change in ship design. One area of change is ship board accommodation for junior rates. At the turn of the 20th century all junior rates slept in hammocks that were eight feet long and 18 inches wide. At this time it would have been possible to sleep more than one hundred ratings in one mess deck. In the 1970's hammocks were phased out in favour of the presently used three tier bunks. This was introduced as a way of saving space and therefore weight for each ship.

This way of sleeping people has disadvantages. Firstly, to fit three layers of bunks in the restricted ceiling height means that the vertical distance between beds is only 56cm per person, about 15cm more than a coffin. Effectively, there is not enough room for even the smallest recruit to sit up in bed and for some, there is not enough space to even roll over. Further more, this small amount of headroom causes a lack of personal space.

The need for 'personal space' has increased over the last two decades as the attitude of recruits has changed. The modern recruit has a higher standard of living at home and therefore expects a higher standard of life on board than expected previously by recruits. This high expectation has lead to an increasing number of recruits becoming dissatisfied with their working and more importantly recreational/resting environments.

When three tier bunks were first introduced, only men served in the Royal Navy (RN), therefore, large numbers of ratings could sleep in the same mess deck. Thus, each ship required only a few mess decks, split according to traditional rank structure. However, since 1991 women have served at sea and with most ships' company being a minimum of 10% women, existing accommodation is not versatile enough to maintain both the traditional mess and separate male/female accommodation. With few small messes available and a lack of flexibility with the space in the large messes, housing the changing mix of men and women has become a major problem.

Habitability

A review of RN shipboard accommodation revealed that Junior Rates were the least happy with their living conditions. The areas of least satisfaction were found to be 'privacy and relaxation' and sleeping accommodation. Subjects were asked to rank selected habitability items in order of importance. The most important aspect was found to be sleeping facilities. From these results it is clear that new designs for accommodation need to concentrate on improving sleeping areas so that they become comfortable and private whilst remaining versatile enough to accommodate changing numbers of men and women throughout the ships life.

This paper discusses the development of a concept for sleeping accommodation and the ergonomics issues raised during the course of its innovation.

Ergonomic Solutions

For a new design of sleeping accommodation to succeed it must replicate as many aspects of the habitability of a bedroom as possible, and where this is not feasible, the design must be novel and innovative with respect to the provision of features which enhance habitability. Through the use of focus groups and ship visits the two main criteria for success of a new design were established as 'being able to sit up in bed' and stowage.

Headroom

If new accommodation is to be versatile and comfortable it needs firstly to allow people to sit up in bed, because this makes it possible to read, write and use a laptop or computer game comfortably. Bunks, therefore, need to be reduced from three tiers to two tiers, with the space used more efficiently so that the same number of people can be housed in the same floor area. Headroom on ships is limited by the total 'clear deck height' allocated to sleeping spaces. This is estimated to be 2.1m for future ships. The bottom bunk needs to be raised, ideally 20cm from the floor. If the mattress is 10cm thick the bottom bunk could be raised by 10cm, affording the 20cm height from the floor, leaving 90cm (before compression) for headroom per person. Although 90cm is not optimal for headroom, it is functional for reading as an anticipated angle of repose is $15-20°$ and a maximum anticipated sitting height (at $0°$) approximately 97cm.

Storage

Storage space has to be useful. A small space that is well designed is more valuable than a large, unusable space. An initial concept to increase storage was to utilise the dead .

space over the feet of a sleeping person, making use of space that would otherwise be redundant. If all this dead space is usable then more room could be provided for storage than is currently available to ratings. Integral to this storage unit could be a flat screen and computer link that would allow for access to the Internet, intranet and email facilities. A desk or writing surface could be attached to this unit, in doing this less space would need to be allocated to a separate area for desks and workspaces, some of this space could then be allocated to sleeping accommodation. However, whilst the general principle behind these ideas is good, there are aspects of the design that make these units almost impossible to use.

Design Criteria

The layouts for bunk arrangement are restricted to run fore and aft along the ship as opposed to athwart (across) ships. This is because war ships roll from side to side more than from end to end and it is better for a person in a bunk to roll than experience the vertical accelerations that are experienced by movements along the head to feet axis.

Storage
There are two ways of accessing storage, both of which have design implications:

- External Access: this appeared to be the most obvious solution, traditionally lockers are accessed from outside. However, with a maximum height of 2.1m, top bunk storage units would be largely out of reach and/or difficult to use if they were to be accessed externally.

- Internal Access: mock-ups showed that due to the required depth of the unit (for it to be usable) accessing all the space would require moving parts in order that sections could be pulled to the front. These moving sections would be difficult to design, intrusive, expensive to build, and maintain and more likely to break. The alternative being that the amount of space taken from above the feet is reduced and offers little advantage.

These types of storage units assume that side entry into the bunk is required. Thus large amounts of floor area (and as such, volumes) are allocated for ingress and egress to each bunk. Also if the storage is accessed from the outside then space needs to be allocated to allow for the doors of the unit to open and for people to access the lockers.

Full scale mock-ups of the designs revealed a further potential problem in the feeling of claustrophobia that could be caused by having a storage unit over the feet. On paper, the dimensions appeared to be generous, but in reality, for any form of useful storage to be provided, the size of the unit would be intrusive and difficult to access.

The POD

A development of the initial concept lead to a solution, which has come to be known as the POD. The POD is an enclosed sleeping capsule with front-end entry that is anthropometrically designed to allow for both comfortable sleeping and sitting postures.

It is designed so that a UK male of 99th percentile stature in 2030 (as predicted by the secular trend) could lie and sit up comfortably in his bunk.

Solutions to Original Problems

The POD is designed to act as a two tiered accommodation unit with storage units running down one side. These units would be accessed from the inside of the POD avoiding problems of accessing the top bunk storage unit. Having the storage accessible from the inside means that space does not have to be allocated in passageways for access to lockers. As the storage units are on the side of the POD there is less chance of claustrophobia occurring because there is a large open area in front of the occupant that will give a feeling of space. The storage units are similar to wardrobes, being accessed from the widest side of the storage space, thus the same volume is more usable as the units are not too deep, allowing the space to be easily accessed. This allows the design to remain simple and therefore low cost, easy to maintain and easy to build. The locker design also utilises the entire floor to ceiling volume that is partly wasted with traditional locker designs.

As the POD would be accessed from the front end, less space is wasted with access to the bunk. This is because the front of each POD can make up the side of the corridors. As a result of this, the space saved can be allocated as space for the occupants.

Dimensions

The three tier bunks in current use measure 1900mm * 700mm * 560mm (L*W*H). This means that only a few individuals can stretch out in their bunk, even fewer can roll over and nobody can sit up in bed. The anthropometric-optimum dimensions are 2100mm * 991mm * 1200mm. These dimensions would allow even the largest occupant enough room to sit up in bed, roll over and lie out flat. Standard bed sizes are less generous in length at 183mm, but have no constraint on sitting height. Ship deck heights drive the maximum headroom available, and it is argued that a bed that is too wide may be troublesome in rough seas. In order to establish 'habitability' of PODs it is necessary to base initial assessments on the minimum likely dimensions. Thus if the design is acceptable at minimum dimensions it can be presumed that larger designs will also be acceptable.

Storage

For a POD made to minimum dimensions, the storage units would total 2100mm * 450mm* 1000mm along the length of the sleeping space. This volume is double that currently available to junior rates. The type of storage (shelves, rails, drawers etc.), required in PODs is open to debate. From focus groups conducted at INM (n = 60) with a full-size POD prototype it became obvious that different RN personnel have different requirements. For example a Royal Marine (RM) will require different types of storage to Royal Naval personnel. RMs, require enough space to accommodate a fully packed backpack (Bergen), but have relatively few requirements for hanging space, RN ratings have a mixed requirement for rails and drawers. It is concluded that storage design should be flexible enough to enable the occupant of the POD to choose his or her own preferred combinations of shelving, hanging rails and drawers.

Technology

Initial accommodation concepts focused on advancing the technology (in the form of consumer electronics) provided to RN personnel. From the focus groups it was determined that this was a misconception. Limiting technology in the individuals units allows the ratings to use and update technology if and when they want to. For example, laptops and DVD players are not provided to Junior Rates, yet most have bought their own. RN funded technology would quickly become dated and therefore, either become redundant or costly to replace. Currently, individuals do not have their own dedicated power points or intranet connections. Provision of two power sockets and a LAN connection to each POD would allow future technology to be used and for laptops to be connected to the Ships' Intranet. The design philosophy is to provide a space that will support the use of a variety of user-selected products.

Ingress/Egress

Ingress and egress are a concern with any design of RN bunk, particularly speed of egress, for emergency purposes. However, since the POD is two tier and current bunks are three tier, many aspects of egress would be improved (fewer people escaping into the same space). Trials need to be conducted to establish whether a person can turn round once they are inside a POD and where ladders and grab handles should be situated. The intended entrance cover will be a curtain , chosen for their easy exit in emergencies, as opposed to a hard roller blind or sliding door. Also curtains are inexpensive, easy to repair and take up little space when stored.

Security

The POD allows for improved security of possessions. Potential designs include the front of each storage unit consisting of lockable sliding doors, and one of the storage units could house a built in security box or safe that could be used for more valuable items such as money or credit cards. The main advantage over current layouts is that possessions are located in the sleeping space, in effect creating a mini-cabin.

Privacy

On future ships, the sleeping accommodation will be situated away from the noisy mess area. It is in this quiet area that the PODs could be sited. The POD would therefore be a private space, as personal as the occupants' own bedroom.

Modularity

The modularity refers to the way in which the POD will be constructed. It is imagined that the POD would be made out of a number of sections. This allows for easy fitting and refitting, replacement of one section instead of the whole POD if damage occurs, more convenient storage, easy adaptation to the changing number and type of personnel on board at any given time. Potentially, PODs could be placed in a space and arranged into any required mess size, in theory the PODs could be varied every time there is a change of personnel.

Environmental Conditions

One of the most important aspects of any habitable space is its environment. As with any environmental assessment, there are four areas for consideration in assessment of the POD, lighting, atmosphere, in-POD climate and noise and vibration.

In terms of lighting arrangements, after consultation with user groups, it is understood that two lights are needed. The main light should be an ambient light source positioned in the ceiling, controlled by on/off switches at both ends of the POD. A second, task light is required for reading and writing and would be operated from the head end of the POD.

It would be preferable for PODs to have air-conditioning units, so that the POD climate can be individually controlled. However, these units are difficult and expensive to maintain. It is therefore essential to establish the requirement for ventilation and air conditioning. The temperature in the corridor to the PODs will be controlled and from that the temperature in the POD will be regulated. However, the way air flows into the POD needs to be investigated to establish whether this is adequate or whether there is too much heat build up.

Trials could also assess the insulation of PODs to noise, vibration and atmospheric conditions, specifically the oxygen and carbon dioxide content of the air when the POD is occupied.

Psychosocial Factors

PODs would be installed on future generation warships as part of an overall design philosophy to enhance quality of life onboard. It is assumed that such ships will be fitted with separate gym and recreational facilities and dedicated recreational/social areas. The POD philosophy attempts to improve the quality of *private* spaces and is not a replacement or a substitute for social spaces. The enhancement of social spaces to better support social interaction is beyond the scope of this paper.

The Way Forward

Trials are planned to assess two working-prototype PODs, manufactured by Dexterity Research Ltd, Loughborough, for their habitability. These trials will determine if the minimum size PODs without air-conditioning or ventilation are acceptable for human living, in terms of user acceptance and environmental issues. Subject to satisfactory results, further development of PODs will continue in the areas of cost, weight, material selection, community architecture and storage solutions.

MULTIVARIATE ANTHROPOMETRIC MODELS FOR SEATED WORKSTATION DESIGN

Claire C. Gordon

U.S. Army Natick Soldier Center
Natick, Massachusetts, USA   01760-5020

Multivariate statistical methods were used to establish anthropometric design criteria for a seated workstation intended to accommodate 90% of US users. Subjects from an Army database were statistically weighted to match US adult demographic distributions. Eight body dimensions critical to seated workstation design were subject to Principal Components Analysis (PCA), and equal frequency ellipsoids capturing 90% of the subjects were fit to the male and female sample distributions in PCA space. Boundary models located on the ellipsoid surfaces at axis intersections and at midpoints on the ellipsoid traces between axes were used to establish design ranges and limits for operator seating, clearances under the workstation, and work surface heights.

Introduction

Improved ergonomic guidelines for configuring office workstations have challenged product designers to treat a wide variety of components as a single system, and to function effectively in an ergonomic sense, the size, location, and orientation of workstation components must closely relate to the geometry of the user's body. Although experts differ in their concepts of what constitutes optimal user-workstation geometry (e.g. desirable seat pan angles), and individual users differ on what they consider to be comfortable, it is common to estimate ergonomically desirable workstation dimensions using equations based on the user's body dimensions. These equations generally also include constants representing clothing allowances, clearances for comfort, and leeway for postural adjustments.   Figure 1 illustrates some body dimensions relevant to workstations and provides equations from Pheasant (1996) relating them to workstation design criteria.

Once the designer has established a particular workstation's concept of use and major functional components, body dimensions are the only unknowns remaining in the equations describing user to workstation relationships. At this point, it is common for designers to refer to tabled percentile values for user body dimensions, and to substitute a 5th percentile value for body dimensions requiring a user minimum or a 95th percentile

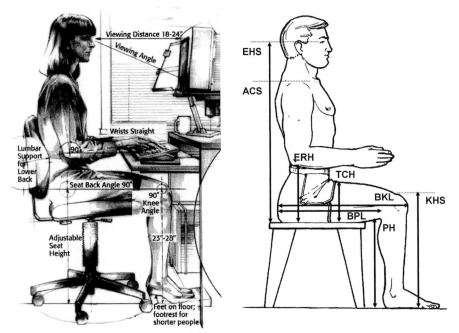

Figure 1. An ergonomic workstation[1] and some relevant body dimensions[2]

Seat Pan Height (*SPH*)	Smallest to Largest **PH** + c (shoe heel) – c (comfort)
Seat Pan Depth	Smallest **BPL** – c (comfort)
Seat Pan Width	Largest **HBS** + c (clothing) + c (leeway to move)
Backrest Top Range	Smallest to Largest **ACS** + c (seat cushion)
Lumbar Support Range	Smallest to Largest **ERH** + c (seat cushion)
Armrest Height Range	Smallest to Largest **ERH** + c (seat cushion)
Input Device Height Range	Smallest to Largest (*SPH* +**ERH**) + c (seat cushion)
Monitor Height Range	Smallest to Largest (*SPH* +**EHS**) + c (seat cushion)
Kneehole Clearance Depth	Largest (**BKL-AED**) + $\sqrt{}$ (**PH**^{2} – **SPH**^{2}) + c (shod foot)
Kneehole Clearance Height	Largest (**PH+TCH**) + c (shoe heel) + c (cushion)

---

[1] Workstation drawing reproduced with permission of the E.O. Lawrence Berkeley National Laboratory, Berkeley, CA.

[2] Abbreviations: **AED**, Abdominal Extension Depth (not shown); **ACS**, Acromion Height Seated; **BKL**, Buttock Knee Length; **BPL**, Buttock Popliteal Length; **EHS**, Eye Height Seated; **ERH**, Elbow Rest Height; **HBS**, Hip Breadth Seated (not shown); **KHS**, Knee Height Seated; **PH**, Popliteal Height; **TCH**, Thigh Clearance Height. Protocols are published in Gordon *et al* (1989).

value for body dimensions requiring a user maximum (ANSI/HFS, 1988; ISO 9241-5, 1998; ISO/FDIS 14738, 2001).   The $5^{th}/95^{th}$ percentile approach is thought to be "conservative" because it is unlikely for a single user to have more than a few body dimensions that are extremely large or extremely small, and so workstation design values that rely on user maxima and minima should provide "generous" estimates for the needs of most users (Pheasant, 1996). The percentile approach is easy and straightforward to implement because tabled percentile values of national user groups and their male and female subgroups are widely available.

However, today's office workstations are comprised of many separate components with their own adjustment mechanisms, including seat pans, seat backs, and armrests, data entry surfaces, writing surfaces, and display support surfaces. To ensure that the components work together as a functional system, user variation in more than 10 body dimensions must be accommodated *simultaneously* to achieve the desired concordance between workstation and user geometry.   In these more complex systems, percentile approaches may cause unanticipated design difficulties for several reasons. Firstly, generous overestimation/underestimation of dimensions caused by adding percentiles together (McConville and Churchill, 1976; Churchill, 1978) may not be tolerable in multiple workstation components because a tighter integration between workstation geometry and user geometry is required.   In addition, many workstations involve adjustment mechanisms whose interactions influence one another and the relationship of the user's body to fixed components in the system. In such cases, individuals with unusual body proportions may constitute the designer's worst case for adjustment rather than uniformly small or large individuals (e.g. torso heights and limb lengths; Zehner et al., 1992).   Finally, and most importantly, when more than one range of adjustment is defined for a single functional system (e.g. seat, armrest, input device, and monitor height), the fact that different body dimensions are not perfectly correlated with one another causes significant reduction in the percentage of users captured by univariate percentile ranges (Moroney and Smith, 1972).

To avoid the problems caused by applying univariate percentiles to functional systems that are multivariate in nature, this paper illustrates an alternative approach developed from earlier work by Bittner (1987) and Zehner (1992) on aviation systems, and by Gordon (1997) on body armor and load carriage systems. Multivariate statistical methods are utilized in the definition of realistic body forms whose dimensions describe the extremes of multivariate body size and shape expected in a centrally located 90% subset of the American population. The body dimensions of these extreme forms are implemented using the equations in Figure 1, and compared with the results achieved by substituting the appropriate 5th and/or 95th percentile values in the same equations. The advantages and disadvantages of this multivariate approach are discussed.

Materials and Methods

Anthropometric data from 5,477 males and 3,469 females measured in the 1988 US Army Anthropometric Survey (Gordon *et al*, 1989) were used in this study.   Subjects were weighted prior to statistical analysis to match prevailing US civilian adult age, sex, and race distributions (Gordon, 2000). However, weighting techniques cannot correct for the fact that military body fat and physical fitness requirements preclude overweight individuals from Army samples, whereas overweight men and women are common in the US civilian population. The absence of overweight subjects in the Army database primarily affects two dimensions in this study: Abdominal Extension Depth and Hip Breadth Sitting. Other study dimensions are primarily related to height, and Army height criteria eliminate less than 2% of civilian adults (Gordon and Friedl, 1994), and can reasonably represent civilian distributions after demographic weighting.

Principal Components Analysis (PCA) was used to reduce variation in 10 body dimensions relevant to workstation design (see Figure 1) to 3 orthogonal components comprised of linear combinations of the original body measurements. Subjects were scored on the PCA eigenvectors, plotted in 3-dimensional PCA space, and an equal frequency ellipsoid capturing 90% of subjects was fit to the distribution of PC scores (see Figure 2). All statistical analyses were conducted in Stata 6.0 (StataCorp, 1999), and male and female subjects were analyzed separately to avoid the biases incurred by force-fitting a multivariate normal model over sexually dimorphic (essentially bimodal) distributions (Gordon *et al*, 1997).

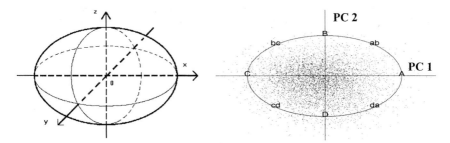

Figure 2. A 3-dimensional ellipsoid and its XY trace

The ellipsoid surface thus represents a 90% accommodation boundary in PCA space. To capture body size and shape extremes represented by the ellipsoid surface, 6 boundary forms are defined at major axis intersections with the surface, and 20 additional forms are located on the surface at arc midpoints using numerical integration. Figure 2 illustrates boundary forms located in the XY (PC1 PC2) plane of the ellipsoid. Once boundary forms are located on the surface of the ellipsoid, the product of their PC scores and eigenvector coefficients are used to determine how far, and in what direction, their body dimensions are located relative to the sample means (Harris, 1975). This process results in a set of 26 extreme forms whose body dimensions are engineering "worst case" scenarios for the 90% of subjects captured by the ellipsoid.

To establish the anthropometric values in the design criteria of Figure 1, the relevant dimensions of each boundary form are substituted in the equations, and statistical software is used to identify the largest and/or smallest values among the 90% boundary forms. In the final step, body dimensions of each subject in the database are compared to the anthropometric limits established using boundary forms. A subject is scored as accommodated only if his/her values are within the design limits for *every* design parameter listed in Figure 1. The proportion of database subjects scored as accommodated in this exercise serves as a check that the design limits derived using boundary forms indeed capture the desired 90% of the user population. Finally, the proportion of subjects captured by the multivariate anthropometric criteria is compared to the proportion of subjects captured if we had substituted 5th and 95th percentile values in the equations in Figure 1.

Results

Results of Principal Components Analysis are presented in Tables 1 and 2. Due to space limitations, only the first 3 PC's are shown and eigenvectors are presented horizontally.

Table 1. Principal Components Analysis of 5,477 males

Component	Eigenvalue	Difference	Proportion	Cumulative
PC 1	4.90161	2.65238	0.4902	0.4902
PC 2	2.24923	0.70385	0.2249	0.7151
PC 3	1.54538	1.11237	0.1545	0.8696

					Scoring Coefficients					
	AED	AHS	BKL	BPL	EHS	ERH	HBS	KHS	PH	TCH
PC 1	0.2159	0.3259	0.4037	0.3795	0.3125	0.1153	0.3248	0.3900	0.3323	0.2473
PC 2	0.2612	0.3417	-0.2332	-0.2871	0.2409	0.5721	0.1983	-0.2714	-0.3604	0.2236
PC 3	0.5051	-0.3358	0.0896	0.0421	-0.4140	-0.2773	0.3456	-0.0917	-0.2282	0.4405

The first 3 Principal Components accounted for 87% of the variation present in the male sample, and for 86% of the variation in females. The patterns of variable loadings in the sexes were identical: PC 1 (accounting for 49% of male and 47% of female variation) represents overall size; PC 2 (22.5% of male variation; 22% of female variation) contrasts lower limb lengths with trunk heights; PC 3 (15.5% of male variation; 17% of female variation) contrasts breadths and depths with trunk heights. The remaining PC's were not retained because they contributed little additional information to the model.

Table 2. Principal Components Analysis of 3,479 females

Component	Eigenvalue	Difference	Proportion	Cumulative
PC 1	4.72338	2.55678	0.4723	0.4723
PC 2	2.16660	0.46066	0.2167	0.6890
PC 3	1.70594	1.22601	0.1706	0.8596

					Scoring Coefficients					
	AED	AHS	BKL	BPL	EHS	ERH	HBS	KHS	PH	TCH
PC 1	0.2510	0.2864	0.4197	0.3958	0.3037	0.0758	0.3149	0.4007	0.3066	0.2654
PC 2	-0.0083	0.4860	-0.2162	-0.2473	0.3989	0.6430	0.0850	-0.1861	-0.2007	-0.0338
PC 3	0.5077	-0.1706	0.0110	-0.0788	-0.2226	-0.0011	0.4310	-0.2268	-0.4759	0.4400

Subjects were scored on each of the first 3 PC's using their sex-specific coefficients. Ninety percent ellipsoids centered on the sex-specific mean values were fit to the three-dimensional PCA distributions of male and female subjects. The sex-specific distributions did not completely overlap (see Figure 3), so some male outliers were captured by the female ellipsoid and vice-versa. Together the two ellipsoids captured 92% of the total sample.

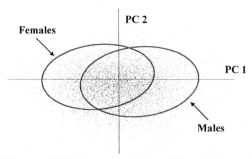

Figure 3. XY Traces of 90% Accommodation Ellipsoids for males and females

As described earlier, body dimensions for each of the 26 boundary forms on the ellipsoid surface were calculated from their PC coordinates and eigenvectors. These boundary forms represent engineering "worst case" scenarios for 90% of male and female users, and they include extremes of body size and of body proportion. To use the boundary forms in estimating the design specifications of Figure 1, each boundary form's body dimensions were substituted into the equations and then analyzed by computer to determine the most extreme design values among the boundary forms. The results of this process are shown below in Table 3, and compared with values derived by substituting the appropriate 5th and/or 95th percentile value in the Figure 1 equations. For comparative purposes, we can ignore the constants representing clothing allowances, comfort space, and compressed seat cushion thickness.

Table 3. Anthropometric Values for Design Specifications, in mm

Design Specification[3]	Multivariate Result	Percentile Result
Seat Pan Height	330 – 484	347 - 472
Seat Pan Depth	421	436
Seat Pan Width	463	450
Backrest Top	497 – 671	514 - 653
Armrest & Lumbar Supports	171 – 294	186 - 280
Input Device Height	543 – 737	560 - 720
Monitor Height	1026 – 1341	1056 - 1307
Kneehole Depth	436	420
Kneehole Height	659	646

[3] For comparative purposes, constants and geometric corrections for seat angle have been ignored. Only anthropometric contributions to the design values are shown.

In the last step of this analysis, theoretical workstation dimensions were computed for every subject in the database using the equations in Figure 1, and the results for each subject were compared against the design criteria derived from boundary form data and univariate percentiles. A subject was scored as accommodated by the design criteria if his or her individual results were within the design limits for *every* design specification in Figure 1. The results of this exercise are shown below in Table 4.

Table 4. Accommodation Rates for Univariate vs. Multivariate Methods

	Percentile Method	Multivariate Method
Males (n=5477)	79.9 %	93.2 %
Females (n=3479)	77.9 %	94.0 %
Total	78.9 %	93.6%

Discussion

As can be seen above, the univariate percentile method is not as conservative as one might have thought. The intended accommodation rate was 90%, and yet only about 80% of workstation users are captured when univariate percentiles for these 10 dimensions were substituted in our workstation design criteria equations. The multivariate method, on the other hand, exceeded the intended 90% accommodation rate, at least in because some of the design criteria required only one-sided limits instead of two-sided ranges.

The multivariate method used here has several advantages. It identifies extreme forms of size and shape that are both realistically proportioned and closely related to the intended accommodation rate of the design. In addition, the anthropometric characteristics of the boundary forms are independent of engineering design details. They describe the extremes of size and shape present in the central 90% of the database population for those body dimensions submitted to the PCA. The designer can change seat cushions, add or remove adjustment mechanisms, change clothing allowances and still use these same extreme forms to establish design criteria, giving true meaning to the term "human centered design".

On the other hand, the multivariate ellipsoid method requires considerable expertise in human biology and biostatistics to exercise fruitfully. The PCA model upon which the ellipsoid is based must capture a large percentage of the variation present in the original design dimensions, and all the critical dimensions must load strongly on at least one of the PC's retained in the model. This requires thoughtful selection of the original body dimensions for consideration, and careful attention to demographic subgroup differences in body size and shape.

Finally, we should note that accurate definition of extreme body sizes and shapes is no guarantee that a design will be successful. Workstation designs that attempt to accommodate poorly correlated body dimensions with a single adjustment mechanism (e.g aircraft and automobile seat rails that use inclines to simultaneously change seat height and proximity to hand/foot controls) require special statistical attention as do adjustment mechanisms that have "stops" at predetermined intervals instead of continuous adjustment. In both cases, accommodating the extremes of size and shape is no guarantee that everyone within the extreme boundaries will be accommodated.

References

ANSI/HFS 100, 1988, *American National Standard for human factors engineering of visual display terminal workstations.* (Human Factors Society, Santa Monica, CA)

Bittner, A.C., Glenn, F.A., Harris, R.M., Iavecchia, H.P., and Wherry, R.J., 1987, CADRE: A Family of Mannikins for Workstation Design. In S.S. Asfour (ed.) *Trends in Ergonomics/Human Factors IV,* (Elsevier, North Holland), 733-740

Churchill, E., 1978, Statistical Considerations. In *Anthropometric Source Book, Volume I. Anthropometry for Designers,* NASA Reference Publication 1024, (National Aeronautics and Space Administration, Washington, D.C.), Chapter IX

Gordon, C.C., Churchill, T., Clauser, C.E., Bradtmiller, B., McConville, J.T., Tebbetts, I. & Walker, R.A., 1989, *1988 Anthropometric Survey of U.S. Army Personnel: Methods and Summary Statistics.* NATICK/TR-89/044, AD A225 094, (US Army Natick RD&E Center, Natick, MA)

Gordon, C.C. and Friedl, K., 1994, Anthropometry in the U.S. Armed Forces. In S.J. Ulijaszek and C.G.N. Mascie-Taylor (eds.) *Anthropometry. The Individual and the Population,* (Cambridge University Press, Cambridge, UK), 178-210

Gordon, C.C., Corner, B.D., and Brantley, J.D., 1997, *Defining Extreme Sizes and Shapes of U.S. Army Personnel for the Design of Body Armor and Load-Bearing Systems,* NATICK/TR-97/012, AD A324 730, (US Army Natick RD&E Center, Natick, MA)

Gordon, C.C., 2000, Case Studies in Statistical Weighting of Anthropometric Databases, Paper presented at the XIVth Triennial Congress of the International Ergonomics Association, San Diego, CA, 3 August 2000

Harris, R.J., 1975, *A Primer of Multivariate Statistics,* (Academic Press, New York, NY)

ISO 9241-5, 1998, *Ergonomic requirements for office work with visual display terminals (VDTs) – Part 5: Workstation layout and postural requirements.* (International Organization for Standardization, Geneva)

ISO/FDIS 14738, 2001, *Safety of machinery – Anthropometric requirements for the design of workstations at machinery.* (International Organization for Standardization, Geneva)

McConville, J.T. and Churchill, E., 1976, *Statistical Concepts in Design,* AMRL-TR-76-29, AD A025 750, (Aerospace Medical Research Laboratory, Wright-Patterson Air Force Base, OH)

Moroney, W.F. and Smith, M.J., 1972, *Empirical Reduction in Potential User Population as the Results of Imposed Multivariate Anthropometric Limits,* NAMRL-1164, AD 752 032, (Naval Aerospace Medical Research Laboratory, Pensacola, FL)

Pheasant, S., 1996, *Bodyspace. Anthropometry, Ergonomics and the Design of Work, Second Edition,* (London: Taylor & Francis, Ltd)

StataCorp, 1999, *Stata Statistical Software 6.0,* (Stata Corporation, College Station, TX)

Zehner G.F., Meindl R.S., and Hudson J.A., 1992, *A Multivariate Anthropometric Method for Crew Station Design: Abridged,* AL-TR-1992-0164, AD A274 588, (Armstrong Laboratory, Wright-Patterson Air Force Base, OH)

ON HUMAN FACTORS STANDARDS FOR DISPLAYS WITHIN ROYAL NAVY COMMAND SYSTEMS

Dr M A Tainsh, F. Erg. S.

Centre of Human Sciences
QinetiQ, Cody Technology Park
Farnborough, Hants GU14 0LX

Current display technology has improved the potential of graphical designs for the Human-Computer Interaction (HCI) within RN command systems. Hence, the available human factors standards of graphics interfaces were believed to be inadequate to meet the requirements for specification. This programme investigated the graphics requirements for Royal Navy (RN) systems, while the room had low ambient lighting.

The first trial considered background colour, while the second and third investigated alphanumeric characters and symbols against the preferred background. The relative benefits of two types of display technology (cathode ray tubes and flat panel displays - in this case liquid crystal) were also assessed. No important differences were found between the display types. Empirical results are presented for optimal display characteristics.

Introduction

Background

Command systems are complex arrangements of users and equipment (generally computer-based) linking sensors from remote locations to a central workstation which in turn is linked to effectors. The users execute tasks which in general are a form of Human-Computer Interaction (HCI). They tend to involve displays of graphical representations. In the Royal Navy (RN) context, the representations are likely to be map-like, showing sea and land with symbols to represent platforms or other objects of interest. Text will be used to provide supplementary material. Unfortunately the human factors standards available to support design of such HCI now lag behind requirements.

This work was initiated and funded by the Ministry of Defence's (MoD's) Defence Procurement Agency (DPA), and aimed at investigating three sets of standards for graphical displays within RN command systems. However, there are good reasons to believe that the results are more widely applicable.

Specific Issues

Three sets of characteristics for graphics representations were selected following a brief study involving RN users and members of the DPA:
    (a) Preferred background colour
    (b) Text size
    (c) Symbol size
The relative benefits of Cathode Ray Tube technology (CRTs) and Flat Panel Displays (FPDs) were also assessed to help establish that the results were relatively independent of implementation.

The three sets of characteristics all have important implications for the design of graphical representations. The preferred background colours not only influence ease of reading but also reports of "visual comfort". The shape and size of the characters and symbols determine both legibility and the amount of information that can be displayed simultaneously.

Knowledge of the relative benefits of CRTs and FPDs is important as the latter have a substantial size advantage over the former but the scientific case from the User perspective has yet to be made for RN command systems.

Trials Overview

The first trial set out to establish the preferred background colour for a graphical representation, the second the size and shape of the alphanumeric characters to be used and the third the size of the symbols. The equipment consisted of a Silicon Graphics Indigo (SGI) system with a standard CRT display next to a Flat Panel Display (FPD). The CRT was a 19 inch Silicon Graphics monitor and the FPD was a 19 inch BARCO D251 Liquid Crystal Display (LCD). The FPD had comparable dimensions and resolution to the CRT.

All three trials were conducted at low levels of ambient illumination in accordance with Royal Navy requirements (NES 624). The illumination level was set to 60 lux. Trials conducted under 'normal' office levels of illumination may yield different results. The representations on the displays were similar to those available on current and future RN command systems. These were radar, sonar and Tote (an alphanumeric display).

All subjects were RN users of command systems and their vision was assessed as normal using standard tests. The viewing distance was approximately 500mm.

The Trials

Trial 1: The Use of Achromatic/Coloured Backgrounds within the Combat System

The aim of the first trial was to investigate the preferred background colour characteristics for reading foreground symbols and alphanumerics on three types of Combat System display representation. The three representations investigated were radar, sonar and Tote (alphanumeric) representations. The colour characteristics assessed were value, saturation, grey level and texture. Value refers to a measurement of lightness of a surface ranging from 0 (black) to 10 (white). Saturation is the quality that distinguishes a hue from white. Pastel shades are de-saturated and vivid colours are saturated. Grey levels were adjusted by means of their value. Texture is the degree of roughness/smoothness, coarseness/fineness of some material object: it was introduced using a random function. The colours selected were: black, white, four shades of grey and ten shades of blue. These all had textured versions. Hence there were thirty-two conditions in total.

The various conditions were presented in random order associated with a Latin Square design to take account of the various materials and display types. Subjects rated legibility, comfort, ease of use and contrast.

Trial 1: Results

Generally the comfort and other ratings given to a particular background were similar for the CRT and the FPD. This suggests that the displays are the same in their requirements for background colour. However some subjects thought that the colours on the FPD were "more subtle" than on the CRT, which improved some of the ratings given to the FPD. A typical set of results is given in Figure One. The best ratings were given for low value, high saturation, blue. Texture was not found beneficial and in some cases was reported as confusing. Blue backgrounds of low value appeared suitable for all types of representation.

Trial 2: The use of various shapes and sizes of fonts within the command system

The aim of this trial was to assess the minimum character size which was readable and the optimum size acceptable for text on three command system display representations at a viewing distance of 500mm.

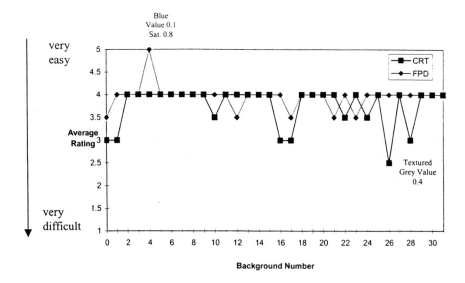

Figure One: Ease of distinguishing symbols presented on the radar display

Two fonts were used: Courier and Helvetica. This was to assess any differences between sans serif (Helvetica) and a serif style font (Courier). Def Stan 00-25 Part 7 recommends that a sans serif font is used for text especially if viewed under adverse conditions. The reason is given as the extra detail of serif font adds to the task. The two font styles used in the trial are exemplified below:

Sans serif style
Serif style

The trial was conducted using fonts of increasing and decreasing sizes and asking the users to rate their ease of reading and preference.

3.4 Trial 2: Results

Subjects can read text that is smaller than the minimum defence standard recommendations. For fonts such as Courier which have a high width to height ratio, text as small as 11 arcmin can be considered 'just readable' for some representations. However for fonts such as Helvetica with a smaller width to height ratio text should be at least 15 arcmin to be considered 'just readable'

When determining the optimally acceptable character size, the combination of character height and width appeared to be more important for subject acceptance than character height alone. See Table One.

Table One: Minimum readable and optimum acceptable sizes of text

Observation made	Display Type	Font	Height	Width	Width to Height Ratio
Minimum Readable Character Size	CRT	Helvetica	15 arcmin	13 arcmin	0.87:1
		Courier	11 arcmin	11 arcmin	1:1
	FPD	Helvetica	15 arcmin	13 arcmin	0.87:1
		Courier	11 arcmin	11 arcmin	1:1
Optimum Acceptable Character Size	CRT	Helvetica	21 arcmin	20 arcmin	0.97:1
		Courier	17 arcmin	19 arcmin	1.1:1
	FPD	Helvetica	21 arcmin	18 arcmin	0.87:1
		Courier	13 arcmin	15 arcmin	1.2:1

Trial 3: The size of symbols for use on graphical representations within the command system
The aim of this trial was to investigate the minimum identifiable symbol size for the set within STANAG 4420, and the optimum acceptable symbol size for normal operation on two combat system representations (CSFAB Style Guide). Subjects were provided with symbols of increasing or decreasing size and required to identify them

Trial 3: Results
The minimum identifiable symbol size for STANAG 4420 symbols to satisfy all symbol requirements on both displays, was 28 arcmin. The optimum acceptable symbol size was 50 arcmin. See Figure Two.

Conclusion

In summary, the following standards can be supported as a result of this work:

(a) The background colour for this class of display, taking account of lighting conditions: highly saturated blues or similar are to be preferred.
(b) The size and shape of characters required is given in Table 1.
(c) The preferred size of the symbol must depend on its construction. Figure 3 indicate the sizes for symbols as given in STANAG 4420 and similar.

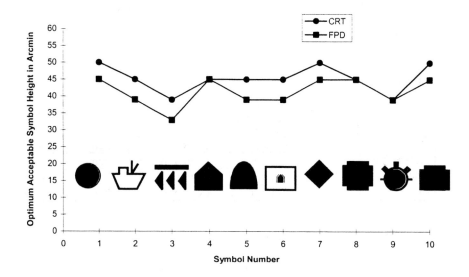

Figure Two: Preferred symbol size

References

1. NES 624 Issue 1 March 1994, Requirements for the Design of an Interface to Electronic Damage Control Information Displays. Ministry of Defence, UK RESTRICTED.
2. Ministry of Defence. Def Stan 00-25 (Part 7) Human Factors for Designers of equipment: Visual Displays. Issue 2, December 1996.
3. NATO, STANAG 4420, Standardisation Agreement - Display Symbology and Colours for NATO Maritime Units, Edition 2.
4. CSFAB. Combat Systems Functional Allocation Board (CSFAB) Human-System Interaction Style Guide Draft Version 1. Website address: http://his.nosc.mil/Publications/CSFAB/index.html

Acknowledgement

The expert guidance of Malcolm Earl of the DPA and Lt Cdr A Hobbs RN was essential to this programme. The important contributions of many colleagues including members of Liveware and Tenet Systems are also acknowledged. The assistance of the US Navy Laboratories in San Diego is gratefully acknowledged.

AUTHOR INDEX

SUBJECT INDEX